T0310213

PUMPS, CHANNELS, AND TRANSPORTERS

CHEMICAL ANALYSIS

A SERIES OF MONOGRAPHS ON ANALYTICAL CHEMISTRY
AND ITS APPLICATIONS

Editor
MARK F. VITHA

VOLUME 183

A complete list of the titles in this series appears at the end of this volume.

PUMPS, CHANNELS, AND TRANSPORTERS
Methods of Functional Analysis

Edited By

RONALD J. CLARKE
School of Chemistry
University of Sydney
Sydney, Australia

MOHAMMED A. A. KHALID
Department of Chemistry
University of Taif
Turabah, Saudi Arabia

Published by John Wiley & Sons, Inc., Hoboken, New Jersey
Published simultaneously in Canada

For general information on our other products and services or for technical support, please contact our Customer Care Department within the United States at (800) 762-2974, outside the United States at (317) 572-3993 or fax (317) 572-4002.

Wiley also publishes its books in a variety of electronic formats. Some content that appears in print may not be available in electronic formats. For more information about Wiley products, visit our web site at www.wiley.com.

Library of Congress Cataloging-in-Publication Data

Pumps, channels, and transporters : methods of functional analysis / edited by Ronald J. Clarke, Mohammed A. A. Khalid.
      pages   cm
   "Ion-transporting membrane proteins permanently embedded in the membranes of cells or cell organelles can be grouped into three broad categories: channels, pumps, and transporters"–Chapter 1, introduction.
   Includes index.
   ISBN 978-1-118-85880-6 (cloth)
1. Carrier proteins.   2. Ion pumps.   3. Ion channels.   4. Membrane proteins.
5. Biological transport.   I. Clarke, Ronald J. (Ronald James), editor.
II. Khalid, Mohammed A. A., editor.
   QP552.C34P86 2015
   572′.696–dc23
                                                                        2015019124

Set in 10/12pt Times by SPi Global, Pondicherry, India

Printed in the United States of America

10   9   8   7   6   5   4   3   2   1

1   2015

CONTENTS

PREFACE

MOHAMMED A. A. KHALID[1] AND RONALD J. CLARKE[2]

[1] *Department of Chemistry, University of Taif, Turabah, Saudi Arabia*
[2] *School of Chemistry, University of Sydney, Sydney, NSW, Australia*

Ion-transporting membrane proteins, which include ion pumps, channels, and transporters, play crucial roles in all cellular life forms. Not only do they provide pathways linking the extracellular medium with the cytoplasm and the cytoplasm with the contents of intracellular organelles, they are also intimately involved in energy transduction in all cells. Furthermore, in organisms with higher levels of cellular differentiation, they provide sophisticated mechanisms for signal transduction, for example, in both electrical and chemical forms of nerve impulse transmission and in muscle contraction and relaxation.

Although probably not the most objective criterion for judging the importance of pumps, channels, and transporters, it is interesting and enlightening to consider the number of Nobel Prizes awarded for research directly related to membrane proteins. The first Nobel Prize ever given for research on a specific membrane protein was awarded to Johann Deisenhofer, Robert Huber, and Hartmut Michel in 1988 for the determination of the first crystal structure of an integral membrane protein, the photosynthetic reaction center of the bacterium *Rhodopseudomonas viridis*. Although the structure determination of membrane proteins is still not routine in comparison to water-soluble proteins, this was a major breakthrough, which paved the way for the crystallization of many other membrane proteins.

A further significant contribution to the field was the development by Erwin Neher and Bert Sakmann of the patch-clamp technique, which enables the activity of single ion channels to be detected. For this important technical advance and its application (described in Chapter 3), together they were awarded the 1991 Nobel Prize in Physiology or Medicine. A few years later, in 1997, Paul Boyer, John Walker, and Jens Christian Skou were awarded the Nobel Prize in Chemistry. Jens Christian Skou won the prize for his discovery of the Na^+,K^+-ATPase. Paul Boyer and John Walker were chosen for their research into the enzymatic mechanism of ATP biosynthesis. The critical contribution of John Walker was his crystallization and structural determination of the cytoplasmic domain of the ATP synthase. Then in 2003, Peter Agre and Roderick MacKinnon were also awarded the Nobel Prize in Chemistry. In the case of Peter Agre, this was for the discovery of aquaporins. Roderick MacKinnon, on the other hand, won the prize for his work on ion channels, in particular the determination of the structure of the K^+ channel of *Streptomyces lividans* using X-ray crystallography.

In this brief overview of Nobel Prizes in the membrane protein field, it is notable that several Nobel Prizes have been awarded for structural determinations via X-ray crystallography. This is certainly an important step toward the complete understanding of any membrane protein. However, crystal structures alone do not immediately explain how a protein functions. After the discovery of the helical structure of DNA and its base pairing by Crick, Watson, Franklin, and Wilkins, the mechanism of information transfer via the genetic code was relatively soon unraveled, and the significance of the structure is now obvious. However, in the case of proteins, often after the publication of a crystal structure one is still none the wiser regarding how they function. To understand how a protein functions, a static picture is insufficient. Time resolution is essential. One needs to observe a protein in action.

Without the crystal structure of a protein or one closely related to it, certainly no meaningful theoretical simulations of the protein's activity via molecular dynamics (MD) methods are possible. Although the timescales of MD simulations are gradually increasing, as described in Chapter 15, many protein conformational changes important for the functions of pumps, channels, and transporters occur over milliseconds or seconds, beyond the reliability of any current MD methods. Therefore, it is clear that experimental methods for the direct detection of the activity of ion-transporting membrane proteins are essential and will likely remain so for many years to come.

The emphasis of this book is, therefore, on experimental methods for resolving the kinetics and dynamics of pumps, channels, and transporters. Structural methods, such as X-ray crystallography or electronmicroscopy, although clearly important for a complete understanding of membrane protein function down to the atomic level, are specifically excluded. The experimental methods treated in the book are divided into three main groups: electrical (Chapters 2–6), spectroscopic (Chapters 7–12), and radioactivity-based and atomic absorption-based flux assays (Chapters 13 and 14). Finally, the book concludes with a chapter on computational techniques (Chapter 15).

Many pumps, channels, and transporters are electrogenic, that is, their activity involves a net transport of charge across the membrane. Hence, it is logical that a variety of techniques have been developed to detect the currents and membrane voltage changes that these proteins produce. For this reason, after an introduction overviewing pumps, channels, and transporters, their energetics, and mechanisms, the following five chapters of the book are devoted to electrical methods.

However, not all ion-transporting membrane proteins are electrogenic. There are many which function in an electroneutral fashion. This is the case when the current produced by the transport of one ion across the membrane is compensated by the transport of another ion with the same charge in the opposite direction. Therefore, electrical techniques alone are insufficient to quantify the activity of all pumps, channels, and transporters. Complementary techniques are necessary. Chapters 7–14 are, therefore, devoted to either spectroscopic or radioactivity-based methods. Apart from allowing the detection of protein activity independent of the protein's electrogenicity, some spectroscopic techniques, in particular IR spectroscopy (Chapter 9) and EPR spectroscopy (Chapter 11), have the added advantage of providing time-resolved structural information, not accessible by any electrical method.

We thank all chapter authors for their contributions to the book, the series editor Mark Vitha for his valuable comments and suggestions and Michael Leventhal of Wiley for his assistance in the preparation of the book for publication. We hope that readers with an interest in dynamic aspects of membrane protein function find the book interesting and of value for their own research.

RONALD J. CLARKE
Berlin, Germany,
February 2015

MOHAMMED A. A. KHALID
Turabah, Saudi Arabia,
February, 2015

LIST OF CONTRIBUTORS

Hans-Jürgen Apell, Department of Biology, University of Constance, Constance, Germany

Peter H. Barry, Department of Physiology, School of Medical Sciences, UNSW Australia (The University of New South Wales), Sydney, New South Wales, Australia

Ingolf Bernhardt, Laboratory of Biophysics, Saarland University, Saarbrücken, Germany

Stephanie Bleicken, Max Planck Institute for Intelligent Systems, Stuttgart, Germany; Interfaculty Institute of Biochemistry, University of Tübingen, Tübingen, Germany

Louise J. Brown, Department of Chemistry and Biomolecular Sciences, Macquarie University, Sydney, New South Wales, Australia

Ronald J. Clarke, School of Chemistry, University of Sydney, Sydney, New South Wales, Australia

Ben Corry, Research School of Biology, Australian National University, Canberra, Australian Capital Territory, Australia

Katia Cosentino, Max Planck Institute for Intelligent Systems, Stuttgart, Germany; Interfaculty Institute of Biochemistry, University of Tübingen, Tübingen, Germany

John S. H. Danial, Department of Chemistry, Chemical Research Laboratory, University of Oxford, Oxford, UK

J. Clive Ellory, Department of Physiology, Anatomy and Genetics, University of Oxford, Oxford, UK

Klaus Fendler, Department of Biophysical Chemistry, Max Planck Institute of Biophysics, Frankfurt/Main, Germany

Thomas Friedrich, Institute of Chemistry, Technical University of Berlin, Berlin, Germany

Ana J. García-Sáez, Max Planck Institute for Intelligent Systems, Stuttgart, Germany; Interfaculty Institute of Biochemistry, University of Tübingen, Tübingen, Germany

Christof Grewer, Department of Chemistry, Binghamton University, Binghamton, NY, USA

Joanna E. Hare, Department of Chemistry and Biomolecular Sciences, Macquarie University, Sydney, New South Wales, Australia

Joachim Heberle, Institute of Experimental Physics, Department of Physics, Free University of Berlin, Berlin, Germany

Mohammed A. A. Khalid, Department of Chemistry, Faculty of Applied Medical Sciences, University of Taif, Turabah, Saudi Arabia

Philip W. Kuchel, School of Molecular Bioscience, University of Sydney, Sydney, New South Wales, Australia

Trevor M. Lewis, Department of Physiology, School of Medical Sciences, UNSW Australia (The University of New South Wales), Sydney, New South Wales, Australia

Andrew J. Moorhouse, Department of Physiology, School of Medical Sciences, UNSW Australia (The University of New South Wales), Sydney, New South Wales, Australia

Valerij S. Sokolov, A. N. Frumkin Institute of Physical Chemistry and Electrocemistry, Russian Academy of Sciences, Moscow, Russia

Francesco Tadini-Buoninsegni, Department of Chemistry "Ugo Schiff", University of Florence, Sesto Fiorentino, Italy

Mark I. Wallace, Department of Chemistry, Chemical Research Laboratory, University of Oxford, Oxford, UK

Eve E. Weatherill, Department of Chemistry, Chemical Research Laboratory, University of Oxford, Oxford, UK

1

INTRODUCTION

MOHAMMED A. A. KHALID[1] AND RONALD J. CLARKE[2]

[1]*Department of Chemistry, Faculty of Applied Medical Sciences, University of Taif, Turabah, Saudi Arabia*
[2]*School of Chemistry, University of Sydney, Sydney, New South Wales, Australia*

1.1 HISTORY

Modern membrane science can be traced back to 1748 to the work of the French priest and physicist Jean-Antoine (Abbé) Nollet who, in the course of an experiment in which he immersed a pig's bladder containing alcohol in water, accidentally discovered the phenomenon of osmosis [1], that is, the movement of water across a semipermeable membrane. The term osmosis was, however, first introduced [2] by another French scientist, Henri Dutrochet, in 1827. The movement of water across a membrane in osmosis is a passive diffusion process driven by the difference in chemical potential (or activity) of water on each side of the membrane. The diffusion of water can be through the predominant matrix of which the membrane is composed, that is, lipid in the case of biological membranes, or through proteins incorporated in the membrane, for example, aquaporins. As the title of this book suggests, here we limit ourselves to a discussion of the movement of ions and metabolites through membranes via proteins embedded in them, rather than of transport through the lipid matrix of biological membranes.

The fact that small ions, in particular Na^+, K^+, and Cl^-, are not evenly distributed across the plasma membrane of cells (see Fig. 1.1) was first recognized by the physiological chemist Carl Schmidt [3] in the early 1850s. Schmidt was investigating the pathology of cholera, which was widespread in his native Russia at the time, and

Pumps, Channels, and Transporters: Methods of Functional Analysis, First Edition. Edited by Ronald J. Clarke and Mohammed A. A. Khalid.
© 2015 John Wiley & Sons, Inc. Published 2015 by John Wiley & Sons, Inc.

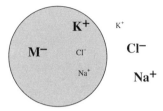

FIGURE 1.1 Ionic distributions across animal cell membranes. M⁻ represents impermeant anions, for example, negatively charged proteins. Typical intracellular (int) and extracellular (ext) concentrations of the small inorganic ions are: $[K^+]_{int} = 140–155\,mM$, $[K^+]_{ext} = 4–5\,mM$, $[Cl^-]_{int} = 4\,mM$, $[Cl^-]_{ext} = 120\,mM$, $[Na^+]_{int} = 12\,mM$, $[Na^+]_{ext} = 145–150\,mM$. (Note: In the special case of red blood cells $[Cl^-]_{ext}$ is lower (98–109 mM) due to exchange with HCO_3^- across the plasma membrane, which is important for CO_2 excretion and the maintenance of blood pH. This exchange is known as the "chloride shift.") Adapted from Ref. 4 with permission from Wiley.

discovered the differences in ion concentrations while comparing the blood from cholera victims and healthy individuals. By the end of the nineteenth century it was clear that these differences in ionic distributions occurred not only in blood, but existed across the plasma membrane of cells from all animal tissues. However, the origin of the concentration differences remained controversial for many years.

In the 1890s at least two scientists, Rudolf Heidenhain [5] (University of Breslau) and Ernest Overton [6] (then at the University of Zürich), both reached the conclusion that the Na⁺ concentration gradient across the membrane was produced by a pump, situated in the cell membrane, which derived its energy from metabolism. Although we know now that this conclusion is entirely correct, it was apparently too far ahead of its time. In 1902, Overton even correctly proposed [7] that an exchange of Na⁺ and K⁺ ions across the cell membrane of muscle—now known to arise from the opening and closing of voltage-sensitive Na⁺ and K⁺ channels—was the origin of the change in electrical voltage leading to muscle contraction. This proposal too was not widely accepted at the time or even totally ignored. It took another 50 years before Overton's hypothesis was rediscovered and finally verified by the work of Hodgkin and Huxley [8], for which they both received the Nobel Prize in Physiology or Medicine in 1963. According to Kleinzeller [9], Andrew Huxley once said that, "If people had listened to what Overton had to say about excitability, the work of Alan [Hodgkin] and myself would have been obsolete."

Unfortunately for Heidenhain and Overton, their work did not conform with the *Zeitgeist* of the early twentieth century. At the time, much fundamental work on the theory of diffusion was being carried out by high-profile physicists and physical chemists, among them van't Hoff, Einstein, Planck, and Nernst. Of particular relevance for the distribution of ions across the cell membranes was the work of the Irish physical chemist Frederick Donnan [10] on the effect of nondialyzable salts. Therefore, it was natural that physiologists of the period would try to explain membrane transport in terms of passive diffusion alone, rather than adopt Overton's and Heidenhain's controversial hypothesis of ion pumping or active transport.

Donnan [10] suggested that if the cytoplasm of cells contained electrolytically dissociated nondialyzable salts (e.g., protein anions), which it does, small permeable ions would distribute themselves across the membrane so as to maintain electroneutrality in both the cytoplasm and the extracellular medium. Thus, the cytoplasm would naturally tend to attract small cations, whereas the extracellular medium would accumulate anions. Referring back to Figure 1.1, one can see that this idea could explain the distribution of K^+ and Cl^- ions across the cell membrane. However, the problem is that the so-called Donnan equilibrium doesn't explain the distribution of Na^+ ions. Based on Donnan's theory, any permeable ion of the same charge should adopt the same distribution across the membrane, but the distributions of Na^+ and K^+ are in fact the opposite of one another. To find a rational explanation for this inconsistency, many physiologists concluded that, whereas cell membranes were permeable to K^+ and Cl^- ions, they must be completely impermeable to Na^+ ions. The logical consequence of this was that the Na^+ concentration gradient should have originated at the first stages of the cell division and persisted throughout each animal's entire life. This view was an accepted doctrine for the next 30 years following the publication of Donnan's theory [11].

In the late 1930s and early 1940s, however, evidence was mounting that the idea of an impermeant Na^+ ion was untenable. In this period, radioisotopes started to become available for research, which greatly increased the accuracy of ion transport measurements. A further stimulus at the time was the development in the United States of blood banks and techniques for blood transfusion, during which researchers were again investigating the distribution of ions across the red blood cell membranes and the effects of cold storage. Researchers in the United States, in particular at the University of Rochester, Yale University, and the State University of Iowa, were now the major players in the field, among them Fenn, Heppel, Steinbach, Peters, Danowski, Harris, and Dean. Details of the experimental evidence that led to the universal discarding of the notion of an impermeant Na^+ ion and the reemergence of the hypothesis of an active Na^+ pump located in the cell membrane of both excitable and nonexcitable cells are described elsewhere [4, 12, 13]. Here it suffices to say that by the middle of the twentieth century the active transport of Na^+ had become an established fact.

The enzyme responsible for active Na^+ transport, the Na^+,K^+-ATPase, which is powered by the energy released from ATP hydrolysis, was isolated by Jens Christian Skou of the University of Aarhus, Denmark, in 1957 [14]. This was the first ever ion-transporting enzyme to be identified. Almost 40 years later, in 1997, when all possible doubt that Skou's Na^+,K^+-ATPase incorporated the complete active transport machinery for sodium and potassium ions and the broad significance of his discovery was clear, he received the Nobel Prize in Chemistry.

One reason why the concept of the active transport of Na^+ took so long to be accepted was probably the perception that it represented a waste of a cell's valuable energy resources. However, rather than think of the pumping of ions across a membrane as energy expenditure, in fact it is more helpful and more accurate to describe it as an energy conversion process. In the case of the Na^+,K^+-ATPase, the energy released by ATP hydrolysis is stored as Na^+ and K^+

electrochemical potential gradients across the membrane. Therefore, the energy can be released again whenever Na^+ or K^+ diffuse passively across the membrane. The Na^+ and K^+ electrochemical potential gradients established by the Na^+,K^+-ATPase across the plasma membrane of all animal cells thus provide the driving force for diffusion of Na^+ and K^+ through all plasma membrane Na^+- and K^+-selective ion channels, which, for example, is the basis of the production of action potentials in nerve and muscle. The Na^+ electrochemical potential gradient created by the Na^+,K^+-ATPase also serves as a secondary source of energy to drive the active uptake or extrusion of other ions or metabolites across the plasma membrane by transporter membrane proteins. For example, the reabsorption of glucose into the bloodstream in the kidney is driven by the energy released by the simultaneous coupled passive flow of Na^+ into the cytoplasm of the epithelial cells lining the kidney collecting tubules. As these examples demonstrate, the realization that the cell membrane is permeable to Na^+ ions and requires a sodium pump to keep the ions out was pivotal for the understanding of membrane transport processes in general, and the change in thinking that this realization generated no doubt contributed to the later discovery of many other membrane protein transport systems, including channels and transporters.

Now we will concentrate for the moment on channels alone. When Hodgkin and Huxley proposed [8] consecutive changes in Na^+ and K^+ membrane permeability of nerve as the origin of the action potential in 1952, their hypothesis was based on the mathematical fitting of kinetic equations to their recorded data. Impressive as their conclusions were, their data still provided no clue as to the molecular origin of the changes in Na^+ or K^+ permeability. A major step forward occurred in 1964 when Narahashi et al. [15] discovered that tetrodotoxin (TTX), a paralytic poison found in some edible (with caution) puffer fish, blocks the action potential in nerve axons by inhibiting the Na^+ conductance but without any effect on the K^+ conductance. This clearly demonstrated that there must be separate pathways or channels for Na^+ and K^+ ions in the membrane. Still the chemical nature of the channels was unclear, but after this discovery TTX became an invaluable tool for the identification of the source of the Na^+ conductance.

The next major advance in the channel field occurred through the application of biochemical purification procedures. The electric eel, *Electrophorus electricus*, is capable of producing voltages as high as 600 V along the whole animal. As one might imagine, its specialized electrical properties made it a prime source for the isolation of the molecules responsible for voltage changes across the cell membranes. In 1978 Agnew et al. [16] succeeded in extracting and purifying a 230 kDa protein that had a high affinity for TTX. After it was shown in 1984 [17] that synthetic vesicles, in which the purified protein had been reconstituted, displayed Na^+ currents that could be inhibited by TTX, there was no longer any doubt that the Na^+ channel had been isolated and that it was indeed a membrane protein. More details on the history of ion channel research including more recent developments can be found in a fine review by Bezanilla [18].

Now, finally in this brief historical overview, we turn our attention to transporters. Particularly in the intestine and in the kidney, many metabolites, including sugars and amino acids, need to be absorbed or reabsorbed, respectively, into the bloodstream. In the early 1960s, shortly after Skou's discovery of the Na^+,K^+-ATPase [14], Robert Crane first suggested [19, 20] that the intestinal absorption of sugar was coupled to the influx of Na^+ into the cell, that is, that the energy released by the passive diffusion of Na^+ into the cell was utilized to absorb sugars. His hypothesis was based in part on the fact that sugar absorption was already known to be dependent on the presence of Na^+ in the medium. Roughly 10 years later, using isolated intestinal epithelial cells, Kimmich [21] showed the sugar uptake system was located in the plasma membrane of the cells and not between the cells of an intact tissue or epithelium. That the Na^+/glucose coupled transport system is in fact a membrane protein was shown in a similar way to that described above for the Na^+ channel, that is, by isolation of the protein from tissue, reconstitution in vesicles, and the demonstration that the reconstituted system carried out Na^+-dependent active transport of glucose across the vesicle membrane [22]. For these experiments, kidney tissue was used because of the higher concentration of the protein that could be isolated in comparison to intestine.

A useful question to ask here is why such coupled transport systems, at least in animals, all utilize the Na^+ gradient across the membrane and not the K^+ gradient. The answer is quite simple. The distribution of K^+ ions across the plasma membrane is quite close to equilibrium, that is, the normal resting electrical potential across the membrane is quite close to what one would calculate theoretically based on the equilibrium theory of Nernst for electrical diffusion potentials. In contrast, the distribution of Na^+ is far from equilibrium. Therefore, the passive diffusion of Na^+ in through a transporter protein releases much more energy that can be used for metabolite uptake or extrusion than the passive diffusion of K^+ out.

At roughly the same time that Crane hypothesized the coupling of the energy stored in the Na^+ gradient to glucose absorption, the idea of energy storage in electrochemical potential gradients was also taken up by Peter Mitchell [23] when he proposed the chemiosmotic theory of oxidative and photosynthetic phosphorylation, for which he received the 1978 Nobel Prize in Chemistry. Central to Mitchell's hypothesis was the existence of a membrane-bound ATPase in mitochondria or chloroplasts that utilized the H^+ gradient built up across their inner membranes for the conversion of ADP to ATP. This enzyme, now known as the ATP synthase or F_0F_1-ATPase, cannot strictly be classified as a transporter, because the energy released as H^+ ions that flows through it across the membrane is not used for the transport of other ions or metabolites, but rather it is converted into chemical energy in the form of ATP. Closely related molecular machines are the bacterial flagellar motors, which also use the energy of an H^+ gradient, but in this case the energy is released in mechanical form as flagellar rotation.

Concluding this historical overview, one can say that the existence of membrane-embedded proteins that act as pumps, channels, and transporters and the means by which they gain their energy to carry out their transport processes were firmly established by the early 1980s. Since that time, further major advances have been made into the details of how they operate. One significant advance was the development of patch-clamp techniques by Neher and Sakmann [24], which enabled the opening and closing of single channels to be directly recorded, and for which they received the Nobel Prize for Physiology or Medicine in 1991. Another major advance has been the resolution of the atomic structure of membrane proteins by X-ray crystallography. The first membrane protein to be crystallized and have its structure determined by X-ray diffraction was that of a bacterial photosynthetic reaction center [25], for which Michel, Deisenhofer, and Huber received the Nobel Prize in Chemistry in 1988. After a slow start, the structures of other membrane proteins at atomic resolution are now being determined at an increasingly rapid rate. With structures becoming available, this has allowed the application of molecular dynamics simulations and other theoretical techniques to obtain an improved chemical understanding of how pumps, channels, and transporters work. Both patch-clamp techniques and molecular simulations are topics of later chapters of this book.

1.2   ENERGETICS OF TRANSPORT

How does one distinguish between pumps, channels, and transporters? The decisive criterion is whether or not energy is required for transport. However, to decide whether the transport of an ion requires energy or not, it is not sufficient to consider its concentration, c, on each side of the membrane. The electrical potential difference, V_m, across the membrane also contributes to the energetics of the process. Therefore, one needs to define the electrochemical potential difference, $\Delta\mu$, which for the transport of an ion into a cell is given by,

$$\Delta\mu = RT \ln \frac{c_{in}}{c_{out}} + zFV_m \tag{1.1}$$

and for the transport of an ion out of a cell is given by,

$$\Delta\mu = RT \ln \frac{c_{out}}{c_{in}} - zFV_m \tag{1.2}$$

In these equations, R is the ideal gas constant, T is the absolute temperature, z is the valence of the ion (e.g., +1 for Na^+, or +2 for Ca^{2+}), F is Faraday's constant, and V_m is the electrical potential difference across the membrane. In both equations V_m is defined as the potential inside the cell minus the potential outside the cell. The movement of an ion across a membrane for which $\Delta\mu$ is calculated to be negative involves a loss of free energy. If the movement is along an electrochemical potential gradient, it is a spontaneous process, and no energy is required. If $\Delta\mu$ is calculated to be positive, on the other hand, then the movement of the ion requires energy, the movement is against an electrochemical potential gradient, the process is non-spontaneous, and a source of energy would be required. Of course, if it is an uncharged metabolite that is moving across the membrane then the second term in Equations 1.1 and 1.2 disappears, and it is only the direction of the chemical potential gradient or the concentration gradient that determines whether or not the transport is spontaneous.

If no energy is required, that is, the transport occurs spontaneously along an electrochemical potential gradient ($\Delta\mu < 0$), then the transport is termed *facilitated diffusion*. In this case, the protein simply provides a pathway for the ions to move more easily through the membrane. This is the situation that occurs with a channel. If energy is required, that is, the transport occurs nonspontaneously up an electrochemical potential gradient ($\Delta\mu > 0$), then the transport is termed *active transport*.

There are a number of possible sources of energy in the case of active transport. If the energy comes directly from light, ATP, or from the energy released in a redox reaction, then this is termed *primary active transport*. All pumps are primary active transporters. If the energy is generated by the flow of an ion down an electrochemical potential gradient created by a pump, then this is termed *secondary active transport*. The Na^+/glucose cotransporter is an example of this. In such a situation, the transport of one species is down an electrochemical potential gradient ($\Delta\mu < 0$) and the transport of the other is up an electrochemical potential gradient ($\Delta\mu > 0$). When summed together, as long as the overall $\Delta\mu$ is negative the transport proceeds. For example, in the case of the Na^+/glucose cotransporter, as long as $\Delta\mu(Na^+) + \Delta\mu(\text{glucose}) < 0$, then both glucose and Na^+ are taken up into the cytoplasm of the cell.

1.3 MECHANISTIC CONSIDERATIONS

Apart from the difference in energetics, there are also important mechanistic differences between active transport and facilitated diffusion processes. In active transport, because the ions are transported against an electrochemical potential gradient, the enzyme's ion-binding sites should not be open to both sides of the membrane simultaneously. If this were to happen the efficiency of pumping would be drastically compromised. The ions should first be bound from one side of the membrane, become occluded within the protein via a conformational change, and then be released to the other side of the membrane via a conformational change. This is in contrast to the mechanism of ion channels, which have their ion-binding sites open to both sides of the membrane at once (see Fig. 1.2). Because of these differences in mechanism, the transport timescales of ion pumps and transporters are very different from those of ion channels.

A channel that is open to both sides of the membrane at once allows a rapid flux of ions across the membrane. For example, the flux through open Na^+ channels of the nerve membrane is approximately 10^7 ions/s, corresponding to an average time for the transport of a single ion of $0.1\,\mu s$. In contrast, ion pumps and transporters function on a much slower timescale. In their case, ion transport requires significant conformational changes to drive the ions or metabolites across the membrane. Because these conformational changes involve a large number of amino acid side chains whose intermolecular interactions need to be broken and formed, they typically have rate constants on the order of $100\,s^{-1}$ or slower. The overall turnover of an ion pump or transporter then usually occurs on a timescale of milliseconds to seconds, that is, four to six orders of magnitude slower than that of ion channels. This has important experimental consequences. Because of the large ion fluxes that they produce, the

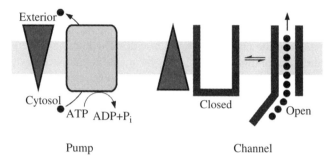

FIGURE 1.2 Ion-transporting membrane proteins. Channels can exist in an open state, in which ions move *down* an electrochemical potential gradient. No energy is required—the transport is termed facilitated diffusion. Pumps transport ions *against* an electrochemical potential gradient. The ion-binding sites are open alternately to the cytosol and the exterior. Energy is required—the transport is termed active transport. In the example shown the energy is derived from ATP hydrolysis. Reproduced from Ref. 26 with permission from Wiley.

opening and closing of single channels can be observed via the patch-clamp technique. Typically the observed currents are in the picoampere range. However, single ion pumps or transporters produce only very small currents across the membranes that are exceedingly difficult if not impossible to measure by electrophysiological means. An alternative approach for pumps is to use the whole-cell patch-clamp technique, whereby the ion flux through many pumps or transporters is recorded simultaneously.

The important point which we would like to make here is that because of these mechanistic and timescale differences, experimental techniques designed for the investigation of ion channels can often not be applied to pumps or transporters. In a similar fashion, techniques devised for research on pumps or transporters cannot generally be directly applied to ion channels. Therefore, in the following chapters one will find some experimental techniques that have been specifically designed with ion channel investigations in mind and others with ion pumps or transporters in mind.

1.4 ION CHANNELS

Ion channels can be classified according to what it is that causes them to open, that is, their gating mechanism. For a channel to allow ions to pass, it must undergo a conformational change from one or more inactive closed states. There are a number of mechanisms by which this conformational change might come about. Below we consider the variety of possible mechanisms one by one.

1.4.1 Voltage-Gated

First we consider voltage-gated channels, for example, the Na^+- and K^+-channels of nerve and muscle responsible for the action potential. If a channel contains movable charged or dipolar amino acid residues, then a change in voltage across the membrane would cause a change in the electric field across the protein. The change in field strength could then induce a translational or rotational motion of the charged or dipolar regions of the protein leading to a conformational change that opens the channel. The charged or dipolar regions are termed the voltage sensor of the channel.

1.4.2   Ligand-Gated

Another important class of ion channels is those that are ligand-gated. In this case, the binding of a ligand to the channel induces a conformational change that leads to channel opening. Classic examples of this type of channel are the nicotinic acetyl choline receptors, which are located in the plasma membrane of nerve and muscle cells. Binding of the neurotransmitter acetyl choline to the receptor stimulates its channel activity allowing Na^+ and K^+ to flow through it. This causes an increase in the membrane potential (termed depolarization), which subsequently causes the opening of voltage-gated Na^+ channels, and the production of the action potential necessary for muscle contraction.

1.4.3   Mechanosensitive

Mechanosensitive channels are a further important class of ion channels. These respond to mechanical deformations of the membrane in which they are embedded, for example, changes in membrane tension, thickness, or curvature. They can be found in all forms of cellular life. An example is the stretch-activated large conductance mechanosensitive channels (MscL) of bacteria, which were first discovered in *Escherichia coli* by Martinac et al. [27] in 1987. These channels allow the passage of ions, water, and small proteins. For bacteria they act as an osmotic emergency release valve when the cells find themselves in a hypotonic solution, for example, through the addition of fresh water to their bathing solution. Under such circumstances water would flow into the cell to re-establish osmotic equilibrium and the cells would swell. As the cells swell the stretching of the membrane activates the opening of the MscL channels allowing ions and small proteins to diffuse out. This inhibits the further influx of water and prevents the bacterial cells from bursting. In animal cells, the osmotic equilibrium across the plasma membrane and cell volume is maintained by the Na^+,K^+-ATPase, which continually pumps Na^+ out of the cell, thus preventing water influx. However, the Na^+,K^+-ATPase isn't present in bacterial cells, hence their requirement for mechanosensitive channels.

1.4.4   Light-Gated

Light-gated channels are a recently discovered class of ion channels. At this stage, only two naturally occurring light-gated ion channels are known to exist—channelrhodopsin-1 and channelrhodopsin-2. In 2002, channelrhodopsin-1 was shown by Nagel et al. [28] to act as an H^+ channel via electrophysiological measurements after expression in *Xenopus* oocytes. In the following year, the same group showed [29] that channelrhodopsin-2 is a cation-selective ion channel. Both of these channels contain the light-isomerizable chromophore all-*trans*-retinal, which is linked to the polypeptide chain of the protein via a protonated Schiff base. The chromophore undergoes a photoisomerization from the all-*trans* to the 13-*cis* state on absorption of a photon of the appropriate wavelength. This induces a conformational change of the protein causing the channel to open. The outward flow of ions then causes a rapid depolarization of the cell. The retinal can relax spontaneously back to the all-*trans* state within milliseconds, closing the channel and stopping the further flow of ions.

In unicellular green algae, where they occur naturally, the channelrhodopsins are involved in the control of phototaxis, that is, movement in response to light. However, their discovery has had much more far-reaching consequences than simply providing a better understanding of algal phototaxis. The expression of channelrhodopsins in other cells is proving to be a valuable tool allowing researchers to optically control the stimulation of excitable cells such as neurons. Previously, this could only be done using microelectrodes. This new field of optically controlling genetically modified cells has been given the name Optogenetics.

1.5   ION PUMPS

Whereas ion channels can be classified in terms of the gating mechanism, ion pumps can be grouped by the source of energy they utilize for ion pumping. In principle, any reaction that releases energy could be coupled to a membrane protein for the uphill pumping of ions. Here we consider only the three most widely represented sources of energy. Others do exist.

1.5.1   ATP-Activated

ATP-activated ion pumps utilize the energy released by the hydrolysis of ATP to ADP and inorganic phosphate to pump ions across membranes against their electrochemical potential gradients. The classic example is the Na^+,K^+-ATPase, which belongs to the P-type class of ion pumps. Other ion pumps belonging to this class include the Ca^{2+}-ATPase of sarcoplasmic reticulum, which plays a crucial role in muscle relaxation; the H^+,K^+-ATPase, which is responsible for the acidification of the stomach; and the H^+-ATPases in the plasma membrane of fungi and plants, which generate an H^+ electrochemical gradient across the membrane for later use as an energy source in nutrient uptake.

Depending on the relative concentrations of ATP, ADP, and inorganic phosphate, the free energy change, ΔG, for ATP hydrolysis is between -50 and $-63\,kJ\,mol^{-1}$ [30]. These values indicate that it is a very favorable reaction. However, if the reaction proceeded in isolation, then this energy would be released predominantly as heat. To be utilized for the work of ion pumping, the ATPases need to possess a mechanism for the coupling of ATP hydrolysis to ion transport. The way that this occurs is by phosphoryl transfer from ATP to the protein. Thus, all the P-type ATPases possess phosphorylated intermediates as part of their reaction cycle. After ions have already bound to the protein, the intermediate which is initially produced by reaction with ATP is a high energy state, termed E1P or E1~P. This state is able to undergo a conformational change by which it relaxes to a lower energy state, termed E2P, but in the process of this conformational change the bound ions are relocated across the membrane. Subsequently, the phosphorylated intermediate E2P releases inorganic phosphate to the surrounding aqueous medium, so that the net reaction is ATP→ADP+Pi and there is no net change in structure of the protein, but the result is that ions are pumped across the membrane.

1.5.2 Light-Activated

Light-activated ion pumps directly use energy from sunlight to pump ions across a membrane. The most prominent member of this class of ion pumps is bacteriorhodopsin, which pumps H^+ ions out of the cytoplasm of photosynthetic *Halobacteria*. Another is halorhodopsin, a light-driven chloride pump.

Bacteriorhodopsin was discovered by Walther Stoeckenius and Dieter Oesterhelt, both then working at the University of California. In 1967, Stoeckenius and Rowen [31] described the isolation of purple membrane patches from the salt-loving bacterium *Halobacterium salinarum* (formerly termed *H. halobium*), which thrives in the Californian salt lakes. A few years later, in 1971, Oesterhelt and Stoeckenius [32] showed that the purple patches contain a single protein, which they named bacteriorhodopsin because it contains the chromophore retinal, as does the visual pigment rhodopsin of humans and other animals. Furthermore, in 1973 they proposed [33] that bacteriorhodopsin acts as a light-driven H^+ pump and that the H^+ electrochemical potential gradient which the protein builds up is utilized by the cell for ATP synthesis, in agreement with Mitchell's chemiosmotic theory [23] which at the time was still controversial. This proposition was confirmed shortly afterwards by Racker and Stoeckenius [34], who reconstituted bacteriorhodopsin in vesicles together with mitochondrial ATP synthase and were able to demonstrate light-driven synthesis of ATP. These early experiments with bacteriorhodopsin were decisive in promoting the general acceptance of Mitchell's chemiosmotic theory for which he won the Nobel Prize in Chemistry only 5 years later.

The cyclic photochemical reaction that bacteriorhodopsin undergoes is very similar to that already described for the channelrhodopsins. The chromophore all-*trans* retinal is attached as a protonated Schiff base to an amino side chain of a lysine residue of the protein. On absorption of a photon the chromophore undergoes a photochemical conversion to the 13-*cis* state linked to a conformational change of the protein. The crucial point is that photon absorption causes a massive drop in the pK_a of the Schiff base, that is, a significant increase in acidity. In the dark, the pK_a of the Schiff base is around 12 and thus protonated at neutral and even slightly basic pH. After photon absorption and the conversion of retinal to the 13-*cis* state, the pK_a of the Schiff base drops by 4–5 pH units, thus promoting the loss of a proton. The drop in pK_a of the Schiff base is due to a change in its electrostatic environment caused by the protein's conformational change, that is, the proximity to other charged amino acid side chains. Subsequent to the loss of the proton, bacteriorhodopsin undergoes further conformational relaxation whereby the pK_a of the Schiff base increases so that a proton is again taken up. After about 6 ms, bacteriorhodopsin returns to its initial conformation with retinal in its all-*trans* conformation and a protonated Schiff base.

The trick in this entire photocycle is that, because of bacteriorhodopsin's location in the membrane and the access it has to the extracellular fluid and the cytoplasm in its different conformations, the H^+ ion that is released after photon absorption is released to the extracellular fluid, whereas the H^+ that is taken up after protein conformational relaxation is taken up from the cytoplasm. Therefore, in each photocycle of bacteriorhodopsin there is a net transport of one H^+ ion out of the cell and the protein acts as a proton pump.

When H^+ ions travel back through the ATP synthase into the cell cytoplasm, the energy released is used to convert ADP to ATP. Thus, bacteriorhodopsin allows the cell to carry out a very simple form of photosynthesis, mechanistically very different from the photosynthesis of green plants, although in both energy storage in the form of an H^+ electrochemical potential gradient is used.

1.5.3   Redox-Linked

The final group of ion pumps we discuss here are those in which the energy required for ion pumping derives from an electron transfer reaction, that is, a redox reaction. The best example is probably the cytochrome-*c* oxidase, the final enzyme of the respiratory electron transport chain of mitochondria, which is a redox-linked proton pump.

The cytochrome-*c* oxidase is situated in the inner mitochondrial membrane of eukaryotic cells and in the plasma membrane of some aerobic bacteria. It catalyzes the transfer of electrons from the reduced form of cytochrome-*c* to oxygen. The overall reaction can be written as follows:

$$4\text{cyt-}c\,(\text{Fe}^{2+}) + 4\text{H}^+ + \text{O}_2 \rightarrow 4\text{cyt-}c\,(\text{Fe}^{3+}) + 2\text{H}_2\text{O}$$

Because O_2 has a higher reduction potential than cyt-c(Fe^{2+}), the reaction is spontaneous, that is, ΔG is negative. However, as described earlier for ATP hydrolysis, if this reaction were carried out in a test tube in the absence of the cytochrome-*c* oxidase, then heat would simply be generated. The cytochrome-*c* oxidase not only catalyzes the reaction, but also couples the energy released to the pumping of H^+ ions across the membrane. Therefore, rather than release energy as heat, the free energy change is used to do the work of ion pumping and store the energy in the form of an H^+ electrochemical potential gradient. When the H^+ ions then move back across the membrane through the ATP synthase and release their energy once again, ADP is converted to ATP.

In effect, the cytochrome-*c* oxidase in eukaryotic cells is performing the same function as bacteriorhodopsin in *Halobacteria*. Both generate an H^+ electrochemical potential gradient for the synthesis of ATP. They simply use different mechanisms for generating the H^+ gradient. Whereas the *Halobacteria* derive their energy for H^+ pumping directly from sunlight, eukaryotic cells derive their energy from a redox reaction.

Ultimately, the energy for the production of ATP in eukaryotic cells comes from their food intake. The oxidative phosphorylation reaction of the electron transport chain of the mitochondria utilizes electron-rich reduced nicotinamide adenine dinucleotide, NADH, produced as a result of glycolysis and the citric acid cycle. In the electron transport chain electrons are transferred in three steps from NADH to O_2, each of which involves the storage of the energy released in the form of the H^+ electrochemical gradient. The cytochrome-c oxidase is the enzyme which catalyzes the final step of the chain. The enzymes catalyzing the two previous steps are the NADH-coenzyme-Q reductase and the coenzyme-Q-cytochrome-c oxidoreductase, which catalyze the transfer of electrons from NADH to coenzyme-Q and from coenzyme-Q to cytochrome-c, respectively. Both of these are simultaneously proton pumps, similar to the cytochrome-c oxidase.

1.6   TRANSPORTERS

In principle there are two different ways in which transporters can be classified. One is according to the direction in which ions are transported. Transporters transport one ion down an electrochemical potential gradient, that is, downhill in energy, and another ion or metabolite up an electrochemical potential gradient, that is, uphill in energy. Depending on the directions of the gradients across the membrane, both transported species could move in the same direction across the membrane or they could move in opposite directions. If the directions are the same, the transporter is termed a *symporter*. If the directions are opposite, the transporter is termed an *antiporter*.

Another way of classifying transporters is according to which ion moves downhill in energy and is therefore the ion that provides the energy for the active transport of the other ion or metabolite. Different ions are used by different forms of life and in different cells or organelles, with Na^+ and H^+ ion being the most common.

1.6.1   Symporters and Antiporters

Symporters, also called cotransporters, are used by cells for the uptake of essential metabolites or nutrients into the cytoplasm. The first symporter to be identified was the Na^+/glucose cotransporter (see Section 1.1), which is present in intestinal and kidney tubule cells. Because glucose provides the cells with energy, it needs to be concentrated into the cytoplasm. The Na^+/glucose cotransporter performs this function by coupling glucose uptake to the passive diffusion of Na^+ ions through the transporter as an energy source. The high electrochemical potential gradient for Na^+ across the membrane, with high concentration in the extracellular fluid and low concentration in the cytoplasm, is created by the outward pumping of Na^+ by the Na^+,K^+-ATPase. Many other cotransporters are known for other nutrients or essential metabolites, for example, amino acids, other sugars, other organic compounds, and inorganic ions.

Antiporters, also called exchangers, are used by cells for the extrusion of ions or organic compounds from the cytoplasm into the extracellular fluid. A good example is the Na^+/Ca^{2+} exchanger, found in the plasma membrane of heart muscle cells. Evidence that such an exchange system existed was first presented by Baker et al. [35] in 1969. In muscle cells, Ca^{2+} is a signaling agent required for muscle contraction. Therefore, after each action potential the Ca^{2+} concentration in the cytoplasm must be lowered to allow muscle relaxation. By coupling Ca^{2+} extrusion from the cell to the inward flow of Na^+ ions, the Na^+/Ca^{2+} exchanger provides an effective mechanism for rapidly reducing the cytoplasmic Ca^{2+} concentration. This is, however, not the only mechanism muscle cells possess for lowering the Ca^{2+} concentration. The sarcoplasmic reticulum Ca^{2+}-ATPase also actively pumps Ca^{2+} ions into the stores of the sarcoplasmic reticulum by using the energy of ATP hydrolysis. Thus, Ca^{2+} is available for release at the next action potential and the stimulation of muscle contraction.

Another important antiporter is the Na^+/H^+ exchanger. The activity of the transporter is reversible and depends on the polarity of the electrochemical potential gradients across the membrane. In animals it utilizes the influx of Na^+ into the cytoplasm to drive the efflux of H^+ ions and thus to control the intracellular pH. In archaea, bacteria, yeast, and plants the direction of transport is the opposite such that the transporter utilizes the influx of H^+ to drive the efflux of Na^+ ions and thus to increase salt tolerance. The reason for the different directions of transport is due to the different ions that are actively pumped across the membrane in different forms of life. This is discussed in the following section.

1.6.2 Na^+-Linked and H^+-Linked

All multicellular animals possess a Na^+,K^+-ATPase in their plasma membrane. As discussed earlier (Section 1.1) this creates a Na^+ electrochemical potential gradient across the membrane which is far from equilibrium. Therefore, transporters of animal cells make use of this Na^+ gradient to provide the energy for the transport of other ions or metabolites, for example, sugars or amino acids.

Archaea, plants, fungi, and bacteria, however, generally don't possess a Na^+,K^+-ATPase. Therefore, they need other mechanisms for absorbing nutrients or extruding other ions. Usually, their transporters are coupled to an H^+ electrochemical potential gradient across the membrane. For example, the H^+ gradient built up across the plasma membrane of *H. salinarum* by the light-driven H^+-pumping of bacteriorhodopsin is used not only to drive the synthesis of ATP by the ATP synthase (described in Section 1.5.2), it is also used to drive the extrusion of Na^+ from the cytoplasm of the bacterium by a H^+-linked transporter located in the plasma membrane. In other bacteria, fungi, and plants other mechanisms for creating the H^+ gradient are employed, but it is still generally the H^+ gradient which provides the energy for nutrient uptake.

Plants and fungi possess an H^+-ATPase in their plasma membranes [36]. This is a P-type ATPase, related to the Na^+,K^+-ATPase, and which performs similar roles to the Na^+,K^+-ATPase in animals. Instead of the H^+ pumping being light-driven as it is in *Halobacteria*, the energy from ATP hydrolysis drives the pumping process. Most of the transporters identified in plants are energized by the H^+ electrochemical gradient created by H^+-ATPases, for example, transporters for sugars, amino acids, peptides, nucleotides, and inorganic anions and cations.

Although animals don't possess H^+-ATPases in their plasma membranes, H^+ electrochemical potential gradients are produced across the inner mitochondrial membrane of every animal cell via the electron transport chain, which then drives ATP synthesis. This is one piece of evidence, amongst many others, that mitochondria evolved from ancestral bacteria (purple bacteria), that is, they were engulfed by eukaryotic cells to perform the function of ATP production. This is termed the endosymbiotic theory. In a similar fashion, the chloroplasts of green plants are thought to have evolved from ancestral cyanobacteria.

1.7 DISEASES OF ION CHANNELS, PUMPS, AND TRANSPORTERS

Diseases caused by ion channels, pumps, and transporters are widespread and as diverse as that of ion channels, pumps, and transporters themselves. Therefore, it is impossible here to provide comprehensive coverage of all the diseases with which they are associated. Instead, by selecting a small number of specific case studies, we hope to emphasize the importance of channels, pumps, and transporters for human health.

1.7.1 Channelopathies

Channelopathy is a term used in pathophysiology to describe any dysfunction of an ion channel. Many ion channel dysfunctions cause diseases in the neuromuscular system, such as epilepsy, ataxia, myotonia, and cardiac arrhythmia. Most channelopathies are inherited disorders, that is, they are the result of mutation in genes encoding channel proteins; while others are autoimmune diseases, meaning the body produces antibodies to its own channel molecules.

Cystic fibrosis was the first disorder to be discovered where the disease could be directly linked to a defect in an ion channel. The link between the channel and the disease was established by Tsui in 1989 [37]. Since then the number of disorders identified to be associated with ion channels has increased considerably. The study of ion channel diseases mostly occurs through identification of the chromosome locus of the disease and the protein coded by that gene. Once the gene has been identified, the next step is to express the mutant channel gene in, for example, HEK (human embryonic kidney cells) or *Xenopus* oocytes and study its activity using electrophysiological techniques (see Chapter 3) [37].

Table 1.1 shows that epilepsy (a disease affecting up to 1% of the world population, and which causes significant morbidity in humans) is caused by the dysfunction of several types of channels. Because of its widespread occurrence, we concentrate here on this disease [38].

Altering neuronal excitability in the brain by a mutation in an ion channel could drive a neuronal network into bursts of synchronized action potentials which

TABLE 1.1 Human Diseases Associated with Ion Channels and ABC Transporters.

Diseases Caused by Channel Disorders [38, 39]

Channel Type	Diseases				
Voltage-gated sodium channels	Epilepsy	Cardiac disorders	Muscle disorders	Malignant hyperthermia	
Voltage-gated and inwardly rectifying potassium channels	Epilepsy	Congenital hearing loss	Episodic ataxia with myokymia syndrome	Bartter syndrome	Hyperinsulinemic hypoglycemia of infancy
Nicotinic acetylcholine receptor channels	Epilepsy	Myasthenia gravis	Congenital myasthenia	Other human central nervous system disorders	
Voltage-gated calcium channels	Hypokalemic periodic paralysis	Episodic ataxia	Familial hemiplegic migraine	Congenital stationary night blindness	Malignant hyperthermia

Diseases Caused by Transporter Disorders [40]

Transporter Type	Diseases
ABCB1 (MDR1)	Cancer
ABCC1 (MRP1)	Cancer
ABCG2 (MXR)	Cancer
ABCB11 (SPGP)	Progressive familial intrahepatic cholestasis
ABCB4 (MDR2)	Progressive familial intrahepatic cholestasis

ultimately lead to an epileptic seizure; evidence of this is supported by the fact that most anticonvulsants used in clinical practice for treating epileptic seizures affect ion channels [38]. From Table 1.1 one can see three channels associated with epilepsy: voltage-gated sodium channels, brain voltage-gated potassium channels, and the neuronal nicotinic acetyl choline receptor.

In the case of the voltage-gated sodium channel, two genes, SCN1A and SCN1B, encode the α and β subunits of the channel, respectively. Mutations in both genes have been identified as possible candidates for the disorder [38, 39]. For example, generalized epilepsy with febrile seizures is caused by mutations in the SCN1A α subunit and in the SCN1B β subunit [41, 42].

Another form of epilepsy, benign neonatal epilepsy, is caused by a disorder in the function of brain voltage-gated potassium channels. In this case mutations to genes encoding two α subunits of the channels, KCNQ2 and KCNQ3, are associated with the disorder. After expression of the mutant channel in *Xenopus* oocytes, either no measurable current at all [43] or a severely reduced potassium current could be detected [44].

Nocturnal frontal lobe epilepsy is caused by a single mutation in the neuronal nicotinic acetyl choline receptor (nAChR)-α4 subunit. Expression of the mutant channel in *Xenopus* oocytes causes faster desensitization to acetyl choline and slower recovery from desensitization than that of the wild type channel [45, 46]. Mutation in the channel results in a decrease of the channel open time, a reduction of the single channel conductance, and an increase of the rate of desensitization. The possible connection between these activity changes and epilepsy is that the mutant channel might mediate the release of the inhibitory neurotransmitter GABA (γ-amino butyric acid). A reduction in nAChR function would then result in enhanced excitability of postsynaptic neurons and lower the seizure threshold [47, 48].

1.7.2 Pump Dysfunction

As an example of ion pump-related disease, we consider here disorder in the activity of the plasma membrane calcium ATPase (PMCA). The PMCA family of pumps is located in the plasma membrane and exports calcium ions from the cell; PMCA function is vital for regulating the amount of calcium within all eukaryotic cells. Improperly functioning PMCA proteins have been found to be associated with diseases, such as sensorineural deafness, diabetes, and hypertension [49].

In mammals four genes (ATP2B1–ATP2B4) encoding four proteins (PMCA1–PMCA4) have been found to be associated with some diseases. It has been proven that mutation in PMCA1, 2, and 4 is associated with disease in mice and man. For example, male mice lacking PMCA4 were found to be infertile due to loss of sperm motility. PMCA1 has a role in hypertension due to its function in the regulation of blood pressure through the alteration of calcium handling and vasoconstriction in vascular smooth muscle cells [50]. On the other hand, PMCA2 plays a role in causing ataxia and hearing loss. This was shown via the varying degrees of severity in mutants with different degrees of loss of PMCA2 function [51]. Furthermore, it was found that changes in PMCA expression are also involved in the development of other diseases including cataract formation, carcinogenesis, diabetes, cardiac hypertension and hypertrophy, and that the severity of these diseases may be associated with subtle changes in the expression level of the PMCA isoforms in those tissues [51].

1.7.3 Transporter Dysfunction

As an example of the effect of transporter dysfunction on human health, we concentrate here on the ATP-binding cassette (ABC) transporters. ABC transporters are found in all known organisms. Approximately 1100 different transporters belonging to this family have so far been described in the literature [52]. ABC transporters typically consist of two transmembrane domains and two nucleotide-binding domains [53]. There are approximately 50 known ABC transporters in humans, defects in 14 of which can cause 13 genetic diseases (including cystic fibrosis, Stargardt disease, adrenoleukodystrophy, and Tangier disease). ABCB4 is a member of the P-glycoprotein family of multidrug resistance transporters (see Table 1.1). Six liver diseases have been found to be associated with defects in the ABCB4 gene. These are drug-induced cholestasis, adult biliary cirrhosis, phospholipid-associated cholelithiasis syndrome, progressive familial intrahepatic cholestasis type 3, transient neonatal cholestasis, and intrahepatic cholestasis of pregnancy [40].

1.8 CONCLUSION

Now that we have introduced the main categories of channels, pumps, and transporters, explained the fundamental differences in their energetics and mechanisms, and described their importance in human health, the chapters that follow are each devoted to an experimental or theoretical method for researching the mechanisms by which these important proteins function on a molecular or atomic level. The experimental methods are divided into three main groups: electrical (Chapters 2–6), spectroscopic (Chapters 7–12), radioactivity- and atomic absorption-based flux assays (Chapters 13 and 14). Finally, we conclude with a chapter on computational techniques (Chapter 15).

We begin with electrical techniques, not only because they are probably the most widely used techniques to study pump, channel, and transporter function, but also for historical reasons. The realization that muscular activity has a fundamental electrical basis goes back to the experiments of the Italian scientist Luigi Galvani [54], who in the late eighteenth century made the revolutionary discovery that contraction of the legs of frogs could be stimulated by the application of an electrical potential via metal electrodes. Although Galvani could not have known at the time that what he was actually doing was inducing the opening of voltage-gated ion channels, his experiments provided the impetus for further electrical-based experiments and he can now be considered the father of the field of electrophysiology.

Following the description of electrical methods, we turn to spectroscopic methods that are increasingly becoming more widely used methods to study pumps, channels, and transporters, sometimes in combination with electrical methods, as in the case of voltage clamp fluorometry (discussed in Chapter 4), but also by themselves in systems for which electrical methods aren't applicable, for example, membrane fragments or small cell organelles, or in obtaining complementary mechanistic information not achievable by electrical means.

Following spectroscopic techniques, we direct our attention to the measurement of fluxes of ions and substrates across the cell plasma membranes based on the detection of radioactivity (Chapter 13) or atomic absorption (Chapter 14). No book on ion transport function would be complete without a discussion of the use of radioactivity. As described in Section 1.1, the use of radioisotopes, which became possible in the late 1930s, enabled rapid major advances in the ion transport field due to the far greater accuracy that could be achieved. In particular, radioisotopes allowed the hypothesis of Na^+ impermeability of cell membranes to be finally buried, a hypothesis which had been hampering progress for decades. As will be described in Chapters 13 and 14, radioactivity-based and atomic absorption-based techniques are still extremely powerful research tools and are particularly useful in the study of charge-neutral transport, when ion transport generates no electrical potential difference across the membrane. Finally, Chapter 15 describes the application of theoretical methods, which have only recently become feasible due to major advances in computer hardware and simulation procedures together with the increasing availability of high-resolution crystal structures of membrane proteins.

ACKNOWLEDGMENTS

R.J.C. received financial support from the Australian Research Council (Discovery Grants DP-12003548 and DP-150103518) and the Alexander von Humboldt Foundation.

REFERENCES

1. J.A. Nollet, Recherches sur les causes du Bouillonnement des Liquides, Histoire de l'Académie Royale des Sciences, Année MDCCXLVIII, (Paris, 1752) 57–104 (reprinted as an English translation in *J. Membr. Sci.* 100 (1995) 1–3).
2. R.J.H. Dutrochet, Nouvelles Observations sur l'Endosmose et l'Exosmose, et sur la cause de ce double phénomème, *Ann. Chim. Phys.* 35 (1827) 393–400.
3. S.S. Zaleski, Carl Schmidt, *Chem. Ber.* 27 (1894) 963–978.
4. R.J. Clarke, X. Fan, Pumping ions, *Clin. Exp. Pharmacol. Physiol.* 38 (2011) 726–733.
5. R. Heidenhain, Neue Versuche über die Aufsaugung im Dünndarm, *Pflugers Arch.* 56 (1894) 579–631.
6. E. Overton, Ueber die allgemeinen osmotischen Eigenschaften der Zelle, ihre vermutlichen Ursachen und ihre Bedeutung für die Physiologie, *Vierteljahrsschr. Naturforsch. Ges. Zürich* 44 (1899) 88–135.
7. E. Overton, Beiträge zur allgemeinen Muskel- und Nervenphysiologie. II. Ueber die Unentbehrlichkeit von Natrium- (oder Lithium-) Ionen für den Contractionsact des Muskels, *Pflugers Arch.* 92 (1902) 346–386.

8. A.L. Hodgkin, A.F. Huxley, A quantitative description of membrane current and its application to conduction and excitation in nerve, *J. Physiol.* 117 (1952) 500–544.

9. A. Kleinzeller, Ernest Overton's contribution to the cell membrane concept: A centennial appreciation, *News Physiol. Sci.* 12 (1997) 49–53.

10. F.G. Donnan, Theorie der Membrangleichgewichte und Membranpotentiale bei Vorhandensein von nicht dialysierenden Elektrolyten. Ein Beitrag zur physikalisch-chemischen Physiologie, *Z. Elektrochem. Angew. Phys. Chem.* 17 (1911) 572–581.

11. P.J. Boyle, E.J. Conway, Potassium accumulation in muscle and associated changes, *J. Physiol.* 100 (1941) 1–63.

12. J.D. Robinson, Steps to the Na^+-K^+ pump and Na^+-K^+-ATPase (1939–62), *News Physiol. Sci.* 10 (1995) 184–188.

13. P. De Weer, A century of thinking about cell membranes, *Annu. Rev. Physiol.* 62 (2000) 919–926.

14. J.C. Skou, The influence of some cations on an adenosine triphosphatase from peripheral nerves, *Biochim. Biophys. Acta* 23 (1957) 394–401.

15. T. Narahashi, J.W. Moore, W.R. Scott, Tetrodotoxin blockage of sodium conductance increase in lobster giant axons, *J. Gen. Physiol.* 47 (1964) 965–974.

16. W.S. Agnew, S.R. Levinson, J.S. Brabson, M.A. Raftery, Purification of the tetrodotoxin-binding component associated with the voltage-sensitive sodium channel from *Electrophorus electricus* electroplax membranes, *Proc. Natl. Acad. Sci. U. S. A.* 75 (1978) 2606–2610.

17. R.L. Rosenberg, S.A. Tomiko, W.S. Agnew, Single-channel properties of the reconstituted voltage-regulated Na channel isolated from the electroplax of *Electrophorus electricus*, *Proc. Natl. Acad. Sci. U. S. A.* 81 (1984) 5594–5598.

18. F. Bezanilla, Ion channels: From conductance to structure, *Neuron* 60 (2008) 456–468.

19. R.K. Crane, Intestinal absorption of sugars, *Physiol. Rev.* 40 (1960) 789–825.

20. R.K. Crane, Hypothesis of mechanism of intestinal active transport of sugars, *Fed. Proc. Fed. Am. Soc. Exp. Biol.* 21 (1962) 891–895.

21. G.A. Kimmich, Preparation and characterization of isolated intestinal epithelial cells and their use in studying intestinal transport, *Methods Membr. Biol.* 5 (1975) 51–115.

22. R.K. Crane, P. Malathi, H. Preiser, Reconstitution of specific Na^+-dependent D-glucose transport in liposomes by triton X-100-extracted proteins from purified brush border membranes of rabbit kidney cortex, *FEBS Lett.* 67 (1976) 214–216.

23. P. Mitchell, Coupling of phosphorylation to electron and hydrogen transfer by a chemiosmotic type of mechanism, *Nature* 191 (1961) 144–148.

24. B. Sakmann, E. Neher, Patch clamp techniques for studying ionic channels in excitable membranes, *Annu. Rev. Physiol.* 46 (1984) 455–472.

25. J. Deisenhofer, O. Epp, K. Miki, R. Huber, H. Michel, Structure of the protein subunits in the photosynthetic reaction centre of *Rhodopseudomonas viridis* at 3 Å resolution, *Nature* 318 (1985) 618–624.

26. R.J. Clarke, Probing kinetics of ion pumps via voltage-sensitive fluorescent dyes, in: E.M. Goldys (Ed.), *Fluorescence Applications in Biotechnology and the Life Sciences*, Wiley-Blackwell, Hoboken, NJ, 2009, pp. 349–363.

27. B. Martinac, M. Buechner, A.H. Delcour, J. Adler, C. Kung, Pressure-sensitive ion channel in *Escherichia coli*, *Proc. Natl. Acad. Sci. U. S. A.* 84 (1987) 2297–2301.

28. G. Nagel, D. Ollig, M. Fuhrmann, S. Kateriya, A.M. Musti, E. Bamberg, P. Hegemann, Channelrhodopsin-1: A light-gated proton channel in green algae, *Science* 296 (2002) 2395–2398.

29. G. Nagel, T. Szellas, W. Huhn, S. Kateriya, N. Adeishvili, P. Berthold, D. Ollig, P. Hegemann, E. Bamberg, Channelrhodopsin-2, a directly light-gated cation-selective membrane channel, *Proc. Natl. Acad. Sci. U. S. A.* 100 (2003) 13940–13945.

30. P. Läuger, *Electrogenic Ion Pumps*, Sinauer, Sunderland, MA, 1991, pp. 45–48.

31. W. Stoeckenius, R. Rowen, A morphological study of *Halobacterium halobium* and its lysis in media of low salt concentration. *J. Cell Biol.* 34 (1967) 365–393.

32. D. Oesterhelt, W. Stoeckenius, Rhodopsin-like protein from the purple membrane of *Halobacterium halobium*, *Nat. New Biol.* 233 (1971) 149–152.

33. D. Oesterhelt, W. Stoeckenius, Functions of a new photoreceptor membrane, *Proc. Natl. Acad. Sci. U. S. A.* 70 (1973) 2853–2857.

34. E. Racker, W. Stoeckenius, Reconstitution of purple membrane vesicles catalyzing light-driven proton uptake and adenosine triphosphate formation, *J. Biol. Chem.* 249 (1974) 662–663.

35. P.F. Baker, M.P. Blaustein, A.L. Hodgkin, R.A. Steinhardt, The influence of calcium on sodium efflux in squid axons, *J. Physiol.* 200 (1969) 431–458.

36. M.G. Palmgren, Plant plasma membrane H^+-ATPases: Powerhouses for nutrient uptake, *Annu. Rev. Plant. Physiol. Plant. Mol. Biol.* 52 (2001) 817–845.

37. L.C. Tsui, The spectrum of cystic fibrosis mutations, *Trends Genet.* 8 (1992) 392–398.

38. B. Dworakowska, K. Dolowy, Ion channels-related diseases, *Acta Biochim. Pol.* 47 (2000) 685–703.

39. D.C. Camerino, D. Tricarico, J.-F. Desaphy, Ion channel pharmacology, *Neurotherapeutics* 4 (2007) 184–198.

40. M.M. Gottesman, S.V. Ambudkar, Overview: ABC transporters and human disease, *J. Bioenerg. Biomembr.* 33 (2001) 453–458.

41. A. Escayg, B.T. MacDonald, M.H. Meisler, S. Baulac, G. Huberfeld, I. An-Gourfinkel, A. Brice, E. LeGuern, B. Moulard, D. Chaigne, C. Buresi, A. Malafosse, Mutations of SCN1A, encoding a neuronal sodium channel, in two families with GEFS+2, *Nat. Genet.* 24 (2000) 343–345.

42. R.H. Wallace, D.W. Wang, R. Singh, I.E. Scheffer, A.L. George Jr., H.A. Phillips, K. Saar, A. Reis, E.W. Johnson, G.R. Sutherland, S.F. Berkovic, J.C. Mulley, Febrile seizures and generalized epilepsy associated with a mutation in the Na^+-channel beta1 subunit gene SCN1B, *Nat. Genet.* 19 (1998) 366–370.

43. C. Biervert, B.C. Schroeder, C. Kubisch, S.F. Berkovic, P. Propping, T.J. Jentsch, O.K. Steinlein, A potassium channel mutation in neonatal human epilepsy, *Science* 279 (1998) 406–409.

44. H. Lerche, C. Biervert, A.K. Alekov, L. Schleithoff, M. Lindner, W. Klinger, F. Bretschneider, N. Mitrovic, K. Jurkat-Rott, H. Bode, F. Lehmann-Horn, O.K. Steinlein, A reduced K^+ current due to a novel mutation in KCNQ2 causes neonatal convulsions, *Ann. Neurol.* 46 (1999) 305–312.

45. S. Weiland, V. Witzemann, A. Villarroel, P. Propping, O. Steinlein, An amino acid exchange in the second transmembrane segment of a neuronal nicotinic receptor causes partial epilepsy by altering its desensitization kinetics, *FEBS Lett.* 398 (1996) 91–96.

46. A. Kuryatov, V. Gerzanich, M. Nelson, F. Olale, J. Lindstrom, Mutation causing autosomal dominant nocturnal frontal lobe epilepsy alters Ca^{2+} permeability, conductance, and gating of human alpha4beta2 nicotinic acetylcholine receptors, *J. Neurosci.* 17 (1997) 9035–9047.

47. C.A. Hubner, T.J. Jentsch, Ion channel diseases, *Hum. Mol. Genet.* 11 (2002) 2435–2445.

48. T.D. Graves, Ion channels and epilepsy, *Q. J. Med.* 99 (2006) 201–217.

49. E.F. Talarico, B.G. Kennedy, C.F. Marfurt, K.U. Loeffler, N.J. Mangini, Expression and immunolocalization of plasma membrane calcium ATPase isoforms in human corneal epithelium, *Mol. Vis.* 11 (2005) 169–178.

50. M. Brini, T. Cali, D. Ottolini, E. Carafoli, The plasma membrane calcium pump in health and disease, *FEBS J.* 280 (2013) 5385–5397.

51. B.L. Tempel, D.J. Shilling, The plasma membrane calcium ATPase and disease, in: E. Carafoli and M. Brini (Eds.), *Calcium Signaling and Disease: Molecular Pathology of Calcium*, Springer Science, New York, 2007, pp. 365–383.

52. S. Lorkowski, P. Cullen, ABCG subfamily of human ATP-binding cassette proteins, *Pure Appl. Chem.* 74 (2002) 2057–2081.

53. J. Stefková, R. Poledne, J.A. Hubácek, ATP-binding cassette (ABC) transporters in human metabolism and diseases, *Physiol. Res.* 53 (2004) 235–243.

54. L. Galvani, *De viribus electricitatis in motu musculari commentarius*, Accademia delle Scienza, Bologna, 1791.

2

STUDY OF ION PUMP ACTIVITY USING BLACK LIPID MEMBRANES

HANS-JÜRGEN APELL[1] AND VALERIJ S. SOKOLOV[2]

[1]*Department of Biology, University of Constance, Constance, Germany*
[2]*A. N. Frumkin Institute of Physical Chemistry and Electrochemistry, Russian Academy of Sciences, Moscow, Russia*

2.1 INTRODUCTION

Living cells depend vitally on the existence of active ion-transport proteins in their membranes to uphold the electrochemical potential gradients for various ion species such as Na^+, K^+, H^+, and Ca^{2+}. These proteins are typically named "ion pumps," and a detailed understanding of their function and transport mechanism is essential to gain insight into cellular metabolism. In terms of physics, the translocation of ions across the membrane is a charge movement through a medium with a low dielectric constant between two electrically conducting phases, and therefore, can be measured in an external measuring circuit as electric current and can be affected by the membrane potential, the electric potential difference across the membrane. This property of a transporter is called electrogenicity. When an ion pump transfers net charges per pump cycle, it is called electrogenic. If the same number of charges are transported in both directions during the pump cycle, it is called (overall) electroneutral. Nevertheless, even in this case, single reaction steps of the pump cycle comprise net charge movements within the membrane domain of the protein, which can be detected by methods described in this and other chapters of this book.

Pumps, Channels, and Transporters: Methods of Functional Analysis, First Edition. Edited by Ronald J. Clarke and Mohammed A. A. Khalid.
© 2015 John Wiley & Sons, Inc. Published 2015 by John Wiley & Sons, Inc.

To study the transport and kinetic properties of a particular ion pump, in most cases, it is necessary to discriminate the ion currents generated by the ion pumps in their respective membranes from those fluxes contributed by numerous other ion transporters present in the same membrane. Therefore, various approaches have been introduced to isolate, purify, and reconstitute the ion pumps as the only transport-protein species in a membrane system convenient for detailed functional investigations. In the following, we introduce an approach that makes use of the well-established black lipid membrane (BLM) technique.

2.2 FORMATION OF BLACK LIPID MEMBRANES

Artificial black lipid membranes with diameters of up to several millimeters and a thickness of less than 10 nm can be formed by a technique introduced by Mueller and Rudin [1, 2]. A Teflon cell with two compartments separated by a thin diaphragm that contains a small circular opening is filled with an appropriate chloride ion containing electrolyte (see Fig. 2.1a). Silver/silver chloride electrodes are immersed, and by a voltage source and a sensitive current meter a complete electric circuit is set up. A small aliquot of a lipid solution is "painted" over the hole in the diaphragm, which forms a lamella (see Fig. 2.1b). The lipid solution is typically lecithin dissolved in n-decane (1% w/v). This lamella becomes gradually thinner and shows bright interference colors when the thickness reaches the range below 1 μm. After some time, a discontinuous transition occurs to a layer thickness below 10 nm, where light reflection vanishes almost completely. These "black" spots increase in size until they cover the entire hole (but for a small torus at its edge). The thickness of the lipid layer

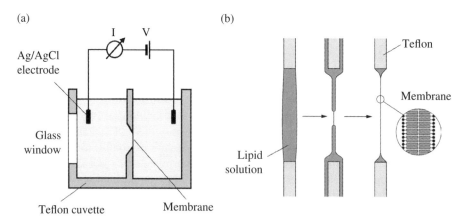

FIGURE 2.1   Setup for experiments with black lipid membranes. (a) Teflon cell and electric circuit to use BLMs for membrane-conductance and charge-dislocation experiments. (b) Formation of a BLM after the hole in the compartment separating diaphragm has been covered with a lamella of a lipid solution. Thinning of the lamella and formation of the BLM is a spontaneous process.

can be determined by electric measurements [3] to be in the range 5–7 nm depending on the choice of the lipid and solvent [4]. This range is close to the length of twice the length of a fully stretched lecithin molecule (see Fig. 2.1b), that is, consistent with the thickness of a lipid bilayer. To perform experiments with adsorbed membrane fragments or vesicles, modifications of the Teflon cells were introduced to allow for the application of UV-light flashes to produce ATP-concentration jumps (see Section 2.6.1) or to reduce the volumes of the electrolyte compartments [5, 6].

2.3 RECONSTITUTION IN BLACK LIPID MEMBRANES

The BLMs were initially developed as a model of biological membranes that allows reconstitution of proteins and study of their functions. The advantages of such an approach are convenient application of electrical measurements, direct access to the aqueous compartments on both sides of the membrane, and the possibility of investigating the functions of a reconstituted protein under well-controlled conditions such as lipid composition of the membrane and the choice of the electrolytes. The first attempt to reconstitute membrane proteins into the membrane was performed with an extract of proteins from nerve cells, which were assumed to be responsible for excitability of the membrane and are therefore referred to as "excitation inducing material" or EIM. This application was presented in the first publication on BLMs in 1962 [1].

2.3.1 Reconstitution of Na$^+$,K$^+$-ATPase in Black Lipid Membranes

Initially, the BLM was considered by many investigators as an ideal system to study the properties of membrane transport proteins, but later the enthusiasm decreased. Small molecules such as polypeptides, pore formers, and ion carriers were studied quite successfully in an abundance of publications. Larger membrane proteins such as ion channels and ion pumps refused to "cooperate." The reason is related to the conditions of reconstitution in the BLM. A widely used reconstitution method consists of two steps. First the protein is biochemically isolated from its native membrane by an appropriate detergent treatment, and subsequently reconstituted into liposomes by lipid addition and detergent removal. For the majority of transporters of interest, a method could be developed that resulted in active transporters in the vesicle membrane. In the second step, the transport proteins have to be transferred into the BLM. In principle, two different methods can be applied—either the proteins are transferred into a BLM by fusion of liposomes with a preformed membrane, or the vesicles are spread on the surface of the aqueous phase and a bilayer is formed from two monolayers containing the protein according to the so-called Mueller–Montal method [7]. The monolayer can be formed by spreading lipids with protein due to simultaneous opening of liposomes at the water–air interface, as it was shown by Schindler [8, 9]. However, these attempts to reconstitute large proteins or protein complexes responsible for active ion transport across the membrane were mostly unsuccessful. The most probable reason is that the unavoidable large fluctuations in the lateral pressure of the lipid monolayer promote denaturation of the tertiary or quaternary structure of the proteins. When ion pumps are studied, an additional constraint has to be taken into account. The typical electrical current that can be detected reliably by an experimental setup is about 10^{-12} A, and such a current is typically produced by single ion channel proteins. In contrast, the (net) current generated by an ion pump, for example, by a single Na$^+$,K$^+$-ATPase, is one elementary charge per enzymatic cycle during a period of about 10 ms. This corresponds to a current of about 10^{-17} A, which is five orders of magnitude below the current level needed. This means that about 10^5 or more active pump proteins have to be reconstituted in a BLM to measure a current above the noise level.

Attempts to reconstitute the Na$^+$,K$^+$-ATPase into BLMs began immediately after purified Na$^+$,K$^+$-ATPase preparations became available. A widely used procedure to isolate and purify the Na$^+$,K$^+$-ATPase was introduced by Jørgensen [10], and yields open membrane fragments of about 0.5 μm in diameter containing the purified protein in a high concentration of 5000–7000 molecules/μm^2. Suspensions of fragments are stable in aqueous solution for several days. The most widely used method for Na$^+$,K$^+$-ATPase reconstitution in liposomes is based on solubilization of the ion pumps in a mild detergent. After mixing with lipid solubilized in the same detergent, the detergent is removed by dialysis [11, 12] and an almost homogeneous population of unilamellar proteoliposomes is obtained [13].

The activity of the reconstituted ATPase depends on the lipid composition [12, 14]. The membrane thickness is an important parameter that affects the enzyme activity. There is an optimal thickness providing for maximal enzyme activity [14]. Usually, a BLM formed by the Mueller–Rudin technique [2] from a lipid solution in decane has a significantly larger thickness than that of a solvent free or biological membrane [4]. Therefore, an appropriate choice of lipids has to be made to obtain sufficient ATPase activity upon reconstitution. Early attempts to incorporate ATPase in a BLM by fusion of proteoliposomes with the membrane led, in many cases, only to fluctuations of the membrane conductance resembling the behavior of single ion channels [15]. The membrane conductance increased upon cleavage of the ATPase by chymotrypsin [16]. Similar ATP-dependent fluctuations were also observed upon insertion of a water-soluble fragment of the protein α-subunit [17]. These publications show that the method allows recording of electric current of denatured protein rather than of native ATPase. More successful was the ATPase reconstitution into solvent-free BLMs formed from monolayers and having a thickness closer to that of biological membranes [18–20]. A non-steady-state electric current was observed through such membranes upon fast release of ATP from caged ATP. The amplitude of the current initiated by caged ATP photolysis depended on the voltage applied to the membrane. However, the detected electric currents were small and not well reproducible. Therefore, this reconstitution technique has not been applied further for intensive functional studies of the Na$^+$,K$^+$-ATPase.

2.3.2 Recording Transient Currents with Membrane Fragments Adsorbed to a Black Lipid Membrane

The most successful assay to record electric signals of the Na$^+$,K$^+$-ATPase using a BLM was developed in the 1980s, and is based on the aligned and tight adsorption of open membrane fragments containing the ATPases in high density [5, 21–26].

In the presence of Mg^{2+} ions the membrane fragments adsorb in an oriented manner with the extracellular side of the membrane fragments facing the BLM surface. This approach allows the detection of electric currents in the external measuring circuit produced by the pump action of the Na^+,K^+-ATPase in the adsorbed membrane fragments, and avoids protein denaturation. This method was introduced in 1976 by Drachev et al., who were the first to record pump currents generated by bacteriorhodopsin [27]. It is based on a capacitive coupling of both membranes (see Section 2.4 and, for its application to solid supported membranes, Section 6.3.1). To trigger Na^+,K^+-ATPase-induced currents, the ion pumps have to be activated simultaneously. This process can be initiated by a concentration jump of a substrate species that shifts the ion pump from its initial equilibrium state. The most widely used substrate is ATP released from its inactive precursor, caged ATP, by a UV-light flash induced photolysis. The transient current generated in response to the ATP-concentration jump is a transfer of positive charge through the Na^+,K^+-ATPase in the membrane fragments, as has been shown in the first publications using this method [5, 21].

In the case of ion pumps that cannot be obtained in flat membrane fragments, which adsorb to a BLM, a modified experimental approach was developed. Vesicular membrane preparation isolated from cell membranes containing ion pumps in a high concentration were adsorbed to the BLM surface instead. These vesicles were also capacitively coupled to the BLM so that ATP-induced current transients could be detected in the external measuring circuit. This method has been applied in studies of the sarcoplasmic reticulum Ca^{2+}-ATPase [28, 29] and the gastric H^+,K^+-ATPase [30]. If an ion pump of interest is not available in a sufficiently high density in native membranes, it can be isolated and functionally reconstituted in lipid vesicles which are subsequently adsorbed to the BLM. This was performed with the K^+-translocating KdpFABC complex from *Escherichia coli* [31, 32].

In the case of Ca^{2+}-ATPase, the transient currents can also be observed upon release of Ca^{2+} ions from caged calcium, a photosensitive chelator [33]. Unfortunately, the methods of photolysis of caged compounds have definite limitations. For example, no caged compounds exist that can be used to produce concentration jumps of sodium or potassium ions. For these substrates, a more promising method is one in which the BLM is replaced by a stable supported membrane, where concentration jumps are performed by fast exchange of the solution [34, 35] (see Chapter 6).

Lipid bilayers with adsorbed membrane fragments can also be used for measurements of electrical signals other than substrate concentration jump induced transient currents. It has been shown that electrogenic ion movements in the Na^+,K^+-ATPase can be detected by measuring changes of the electric potential across the BLM detected by a voltage amplifier of virtually infinite impedance [5, 6]. Both methods of signal recording, current and voltage, are, however, equivalent to each other, because the current is proportional to the time derivative of the potential [5]. The effect of externally applied voltage modulations on the ion transport can be also studied in such a setup by detecting small changes of membrane capacitance and conductance (see Section 2.7).

2.4   THE PRINCIPLES OF CAPACITIVE COUPLING

The principle of current measurement across membrane fragments tightly adsorbed to a BLM can be represented by the equivalent circuit shown in Figure 2.2. Because conductance and capacitance of BLM and membrane fragments are connected in series, such a system can only be used to measure transient currents due to recharging the BLM and fragment capacitances. Under these conditions, the BLM serves as a

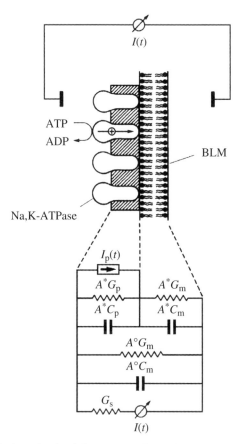

FIGURE 2.2   Equivalent circuit of the compound membrane system consisting of a black lipid membrane with adsorbed membrane fragments containing Na$^+$,K$^+$-ATPase in a high density. Upon an ATP-concentration jump the ion pumps generate charge translocations that sum to the pump current $I_p(t)$. This current induces a current $I(t)$ in the external measuring circuit that is recorded. G_p and C_p are the specific conductance and specific capacitance of the membrane fragments, respectively. G_m and C_m are the corresponding values of the BLM. A^* and A° are the areas of the covered and uncovered parts of the BLM. G_s is the inverse of the series resistance of the external detection circuit, including the resistance of the electrolyte, the electrodes and the amplifier. Adapted from Ref. 5 with kind permission from Springer Science and Business Media.

"capacitive electrode," and couples the charge transfer in the protein to a compensatory charge movement between both compartments separated by the BLM. The relationship between the pump current, $I_p(t)$, flowing across the membrane fragment and the current, $I(t)$, measured in the external circuit can be determined by an analysis of the circuit. In the simplest case, if the (leak) conductances of the gap between the BLM and fragment, G_p, can be neglected, the detected current, $I(t)$, is proportional to the pump current, $I_p(t)$, with a coefficient, α_C, specified by the ratio of the capacitances of BLM and membrane fragment:

$$I(t) = \alpha_C I_p(t); \quad \alpha_C = \frac{C_m}{(C_p + C_m)}, \tag{2.1}$$

where C_p and C_m are the specific capacitances of the membrane fragment and lipid bilayer, respectively. The coefficient can be easily evaluated. If the lipid bilayer is formed by the Mueller–Rudin technique, it contains solvent (usually decane), and its thickness is twice that of a solvent-free lipid bilayer (or the membrane fragments) [4]. In this case holds $C_p \approx 2C_m$, and the coefficient α_C is equal to ~1/3. This is, however, the simplest case, and this approximation is only valid if the duration of the current generated by the pump is short in comparison to the time constant of the discharge of the capacitance of the gap between the membrane fragment and BLM. In the more general case, the relationship between the current measured in the external circuit, $I(t)$ and the pump current, $I_p(t)$, is more complex [5]. It can be expressed by the equation

$$I_p(t) = \left(1 + \frac{C_p}{C_m}\right)\left[I(t) + \frac{G_p}{C_m + C_p}\int_0^t I(\tau)\,d\tau\right] \tag{2.2}$$

where G_p is the leakage conductance of the gap between the membrane fragment and the BLM.

2.4.1 Dielectric Coefficients

A detailed study of electrogenic pump currents requires the introduction of a parameter that quantifies the ion movements through the protein in different transport steps. It turned out to be practical and graphic to use the (dimensionless) fraction of the dielectric thickness of the membrane through which an ion or a charged group of the protein moves in each single step of the ion transport cycle. This parameter is referred to as a dielectric coefficient [36]. If the voltage across the membrane is constant, the dielectric coefficient can easily be determined because any charge movement inside the membrane dielectric will be compensated by a corresponding charge flow through the external measuring circuit. It is normalized so that the dielectric coefficient has the value of 1 when one elementary charge, e_0, is moved from one side of the membrane to the other. Values below 1 indicate that either the translocation takes place over a shorter distance or the ion movement is coupled to a simultaneous counter movement of charged groups or dipole rotations in the protein. When during a single turnover of an ion pump n ions of valency z are transported across the membrane, the sum of all dielectric coefficients is $n \cdot z$, and the total charge moved across the membrane is $Q = nze_0$.

There are several ways to determine dielectric coefficients, as can be illustrated by a simple example [36]. The protein together with the surrounding membrane is replaced by a capacitor with a homogeneous dielectric (see Fig. 2.3). Suppose the reaction translocates the charge q_{in} in the membrane. Under short-circuit conditions, the electric potential difference across the membrane remains zero and is maintained by the compensating charge, q_{out}, transferred through the external circuit. (q_{out} is the time integral of the measured short-circuit current.) An analysis of an equivalent circuit shows that q_{out} is smaller than q_{in}, and their ratio, α, is the dielectric coefficient. In the equivalent circuit shown (see Fig. 2.3), the dielectric coefficient can be determined by the relationship (Eq. 2.3) between the effective capacitance of the medium, C_{in}, in which the charge transfer takes place and that of the total membrane capacitance [37],

$$\alpha = \frac{q_{out}}{q_{in}} = \frac{1/C_{in}}{1/C_{in} + 1/C_{add}} = \frac{\delta_{in}/\varepsilon_{in}}{\delta_{in}/\varepsilon_{in} + \delta_{add}/\varepsilon_{add}}, \quad (2.3)$$

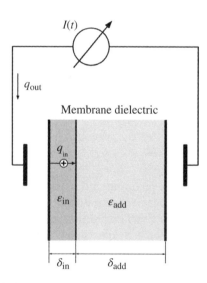

FIGURE 2.3 Schematic representation of hydrophobic interior of a BLM in which a charge, q_{in}, is moved across a distance of δ_{in} with the dielectric constant, ε_{in}. Under short-circuit condition in the external measuring circuit the charge, q_{out}, is transferred to compensate the intramembrane charge shift. With these numbers and the knowledge of the membrane thickness, $d = \delta_{in} + \delta_{add}$ and its dielectric constant, ε_{add}, the dielectric coefficient, α, can be calculated. In the simple case that $\varepsilon_{in} = \varepsilon_{add}$ the value of α is $\delta_{in}/d = q_{out}/q_{in}$.

where, C_{in}, δ_{in}, and ε_{in} are the capacitance, thickness, and dielectric permittivity, respectively, of the layer inside the membrane over which charge translocation takes place. C_{add}, δ_{add}, ε_{add} are the respective quantities of the remaining layer (see Fig. 2.3). The dielectric coefficient may also be determined by a different method in which the effect of an externally applied voltage, φ, on the charge transfer within the membrane is studied. When in such a process charge is transferred through a part of the membrane dielectric, only a fraction of the applied voltage, namely $\alpha \cdot \varphi$, will affect the transfer. The dielectric coefficient is defined as the fraction of the applied voltage that affects the charge transfer. It can be shown that it depends on the ratio of capacitances, C_{in} and C_{add}, in accordance to Equation 2.3, and is also equal to the coefficient α. Thus, both methods yield the same dielectric coefficient.

In the example described earlier, the relationship between the intramembrane charge transfer and the measured electric current in the external circuit was evident. The Na^+,K^+-ATPase performs, however, a complex transport process, and ion translocation may be accompanied by changes in the dielectric permittivity along the translocation pathway (e.g., upon the opening and closing of the ion access channel) and by movements of charged groups in the protein [36]. To characterize the transport electrogenicity in such a system, a more general definition of dielectric coefficient is required. It can be obtained on the basis of thermodynamics as the ratio of electric work in each separate step of the ion transfer through the membrane and the total electric work necessary to transfer an ion from one solution to the other. The total electric work of the ion transfer across the membrane is determined on the basis of thermodynamic considerations, and hence, is independent of the transfer mechanism. It is equal to the product of the ion charge transferred across the membrane and the electric potential difference between both solutions.

2.5 THE GATED-CHANNEL CONCEPT

When mechanistic concepts for ion pumps were developed to allow a quantitative description, it was soon generally accepted that enzymatic activity, such as ATP hydrolysis and ion transport, are not located together within the protein. Indeed, this has been confirmed in recent years by published protein structures with atomic resolution for a number of ion pumps. The first attempts to describe molecular mechanisms of pump function were published about 35 years ago. They introduced a "rotating carrier mechanism" [38] and a "channel mechanism for electrogenic ion pumps" [39]. The latter concept was extensively elaborated by Peter Läuger and applied to ATPases [36]. A crucial condition of the channel concept is the existence of two strictly regulated gates that encase the ion-binding sites. This requirement secures that (i) the access to the binding sites is possible alternately only from one of the two sides of the membrane, and that (ii) an occluded state is generated transiently in which the bound ions cannot exchange with either side. This mechanism ensures that a membrane-spanning ion pathway is never open end-to-end, and thus, no "short circuit" can occur. In fact, the pathway is divided into two alternatingly opened access channels (or half channels) that connect the binding sites to the respective aqueous phases of the membrane. A schematic representation of this concept is shown in Figure 2.4 in terms of a minimal reaction scheme. The fixed structural arrangement of the binding sites inside the membrane domain, in an environment of a low dielectric constant, leads to the property of electrogenic ion movements in the access channels; while all the other reaction steps of the pump cycle, such as enzyme phosphorylation/dephosphorylation and conformation transitions, are virtually electroneutral. As mentioned earlier, the electrogenic reaction steps are detectable in an external electric measuring circuit. The translocation of an ion through an ion pump is performed by a defined and ordered sequence of steps as has been shown for the Na^+,K^+-ATPase (see Fig. 2.5) in detailed analyses [40, 41] based on the so-called Post-Albers cycle [42, 43]. First, three Na^+ ions migrate through a (narrow) access channel which requires, at least partially, dehydration of the ion. This movement is electrogenic and diffusion controlled. Subsequent ion binding to a corresponding site is accompanied by dehydration of the ion and coordination by

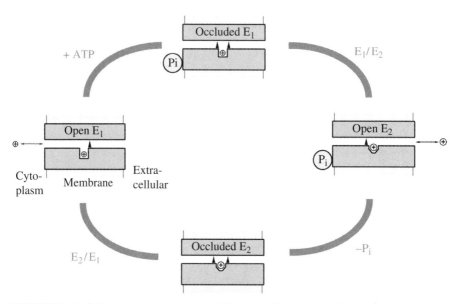

FIGURE 2.4　Schematic representation of the gated-channel model of a prototype P-type ATPase that transports two ion species in a ping-pong mode across the membrane. The pump cycle is reduced to four states. In the E_1 conformation the ion-binding sites are accessible from the intracellular side of the membrane. Phosphorylation by ATP triggers occlusion of the bound ion, and the binding site is closed off from both access channels before the conformational transition into the E_2 conformation takes place and the extracellular access channel is opened. Now both the ion species can exchange, usually promoted by a pronounced difference in binding affinity of both ion species. After the exchange, the ion-binding site is occluded again and the enzyme is dephosphorylated. After the conformational transition back to the E_1 conformation both ion species are exchanged again, and the next cycle starts. Adapted from Ref. 40 with permission from Elsevier.

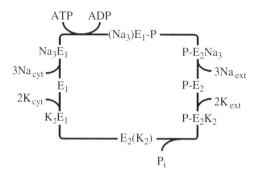

FIGURE 2.5 The Post-Albers scheme describes the pump cycle of the Na^+,K^+-ATPase under physiological conditions [42, 43]. In the E_1 conformation the ion-binding sites are accessible to the cytoplasm and allow exchange of two potassium ions for three sodium ions. In the E_2P conformation the reverse exchange is performed with the extracellular medium. In the "occluded" states, $(Na_3)E_1$-P and $E_2(K_2)$, the ions bound are trapped inside the membrane domain, unable to exchange with either aqueous phase. The upper half cycle is ATP driven and transports sodium ions outside the cell. The lower half cycle is controlled by enzyme dephosphorylation and transfers K^+ ions into the cytoplasm. The stoichiometry of the Na^+,K^+-ATPase is 3 Na^+/2 K^+/1 ATP. Therefore it is an electrogenic ion pump. Reproduced from Ref. 44 with permission from Elsevier.

oxygen atoms of amino acids that form a high-affinity environment. This process is electroneutral. When all the binding sites are occupied, the ion pump is phosphorylated by ATP and simultaneously the ions are occluded. This process is only transient and followed by a major transition from the P-E_1 into the P-E_2 conformation. Both processes are electroneutral. Then occlusion is suspended and the Na^+ ions are released sequentially with different electrogenicity. These features reveal that, on the one hand, the electrogenicity is related exclusively to the ion movements inside the access channels and that, on the other hand, these movements are much faster (in the microsecond range) compared to conformational transitions that occur in the millisecond range. The consequence is that reaction steps that precede the electrogenic step control the kinetics of the detected electric signal. This fact was exploited in numerous investigations, as will be shown later. A corresponding reaction sequence transfers K^+ ions in the opposite direction through the membrane. The conformational transition back to the E_1 conformation is triggered by enzyme dephosphorylation.

To obtain kinetic information on the molecular processes generating ion-translocation in the Na^+,K^+-ATPase, two different approaches may be applied—the relaxation technique and the periodic perturbation technique. In relaxation experiments, the ion pumps are initially in an equilibrium (or steady) state when an external variable such as voltage or a substrate concentration is abruptly changed and the equilibrium is perturbed. The time course of the approach to (or relaxation into) a new equilibrium state is detected, and the analysis of its time dependence reveals kinetic parameters of the pump's ion-transport mechanism. This technique has been widely used to study various ion transporters (see Chapter 5), carriers, and ion channels [45] but also the Na$^+$,K$^+$-ATPase [5, 21, 22, 24, 46]. An alternative to this approach is the periodic perturbation technique in which the application of a small external voltage in the form of a sine wave in the millivolts range produces minor periodic deviations from the equilibrium state. These small changes affect the electrogenic reaction steps, and cause modifications of the externally detectable admittance of the membrane containing the Na$^+$,K$^+$-ATPase [37, 47, 48]. It has been shown that both techniques allow the determination of the same kinetic parameters of the ion transport [49, 50]. These techniques and their application to the Na$^+$,K$^+$-ATPase are discussed in the following sections.

2.6   RELAXATION TECHNIQUES

2.6.1   Concentration-Jump Methods

Two different concentration-jump methods have mainly been applied to study the properties of the Na$^+$,K$^+$-ATPase: the stopped-flow technique and substrate release from caged compounds. The first method is presented and discussed in Chapter 7. The application of the latter technique is introduced here (see Chapter 5 for its application to secondary transporters). The predominantly used caged compound is caged ATP in the case of P-type ATPases [5, 21, 46, 49, 51–54]. However, caged phosphate has also been used to study specific parts of the pump cycle [55]. Caged ATP is a derivative of ATP with a photolabile blocking group attached to the terminal γ-phosphate of the molecule. Two different species of caged ATP have so far been applied—the P^3-1-(2-nitrophenyl)ethyl ester of ATP (NPE-caged ATP) [51, 52] and the P^3-(1-(3′,5′-dimethoxyphenyl)-2-phenyl-2-oxo)ethyl ester of ATP (DMB-caged ATP) [46, 56]. The commercially available NPE-caged ATP has pH-dependent release kinetics according to [6, 52]:

$$c_{\mathrm{ATP}}\left(t\right)=\vartheta\cdot c_{\mathrm{caged\ ATP}}\cdot\left(1-e^{-\lambda t}\right),\quad \lambda=2.2\cdot10^9\cdot10^{-\mathrm{pH}}\ \mathrm{s}^{-1}, \tag{2.4}$$

where ϑ is the yield of ATP release from caged ATP, which is dependent on the intensity of the UV flash (210–240 nm) and λ is the pH-dependent rate constant of the release reaction. The release rate of ATP from DMB-caged ATP is pH independent ($\lambda>10^5\,\mathrm{s}^{-1}$). Unfortunately, caged compounds of the transported ions, Na$^+$ and K$^+$, are not yet available. They would be extremely interesting compounds to reveal additional details of the ion-transport mechanism in the Na$^+$,K$^+$-ATPase. In contrast, caged calcium and caged proton, the ion species transported by the Ca^{2+}-ATPase of the sarcoplasmic reticulum, are available and have been applied to study kinetic properties of this closely related P-type ATPase [57–59].

When flat membrane fragments containing Na$^+$,K$^+$-ATPase are adsorbed to planar lipid bilayers, capacitive coupling allows a detection of charge movements in the ion pumps (cf. Section 2.2). When the Na$^+$,K$^+$-ATPase is equilibrated in the presence of a saturating NaCl concentration, almost all ion pumps are present in a state with 3 Na$^+$ bound to their sites in the E$_1$ conformation. After a UV-flash-induced fast release of ATP from its caged precursor, an ATP-concentration jump from 0 to 100 μM takes place and this process triggers a synchronized reaction sequence of all ion pumps present in the Na_3E_1 state (see Fig. 2.5),

$$Na_3E_1 + ATP \rightarrow (Na_3)E_1 \sim P + ADP \rightarrow P\text{-}E_2Na_3 \rightarrow P\text{-}E_2 + 3Na^+$$

A typical current response detected in the external measuring circuit is shown in Figure 2.6, trace A. The Na^+,K^+-ATPase was equilibrated in 150 mM NaCl, 10 mM $MgCl_2$, 0.5 mM caged ATP and pH 7. A transient current was observed with a peak amplitude of 350 pA about 26 ms after the UV flash. At $t > 140$ ms after the UV flash a small negative current was detected that faded out within about 1 s. Enzyme phosphorylation was inhibited when the experiment was repeated in the absence of Mg^{2+}, and the remaining current signal shows only a minor current artifact (trace B) that decayed within less than 10 ms, and therefore, did not significantly affect the recorded charge movements performed by the Na^+,K^+-ATPase. This current spike was caused most likely by an electrostatic interaction of the released ATP with the BLM surface.

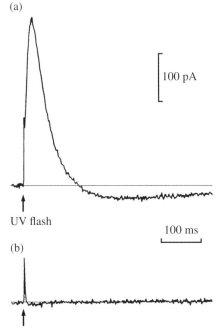

(a)

100 pA

UV flash

100 ms

(b)

FIGURE 2.6 Current signals produced by the Na^+,K^+-ATPase upon an ATP-concentration jump induced by photolytic release from its caged precursor. The membrane fragments with a high density of ion pumps were adsorbed to a BLM in a buffer containing 150 mM NaCl and 0.5 mM caged ATP, pH 7. The current was recorded in the presence of 10 mM $MgCl_2$ (trace A) and in the absence of $MgCl_2$ (trace B). Adapted from Ref. 5 with kind permission from Springer Science and Business Media.

When the experiment is performed in an electrolyte without K$^+$, the ATP-driven half cycle is fast while the subsequent return from the P-E$_2$ state to the E$_1$ conformation is slow [5]. Because, at a concentration of 0.5 mM caged ATP, much more ATP is released than necessary for a single turnover of all pumps present in the experiment, multiple turnovers are possible. When a second UV flash was applied to start another ATP-concentration jump at times greater than 1 s after the first flash almost no current transient was detected over a period of up to 100 s (see Fig. 2.7b, dashed line).

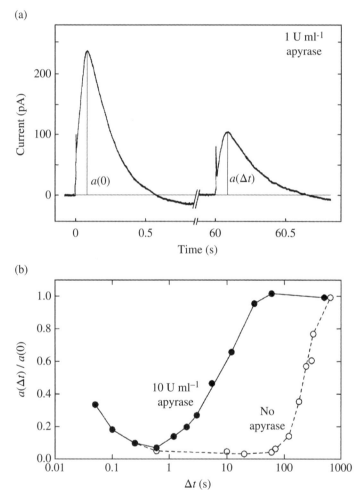

FIGURE 2.7   Recovery of the photo response in double-flash experiments in the absence and presence of apyrase VI in the electrolyte. (a) In the presence of 1 U/ml apyrase VI the maximum current amplitude, $a(\Delta t = 60\,\mathrm{s})$ is about 40% of the maximum current of the first flash, $a(0)$. (b) Relative current recovery in double-flash experiments in dependence of the waiting time, Δt. The recovery process is accelerated by a factor of 20 in the presence of 10 U/ml apyrase. At 100 U/ml it was even a factor of 200 [5]. Adapted from Ref. 5 with kind permission from Springer Science and Business Media.

To reduce the recovery period, apyrase VI, a soluble ATPase, may be added that degrades ATP but not caged ATP. In Figure 2.7a an experiment in the presence of 1 U/ml apyrase is shown. Under this condition the recovery was approximately 40% after 1 min. In Figure 2.7b a comparison of the relative recovery of the relative peak-current amplitude, $a(\Delta t)/a(0)$ is shown as a function of the waiting time, Δt, in the absence and presence of 10 U/ml apyrase VI. The latter condition allows a repetition of experiments every 100 s.

The detected current, $I(t)$, in the external measuring circuit is uniquely related to the pump current, $I_p(t)$, generated by the Na$^+$,K$^+$-ATPase upon the photochemical release of ATP. Analyzing the equivalent circuit representing the compound system of membrane fragments adsorbed to the BLM (Fig. 2.2) allows the calculation of $I_p(t)$ as introduced earlier (Eq. 2.2) with independently obtained values of C_p, C_m, and G_p. Applying Equation 2.2 to the recorded current, $I_{detected}$, transforms it to the pump current I_p as shown in Figure 2.8. As can be seen, the pump current is at its peak value about 3.7-fold larger than the detected current, $I(t)$, and at times larger than 400 ms a steady-state current was present that reflects repetitive turnovers of the pumps, observable as long as ATP is still present. The rather low steady-state amplitude obtained has to be expected because the experiments were performed in the absence of K$^+$ so that the reaction sequence from the state P-E$_2$ back to E$_1$ is rate limiting. In the presence of Na$^+$ and K$^+$ significantly larger currents could be induced [5, 21]. The biphasic shape of the time course of the pump current, I_p, can be well approximated by a phenomenological function,

$$I_p(t) = I_1 \cdot \exp(-k_1 t) + I_2 \cdot \exp(-k_2 t) + I_p^\infty \tag{2.5}$$

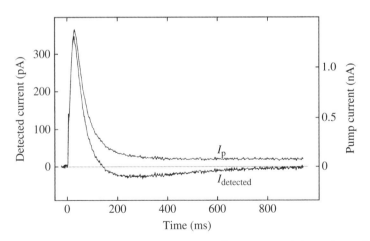

FIGURE 2.8 Intrinsic pump current $I_p(t)$, calculated from the recorded experimental current signal according to Equation 2.2. The parameters needed for the transformation, the ratio of $C_p/C_m \approx 2.7$, and $G_p/(C_m + C_p) = 2.5\,\mathrm{s}^{-1}$, were determined in independent experiments [5]. The steady state current found at $t > 400$ ms represents the slow turnover of the pumps in the Na$^+$-only mode of the Na$^+$,K$^+$-ATPase due to the pool of free ATP which accumulates after its photochemical release from caged ATP (before its complete degradation by apyrase VI). Adapted from Ref. 5 with kind permission from Springer Science and Business Media.

The current components, I_1, I_2, and I_p^∞, are time-independent constants. An optimum fit of I_p in Figure 2.8 was obtained with $k_1 = 67.3\,\text{s}^{-1}$ and $k_2 = 24.2\,\text{s}^{-1}$. The assignment of these kinetic parameters is model dependent and has been discussed elsewhere in detail [22, 36]. Besides such a phenomenological approach, efforts have been made to simulate the kinetic behavior on the basis of a detailed Post–Albers cycle and to assign rate constants in the reaction scheme on the basis of the Na^+ and ATP concentration dependences [22, 25, 36].

This method has so far been the only one that has allowed the proof that phosphorylation by ATP, occlusion of Na^+, and the ATP–ADP exchange are electroneutral, that is, they are not accompanied by (net) charge movements across the membrane domain of the Na^+,K^+-ATPase [5]. In these experiments, a known biochemical property of α-chymotrypsin was used. This protease cleaves a specific peptide bond of the Na^+,K^+-ATPase [60] that prevents the transition $(Na_3)E_1 \sim P \rightarrow P\text{-}E_2$ (Na_3) but not enzyme phosphorylation and Na^+ occlusion [61, 62]. The progress of cleavage was monitored as a decreasing-current amplitude, whereas the shape of the transient remained unaltered (see Fig. 2.9). This finding was supported recently when structures of the Na^+,K^+-ATPase were resolved with a resolution of a few Ångströms, and the ion-binding sites were located in the E_1 and E_2 conformation at the same positions in the membrane domain of the Na^+,K^+-ATPase [63–65].

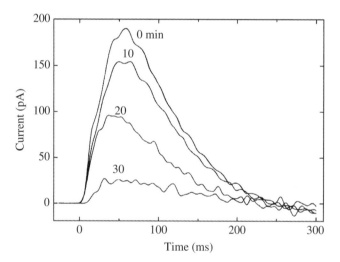

FIGURE 2.9 Effect of α-chymotrypsin cleavage on the Na^+,K^+-ATPase in membrane fragments adsorbed to a BLM. At $t = 0$, ATP was released from 0.5 mM caged ATP in buffer containing 10 mM NaCl, 50 mM tris, 5 mM $MgCl_2$, 20 mM dithiothreitol, pH 7 at 20°C. The current transient measured immediately before addition of 5 mg/ml protease is labeled "0 min." Current responses upon ATP-concentration jumps were recorded again after 10, 20, and 30 min. While the amplitude decreased, the shape of the transients remained constant. This indicates that the cleavage decreased the number of active Na^+,K^+-ATPase molecules but had no effects on their kinetics. Adapted from Ref. 5 with kind permission from Springer Science and Business Media.

2.6.2 Charge-Pulse Method

As mentioned earlier, release of ATP from the commercially available NPE-caged ATP has pH-dependent kinetics [5, 52] with a rate constant of $220\,s^{-1}$ at pH 7 and $2200\,s^{-1}$ at pH 6 (Eq. 2.4). This property limits the time resolution of the investigated electrogenic reaction steps of the Na^+,K^+-ATPase. This constraint is not severe when the rate-limiting conformational transition is studied. Ion-binding and release steps are, however, significantly faster. To overcome this restriction, the so-called charge-pulse method [45, 66] may be applied. This allows the detection of electrogenic reactions with time constants in the order of 50 ns. The method was applied to investigate Na^+ movements in the P-E_2 conformation of the Na,K-ATPase in membrane fragments adsorbed to a BLM [6]. It has been shown that the release of the first Na^+ ion is the major electrogenic process while the release of both subsequent Na^+ ions is only weakly electrogenic, due to a conformational relaxation during which the access channel is widened and filled with water molecules [6, 67].

2.7 ADMITTANCE MEASUREMENTS

Admittance measurements yield the same information on charge transfer in ion pumps as obtained by voltage clamp measurements with biological membranes. The equivalence of current-relaxation and admittance techniques was first shown in studies of the kinetics of passive ion transport through a BLM in the presence of ionophores or hydrophobic ions [68]. Transient currents of the Na^+,K^+-ATPase in response to voltage steps were measured in various biological membranes as current differences recorded in the absence and presence of ouabain, a Na^+, K^+-ATPase-specific inhibitor, and its analogs [69]. The transient currents were described by a simple model that considers sodium ion-movements in a narrow access channel, affected by the electric field, and (electroneutral) ion binding to a site inside the protein (see Fig. 2.10a). This model [70] describes the transient currents initiated by stepwise changes in the membrane potential as a function of both the potential and the concentration of sodium ions (see Fig. 2.10b). A corresponding model that describes the change in admittance of the same sodium ion movement in the access channel induced by a sine-wave voltage is shown for comparison in Figure 2.10c. In Sokolov et al. [49] the relationship between current relaxations and the admittance changes has been discussed.

The time dependence of the transient current initiated by a stepwise change of the voltage can be approximated by an exponential function (see Fig. 2.10b),

$$I(t) = I_0 \exp\left(\frac{-t}{\tau}\right), \tag{2.6}$$

The characteristic parameters, I_0 and τ, can also be determined by measuring changes of the membrane admittance. The measured current response to an externally applied sine-wave voltage is also sinusoidal with an amplitude and phase shift dependent on

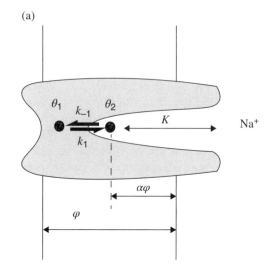

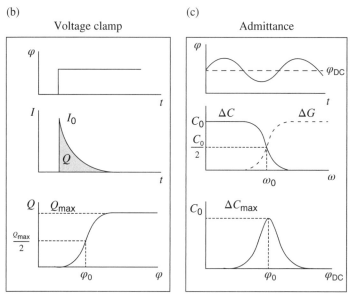

FIGURE 2.10 Comparison of two experimental methods for the study of electrogenic transport by the Na⁺,K⁺-ATPase with a simple access channel model. (a) Schematic representation of the ion pump. (b) In the voltage-clamp technique the voltage, φ, is changed stepwise causing a shift of the ion distribution within the access channel, which is measured as a transient current, $I(t)$. The integral of this current is the charge, Q, transferred across the membrane. The dependence of Q on the size of the voltage step, φ, is described by Equation 2.9. (c) In the admittance technique a sine-wave voltage is applied to the membrane. The detected alternating current arises from ion movement within the access channel and contributes to the membrane capacitance and conductance. Both are frequency dependent as shown in Equation 2.7. The magnitude of the capacitance increase, C_0, depends on the (constant) voltage offset, φ_{DC}, and is related to the charge movement, $Q(\varphi)$, in voltage-clamp experiments, $C_0(\varphi) = dQ(\varphi)/d\varphi$. Adapted from Ref. 49 with kind permission from Springer Science and Business Media.

the frequency of the applied voltage. The current can be represented as a sum of the real and imaginary part of the sine wave [49]. Their magnitudes can be derived from the exponential response (Eq. 2.6) by Fourier transformation without any model-dependent assumptions. For visualization, one can also derive them from the kinetic model of ion movements in the ion access channel of Figure 2.10a [49]. These parameters are transformed into more convenient quantities, the shifts of the membrane capacitance ΔC and conductance ΔG, which are obtained directly from measurements. Their dependencies on frequency can be expressed by so-called Lorentz functions

$$\Delta C = C_0 \frac{\omega_0^2}{\omega^2 + \omega_0^2}, \quad \Delta G = C_0 \omega_0 \frac{\omega^2}{\omega^2 + \omega_0^2}, \tag{2.7}$$

where ω is the cyclic frequency and ω_0 the corner frequency, which is related to the relaxation time, τ, of the exponential decrease of the current (Eq. 2.6), by

$$\omega_0 = \frac{1}{\tau}. \tag{2.8}$$

Equation 2.8 shows that the time constant, τ, and the characteristic circular frequency of the Lorentz function, ω_0, are equivalent. Other useful parameters can be determined from the current relaxation technique by the dependence of the transferred charge, Q, on the applied voltage, φ [70]. The charge is calculated as the integral of the transient current. This dependence is represented by the following function describing the charge distribution equilibrium between the aqueous solution and binding sites inside the ion pump (see Fig. 2.10b).

$$Q(\varphi) = \frac{Q_{max}}{1 + \exp\left[\alpha\beta(\varphi_0 - \varphi)\right]}, \tag{2.9}$$

where $\beta = F/RT$, F is the Faraday constant, R is the universal gas constant, and T is the absolute temperature. This equation allows the determination of the midpoint potential, φ_0, the maximal charge that can be transferred through the membrane, Q_{max}, and the dielectric coefficient, α, that determines which fraction of the applied voltage affects the charge transfer. The same parameters can also be determined via the admittance method when the dependence of the capacity increment, C_0, is measured at a low frequency with a voltage bias, φ_{DC}, (see Fig. 2.10c). As has been shown in Ref. 49, the voltage dependence of C_0 is equivalent to the dependence of ΔQ on φ, since it is the derivative of Q with respect to φ,

$$C_0(\varphi) = \frac{d}{d\varphi}(\Delta Q) = \frac{\alpha\beta Q_{max}}{4 \cdot \left[\cosh\left(\alpha\beta \frac{\varphi_0 - \varphi}{2}\right)\right]^2}. \tag{2.10}$$

The midpoint potential, φ_0, is the voltage of the maximal slope of $\Delta Q(\varphi)$ in current-relaxation measurements. In the admittance technique it is the voltage of the maximal capacitance increase. The dependence of the capacitance change on the voltage bias cannot be measured with Na^+,K^+-ATPase-containing membrane fragments adsorbed to a lipid bilayer. However, it was observed in giant-patch experiments with Na^+,K^+-ATPase, and this dependence was bell-shaped with a maximum capacitance dependent on the Na^+ concentration [50]. The maximal capacitance change, C_0, is obtained when sodium ions occupy half of the binding sites available. As is shown later, this property was used to determine the Na^+-dissociation constant by varying the ion concentration in the electrolyte. This approach can also be used to study transport processes in which several electrogenic steps are contributing.

2.8   THE INVESTIGATION OF CYTOPLASMIC AND EXTRACELLULAR ION ACCESS CHANNELS IN THE Na^+,K^+-ATPase

The model introduced earlier (see Fig. 2.10a) describes only the extracellular ion access channel. But this model can be generalized for more complex cases. If the transport process includes several steps that are characterized by dielectric coefficients, α_i, and time constants, τ_i, the frequency dependence of the capacitance and conductance are described by a sum of Lorentz functions with corresponding characteristic frequencies, ω_i [47].

The separate steps of non-steady-state sodium ion transport by the Na^+,K^+-ATPase were studied by measuring the ATP-driven changes of the membrane capacitance and conductance at various frequencies of the alternating voltage. This approach allows the study of ion movements on both the extracellular side of the Na^+,K^+-ATPase (after enzyme phosphorylation and at high Na^+ concentrations in the range of their dissociation constant of around 500 mM [54]) and on the cytoplasmic side (at low Na^+ concentrations near their dissociation constant of about 1–10 mM [71]).

At high sodium ion concentrations in the $P-E_2$ conformation, the main contribution to the admittance is due to ion transfer in the extracellular channel. When ion movements on the extracellular side of the Na^+,K^+-ATPase are studied with membrane fragments adsorbed to a BLM, one has to keep in mind that the extracellular membrane interface is facing the BLM surface, and the aqueous phase in the narrow cleft is not well defined. In early experiments the frequency dependencies of the capacitance and conductance change were approximated by a single Lorentz function with a frequency-independent component of the capacitance, indicating that the process consists of two electrogenic steps, that is, a slow first one and a fast second one [49, 72]. This was explained by a voltage-driven redistribution of sodium ions in the extracellular access channel coupled to the conformational transition which controls the observed slow kinetics. Later, more detailed measurements revealed that an additional fast component of the signal indicates an intermediate component of sodium-ion transport [48]. This component was assigned to the sodium ion binding/release process. Sodium ion transport through the extracellular channel has been studied in greater detail by the voltage-clamp technique on cells and giant patch membranes with well-defined access to the extracellular side. The non-steady-state currents generated by the Na^+,K^+-ATPase in the $P-E_2$ conformation in the presence of Na^+ but absence of K^+ were measured on mice cardiac myocytes [73], frog oocytes [70], and squid axons [74]. The most detailed investigation on squid giant axons revealed three current components: a slow one, assigned to the conformational transition E_1P/E_2P, an intermediate one, its rate controlled by binding/release of the third sodium ion, and an ultrafast step, attributed to the movement of two other sodium ions with lower electrogenicity in a wider access channel [67].

When experiments were performed with adsorbed membrane fragments at low sodium ion concentrations, a new transport step was detected, which was not detected by the current-relaxation method in electrophysiological investigations. It was found that the detected capacitance and conductance changes may have a negative sign after an ATP-concentration jump [47, 75], as shown in Figure 2.11a. The negative capacitance change was observed only at low Na^+ concentrations (<10 mM) and in the frequency range between 10 and 100 Hz (see Fig. 2.11c). It has to be assigned to an electrogenic process that disappears (or is significantly decreased) after enzyme phosphorylation and the concomitant Na^+ occlusion occurs. A viable candidate for this process is the electrogenic movement of the third Na^+ in the cytoplasmic access channel of the ion pump. At high Na^+ concentrations (e.g., 150 mM) the contribution of this process to the membrane capacitance decreased significantly because the cytoplasmic binding sites were all permanently occupied before the transition to the $P-E_2$ conformation, and the electrogenic movement of Na^+ in the extracellular access channel begins to contribute to the current signal (see Fig. 2.11b). As discussed earlier, the maximal effect on capacitance and conductance changes is expected at Na^+ concentrations in the half-saturating concentration range of the binding sites. In the E_1 conformation (on the cytoplasmic side) it is known to be 4 mM (at 5 mM Mg^{2+}, pH 7.2 [71]), and 8 mM (at 10 mM Mg^{2+}, pH 7.2 [76]). The experiments demonstrated that the negative capacitance changes disappeared at higher concentrations (see Fig. 2.11d). These experiments were satisfactorily explained by a theoretical model [47]. It allows the determination of the Na^+ dissociation constants of the cytoplasmic and extracellular binding sites of the Na^+,K^+-ATPase and the corner frequencies characterizing the rate constants of sodium-ion exchange on both sides of the membrane. The characteristic parameters of sodium ion transport on the extracellular side were similar to that determined earlier [48], while the rate constants of sodium ion transport in cytoplasmic access channels were determined for the first time. From the experimentally obtained amplitudes it can be concluded that the cytoplasmic access channel is less deep than the extracellular one, in agreement with other findings in the literature [6, 77, 78].

2.9 CONCLUSIONS

Electrogenic activities of ion pumps isolated in flat or vesicular membranes or reconstituted in proteoliposomes can be easily investigated in a compound membrane system formed by adsorption of the protein-containing membranes to a planar lipid bilayer acting as a capacitive electrode. Because charge movements in a single ion pump are far too small to be detected in any external measuring device, a synchronized

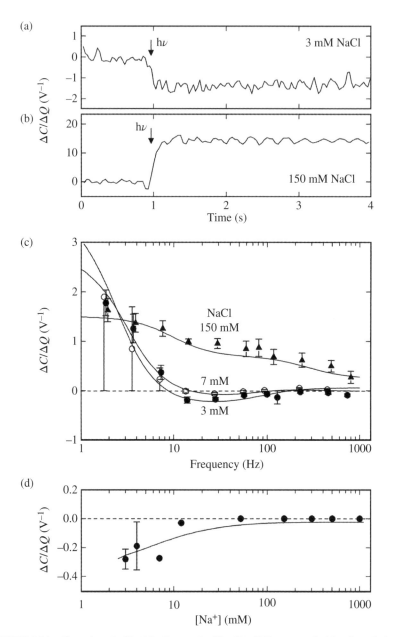

FIGURE 2.11 Cytoplasmic Na$^+$ binding to the Na$^+$,K$^+$-ATPase revealed by the admittance method. The aqueous solution contained 10 mM MgCl$_2$, 1 mM EDTA, 30 mM imidazole, pH 6.5, 100 μM caged ATP and the indicated concentrations of NaCl, (a) At a low sodium concentration of 3 mM, a decrease in the effective membrane capacitance was observed upon an ATP-induced transfer of the ion pumps due to the conformational transition E$_1$ → P-E$_2$. (The arrow indicates when ATP was released from the inactive caged ATP). (b) In the presence of 150 mM Na$^+$ the effective capacitance increased due to the onset of ion movements in the extracellular access channel. (c) Frequency dependence of the capacitance increments, ΔC, normalized by the total amount of charge transferred upon the ATP-concentration jump for three different Na$^+$ concentrations. (d) Sodium ion-concentration dependence of the negative component of the capacitance increment. The contributions are significant only below 10 mM Na$^+$. Adapted from Ref. 47 with kind permission from Springer Science and Business Media.

action involving virtually all ion pumps has to be triggered. Two different approaches have been used to accomplish this task. The first approach is the relaxation technique in which the equilibrium state of the ion pump is perturbed by a fast concentration jump of a substrate. The most frequently used substrate is ATP. Another possibility is the application of a charge pulse by which the membrane potential across the membrane is modulated within microseconds. In both cases the ion movements within the ion pump are detected when they relax into a new equilibrium state. The second approach is the periodic perturbation method which does not need significant deviations from the equilibrium state, but exploits the current response to an externally applied minor cyclic voltage. The analyses of the results from both methods provide the same kinetic information on elementary processes of the ion transport through ion pumps, and have been applied to analyze various reaction steps of the Na^+,K^+-ATPase pump cycle. Advantages of the BLM-related method are that isolated and purified preparations of ion pumps can be studied which (i) do not require subtraction of current contributions from other ion transporters, (ii) provide excellent reproducibility of the results, and (iii) allow the analysis of reaction steps in the pump cycle of minor electrogenicity that are not accessible to other (electrophysiological) methods. Handicaps of the method are (i) the composition of the aqueous solution on the extracellular side of the membrane is not easily accessible or alterable (i.e., in the narrow cleft between membrane fragments and BLM or the intravesicular volume), (ii) the possible fragility of the compound membrane, and (iii) the limited choice of substrates that may be used in fast concentration jump relaxation experiments.

ACKNOWLEDGMENTS

V.S.S. was financially supported by the Russian Foundation for Basic Research (project no. 13-04-01624).

REFERENCES

1. P. Mueller, D.O. Rudin, H.T. Tien, W.C. Wescott, Reconstitution of cell membrane structure in vitro and its transformation into an excitable system, *Nature* 194 (1962) 979–980.
2. P. Mueller, D.O. Rudin, H.T. Tien, W.C. Wescott, Methods for the formation of single bimolecular lipid membranes in aqueous solution, *J. Phys. Chem.* 67 (1963) 534–535.
3. T. Hanai, D.A. Haydon, J. Taylor, An investigation by electrical methods of lecithin-in-hydrocarbon films in aqueous solutions, *Proc. R. Soc. Lond. A* 281 (1964) 377–391.
4. R. Benz, O. Fröhlich, P. Läuger, M. Montal, Electrical capacity of black lipid films and of lipid bilayers made from monolayers, *Biochim. Biophys. Acta-Biomembr.* 394 (1975) 323–334.
5. R. Borlinghaus, H.-J. Apell, P. Läuger, Fast charge translocations associated with partial reactions of the Na, K-pump: I. Current and voltage transients after photochemical release of ATP, *J. Membr. Biol.* 97 (1987) 161–178.
6. I. Wuddel, H.-J. Apell, Electrogenicity of the sodium transport pathway in the Na, K-ATPase probed by charge-pulse experiments, *Biophys. J.* 69 (1995) 909–921.
7. M. Montal, P. Mueller, Formation of bimolecular membranes from lipid monolayers and a study of their electrical properties, *Proc. Natl. Acad. Sci. U. S. A.* 69 (1972) 3561–3566.

8. H. Schindler, Exchange and interactions between lipid layers at the surface of a liposome solution, *Biochim. Biophys. Acta-Biomembr.* 555 (1979) 316–336.

9. H. Schindler, U. Quast, Formation of planar membranes from natural liposomes. Application to acetylcholine receptor from *Torpedo*, *Ann. N. Y. Acad. Sci.* 358 (1980) 361.

10. P.L. Jørgensen, Isolation of $(Na^{+}+K^{+})$-ATPase, *Methods Enzymol.* 32 (1974) 277–290.

11. B.M. Anner, M. Moosmayer, On the kinetics of the Na:K exchange in the initial and final phase of sodium pump activity in liposomes, *J. Membr. Sci.* 11 (1982) 27–37.

12. F. Cornelius, Incorporation of $C_{12}E_{8}$-solubilized Na^{+},K^{+}-ATPase into liposomes: Determination of sidedness and orientation, *Methods Enzymol.* 156 (1988) 156–167.

13. E. Skriver, A.B. Maunsbach, P.L. Jørgensen, Ultrastructure of Na,K-transport vesicles reconstituted with purified renal Na,K-ATPase, *J. Cell Biol.* 86 (1980) 746–754.

14. M.M. Marcus, H.-J. Apell, M. Roudna, R.A. Schwendener, H.G. Weder, P. Läuger, $(Na^{+}+K^{+})$-ATPase in artificial lipid vesicles: Influence of lipid structure on pumping rate, *Biochim. Biophys. Acta-Biomembr.* 854 (1986) 270–278.

15. E.S. Hyman, Electrogenesis from an ATPase-ATP-sodium pseudo pump, *J. Membr. Biol.* 37 (1977) 263–275.

16. R. Reinhardt, B. Lindemann, B.M. Anner, Leakage-channel conductance of single $(Na^{+}+K^{+})$-ATPase molecules incorporated into planar bilayers by fusion of liposomes, *Biochim. Biophys. Acta-Biomembr.* 774 (1984) 147–150.

17. G.D. Mironova, N.I. Bocharnikova, N.M. Mirsalikhova, G.P. Mironov, Ion-transporting properties and ATPase activity of $(Na^{+}+K^{+})$-ATPase large subunit incorporated into bilayer lipid membranes, *Biochim. Biophys. Acta-Biomembr.* 861 (1986) 224–236.

18. M. Stengelin, A. Eisenrauch, K. Fendler, G. Nagel, H.T. van der Hijden, J.J. de Pont, E. Grell, E. Bamberg, Charge translocation of H,K-ATPase and Na,K-ATPase, *Ann. N. Y. Acad. Sci.* 671 (1992) 170–188.

19. A. Eisenrauch, E. Grell, E. Bamberg, Voltage dependence of the Na,K-ATPase incorporated into planar lipid membranes, *Soc. Gen. Physiol. Ser.* 46 (1991) 317–326.

20. E. Bamberg, H.J. Butt, A. Eisenrauch, K. Fendler, Charge transport of ion pumps on lipid bilayer membranes, *Q. Rev. Biophys.* 26 (1993) 1–25.

21. K. Fendler, E. Grell, M. Haubs, E. Bamberg, Pump currents generated by the purified $Na^{+}K^{+}$-ATPase from kidney on black lipid membranes, *EMBO J.* 4 (1985) 3079–3085.

22. H.-J. Apell, R. Borlinghaus, P. Läuger, Fast charge translocations associated with partial reactions of the Na,K-pump: II. Microscopic analysis of transient currents, *J. Membr. Biol.* 97 (1987) 179–191.

23. K. Fendler, E. Grell, E. Bamberg, Kinetics of pump currents generated by the Na^{+},K^{+}-ATPase, *FEBS Lett.* 224 (1987) 83–88.

24. G. Nagel, K. Fendler, E. Grell, E. Bamberg, Na^{+} currents generated by the purified $(Na^{+}+K^{+})$-ATPase on planar lipid membranes, *Biochim. Biophys. Acta-Biomembr.* 901 (1987) 239–249.

25. R. Borlinghaus, H.-J. Apell, Current transients generated by the Na^{+}/K^{+}-ATPase after an ATP concentration jump: Dependence on sodium and ATP concentration, *Biochim. Biophys. Acta-Biomembr.* 939 (1988) 197–206.

26. R.T. Borlinghaus, H.-J. Apell, P. Läuger, Fast charge translocations associated with partial reactions of Na,K-ATPase induced by ATP concentration-jump, *Prog. Clin. Biol. Res.* 268A (1988) 477–484.

27. L.A. Drachev, V.N. Frolov, A.D. Kaulen, E.A. Liberman, S.A. Ostroumov, V.G. Plakunova, A.Y. Semenov, V.P. Skulachev, Reconstitution of biological molecular generators of electric current. Bacteriorhodopsin, *J. Biol. Chem.* 251 (1976) 7059–7065.

28. K. Hartung, E. Grell, W. Hasselbach, E. Bamberg, Electrical pump currents generated by the Ca^{2+}-ATPase of sarcoplasmic reticulum vesicles adsorbed on black lipid membranes, *Biochim. Biophys. Acta-Biomembr.* 900 (1987) 209–220.

29. K. Hartung, J.P. Froehlich, K. Fendler, Time-resolved charge translocation by the Ca-ATPase from sarcoplasmic reticulum after an ATP concentration jump, *Biophys. J.* 72 (1997) 2503–2514.

30. M. Stengelin, K. Fendler, E. Bamberg, Kinetics of transient pump currents generated by the (H,K)-ATPase after an ATP concentration jump, *J. Membr. Biol.* 132 (1993) 211–227.

31. K. Fendler, S. Dröse, K. Altendorf, E. Bamberg, Electrogenic K^+ transport by the Kdp-ATPase of *Escherichia coli*, *Biochemistry* 35 (1996) 8009–8017.

32. D. Becker, K. Fendler, K. Altendorf, J.C. Greie, The conserved dipole in transmembrane helix 5 of KdpB in the *Escherichia coli* KdpFABC P-type ATPase is crucial for coupling and the electrogenic K^+-translocation step, *Biochemistry* 46 (2007) 13920–13928.

33. J.H. Kaplan, G.C. Ellis-Davies, Photolabile chelators for the rapid photorelease of divalent cations, *Proc. Natl. Acad. Sci. U. S. A.* 85 (1988) 6571–6575.

34. J. Pintschovius, K. Fendler, Charge translocation by the Na^+/K^+-ATPase investigated on solid supported membranes: Rapid solution exchange with a new technique, *Biophys. J.* 76 (1999) 814–826.

35. F. Tadini-Buoninsegni, G. Bartolommei, M.R. Moncelli, K. Fendler, Charge transfer in P-type ATPases investigated on planar membranes, *Arch. Biochem. Biophys.* 476 (2008) 75–86.

36. P. Läuger, *Electrogenic Ion Pumps*, Sinauer Associates, Sunderland, MA, 1991.

37. K.V. Pavlov, V.S. Sokolov, Electrogenic ion transport by Na^+,K^+-ATPase, *Membr. Cell Biol.* 13 (2000) 745–788.

38. A. Dutton, E.D. Rees, S.J. Singer, An experiment eliminating the rotating carrier mechanism for the active transport of Ca ion in sarcoplasmic reticulum membranes, *Proc. Natl. Acad. Sci. U. S. A.* 73 (1976) 1532–1536.

39. P. Läuger, A channel mechanism for electrogenic ion pumps, *Biochim. Biophys. Acta-Biomembr.* 552 (1979) 143–161.

40. H.-J. Apell, How do P-type ATPases transport ions?, *Bioelectrochemistry* 63 (2004) 149–156.

41. H.-J. Apell, Kinetic and energetic aspects of Na^+/K^+-transport cycle steps, *Ann. N. Y. Acad. Sci.* 834 (1997) 221–230.

42. R.W. Albers, Biochemical aspects of active transport, *Annu. Rev. Biochem.* 36 (1967) 727–756.

43. R.L. Post, C. Hegyvary, S. Kume, Activation by adenosine triphosphate in the phosphorylation kinetics of sodium and potassium ion transport adenosine triphosphatase, *J. Biol. Chem.* 247 (1972) 6530–6540.

44. G. Bartolommei, N. Devaux, F. Tadini-Buoninsegni, M.R. Moncelli, H.J. Apell, Effect of clotrimazole on the pump cycle of the Na,K-ATPase, *Biophys. J.* 95 (2008) 1813–1825.

45. P. Läuger, R. Benz, G. Stark, E. Bamberg, P.C. Jordan, A. Fahr, W. Brock, Relaxation studies of ion transport systems in lipid bilayer membranes, *Q. Rev. Biophys.* 14 (1981) 513–598.

46. V.S. Sokolov, H.-J. Apell, J.E. Corrie, D.R. Trentham, Fast transient currents in Na,K-ATPase induced by ATP concentration jumps from the P3-[1-(3′,5′-dimethoxyphenyl)-2-phenyl-2-oxo]ethyl ester of ATP, *Biophys. J.* 74 (1998) 2285–2298.

47. V.S. Sokolov, A.A. Scherbakov, A.A. Lenz, Yu.A. Chizmadzhev, H.-J. Apell, Electrogenic transport of sodium ions in cytoplasmic and extracellular ion access channels of Na⁺,K⁺-ATPase probed by admittance measurement technique, *Biochem. Mos. Suppl. Ser. A* 2 (2008) 161–180.

48. V.S. Sokolov, A.G. Ayuan, H.-J. Apell, Assignment of charge movements to electrogenic reaction steps of the Na,K-ATPase by analysis of salt effects on the kinetics of charge movements, *Eur. Biophys. J.* 30 (2001) 515–527.

49. V.S. Sokolov, S.M. Stukolov, A.S. Darmostuk, H.-J. Apell, Influence of sodium concentration on changes of membrane capacitance associated with the electrogenic ion transport by the Na,K-ATPase, *Eur. Biophys. J.* 27 (1998) 605–617.

50. C.-C. Lu, A.Y. Kabakov, V.S. Markin, S. Mager, G.A. Frazier, D.W. Hilgemann, Membrane transport mechanisms probed by capacitance measurements with megahertz voltage clamp, *Proc. Natl. Acad. Sci. U. S. A.* 92 (1995) 11220–11224.

51. J.H. Kaplan, B. Forbush, III, J.F. Hoffman, Rapid photolytic release of adenosine 5′-triphosphate from a protected analogue: Utilization by the Na:K pump of human red blood cell ghosts, *Biochemistry* 17 (1978) 1929–1935.

52. J.A. McCray, L. Herbette, T. Kihara, D.R. Trentham, A new approach to time-resolved studies of ATP-requiring biological systems; laser flash photolysis of caged ATP, *Proc. Natl. Acad. Sci. U. S. A.* 77 (1980) 7237–7241.

53. I. Wuddel, W. Stürmer, H.-J. Apell, Dielectric coefficients of the extracellular release of sodium ions, in: E. Bamberg and W. Schoner (Eds.), *The Sodium Pump*, Steinkopf, Darmstadt, 1994, pp. 577–580.

54. S. Heyse, I. Wuddel, H.-J. Apell, W. Stürmer, Partial reactions of the Na,K-ATPase: Determination of rate constants, *J. Gen. Physiol.* 104 (1994) 197–240.

55. H.-J. Apell, M. Roudna, J.E. Corrie, D.R. Trentham, Kinetics of the phosphorylation of Na,K-ATPase by inorganic phosphate detected by a fluorescence method, *Biochemistry* 35 (1996) 10922–10930.

56. H. Thirlwell, J.E.T. Corrie, G.P. Reid, D.R. Trentham, M.A. Ferenczi, Kinetics of relaxation from rigor of permeabilized fast-twitch skeletal fibers from the rabbit using a novel caged ATP and apyrase, *Biophys. J.* 67 (1994) 2436–2447.

57. C. Peinelt, H.-J. Apell, Time-resolved charge movements in the sarcoplasmatic reticulum Ca-ATPase, *Biophys. J.* 86 (2004) 815–824.

58. C. Peinelt, H.-J. Apell, Kinetics of Ca²⁺ binding to the SR Ca-ATPase in the E₁ state, *Biophys. J.* 89 (2005) 2427–2433.

59. A. Fibich, K. Janko, H.-J. Apell, Kinetics of proton binding to the sarcoplasmic reticulum Ca-ATPase in the E₁ state, *Biophys. J.* 93 (2007) 3092–3104.

60. P.L. Jørgensen, J.H. Collins, Tryptic and chymotryptic cleavage sites in sequence of alpha-subunit of (Na⁺ + K⁺)-ATPase from outer medulla of mammalian kidney, *Biochim. Biophys. Acta-Biomembr.* 860 (1986) 570–576.

61. I.M. Glynn, Y. Hara, D.E. Richards, The occlusion of sodium ions within the mammalian sodium-potassium pump: Its role in sodium transport, *J. Physiol.* 351 (1984) 531–547.

62. P.L. Jørgensen, J. Petersen, Chymotryptic cleavage of α-subunit in E₁-forms of renal (Na⁺+K⁺)-ATPase: Effects on enzymatic properties, ligand binding and cation exchange, *Biochim. Biophys. Acta-Biomembr.* 821 (1985) 319–333.

63. J.P. Morth, B.P. Pedersen, M.S. Toustrup-Jensen, T.L. Sorensen, J. Petersen, J.P. Andersen, B. Vilsen, P. Nissen, Crystal structure of the sodium-potassium pump, *Nature* 450 (2007) 1043–1049.

64. T. Shinoda, H. Ogawa, F. Cornelius, C. Toyoshima, Crystal structure of the sodium-potassium pump at 2.4 A resolution, *Nature* 459 (2009) 446–450.

65. R. Kanai, H. Ogawa, B. Vilsen, F. Cornelius, C. Toyoshima, Crystal structure of a Na^+-bound Na^+,K^+-ATPase preceding the E1P state, *Nature* 502 (2013) 201–206.

66. C. Barth, H. Bihler, M. Wilhelm, G. Stark, Application of a fast charge-pulse technique to study the effect of the dipolar substance 2,4-dichlorophenoxyacetic acid on the kinetics of valinomycin mediated K^+-transport across monoolein membranes, *Biophys. Chem.* 54 (1995) 127–136.

67. D.C. Gadsby, F. Bezanilla, R.F. Rakowski, P. de Weer, M. Holmgren, The dynamic relationships between the three events that release individual Na^+ ions from the Na^+/K^+-ATPase, *Nat. Commun.* 3 (2012) 669.

68. S.B. Hladky, Ion transport and displacement currents with membrane-bound carriers: The theory for voltage-clamp currents, charge-pulse transients and admittance for symmetrical systems, *J. Membr. Biol.* 46 (1979) 213–237.

69. R.F. Rakowski, F. Bezanilla, P. de Weer, D.C. Gadsby, M. Holmgren, J. Wagg, Charge translocation by the Na/K pump, *Ann. N. Y. Acad. Sci.* 834 (1997) 231–243.

70. R.F. Rakowski, Charge movement by the Na/K pump in Xenopus oocytes, *J. Gen. Physiol.* 101 (1993) 117–144.

71. A. Schneeberger, H.-J. Apell, Ion selectivity of the cytoplasmic binding sites of the Na,K-ATPase: I. Sodium binding is associated with a conformational rearrangement, *J. Membr. Biol.* 168 (1999) 221–228.

72. A. Babes, K. Fendler, Na^+ transport, and the E_1P-E_2P conformational transition of the Na^+/K^+-ATPase, *Biophys. J.* 79 (2000) 2557–2571.

73. M. Nakao, D.C. Gadsby, Voltage dependence of Na translocation by the Na/K pump, *Nature* 323 (1986) 628–630.

74. M. Holmgren, J. Wagg, F. Bezanilla, R.F. Rakowski, P. de Weer, D.C. Gadsby, Three distinct and sequential steps in the release of sodium ions by the Na^+/K^+-ATPase, *Nature* 403 (2000) 898–901.

75. V.S. Sokolov, A.A. Lenz, H.-J. Apell, Negative changes of the membrane capacitance due to electrogenic Na transport by the Na,K-ATPase, *Ann. N. Y. Acad. Sci.* 986 (2003) 229–231.

76. A. Schneeberger, H.-J. Apell, Properties of the sodium-pump ion binding sites in state E_1, in: K. Taniguchi and S. Kaya (Eds.), *Na/K-ATPase and Related ATPases, Excerpta Medica Internaitonal Congress Series*, Elsevier Science B.V., Amsterdam, 2000, pp. 319–326.

77. E. Or, E.D. Goldshleger, D.M. Tal, S.J. Karlish, Solubilization of a complex of tryptic fragments of Na, K-ATPase containing occluded Rb ions and bound ouabain, *Biochemistry* 35 (1996) 6853–6864.

78. W. Domaszewicz, H.-J. Apell, Binding of the third Na^+ ion to the cytoplasmic side of the Na,K-ATPase is electrogenic, *FEBS Lett.* 458 (1999) 241–246.

3

ANALYZING ION PERMEATION IN CHANNELS AND PUMPS USING PATCH-CLAMP RECORDING

Andrew J. Moorhouse, Trevor M. Lewis, and Peter H. Barry

Department of Physiology, School of Medical Sciences, UNSW Australia (The University of New South Wales), Sydney, New South Wales, Australia

3.1 INTRODUCTION

The selective movement of ions across biological membranes is critical for cell and organ function, and for life to exist. Ions traverse cell membranes via three broad classes of membrane transport protein complexes: those that facilitate diffusion of solutes down their energy gradients (channels and carriers), those that move solutes against their energy gradients (primary active transporters or "pumps"), and those that move one solute down an energy gradient simultaneously with the transport of another solute against an energy gradient (secondary active transporters). In this chapter we concentrate on channels and pumps, which both act in unison to mediate cell function. Pumps directly use energy to establish gradients of ions across cell membranes. Ion channels dissipate these gradients via ion diffusion through aqueous pores to elicit rapid cell signaling events. This is a key functional thermodynamic distinction between ion translocation by pumps and channels, yet common molecular features are becoming clear [1]. Both channels and pumps typically have gates that exclude access of ions to the interior of the membrane protein, and both have key structural domains where ion selectivity and transport rates can be determined. As described throughout this book, many different approaches can be used to investigate

Pumps, Channels, and Transporters: Methods of Functional Analysis, First Edition. Edited by Ronald J. Clarke and Mohammed A. A. Khalid.
© 2015 John Wiley & Sons, Inc. Published 2015 by John Wiley & Sons, Inc.

functional properties of channels and pumps; in this chapter we focus on the use of the patch-clamp technique to investigate ion permeation in channels and pumps. We begin with a historical overview of the whole-cell patch-clamp technique and then focus on descriptions of ion channel conductance, selectivity and kinetics, and practical approaches to accurately quantify these parameters as derived largely from our own experiences in this area. We then summarize the results from patch-clamp experiments that have elucidated the structural features of the ion permeation pathway in selected examples of channels and pumps.

3.2   DESCRIPTION OF THE PATCH-CLAMP TECHNIQUE

The patch-clamp technique enabled the direct recording of the ionic currents that flow through individual ion channels, and consequently enabled direct measurements of single channel conductance and of the gating transitions between nonconducting and conducting conformations [2]. The different configurations of the patch-clamp recording technique, and the ability to combine this with solution and voltage control across a cell membrane, markedly facilitated the careful characterization of the great diversity of membrane currents and the underlying transporters controlling these currents.

3.2.1   Development of Whole-Cell Dialysis with Voltage-Clamp

In the whole-cell and excised-patch variants of the patch-clamp recording technique [3], the solutions on each side of the membrane are controlled by the experimenter. Furthermore, the membrane can be voltage-clamped, so that the driving force for ion movement and the conformational states of voltage-gated channels can both be dictated. These are critical points to enable the study of the ion permeation properties of channels and pumps. The power of combined voltage-clamp and solution control was appreciated since the intracellular recordings from the perfused squid axon by Hodgkin and Huxley [4, 5], and others after them, which enabled a quantitative description of the mechanisms of the action potential (see Fig. 3.1a). Figure 3.1 illustrates some of these early techniques for cell dialysis coupled with voltage-clamp. Somatic "whole-cell" perfusion was developed for large invertebrate neurons in Kiev and Galveston in the mid-1970s. In Ukraine, Kostyuk and Krishtal [6, 7] combined voltage-clamp and internal dialysis of large amphibian neurons by embedding a cell between two isolated and independently perfused partitions and applying a Ca^{2+}-free external solution, or later negative pressure, to one side, to rupture part of the cell membrane (see Fig. 3.1b). This membrane rupture approach bears resemblance to some of the "modern" automated planar patch-clamp approaches. This approach was also adapted to smaller cultured neurons by drawing the cell into an aperture mounted on a moveable "U-tube" [8] (see Fig. 3.1c). Meanwhile in Texas, Lee et al. [9, 10] sucked a large invertebrate neuron into a "pipette" made from Pyrex tubing, and used a wire inside this suction pipette to both rupture part of the membrane and provide an intracellular current electrode [11]. Figure 3.1d illustrates an example of these "suction

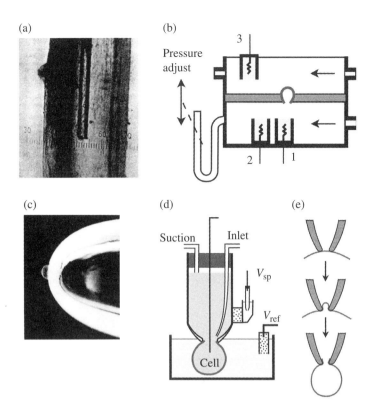

FIGURE 3.1 Examples of early approaches to achieve cell dialysis combined with voltage-clamp, as needed for measurements of channel and pump ion permeation properties. (a) Photo of wire electrode inserted inside single squid axon, which could be perfused with solutions of desired concentrations. Reprinted by permission from Ref. 4, Macmillan Publishers Ltd, © 1939. (b) Dialysis and voltage-clamp of mollusc neurons. Redrawn and adapted from Ref. 7 with permission from John Wiley & Sons. A single neuron was brought into an aperture (sealed with paraffin grease/oil mixture) between two isolated and perfused compartments by mild suction, further suction applied to the lower compartment ruptured the cell membrane to enable dialysis and voltage-clamp. (c) A modification of the method described in panel (b) was used to dialyze and clamp smaller neuroblastoma cells. A small aperture was cut into an independently perfused V-shaped plastic tool mounted on a manipulator and a cell was drawn into this aperture by suction. Further suction ruptured the membrane within the interior of the tool enabling dialysis and voltage-clamping. Reprinted from Ref. 8, © 1978, with permission from Elsevier. (d) In the suction pipette technique, a dispersed snail neuron was drawn by suction into a wide-bore glass pipette. The membrane was typically ruptured by inserting a sharpened wire (central vertical line) through the membrane, to allow dialysis of the cell with pipette solution. The inserted wire could also be used to pass current or record membrane potential. The internal solution could be changed by an inlet tube. The potential of the suction pipette (V_{sp}) could be measured by a calomel half cell via a 3 M KCl/agar salt-bridge and the potential of the bath reference (V_{ref}) via a 3 M KCl/agar salt-bridge and another calomel half cell (not shown). Redrawn and adapted from figure 3 of Ref. 11 with kind permission from Elsevier. (e) The traditional whole-cell patch-clamp technique in which a small-bore fire-polished electrode makes a Giga Ohm seal with a small mammalian neuron. Adapted from parts of figure 9 of Ref. 3, with kind permission from Springer Science and Business Media. Arrows indicate application of suction pressure to go from a low resistance membrane seal to a higher resistance Giga Ohm seal, and finally to rupture the membrane to enable whole-cell voltage-clamp and cell dialysis.

pipette" approaches. These early studies had success in isolating and characterizing different voltage-dependent channels, particularly the inward Ca^{2+} currents that previously were difficult to accurately measure without minimizing other currents. However, these approaches were most applicable to large cells ($\approx 100\,\mu m$ diameter), and the seal resistances were less than $100\,M\Omega$, limiting the applicability for characterizing ion channel permeation properties. These caveats were solved in 1981, when Neher and Sakmann's lab reported [3] the whole-cell patch-clamp technique in small cells using smooth glass pipettes that combined $G\Omega$ seals, voltage-clamping, and cell dialysis (see Fig. 3.1e). Coupled with the ability to record single channels in excised patches [3, 12], the key requirements of current resolution and solution and voltage control were achieved. The stage was set for an avalanche of work to characterize the permeation properties of ion channels.

3.3 PATCH-CLAMP MEASUREMENT AND ANALYSIS OF SINGLE CHANNEL CONDUCTANCE

Defined simply, conductance is a description of how easily current is able to flow between two points in response to an electrical potential difference between them. Here we are considering conductance in terms of ionic current moving across the cell membrane. This is possible when ion channels in the cell membrane are open and allow the selective movement of ions according to the electrochemical potential gradient along the channel. At the single-channel level, conductance is often one of the features used to characterize and identify different ion channels and subtypes. It is an elementary property of an ion channel, because the conductance is influenced by all the physical processes that a permeant ion experiences as it enters the permeation pathway from the bulk solution, traverses the pore, and exits on the other side. Thus, all the structural features of the ion channel that impact on the pathway taken by a permeant ion can have an influence upon conductance (see Section 3.6). Although conductance tends to reflect the ion permeability of a channel, there is an additional contribution to the channel conductance from the ion concentration (see Section 3.3.1).

In this section, we more formally define conductance and illustrate how conductance may be measured and calculated using patch-clamp recordings.

3.3.1   Conductance and Ohm's Law

Conductance is an electrical term that is the reciprocal of the resistance. As such, it can be thought of as an indication of permeability—how readily ions are able to move through the channel. If we take a simple situation with symmetrical concentrations of ions on either side of the cell membrane and open, water-filled pores that allow the selective movement of ions, the conductance, G, can be determined from Ohm's law:

$$I = GV \tag{3.1}$$

where I is the measured ionic current in the presence of the electrical potential difference, V, across the membrane. Under physiological conditions, there is typically a concentration gradient for the permeant ion across the membrane, and hence both the electrical and chemical gradients need to be taken into account. Furthermore, in the case of a single channel with conductance γ, and single channel current i, a modified form of Ohm's law is applicable:

$$i = \gamma \left(V_m - V_{rev} \right) \qquad (3.2)$$

where V_m is the membrane potential and V_{rev} is the reversal (or zero current) potential for the permeant ion, such that $V_m - V_{rev}$ is the electrochemical driving force.

Equation 3.2 describes a linear (or "Ohmic") relationship between current and voltage, with the single channel conductance (γ) as a constant. Under many circumstances, this is the case and thus γ is a useful parameter to characterize ion channels. However, there are circumstances where the single channel current-voltage relationship is not strictly linear. This may be due to asymmetric ion concentrations across the membrane (see below), large driving forces altering the local ion concentrations in the permeation pathway, the influence of divalent ions resulting in rectification [13, 14], or structural elements of the permeation pathway such as charged residues that can also result in rectification [13–15].

It is implicit from Ohm's law that conductance cannot be measured directly, but instead must be calculated from measured values of the membrane potential (V_m) and the amplitude of the single channel current (i). In practical terms, this requires the cell membrane to be voltage-clamped at a known potential and the resultant current measured. In addition, the reversal potential (V_{rev}) of the current either needs to be estimated or experimentally determined. Estimating V_{rev} requires knowledge of the ion channel selectivity and of the concentrations of ions on both sides of the cell membrane. Experimentally, V_{rev} can be determined by recording the current at a range of membrane potentials that span the reversal potential, allowing interpolation, or else by extrapolating from a range of potentials that approach the reversal potential. Once the V_{rev} is known, the conductance is calculated from measurement of single channel current at a particular membrane test voltage (V_m):

$$\gamma_{chord} = \frac{i}{V_m - V_{rev}} \qquad (3.3)$$

This gives the *chord conductance* for the channel (γ_{chord}). Graphically, on a current-voltage plot, this is the slope of the line between the test voltage (V_m) and the point where the driving force is zero ($V_m = V_{rev}$). An alternative method of determining the conductance is the *slope conductance* (γ_{slope}):

$$\gamma_{slope} = \frac{\Delta i}{\Delta V_m} \qquad (3.4)$$

This is calculated by measuring the change in current (Δi) for a known change in the membrane potential (ΔV_m). The slope conductance is useful for providing information of the slope of the current-voltage (I-V) curve across a certain voltage range.

Only if the single channel I-V curve is linear will the chord and slope conductances give the same value. A nonlinear I-V relationship arises when the ions in the pore of a channel do not behave as they do in simple dilute solution due to interactions with each other and with the pore. The influence of ion channel structure on conductance is described in Section 3.6. Furthermore, the interactions between pore and ion are not the same for each ionic species, so not every ion species has the same conductance, even when differences in solution mobility are taken into account. These ion–ion and ion-pore interactions give rise to phenomena such as single channel current rectification and saturation of conductance (both as voltages increase or as bath ion concentrations increase), anomalous mole-fraction behavior, and ion blocking effects. Interested readers could consult a more comprehensive review [16], but it suffices here to point out that even for the same channel the conditions under which one determines the conductance can affect the conductance value. This highlights the importance of reporting these conditions with any conductance value determined.

Besides the conductance of a single ion species, γ_i, another way of expressing the ease with which a particular ion can traverse the membrane is via the permeability coefficient, P_i. In a situation where there is a concentration difference, Δc_i, for an ion across the membrane acting as a driving force for net diffusion of the ion from the side of the membrane with a higher concentration to the side with the lower concentration, the permeability coefficient represents the constant of proportionality between the flux density, Φ_i, that is, the number of moles of ions crossing the membrane per unit area of membrane per unit time, and the concentration difference:

$$\Phi_i = P_i \Delta c_i \tag{3.5}$$

Based on Equation 3.5, the permeability coefficient can also be seen as the flux density of the ion expected per unit concentration difference across the membrane. Because permeability and conductance are both ways of expressing the facility of passive ion transport, it makes sense that there should be some mathematical relationship between them. From the generalized constant field equation [17] for equal concentrations, c_i, of an ion on each side of the membrane it may be shown that the ionic conductance for that ion, γ_i, is related to the permeability coefficient, P_i, by:

$$\gamma_i = \frac{c_i F^2}{RT} P_i \tag{3.6}$$

where F represents Faraday's constant, R the ideal gas constant, and T the absolute temperature.

3.3.2 Conductance of Channels versus Pumps

Ohm's Law can be used as a framework to estimate the maximum conductance a channel may be capable of by considering [18] the resistance of a hypothetical pore constriction of physiological dimensions, for example, 6Å diameter and 5Å length, bathed in physiological salt concentrations with a resistivity (resistance of a unit volume) of $100\,\Omega\,cm$. Because resistance=resistivity times length per unit area, such a hypothetical pore would be expected to have a resistance of $1.8\,G\Omega$ [18].

Including the access resistance of each entrance, this increases to $3.4\,G\Omega$ and the reciprocal equates to a hypothetical maximum conductance of $290\,pS$. The recombinant $\alpha1$ homomeric human glycine receptor (GlyR) has a measured conductance of approximately $90\text{-}100\,pS$ at $20\text{-}23°C$ and at a V_m of $-100\,mV$ [19]. Under a driving force of $100\,mV$, the $90\,pS$ GlyR can pass $9\,pA$ of current, equating to about 50 million ions per second. Hence, fully open ion channels allow ions to flow at rates approaching that of free diffusion, with restrictions imposed partly from the restricted dimensions of their aqueous pores. In contrast, pumps have much lower rates of ion flux. The Na^+,K^+-ATPase, for example, has a maximum turnover rate of only $500\,ions\,s^{-1}$ at $38°C$ [18], but seldom operates physiologically at this maximum. This illustrates a key functional difference between electrodiffusion-mediated ion conductance in channels and conformational kinetics-limited ion translocation in pumps. The net rate of ion transport through an ion channel also depends on the amount of time the channel is actually in the conducting state, hence influenced by the channel's gating kinetics, that is, the rate at which the channel opens and closes. This involves protein conformational changes, as in the case of pumps. However, because effective ion pumping prohibits the protein from allowing ion access to both sides of the membrane simultaneously, it makes sense that the conformational changes associated with ion pump activity should be more extensive and slower than the gating movements of ion channels.

3.3.3 Fluctuation Analysis

The macroscopic conductance determined from the currents of whole cells or macropatches of membrane provides a description of the averaged behavior of the population of channels that contributes to the conductance. Each of the individual channels open and close in a random fashion, and the number of channels open at any point in time varies (fluctuates). This gives rise to a "noisy" current (above any background noise), and fluctuation analysis can be used to extract information about the amplitude and mean lifetime of the unitary channel events that contribute to this noise. Fluctuation analysis was initially applied to the neuromuscular junction to investigate the underlying events of the muscle endplate potential [20–22], at a time before single channel patch-clamp recordings were possible. Fluctuation analysis is now largely surpassed by direct measurement of single channel amplitudes and sophisticated kinetic analysis that is possible with high-resolution single-channel recordings (see Section 3.3.4). The exception to this is when the ion channel has a small conductance, making it difficult to reliably distinguish the single channel events from the background noise. A brief description of the methods is provided here assuming a simple two-state (open and closed) model of ion channel activity as shown in the following Scheme A:

$$C \underset{\alpha}{\overset{\beta}{\rightleftharpoons}} O$$

SCHEME A

where C is the closed state, O is the open state, β is the forward rate constant, and α is the reverse rate constant. A more comprehensive theoretical treatment including consideration of more complex models is available elsewhere [23–25].

Autocovariance. The sample autocovariance function, $C(\Delta t)$, measures the correlation between individual current values separated by a time interval, Δt, in the recorded current, and is given by the equation:

$$C(\Delta t) = \frac{1}{n-1} \sum \left[x(t) - \bar{x} \right]\left[x(t + \Delta t) - \bar{x} \right] \tag{3.7}$$

where $x(t)$ is the value at any time t, $x(t+\Delta t)$ is the value at the time $(t+\Delta t)$ later in the recording, $\bar{x}$ is the mean value of x across the recording, and n is the number of data points sampled in the recording. It is assumed here that the current recording is stationary, meaning that it is a steady-state recording with regards to the clamped membrane potential and other parameters such as the ligand concentration. When $\Delta t = 0$, then the autocovariance function simply gives the variance of the recording. As an aside, if the autocovariance is normalized by dividing by the variance, this gives the autocorrelation function, which is described in Chapter 10 with regards to fluctuation analysis of diffusion of molecules in membranes. Current recordings are typically prepared for analysis by subtracting the background noise (noise present in the absence of the channel activity of interest) and by subtracting the mean current. This gives a signal that varies about a mean of zero ($\bar{x} = 0$) and so Equation 3.7 simplifies to:

$$C(\Delta t) = \sum \frac{x(t)x(t + \Delta t)}{n-1} \tag{3.8}$$

Because these fluctuations are due to the opening and closing of ion channels, this can now be expressed in terms of the activity of a single channel. Assuming a homogeneous population of independent channels, with one conductance level, the autocovariance of the current for the activity of N channels in the membrane is simply the autocovariance of the single channel current multiplied by N. Following Gray [25], each data point, $x(t)$, then equates to the single channel current at that time, $i(t)$, and so $i(t+\Delta t)$ is the current through the same channel at time $t+\Delta t$. Equation 3.8 can then be expressed as:

$$C(\Delta t) = N \sum \frac{i(t)i(t + \Delta t)}{n-1} \tag{3.9}$$

which is simply N times the mean of $[i(t)i(t+\Delta t)]$. The single channel current will be i when the channel is open, and zero when closed. If the channel is open at t and at $t+\Delta t$, then $i(t)$ $i(t+\Delta t)$ will be i^2, but if the channel closes at $t+\Delta t$, then it will be zero. The mean value of $i(t)$ $i(t+\Delta t)$ can then be expressed as the product of i^2 and the probability of the channel being open at t and still open at $t+\Delta t$. If the time interval, Δt, is small, then there is a high probability that the channel will remain open at time $t + \Delta t$ and the correlation $(C(\Delta t))$ will be high. As Δt increases, the correlation will diminish as it becomes less likely that the channel remains open. This can be described by an exponential decay [25, 26] and the covariance function becomes:

$$C(\Delta t) = Ni^2 p_o \exp\left(\frac{-\Delta t}{\tau}\right) \tag{3.10}$$

where p_o is the open probability of the channel and τ is the time constant of the relaxation. The time constant is the reciprocal of the sum of the forward and reverse rate constants in Scheme A. By manipulating the recording conditions so that the open probability is small, the forward rate constant will be negligible and τ will approximate the reciprocal of the reverse rate constant ($=1/\alpha$ in Scheme A), which is the mean open lifetime of the channel. Further, Nip_o gives the mean current, $\bar{I}$, and $i = \gamma (V_m - V_{rev})$ for an ohmic channel (see Section 3.3.1), so:

$$C(\Delta t) = \bar{I}\gamma(V_m - V_{rev})\exp\left(\frac{-\Delta t}{\tau}\right) \tag{3.11}$$

which describes the autocovariance in terms of the single channel conductance, γ, and the mean lifetime of the open state, τ. Typically, this would be presented in the form of a power density spectrum, which is obtained by taking the Fourier transform of the covariance function, thus transforming it into the frequency domain to give a Lorentzian function of the form:

$$S(f) = 4\bar{I}\gamma(V_m - V_{rev})\frac{1}{1 + (f / f_c)} \tag{3.12}$$

where $S(f)$ is the spectral density, f is the frequency, and f_c is the cut-off frequency at which the spectral density is reduced to half of that at zero frequency. The value of f_c is related to the relaxation time constant from the covariance function by the relationship:

$$\tau = \frac{1}{2\pi f_c} \tag{3.13}$$

With more than a simple open and closed state, the covariance and spectral density functions become much more complex. For channels that display n kinetically distinct states, there will be $n - 1$ exponential terms in the covariance function and the same number of Lorentzian components in the spectral density function [23, 24]. The interpretation then becomes much more difficult. In particular, very fast gating kinetics are unable to be resolved. Conversely, very long open times have very few correlations within the recording and the variance will be too low to be detected. By comparison, single-channel recordings can provide much greater temporal resolution and the distribution of open and closed times of the kinetically distinct states can be determined directly (see Section 3.3.4).

Binomial distribution. An alternative statistical approach to interpreting current fluctuations considers the activity in terms of a binomial distribution [24], where the channel is either open or closed, with the simple assumptions of independence in the opening and closing of each of the channels and a homogeneous population of channels. The mean macroscopic conductance $(\bar{G})$ is the product of the number of ion channels in the membrane, N, the associated single channel conductance, γ, and the open probability, p_o, for that population of channels:

$$\bar{G} = \gamma N p_o \qquad (3.14)$$

It can be shown that the variance of the macroscopic conductance, $\sigma^2(G)$, is given by:

$$\sigma^2(G) = \gamma^2 N p_o (1 - p_o) \qquad (3.15)$$

Hence,

$$\frac{\sigma^2(G)}{\bar{G}} = \gamma(1 - p_o) \qquad (3.16)$$

When applied to steady-state recordings, the mean conductance and the variance are calculated over the duration of the record. It is necessary to subtract the background "noise" that is present in the absence of channel activity so as to resolve the variance that is due only to the opening and closing events of the channels. If the recording conditions are carefully chosen such that p_o is very low, then $(1 - p_o)$ approaches 1 and Equation 3.16 simplifies to:

$$\frac{\sigma^2(G)}{\bar{G}} = \gamma \qquad (3.17)$$

which allows the single channel conductance, γ, to be calculated.

These methods are called stationary fluctuation analysis because they are applied to current recordings obtained at a steady state, where the probability of channel opening is assumed not to change with time. However, in the case where the channel activity changes over time, for example, the post-synaptic current following the synaptic release of neurotransmitter, or the voltage dependent sodium current following a step change in the membrane potential, non-stationary or ensemble fluctuation analysis can be used.

Nonstationary fluctuation analysis [27, 28] requires a set of multiple current recordings from the same preparation, in response to the same stimulus. The ensemble average of these current recordings gives the mean total current at each point in time, $\bar{I}(t)$, which by analogy with Equation 3.14, is given by:

$$\bar{I}(t) = i N p_o(t) \qquad (3.18)$$

where i is the single channel current, N is the total number of channels, and $p_o(t)$ is the channel open probability at each point in time. The ensemble variance is then, by analogy with Equation 3.15, given by:

$$\sigma^2\left[I(t)\right] = i^2 N p_o(t)\left[1 - p_o(t)\right] \tag{3.19}$$

By substituting $p_o(t)$ from Equation 3.18 into Equation 3.19, the ensemble variance is then:

$$\sigma^2\left[I(t)\right] = i\bar{I}(t) - \frac{\bar{I}^2(t)}{N} \tag{3.20}$$

which is a parabolic relationship between the total mean current (on the abscissa) and the ensemble variance (on the ordinate). Experimentally, the variance is determined by first calculating the deviation of each individual current recording from the mean total current (simply subtracting the mean current from the individual recordings). The variance, σ^2, is then the sum of the square of these deviations. A plot of the variance against the mean current can then be fitted with Equation 3.20 to estimate the single channel current, i, and the total number of channels in the membrane, N. It is then a simple matter to calculate the chord conductance from the value of i together with the applied membrane potential, V_m, and the measured or calculated reversal potential, V_{rev}, using Equation 3.3.

Summary of fluctuation analysis. Fluctuation analysis has several drawbacks in estimating the single channel conductance. If there are multiple conductance states that the ion channel can adopt, this will influence the estimate of the single channel conductance, which will effectively be a weighted mean of the contributing conductance states. The duration of the events also affects the estimate. Patch-clamp recordings are able to provide a much more direct measure of single channel current and hence conductance, and this is often the method of choice.

3.3.4   Single Channel Recordings

Single channel conductance and information on gating kinetics can be directly determined from current recordings of one or several channels under voltage-clamp conditions. This was initially achieved by inserting model pores, such as the antibiotic gramicidin A, into planar lipid bilayers [29, 30]. This technique allowed easy access by low-resistance electrodes to either side of the membrane for voltage-clamping, but requires the ability to isolate sufficient amounts of the ion channel protein for insertion into the bilayer.

With the advent of the patch-clamp technique [2, 3], voltage-clamp recordings were possible from native and cultured cells expressing ion channels of interest. The key advance with this technique was the ability to form a high resistance seal (~10–100 GΩ; "gigaseal") between the recording electrode and the patch of membrane containing the ion channel. This significantly reduces the background noise levels and consequently improves the temporal resolution of the recording. Here we discuss issues associated with determining single channel conductance and briefly describe the measurement of ion gating kinetics. For other excellent sources that discuss the patch-clamp methodology see, for example, [31, 32].

Single channel recordings can be made using three modes of the patch-clamp technique: the cell-attached patch, inside-out patch, and outside-out patch (see Fig. 3.2).

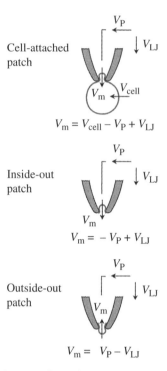

Cell-attached patch

$$V_m = V_{cell} - V_P + V_{LJ}$$

Inside-out patch

$$V_m = -V_P + V_{LJ}$$

Outside-out patch

$$V_m = V_P - V_{LJ}$$

FIGURE 3.2   Three patch-clamp configurations, which may be used for directly measuring single channel currents (cell-attached patch, inside-out patch, and outside-out patch) together with the voltage components that contribute to the measured potential across the membrane patch, V_m (defined as the potential at the inside surface of the membrane relative to the outside one). V_P represents the pipette potential at the input to the patch-clamp amplifier and V_{LJ} represents the LJP (liquid junction potential) offset in the amplifier that arises to balance the LJP between bath solution and the pipette solution when the amplifier is zeroed, prior to the formation of the membrane patch. N.B. The original LJP is replaced by V_m, but the offset V_{LJ} still remains as part of the measured amplifier voltage. The other potential V_{cell}, which only occurs in the intact patch case, represents the membrane potential of the cell (inside relative to outside). Note that the potential components for the outside-out patch are exactly the same ($V_m = V_P - V_{LJ}$) as for the whole cell configuration (see Fig. 3.1e). For the relationships between the voltage components and V_{LJ} refer to [33]. The pipette-membrane configurations were redrawn and adapted from parts of figure 9 of Ref. 3, with kind permission from Springer Science and Business Media.

The cell-attached mode is relatively easy to attain by forming a gigaseal directly with the surface of the intact cell. This means there is no disruption to the intracellular environment of the cell, which can be an advantage if the channel exhibits "run-down" in activity when removed from its intracellular environment. Ligand-gated ion channel activity can be recorded by the inclusion of agonist in the patch electrode solution [34], but there is no opportunity to check for channel activity in the absence of agonist (in the same patch) as is possible with the outside-out patch mode (see below). Taking advantage of the intact cell, receptor-activated second messenger modulation of single channel behavior can also be recorded [35]. However, this configuration has difficulties with knowing the potential across the patch membrane because there is no electrode on the inside of the cell to clamp the cell membrane potential. The command potential applied to the patch electrode is in series with the cell membrane potential, and the patch of membrane under the patch electrode is clamped at a potential, V_m, given by:

$$V_m = V_{cell} - V_p + V_{LJ} \tag{3.21}$$

where V_{cell} is the cell membrane potential, V_p is the command potential applied to the patch electrode, and V_{LJ} is the correction for the liquid junction potential (see Fig. 3.2). There are a couple of solutions to this problem of unknown V_{cell}. If there are sufficiently good estimates of the internal ion concentrations and of the membrane permeability at the resting membrane potential, it may be possible to estimate the cell membrane potential theoretically via the Goldman–Hodgkin–Katz equation. Alternatively, the membrane potential may be manipulated by bathing the entire cell in a high [K$^+$] solution that depolarizes the cell so that $V_{cell} \approx 0\,mV$ and therefore $V_m = -V_p + V_{LJ}$. However, prolonged depolarization may be deleterious to the viability of the cell.

The inside-out patch mode is attained from the cell-attached mode by moving the patch electrode away from the surface of the cell. This leaves a small patch of cell membrane attached to the electrode with the intracellular surface now exposed to the bath solution. Sometimes it may instead form a vesicle on the tip of the electrode that may be observed under high magnification. Alternatively, the vesicle may be inferred by lack of response to intracellular ligands or by non-steady single channel current amplitudes. The vesicle can be easily disrupted by very briefly lifting the electrode out of the bath solution and then returning it to the recording bath leaving the inside-out patch attached. In this mode, the membrane potential is given by:

$$V_m = -V_p + V_{LJ} \tag{3.22}$$

It is commonly used for recordings of single channels that are modulated or activated by intracellular ligands, as there is clear and easy access to the intracellular surface, for example, the investigation of cyclic-nucleotide-gated ion channels [36] or calcium-activated potassium channels [37].

The outside-out patch mode is attained from the whole-cell mode, where the membrane underneath the patch electrode has been disrupted so that the electrode is continuous with the inside of the cell. The electrode is then pulled away from the cell, initially drawing a thin straw of membrane from the cell surface, eventually pinching off to form the outside-out patch. The membrane potential in this case is given by:

$$V_m = V_p - V_{LJ} \tag{3.23}$$

Because the extracellular surface of the membrane remains facing the bath, this mode is commonly used to record single channel activity from extracellular ligand-gated ion channels. Exchanging the bath solution allows for stable, steady-state recordings of single channel activity with different ion or ligand concentrations. The specificity of this activity can be confirmed by returning to the control bath solution conditions.

Single channel conductance. Once the desired mode of patch-clamp recording has been attained and the membrane potential has been correctly determined (i.e., using one of the Equations 3.21–3.23), the conductance can be calculated from the amplitude of the single channel currents obtained from the analyzed record (see below). This may be a chord conductance, using Equation 3.3, or a slope conductance, using Equation 3.4, as described in Section 3.3.1.

Analysis of single channel recordings. Low noise patch-clamp recordings of single channel activity are able to provide direct information on the amplitude of events to less than 1 pA and durations as brief as 10 μs, provided the data are filtered, digitized, and analyzed appropriately [38, 39]. The first step in analysis is usually to convert the data record into an idealized record of the sequence of events, represented as a series of amplitudes and the corresponding dwell time at that amplitude. There are three main methods by which this idealized record can be obtained, described below. The choice of method depends on what parameters are needed from the data and the nature of the channel activity (e.g., presence of multiple conductance states, the duration of the events). Thus, the data record should be carefully inspected beforehand to determine which method is most suitable for the purpose.

Threshold-crossing method. In this method, the baseline (closed level) and the open level of the data record is set by the user, along with a threshold level somewhere between these two levels, which is typically 50% of the amplitude [38, 39]. Single channel events are detected when the channel opens and the current level crosses the threshold. The amplitude and duration of the open level is determined from the data points until the channel closes and the current recrosses the threshold, returning to the baseline. The duration of the closed event is determined until the next opening event crosses the threshold. This analysis can be completed relatively quickly compared to other methods, but it suffers from several limitations. Records with multiple conductance states are not well handled by this analysis, and the time resolution achieved is generally less than can be achieved by other methods. Thus, it is not well suited to detailed kinetic analysis.

Time course fitting method. This method takes account of the filtering of the single channel record and fits channel openings directly with a square pulse that has been filtered with the same cut-off frequency [39]. This allows for fitting of multiple amplitudes in the current record. Significantly, better time resolution is also possible because event durations are determined from the fit, rather than from simply counting the number of data points once a threshold has been crossed. This method is slower, since each event is individually fitted, but it is able to provide a more detailed description of the data for further kinetic analysis.

Hidden Markov models. This method analyzes the single channel events as a series of transitions between a finite number of states: a Markov process. These series of transitions are "hidden" in the noise and distinct kinetic states that share the same conductance cannot be easily distinguished. It has been applied to extract the idealized channel activity, the distribution of amplitudes and dwell times, and the likely microscopic rate constants for an aggregated Markov model of the channel behavior [40–42]. This approach has the capacity to extract information on fast events and small amplitudes from noisy data records that would be impossible with either of the two methods described above [41]. However, it is generally much more computationally intensive. If just the distribution of amplitudes and event dwell times is required from the recording, then the Markov model used need only have the minimum number of states required for the different conductance levels in the data, and this reduces the computation required.

Once the channel recordings have been analyzed, it may be possible to extract information about the underlying kinetics that generates the series of channel openings. For the simple two-state model shown in Scheme A (see Section 3.3.3), with one open state and one closed state, there will be one conductance level. It can be shown that the mean dwell times of the open and closed states are exponentially distributed [26], with the mean dwell times being the reciprocal of the microscopic rate constants that lead away from the state. Thus:

$$\text{Mean open time} = \tau_o = \frac{1}{\alpha} \tag{3.24}$$

$$\text{Mean closed time} = \tau_c = \frac{1}{\beta} \tag{3.25}$$

The distribution of dwell times are described by exponential probability density functions of the form:

$$f(t) = \tau_o^{-1} \exp\left(\frac{-t}{\tau_o}\right), \text{ for open times} \tag{3.26}$$

$$f(t) = \tau_c^{-1} \exp\left(\frac{-t}{\tau_c}\right), \text{ for closed times} \tag{3.27}$$

Thus, in principle, the microscopic rate constants can be estimated from the mean dwell times of the open- and closed-time distributions obtained from experiments. However, real ion channels are much more complicated than this two-state model. There may be more than one conductance level, and there is likely to be more than one closed state or open state. As a consequence, the dwell times are a mixture of exponentials, with the number of components being a lower limit for the number of kinetic states [26]. Scheme A can be extended to include a ligand-binding step to give Scheme B for a simple ligand-gated ion channel [43]:

$$A+R \underset{k_{-1}}{\overset{k_{+1}}{\rightleftharpoons}} AR \underset{\alpha}{\overset{\beta}{\rightleftharpoons}} AR^*$$

SCHEME B

where A is the concentration of agonist (ligand), R is the unbound receptor ion channel, AR is the agonist-bound state (nonconducting), and AR^* is the open state of the channel. The microscopic ligand association and dissociation rate constants are k_{+1} and k_{-1} respectively; the channel opening rate is β and the closing rate is α (as in Scheme A). In such a scheme there is one open state, and so there is a single exponential distribution with a mean open time of $1/\alpha$. For the closed dwell times, two exponential components are expected, but it is not always possible to directly determine the mean dwell time of a particular kinetic state from the probability density function, $f(t)$. This is because the single channel record can only indicate if the channel is conducting (open) or not; it is unable to identify the kinetic state that the channel occupies – this must be inferred from additional analysis. For example, the receptor channel may spend several sojourns in the R and AR states (both closed and nonconducting) as agonist molecules bind and dissociate before eventually transitioning to the open state AR^*, which terminates the closed dwell time. Thus, the dwell time is not determined by a single closed state. Once the channel is open, there may be several transitions between the open AR^* and the closed AR states giving rise to a "burst" of openings separated by short closures that represent the dwell time in AR [44]. In this case, the mean short dwell time, τ_s, is determined by:

$$\tau_s = \frac{1}{\left(k_{-1} + \beta\right)} \tag{3.28}$$

This illustrates how it is necessary to determine something of how the kinetic states are connected, so as to infer the kinetic scheme and the rate constants. This can be approached in an empirical fashion, by determining those dwell times that show dependence upon the stimulus strength (i.e., change with voltage or ligand concentration), and by extracting information on the connectivity of states from conditional distributions or from autocorrelation analysis of dwell times [45–47].

Direct fitting of kinetic schemes to single channel records. A better solution to the empirical approach is to directly fit a kinetic scheme to the single channel record. The advantage of this approach is that it retains all of the temporal information on the sequence of events in the channel record. There are currently two programs available that are capable of directly fitting channel records. These are HJCFIT [47, 48] and MIL, part of the QuB software suite [42, 49, 50]. These are powerful approaches that allow all of the microscopic rate constants in a kinetic scheme to be directly estimated. The predictions from the kinetic scheme can then be compared to the distribution of open and closed dwell times, for example, obtained from the single channel activity. The single channel data still needs to be idealized first and a time resolution imposed on the record before attempting to fit a kinetic scheme. The choice of kinetic scheme will be informed by the interpretation of the dwell time distributions and the analysis of connectivity between the kinetic states. Having single channel records at different stimulus conditions (different agonist concentration for ligand-gated channels, or different membrane potentials for voltage-gated channels) provides important information for determining the kinetic scheme. While these approaches have been very successful in describing important aspects of single channel activity (e.g., [34, 51, 52]), the more complex the behavior of the channel, the more difficult it becomes to ascribe a kinetic scheme.

3.4 DETERMINING ION SELECTIVITY AND RELATIVE PERMEATION IN WHOLE-CELL RECORDINGS

Section 3.3 described the measurement of the channel conductance—the rate of ion flow through channels under a voltage gradient. Ion selectivity relates to the relative permeabilities of different ions passing through the open pore conformation of a channel. It is typically quantified as a relative permeability, P_X/P_Y, where P is the permeability coefficient (see Section 3.3.1) of ions X and Y crossing a cell membrane. The relative permeability is determined from the shift in zero-current reversal potential, V_{rev}, when the electrolyte composition on one side of the channel is changed. In whole-cell recordings, the external bathing solution is typically changed from an initial control solution (ideally with a composition almost the same as the internal solution) and the resulting shift in V_{rev} upon changing the external solution is measured. To determine anion–cation permeability selectivity, so-called dilution potentials are measured as the shift in V_{rev} when the concentration of the bathing solution is diluted. For measurements of relative permeabilities among ions of the same polarity (i.e., both positively or both negatively charged), so-called bi-ionic potentials are measured in which the electrolyte solution is changed to one of a different composition but keeping either the anion or cation common and at the same concentration in both solutions. In each of these approaches, careful control of solutions and voltage is paramount (see Sections 3.4.3 and 3.4.4). Furthermore, appropriate corrections for liquid junction potentials are needed (see Section 3.4.5), and a suitable reference electrode should be used. The ideal reference electrode should be connected to the bath solution via an agar salt-bridge with similar composition to the control extracellular solution, thereby minimizing history-dependent liquid junction potential effects [53, 54]. The measured V_{rev} values are then fitted to an appropriate theoretical equation for the membrane potential to determine the relative permeabilities. We will discuss both the measurement of dilution potentials and bi-ionic potentials below.

3.4.1 Dilution Potential Measurements

Such measurements are the most appropriate for determining anion–cation (e.g., P_{Cl}/P_{Na}) permeability ratios in ion channels. Ideally, the measurements are accomplished by having a simple uni-univalent salt of the appropriate ions at the same concentration on both sides of the membrane (e.g., "150 mM NaCl"). In reality, the intracellular and extracellular solutions are never precisely the same, given the need to balance pH and chelate Ca^{2+}. For example, we typically use a bath solution of 145 mM NaCl, 10 mM HEPES, with approximately 5 mM NaOH added, and an internal solution of 145 mM NaCl, 10 mM HEPES, 2 mM $CaCl_2$, 5 mM EGTA, with

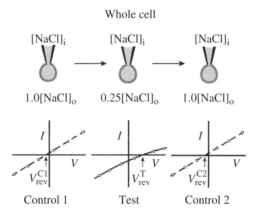

FIGURE 3.3 A schematic illustration of the experimental bracketing procedure (in sequence Control 1 → Test → Control 2) for determining the uncorrected shift in reversal potential (V_{rev}) during a whole-cell patch-clamp dilution experiment. In this example, the external control NaCl solution (1.0 [NaCl]$_o$, see text for full details]) is diluted to approximately 25% of its concentration. The shift in the uncorrected experimental zero-current reversal potential, ΔV_{rev}^{exp}, is given by: $\Delta V_{rev}^{exp} = V_{rev}^{T} - \left(V_{rev}^{C1} + V_{rev}^{C2} \right)/2$. This procedure automatically eliminates any constant pipette tip potential, corrects for minor changes in cell ionic composition, and checks that there are no significant changes in offset between the two control measurements. Note that the same procedure can be used for a bi-ionic potential, in which case the Test solution could be something like 1.0 [NaF]$_o$.

approximately 18 mM NaOH added. The experimental procedure is shown in Figure 3.3. The (uncorrected) experimental zero-current reversal potential, V_{rev}^{C1}, is measured for the control solution. The external bath solution is then replaced by a Test NaCl solution diluted to either 75 mM or 37.5 mM NaCl (each with HEPES and NaOH as above) and the new V_{rev}^{T} is measured. The external solution is changed back to the original "150 mM NaCl" control solution and V_{rev}^{C2} measured again. The shift in uncorrected experimental zero-current reversal potential, ΔV_{rev}^{exp}, is then given by subtracting the averaged control measurement from the test measurement, as indicated in Figure 3.3 and Equation 3.29.

$$\Delta V_{rev}^{exp} = V_{rev}^{T} - \left(V_{rev}^{C1} + V_{rev}^{C2} \right)/2 \qquad (3.29)$$

Then ΔV_{rev}^{exp} can then be corrected for liquid junction potentials (see Section 3.4.5) to give the corrected ΔV_m^c:

$$\Delta V_m^c = \Delta V_{rev}^{exp} - \Delta V_{LJ}^{21} \qquad (3.30)$$

The corrected ΔV_m^c can then be fitted to the appropriate equation to derive the relative permeability (e.g., P_{Cl}/P_{Na}). The most useful and generally used equation for this purpose is the GHK (Goldman–Hodgkin–Katz) equation [55, 56], which is a special zero-current potential case of the general Goldman current equation or constant field equation. For justification of the validity and reliability of the GHK equation, which is much more model independent than the derivation of the original Goldman equation might suggest, see Ref. 57.

In general, for four permeant monovalent ions, for example, Na$^+$, K$^+$, Cl$^-$, and F$^-$, the zero-current reversal potential, V_m, across a cell membrane would be given by the GHK equation as:

$$V_m = \frac{RT}{F} \ln \left(\frac{a_{Na}^o + (P_K / P_{Na}) a_K^o + (P_{Cl} / P_{Na}) a_{Cl}^i + (P_F / P_{Na}) a_F^i}{a_{Na}^i + (P_K / P_{Na}) a_K^i + (P_{Cl} / P_{Na}) a_{Cl}^o + (P_F / P_{Na}) a_F^o} \right) \quad (3.31)$$

in which R, T, and F are the ideal gas constant, absolute temperature, and Faraday constant, respectively; a and P are ion activity and permeability coefficients, respectively, of the ions indicated by the subscript; superscript "o" and "i" refer to external (outside) and internal solutions respectively.

For a simple NaCl dilution experiment, the relative permeability (e.g., P_{Cl}/P_{Na}) can be determined by fitting the corrected ΔV_m^c with the following equation (the difference in GHK equations for test and control conditions):

$$\Delta V_m^c = \frac{RT}{F} \ln \left(\frac{a_{Na}^{oT} + (P_{Cl} / P_{Na}) a_{Cl}^i}{a_{Na}^i + (P_{Cl} / P_{Na}) a_{Cl}^{oT}} \right) - \frac{RT}{F} \ln \left(\frac{a_{Na}^{oC} + (P_{Cl} / P_{Na}) a_{Cl}^i}{a_{Na}^i + (P_{Cl} / P_{Na}) a_{Cl}^{oC}} \right) \quad (3.32)$$

The superscripts "T" and "C" refer to the Test (dilutions) and Control solutions, respectively. If the internal solution is the same as the control solution, the second GHK component is zero.

Equation 3.32 needs to be solved for P_{Cl}/P_{Na} numerically, using appropriate software, from the experimental value of ΔV_m^c and the known ion activities. If there are small quantities of other permeant ions present, then a small correction for their permeability contribution may need to be taken into account. However, if these permeabilities are small (e.g., HEPES$^-$, EGTA^{2-}, and pipette Ca^{2+} for selective ion channels), then the contribution to the zero-current reversal potential should be minimal and could be ignored in Equation 3.32.

3.4.2 Bi-Ionic Potential Measurements

To measure relative permeabilities of cations in a cation-selective channel, or anions in an anion-selective channel, bi-ionic potential measurements are recommended. In such a case, for an anion-selective channel, the concentration of the two different salts, sharing a common cation, should be the same. For example, to measure the relative permeability of F$^-$ to Cl$^-$ in an anion-selective channel, 150 mM NaCl and 150 mM NaF solutions and a procedure similar to that outlined for dilution potentials should be used (see also Fig. 3.3). The shift in the experimental zero-current reversal potential ΔV_{rev}^{exp} for the bi-ionic measurements between the control and test solutions should be first measured and corrected for liquid junction potentials (see Equation 3.30) to obtain ΔV_m^c.

As in the dilution case (see Equation 3.32), for these bi-ionic measurements, the value of P_F/P_{Cl} is determined from the difference in V_m theoretically predicted by the relevant GHK equations:

$$\Delta V_m^c = \frac{RT}{F} \ln \left(\frac{\left(P_{Na}/P_{Cl}\right) a_{Na}^{oT} + a_{Cl}^{i}}{\left(P_{Na}/P_{Cl}\right) a_{Na}^{i} + \left(P_F/P_{Cl}\right) a_F^{oT}} \right) - \frac{RT}{F} \ln \left(\frac{\left(P_{Na}/P_{Cl}\right) a_{Na}^{oC} + a_{Cl}^{i}}{\left(P_{Na}/P_{Cl}\right) a_{Na}^{i} + a_{Cl}^{oC}} \right) \quad (3.33)$$

where again superscripts "T" and "C" refer to the activities of the Test and Control solutions. However, before numerically fitting the bi-ionic experimental values of ΔV_m^c to Equation 3.33 to obtain the value of P_F/P_{Cl}, the permeability ratio P_{Na}/P_{Cl} has to be obtained separately from dilution potential measurements (see Section 3.4.1).

For bi-ionic measurements intended to measure relative cation permeabilities, for example, P_X/P_{Na}, the test cation could be changed to X^+, maybe with a small amount of Na^+ from the NaOH used in buffering. In this case the relevant theoretical equation for ΔV_m^c is:

$$\Delta V_m^c = \frac{RT}{F} \ln \left(\frac{a_{Na}^{oT} + \left(P_X/P_{Na}\right) a_X^{oT} + \left(P_{Cl}/P_{Na}\right) a_{Cl}^{i}}{a_{Na}^{i} + \left(P_{Cl}/P_{Na}\right) a_{Cl}^{oT}} \right) - \frac{RT}{F} \ln \left(\frac{a_{Na}^{oC} + \left(P_{Cl}/P_{Na}\right) a_{Cl}^{i}}{a_{Na}^{i} + \left(P_{Cl}/P_{Na}\right) a_{Cl}^{oC}} \right)$$

$$(3.34)$$

As in the case of anion selectivity measurements, before fitting to Equation 3.34 it is necessary to have a value of P_{Cl}/P_{Na}, which can be determined separately from NaCl dilution potential measurements.

3.4.3 Voltage and Solution Control in Whole-Cell Patch-Clamp Recordings

Control of solution composition and voltage is of utmost importance for ion permeation measurements. To this end, whole-cell patch-clamp recordings require good access to the cell interior. This access resistance (the resistance in series between the electrode and the membrane) can be routinely reduced below $10\,M\Omega$ with perseverance and pipettes of the right shape (short, shallow tapered tips of relatively larger diameters). Such pipettes give resistances of $3–4\,M\Omega$ in the standard solutions and contribute to clean rupturing of the cell membrane under the pipette tip. An access resistance of less than $10\,M\Omega$ is important for two reasons: first to reduce errors in current measurements that arise due to "series resistance errors"—the voltage drop that occurs across the series resistance and the distortion of current due to the filtering effect of this resistance (Note: the circuit response time is given by the product of the resistance and capacitance of the electrical circuit). The series resistance voltage drop depends directly on the magnitude of the current flow across this resistance. This is zero in experiments that measure ion channel selectivity by the shift in V_{rev} (zero current voltage) in the presence of different ionic solutions, but series resistance voltage errors may arise if the reversal potential is inferred from nonzero currents, for example, when subtraction of significant leak or other currents is needed. A second advantage of low access resistance is that it increases the speed and efficiency of dialysis of the cell interior and the ability to maintain a well-defined intracellular solution. For a given cell and pipette solution, the rate of diffusion of pipette solution into a cell is directly related to tip diameter and access resistance [58, 59]. The diffusion of less mobile intracellular ions is expected to be slower. In addition, large intracellular anions may maintain a Donnan potential, with the intracellular solution more negative than the pipette potential (V_p), producing an error of up to 10 mV. Such large impermeable intracellular anions may cause equilibration to take up to 10–20 min as less mobile anions only gradually diffuse from the cell [60, 61]. The transmembrane currents may also lead to a different concentration adjacent to the membrane from that present in the bulk intracellular solution (in equilibrium with pipette solution). An example of such nonuniformity—ion shift effects—is described later (Section 3.4.4). Increasing tip diameters and reducing access resistance minimizes both of these potential errors.

Keeping the solutions bathing a cell membrane as simple as possible facilitates analysis and accurate determination of ionic selectivity in whole-cell patch-clamp recordings by enabling all ionic species to be included in the fitting of V_{rev} to electrodiffusion equations (see Sections 3.4.1 and 3.4.2), as needed to quantify selectivity ratios. Simple solutions also facilitate accurate determination of liquid junction potentials and any other voltage offsets (see Section 3.4.5). The presence of multiple ions, particularly divalent and multivalent ions, can also significantly impact both conductance and relative selectivity measurements [13, 62]. While solutions devoid of more complex constituents with poorly defined mobilities and smaller diffusion constants, such as large anions, multivalent ions like ATP, excess EGTA, or Ca^{2+} and Mg^{2+}, can have negative impacts on cell viability over long recording periods, they simplify selectivity measurements and analysis.

3.4.4 Ion Shift Effects During Whole-Cell Patch-Clamp Experiments

The following example illustrates the potential problem of non-uniform distribution of ion concentrations between the pipette, the bulk intracellular solution, and the solution adjacent to the membrane. The recordings were obtained using standard whole-cell patch-clamp of HEK293 cells expressing recombinant α1 glycine receptors and bathed in approximately symmetrical 150 mM Cl^- solutions (see Ref. 63 for details of experimental conditions). As illustrated in Figure 3.4a, application of 1 mM glycine to a cell voltage-clamped at −30 mV induces a large inward current (Cl^- efflux) that rises rapidly to a peak amplitude and then decays. The polarity of the peak response is directly related to the clamped membrane potential, reversing at about 0 mV (0.7 mV, "I_{gly} at peak" in Fig. 3.4b). Reapplication of the same glycine concentration, followed by voltage ramps (from −20 mV to +10 mV over 1.25 s) to measure V_{rev} during the decay phase, reveals shifts in V_{rev} to progressively more negative values (see Fig. 3.4b). This reflects a marked decrease in the local internal concentrations of Cl^- adjacent to the cell membrane due to the Cl^- efflux evoked by glycine application. Despite uniform concentrations in the pipette and bath, the diffusion rate of Cl^- from the pipette is too slow to maintain the same Cl^- concentration at the membrane. The example illustrated in Figure 3.4 is a "worst case" scenario,

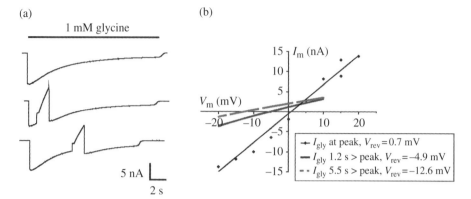

FIGURE 3.4 Illustration of Cl⁻ shift effects potentially complicating measurement of channel current reversal potentials (V_{rev}). (a) Whole-cell currents recorded in a HEK-293 cell expressing α1 glycine receptors, bathed in an approximately symmetrical (~150mM) NaCl solution, and voltage-clamped at −30mV. Three sequential glycine applications evoked large inward currents. In the center and lower traces, a voltage step to −20mV is followed by a voltage ramp to +10mV, resulting in a reversal of the current amplitude. (b) Current-voltage curves from the same cell as in (a). The dark trace (I_{gly} at peak) plots the peak current response as a function of membrane voltages and sequential applications of glycine at each voltage. The thicker solid and dashed curves show the instantaneous current-voltage relationship obtained from voltage ramps applied 1.2 s and 5.5 s after the peak I_{gly}, as shown in the middle and lower panels of (a). Note the shift in V_{rev} from +0.7 mV (peak I_{gly}) to −4.9 and −12.6 mV during the decay phase of glycine receptor responses, consistent with a nonsymmetrical Cl⁻ concentration. The large Cl⁻ efflux associated with the current response at −30mV results in a reduction of Cl⁻ concentration at the intracellular face of the membrane, and a probably smaller increase in Cl⁻ concentration at the extracellular face of the membrane (i.e., a net "Cl⁻ shift" across the membrane).

caused by the very large amplitude of currents in this particular cell, coupled with the unusually large and/or complex cell geometry as indicated by a relatively high cell capacitance of approximately 55 pF. Small currents, small round cells, and low access resistances can minimize these nonuniform concentration distributions. Alternatively, the peak response to rapid channel activation can be used as a more accurate measure of channel current. These Cl⁻ shift effects in large and complex cells such as neurons have been described previously [64].

3.4.5   Liquid Junction Potential Corrections

Whenever two liquids of different compositions are in contact, a potential develops across the junction between them known as a liquid junction potential (LJP). LJPs are especially significant in patch-clamp measurements, where they may be up to 10 mV or even more. The LJP between pipette and bath solution cannot be eliminated by the normal zeroing procedure, whereby residual potentials between the pipette and bath solutions are offset in the patch-clamp amplifier prior to the cell being

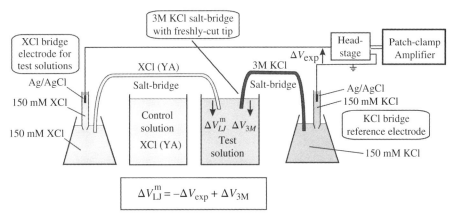

$$\Delta V_{LJ}^{m} = -\Delta V_{exp} + \Delta V_{3M}$$

FIGURE 3.5 The experimental setup for the measurement of liquid junction potentials (LJPs, given by ΔV_{LJ}^{m}; see figure 1 of Ref. 54, but allowing for different anions and cations in the control solution and control salt bridge, XCl/YA). X represents different monovalent cations (e.g., Na^+, K^+, Cs^+, etc.). For typical monovalent control solutions, both control solution and control salt-bridge would be XCl at 150 mM concentration. However, with appropriate care of the salt-bridge, these could be different both in composition and concentration (as YA, e.g., to represent $MgSO_4$ at 50 mM) from the solution in the left conical flask.

patched [65] (see Fig. 3.2 for different patch configurations). An additional LJP at the reference salt bridge also develops when the external bath solution is changed. This latter shift in reference LJP needs to be taken into account to determine the true shift in V_{rev} in either dilution potential or bi-ionic potential measurements (see Equation 3.30). These principles have been previously outlined elsewhere [33, 53, 65]. The values of the LJPs required for correcting these measurements may either be calculated using the generalized Henderson liquid junction potential equation (eq. 12 of [65]), or by measuring them (discussed later in this section) and following the procedures outlined in Ref. 65. Commercially available programs (within pClamp software, Molecular Devices Corp, or stand-alone programs 16-32-bit compatible *JPCalcW* and a new 32-64-bit compatible *JPCalcWin*, see http://web.med.unsw.edu. au/phbsoft/; [66]) exist for both calculating the LJPs and correctly applying them under different patch-clamp and other electrophysiological conditions. The Henderson LJP equation requires the mobility values of all the significant ions in the solution. The commercially available programs include a built-in library of typical ions, and an extended list can be found on the web link http://web.med.unsw.edu.au/phbsoft/mobility_listings.htm. The use of the Henderson LJP equation has been validated for a range of monovalent ions (also in the presence of small concentrations of divalent ions) to an accuracy generally within 0.1–0.2 mV [54]. Where ion mobility data is not available or for further validation of LJP values, they may be experimentally measured [67, 68]. An optimised technique for doing this has recently been described by Barry et al. [54], which includes 3 M KCl reference electrode corrections for greater accuracy (see Fig. 3.5).

As an example to illustrate the critical importance of LJP corrections in ion selectivity measurements, let us consider whole-cell patch-clamp dilution experiments on recombinant α1 GlyRs expressed in HEK293 cells [69]. Both the pipette and bath solutions were initially composed of predominantly 150 mM LiCl solutions. Diluting the outside LiCl solution to about 0.25 of the initial solution, the average shift in the measured zero-current reversal potential, ΔV_{rev}^{exp}, was +19.6±0.4 mV.

While $V_L{}'$—the original LJP between the first bath solution and the pipette solution—is very small, being only 0.1 mV for these two very similar solutions, the only relevant LJP is the LJP shift at the reference electrode between the new and control bath solutions, ΔV_{LJ}^{21}, calculated as −11.1 mV in this example. Thus, the corrected shift in V_m, ΔV_m^c, is predicted to be +30.7 mV (see Equation 3.30). The effects of this on the derived value of P_{Cl}/P_{Li} is to change it from about 5 (without corrections) to well over 30 (with corrections), indicating that LJP corrections can make a huge difference to a resultant permeability ratio (see also Section 3.5).

While it is generally adequate to use concentrations for bi-ionic potentials, where the total ion concentrations in each solution are approximately the same, for dilution potentials greater accuracy is generally obtained by using ion activities, particularly when the two solutions have large total ion concentration differences. This can be seen from the LiCl results of tables 2 and 3 of Ref. 69, where the calculations used ion activities, and the total liquid junction potential correction was −10.3 mV. This contrasts with the value of −11.1 mV described earlier, when ion concentrations were used.

3.5   INFLUENCE OF VOLTAGE CORRECTIONS IN QUANTIFYING ION SELECTIVITY IN CHANNELS

Two examples are given to illustrate the effects of voltage corrections on the quantification of relative ion permeability ratios. The data again come from experiments in recombinant α1 GlyRs, where we can compare the results analyzed with and without such corrections.

3.5.1   Analysis of Counterion Permeation in Glycine Receptor Channels

In two papers [69, 70] we investigated the relative selectivity of the counterions Li$^+$, Na$^+$, and Cs$^+$ in two different anion-selective GlyRs with differing pore dimensions: an α1 WT GlyR channel with an estimated minimum pore diameter of about 5.3Å, and a mutant α1 P-2′Δ GlyR channel with a minimum pore diameter of 6.9Å. The aim of the experiments was to evaluate how relative ion charge selectivity interacted with minimum pore diameter and the size of the counterion cation, and thereby provide insights into how both anions and cations could permeate the same pore. The experiments and analyses followed the procedures outlined in Section 3.4: dilution potential experiments with corrections of the V_{rev} measurements for both LJPs (Eq. 3.30) and for the small but significant offset potential. Offset potentials correspond to the second terms in Equations 3.32, 3.33, and 3.34, and arise due to any differences between internal cell composition and external control salt solution. The recalculated (unpublished) P_{Cl}/P_{cation} values for the raw data showed no clear differences between cations (being 4.6 ± 0.2 ($n=6$), 4.8 ± 0.1 ($n=32$) and 5.1 ± 0.5 ($n=6$) for LiCl, NaCl, and CsCl, respectively) in the smaller α1 WT GlyR channel. For the larger mutant α1 P-2'Δ GlyR channel, the P_{Cl}/P_{cation} values for the raw data had values of 2.4 ± 0.1 ($n=7$), 2.0 ± 0.1 ($n=8$) and 2.0 ± 0.1 ($n=7$) for LiCl, NaCl, and CsCl respectively. Hence, the permeability ratios seemed to indicate that cation size had virtually no effect on anion/cation permeability. The WT channel had a P_{cation}/P_{Cl} of about 0.2 for each of the cations, and the larger mutant channel had a P_{cation}/P_{Cl} of about 0.4–0.5 for each of the cations.

However, when LJPs are taken into account we have a completely different picture. For the small WT GlyR channel, the P_{Cl}/P_{cation} values were 23.4, 10.9, and 5.0 for LiCl, NaCl, and CsCl, respectively (see table 1 of Sugiharto et al. [70]). For the larger mutant GlyR channel, the P_{Cl}/P_{cation} values were 6.0, 3.3, and 1.9 for LiCl, NaCl, and CsCl, respectively (table 1 of Sugiharto et al. [70]). Hence, for both channels $P_{Li} \ll P_{Na} \ll P_{Cs}$, with the range of values from 0.04 to 0.2 in the WT channel and from 0.2 to 0.5 in the larger mutant channel. Accounting for the offset potential was equivalent to using the full version of Equation 3.32 when quantifying the permeability ratio (rather than simply using the first term only of Equation 3.32). For the data from the smaller WT channel, this caused the greatest reduction in permeability for Li$^+$, the ion with the largest hydrated radius ($P_{Cl}/P_{Li}=30.6$, $P_{Li}/P_{Cl}=0.03$), with a smaller reduction for Na$^+$ ($P_{Cl}/P_{Na}=11.8$) and no obvious effect for Cs$^+$, the ion with the smallest hydrated radius ($P_{Cl}/P_{Cs}=5.0$; table 5 of Ref. 69). For the larger mutant channel, only the P_{Cl}/P_{Li} was slightly altered by this offset correction (table 5 of Ref. 69). The picture of the permeability data with these corrections then indicates that anion–cation permeability in both channels correlates well with the hydrated (cation) counterion radius compared to the radius of the channel pore [69, 70], a totally different conclusion than if the LJP and offset corrections had been ignored.

3.5.2 Analysis of Anion–Cation Permeability in Cation-Selective Mutant Glycine Receptor Channels

The second example relates to anion–cation permeability calculations for a series of mutant α1 GlyRs, which we abbreviate as SDM, SDM+271A, and SDM+271E, which all converted the α1 GlyR channel from a normally anion-selective channel ($P_{Cl}/P_{Na}=24.6$ [71]) to a channel that is now predominantly cation-selective [72].

Recalculation of dilution potential measurements ignoring LJP corrections for the above mutant channels radically underestimated the P_{Cl}/P_{Na} values, grossly overestimating the relative P_{Na}/P_{Cl} values. The raw, unpublished ΔV_{rev} (ΔV_{rev}^{exp} in Eq. 3.30) values were obtained by removing the LJP corrections from the corrected ΔV_{rev} values used in Ref. 72. For example, for the SDM+271A mutant α1 GlyR channel the raw ΔV_{rev} value was -25.6 mV instead of a corrected value of -19.3 mV due to a ΔV_{LJ} of -6.3 mV. These calculations indicated that when the LJP corrections were applied to the average ΔV_{rev} values, the P_{Na}/P_{Cl} values correctly dropped from values of 25, 10, and 17 close to values of 7.7, 4.3, and 5.6, (equivalent to the published P_{Cl}/P_{Na} values

0.13, 0.23, and 0.19; table I of Ref. 72) for the SDM, the SDM+271A, and the SDM+271E mutant α1 GlyR channels respectively. These recalculations have shown that had LJP corrections not been applied to these selectivity measurements, the calculated permeabilities for these different mutations would have grossly overestimated the effects of the mutations on the relative shifts from anion to cation selectivity on these mutant α1 GlyR channels and given a misleading picture of the structural changes to the channels caused by the mutations. In fact, our impression is that many people ignore liquid junction potential corrections and of those that try to take them into account, many of them do not do so correctly.

3.6   ION PERMEATION PATHWAYS THROUGH CHANNELS AND PUMPS

Given the number of crystal structures that are now available for a variety of channels and pumps, and the wealth of information that has been assembled from mutation studies on recombinant transporters, it seems natural to think of how the protein structure influences ion transport by channels and pumps. The interplay between patch-clamp electrophysiology and rapid advances in molecular structures of membrane proteins has enabled remarkable insights into how ions are transported across cell membranes. We illustrate below one such example of this—elucidation of the ion permeation pathway of the pentameric ligand-gated ion channels (pLGICs; also known as cys-loop receptor-channels). This family of receptor-channel proteins is comprised, in mammals, of anion-selective glycine and γ-amino-butyric acid (GABA) gated receptor-channels (GlyRs and $GABA_A$Rs) and the cation-selective acetylcholine (ACh) and serotonin (5-HT) gated channels (nicotinic AChRs and $5-HT_3$Rs).

3.6.1   The Ion Permeation Pathway in Pentameric Ligand-Gated Ion Channels

The pLGICs mediate synaptic communication in the brain and peripheral nervous system. They respond to their specific neurotransmitter ligands by undergoing a conformational change that activates the receptor-channel gate to allow permeant ions to flow down their electrochemical potential gradients to excite or inhibit the postsynaptic cell. Key to this functional role is the charge of the permeant ion; influx of anions into the postsynaptic cell results in membrane hyperpolarization and inhibition, while influx of cations results in depolarization and excitation. This functional distinction depends on the structure of the ion permeation pathway in the open conformation of the receptor-channel, which is comprised of distinct but continuous regions of the protein's quarternary structure (see Fig. 3.6). The pLGIC structure is an approximate symmetrical arrangement of five subunits around a central transmembrane pore. Each subunit is comprised of a large extracellular domain harboring the ligand-binding region, four transmembrane spanning α-helices (TM1–TM4), and a large intracellular loop harboring modulatory elements and synaptic attachment sites. The second transmembrane domain (TM2) of the five subunits lines

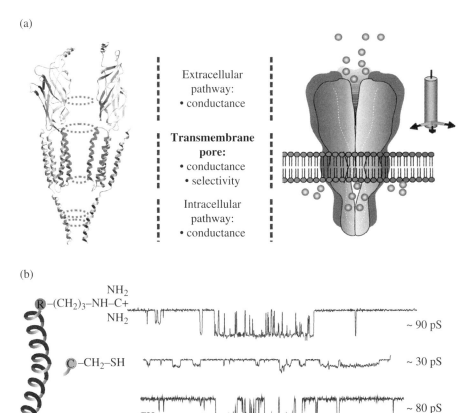

(a)

Extracellular
pathway:
• conductance

**Transmembrane
pore:**
• conductance
• selectivity

Intracellular
pathway:
• conductance

(b)

NH$_2$
R–(CH$_2$)$_3$–NH–C+
NH$_2$

~ 90 pS

C–CH$_2$–SH

~ 30 pS

~ 80 pS

CH$_3$
C–CH$_2$–SS–(CH$_2$)$_2$–N$^\pm$CH$_3$
CH$_3$

3 pA

20 mS

FIGURE 3.6 The ion permeation pathway of pLGICs indicating the discrete domains of the permeation pathway and the influence of charged residues on channel conductance. (a) Opposing α and β subunits of the *Torpedo californica* nAChR generated from PDB ID 2BG9 [73]. Indicated are the approximate locations of rings of charged residues described in the text to influence conductance. The permeation pathway can be naturally divided into extracellular, transmembrane, and intracellular pathways, continuous in the open channel. All domains influence channel conductance with the transmembrane domain additionally influencing ionic selectivity, with the **bold label** to stress its dominant role in permeation. The right-hand panel shows a schematic view of a pLGIC illustrating the pathway of ions (depicted as balls) flowing from extracellular to intracellular solution. Ions enter the transmembrane pore via an extended extracellular funnel, and exit the channel via lateral portals or fenestrations formed in the intracellular domain. The right-hand panel is reproduced with permission from Ref. 74. © the Biochemical Society. (b) Effects of amino acid side-chain charge neutralization on channel conductance. Current traces are from outside-out patches from HEK-293 cells expressing recombinant homomeric α1 GlyRs held at a V_m of −60 mV (for experimental conditions see Ref. 13). The schematic α-helix represents the TM2 domain showing the residue side chain at the TM2 extracellular loop (position 19′). The uppermost current trace represents WT GlyRs, the center trace is from mutant GlyRs where the Arg positive charge has been neutralized by mutation to Cys. The same patch exposed to a methanethiosulfonate reagent covalently labels the Cys to restore the positively charged side chain and returns the conductance back toward the control level (lowermost trace).

the transmembrane pore and contains the activation gate. This basic architecture is conserved across the family of pLGICs, which also share common biophysical and molecular mechanisms of ion permeation. The distinct components of the ion permeation pathway can be divided into extracellular, intracellular, and transmembrane domains (see Fig. 3.6).

3.6.1.1   *Extracellular and Intracellular Components of the Permeation Pathway*
An ion moving from the extracellular to the intracellular solution encounters an extended permeation pathway (see Fig. 3.6), first being drawn through a narrow external funnel from the top of the receptor, through to the top of the transmembrane pore. In the nAChR, the extracellular funnel is approximately 60Å long with an internal diameter of approximately 20Å [75]. If the receptor channel is in an open state, the ion will then completely traverse the approximately 30Å long transmembrane pore, and finally leave the channel via lateral portals or fenestrations in the intracellular domain. Each of these discrete domains contains physical constrictions, down to a diameter of approximately 8Å at the TM2 constriction and lateral portals, sufficient to interact with and influence the permeating ion, and necessitating at least transient and partial dehydration of the ion during permeation. The walls of the extracellular vestibule are lined at various places by charged amino acids, with up to two "rings" of charge formed from the five subunits shown by mutagenesis to affect ion permeation across the pLGIC family [76–78]. In the muscle nAChR, a ring of negative charge is formed by Asp residues in all five subunits (position 97 in the α subunit) located approximately half way along the extracellular funnel. Inverting the charge by mutating all these residues to a positive Lys reduced inward cation conductance by approximately 70% [76]. A similar effect (~70–90% decrease in conductance) is observed when the homologous Asp residue (D127) in the 5-HT$_3$R is mutated to Lys [77] or when the corresponding Lys residue in anionic α1 GlyRs is mutated to Glu (K104E) [78]. Charge neutralization, rather than inversion, reduces conductance to a smaller extent [79]. Mutating additional charged residues in this extracellular funnel can also affect conductance, with the extent of effects seemingly increasing as the residues are located closer to the transmembrane pore [77, 79]. Ion flux from the extracellular to intracellular solutions is affected to a much greater extent by these mutations than ion flux from inside to outside, suggesting these charges concentrate permeant ions in the external vestibule.

    While the structural data is less well resolved for the intracellular domain of the pLGICs (and this domain is largely absent in the crystallized bacterial and *Caenorhabditis elegans* channels), access from the transmembrane pore to the intracellular solution in *Torpedo* nAChR appears to be exclusively via fenestrations or lateral portals [73]. The walls of these portals are also framed by three to four key charged residues in each subunit, the polarity of which differs in the cationic and anionic channels. Inverting the charge of these residues in the anionic and cationic pLGICs similarly impacts on single channel conductance [80–82]. This is of particular physiological relevance for the cationic 5-HT$_3$R receptor splice variants 5-HT$_3$A and 5-HT$_3$B. Channels comprised of homomeric 5-HT$_3$A subunits have portals lined with basic Arg residues at positions R432, R436, and R440, and consequently have a very low single channel conductance of less than 1 pS. In contrast, heteromeric channels comprised of both $5HT_3A$ and $5HT_3B$ subunits, the latter of which has neutral and acidic residues at these positions, have a much larger conductance of around 10–20 pS for inward cation flow under the same conditions [80]. Qualitatively, similar effects of neutralizing or inverting the charge of residues lining the portals are seen in nAChRs and in anionic GlyRs [81, 82]. As with the extracellular vestibule/funnel, the influence on conductance is not symmetrical, but is more pronounced for ion flux from the intracellular to the extracellular side. In the GlyRs, for example, inward conductance is reduced by about 30% when four basic Arg or Lys residues, putatively aligned with the portals, were collectively mutated to Glu, while outward conductance was largely unaffected (<5% decrease [82]). It seems likely that the major effect is to accumulate permeant ions at the intracellular end of the receptor-channels to facilitate rapid ion flux across the transmembrane pore under the appropriate electrochemical potential gradient.

Charged residues flanking the extracellular and intracellular ends of the TM2-demarcated transmembrane pore also markedly affect ion conductance [13, 14]. Mutating the three acidic Glu/Asp residues at the TM2 extracellular end in the α and β subunits of the pentameric $\alpha_2\beta\delta\gamma$ nAChR to Lys results in an approximately 90% reduction in inward cation conductance from an initial value of approximately 80 pS [14]. However, conductance seems even more sensitive to charges at the intracellular end of TM2 (the "intermediate ring" at $-1'$ in the TM2 numbering nomenclature). Neutralizing just two $-1'$ Glu residues in the pentamer results in an approximately 80% reduction in conductance [14]. It's been estimated that each single subunit Glu residue at this $-1'$ site in the pentameric nAChR results in approximately 50 pS additional inward cation conductance [83]. A qualitatively similar effect of TM2 charge mutations is seen for anionic GlyRs, where neutralizing the Arg residue at the extracellular end of the homomeric α1 GlyR TM2 markedly reduces channel conductance [84]; see Fig. 3.6b). Attempts to mutate the intracellular TM2 Arg at $0'$ in GlyRs, corresponding to the intermediate ring of charge, results in nonfunctional channels [85]; although mutations to adjacent residues that may alter the local electrostatics imparted by Arg $0'$ results in very small conductances (<2 pS for the Pro-$2'$Thr mutation [86]). Together, the available results suggest that rings of charge at extracellular and intracellular ends of TM2 have marked effects on channel conductance, with the charge at $0'/-1'$ having a dominant influence. In contrast to the results with the extracellular and intracellular components of the extended permeation pathway, the effects on conductance of TM2 mutations are more symmetrical, especially for this $-1'$ intermediate ring.

3.6.1.2 *The TM2 Pore Is the Primary Ion Selectivity Filter*   The mutations to the intracellular and extracellular components of the extended permeation pathway described above have large effects on conductance, but do not have a marked impact on relative ion selectivity. One caveat to this conclusion is that the extracellular vestibule mutations in $5HT_3Rs$ impact on relative Ca^{2+} permeability, and on Ca^{2+} access to inhibit monovalent cation conductance [77], and likely influence multivalent ion interactions in the other pLGICs. But the same mutations, as well as the charge reversing mutations at the intracellular portals, do not significantly affect

monovalent ion charge selectivity (i.e., $P_{Cl}:P_{Na}$, $P_{Cl}:P_{K}$) [77, 82]. Rather, the transmembrane TM2 pore is the major determinant of ionic selectivity, and again the intracellular end of TM2 is dominant (reviewed in [87]). A single point mutation to replace the neutral Ala with a negative Glu at position −1′ in the α1 GlyR can change the GlyR from being anion-selective to cation-selective [72]. The chloride to sodium permeability ratio changes from a control $P_{Cl}:P_{Na}$ value of approximately 15 to a value of approximately 0.3 in the Ala-1′Glu mutant. $P_{Cl}:P_{Na}$ is even lower (~0.1) if this mutation is combined with the deletion of the adjacent −2′ Pro. Hence, this internal end of TM2 is the primary selectivity filter and the region that has the closest functional interactions with the permeating ion(s). The recent crystal structures of pLGICs confirm that the internal end of TM2 is the most constricted region of the transmembrane pore in the open state [88, 89]. Protonation of basic Arg or Lys residues at 0′ has been proposed as a key to anion selectivity in pLGICs, with this basic 0′ residue present but neutral (deprotonated) in cationic nAChRs. Acidic residues at −1′ in nAChRs dominate pore electrostatics and enable the high cation conductance of these channels [83, 90]. Consistent with the results from the anion–cation selectivity converting mutations described earlier, neutralization of the −1′ Glu or Asp in nAChRs and insertion of a −2′ Pro facilitates protonation of the 0′ Arg and thereby changes the electrostatic profile [83, 90].

In summary, mutagenesis and patch-clamping have revealed the extended permeation pathways of pLGICs, with specific charged residues identified that markedly influence both conductance and selectivity. The transmembrane TM2 delineated pore is the major functional component of the permeation pathway, where both the channel gate and the determinants of ionic selectivity reside.

3.6.2   Ion Permeation Pathways in Pumps Identified Using Patch-Clamp

Ion permeation in pumps is distinctly different from that in ion channels. The membrane-spanning aqueous pores of channels allow high ion flux rates at levels approaching that of free diffusion. Pumps, in contrast, have much lower rates of ion translocation. Hence pumps are not pores with aqueous "ion permeation" pathways, but rather they transport ions via very different mechanisms. The commonly accepted occluded access model of pump transport has ions (and other transported solutes) accessible to intracellular solutions in one conformation, and accessible to the extracellular solution in a different conformation. In between these conformations, ions are occluded within the protein interior. In essence, pumps have two gates: an extracellular gate and an intracellular gate, which occludes access to the surrounding solutions, that open and close separately. Ion "permeation" in pumps therefore includes aspects of how tightly ions bind (and unbind) at specific sites in these different conformations, the selectivity of these binding sites to different ions, and the rates at which the pumps change between conformations to allow ions to pass from one side of the membrane to another. With much slower rates of ion flux, and hence much lower transfer of charge per unit time, the currents carried by ions during a transport cycle are very small (≪0.1 fA) and are not amenable to direct measurement using patch-clamp recordings. Macroscopic pump currents can be measured using patch-clamp electrophysiology, but requires a very high number of functional pumps in a synchronous conformation. The larger recombinant expression levels that can be achieved in *Xenopus* oocytes has favored two-electrode voltage-clamp in these cells for electrophysiological analysis of pumps (see Chapter 4), although this system lacks the solution control inherent in whole-cell recordings.

There are nevertheless many instances where ion transport through pumps has been measured with patch-clamp recordings to shed light on physiological mechanisms. Below we briefly review one interesting example that has illuminated the access or permeation pathway of ions to their binding site(s) within a pump.

3.6.2.1   *Palytoxin Uncouples the Occluded Gates of the Na⁺,K⁺-ATPase*   Palytoxin

is a highly potent marine toxin collected from coral reefs in Hawaii and elsewhere that strongly binds to the Na⁺,K⁺-ATPase (Na⁺ pump) and converts this pump into a non-selective cation channel [91]. Application of saturated palytoxin (100 nm) to lipid membranes expressing the Na⁺ pump and bathed in physiological salt concentrations results in clear single channel current steps of about 10 pS conductance [92]. The mechanism of action of palytoxin appears subtle; the extracellular and intracellular gates can be opened and closed by physiological ligands/phosphorylation as usual, but the coupling appears disrupted such that both gates can sometimes be open simultaneously [91]. The high conductance of this modified pump-channel indicates passive transmembrane ion flow in a channel-like aqueous pore. Such spontaneous opening of both gates simultaneously is unlikely to occur under any physiological conditions. Given that the Na⁺ pump moves Na⁺ and K⁺ ions "uphill" at rates of about $100 \, s^{-1}$, the occasional slip of the pump into a channel mode by having both its gates open, and allowing downhill movement of ions at $10^6 \, s^{-1}$, would be functionally disastrous. Indeed, this channel mode is never observed in intact pumps [91]. The two gates must remain tightly coupled for functional Na⁺ pumping. The toxin exposes an open pore that appears to have the same access pathways as in the functional pump, and has, therefore, proved informative in studying the "permeation pathway" of the pump [91, 93, 94]. Using the substituted cysteine accessibility method, Reyes and Gadbsy [93] probed the permeation pathway by examining effects of positively and negatively charged methanethiosulfonate (MTS) reagents on the single "channel" conductance of palytoxin-modified Na⁺ pumps. The results were mapped on to a homology model of the structurally related Ca²⁺ATPase, crystallized in a conformation with its extracellular gate open and hence the permeation pathway accessible to extracellular solutions [94]. Covalent MTS modification of single Cys residues that affected cation conductance in ways dependent on the MTS charge (negative reagents increasing, positive reagents decreasing) were interpreted as identifying residues located in a wide external vestibule that electrostatically influenced ion permeation. At Cys positions deeper within the permeation pathway, both positive and negative reagents sterically hindered cation flow to reduce conductance. Deeper still within the permeation pathway, only the positive MTS could access the introduced Cys, indicating that these residues were at, or beyond, the "selectivity filter." Substitution of Cys for some of these deep residues thought to coordinate bound cations in the pump (specifically site II residues E336 in TM4 and D813 in TM6) strongly reduced cation to anion selectivity of the pump-channel current [93]. Labelling of the E336C mutation with the positive MTS could even invert the channel selectivity from cationic ($P_{Na}:P_{NO3}=2.8\pm0.03$) to anionic ($P_{Na}:P_{NO3}=0.2\pm0.01$). Hence, acidic residues from the cation-coordinating binding site II form the ion charge selectivity of the modulated pump-channel.

Notably, these elegant results of the Gadsby lab on determinants of conduction and selectivity in palytoxin-modified Na^+ pump-channels show similarity to the role of acidic and basic side chains in the vestibules and selectivity filter of the pLGICs (described in Section 3.6.1). In both, vestibule electrostatics is a key aspect in determining ion conductance, and side chain charges at the selectivity filter can determine ion-charge selectivity. Clearly, the same biophysical and molecular principles operate to attract and select ions to permeate through channels or into and out of pump-binding sites, and in both cases the patch-clamp technique has provided insights into these processes.

3.7   CONCLUSIONS

In this chapter, we have highlighted the utility of patch-clamp recordings to characterize ion channels and pumps and their underlying molecular mechanisms. The pores of ion channels can mediate passive flux rates of well over 10^6 ions per second driven by physiological electrochemical potential gradients that enable the direct measurement of resulting ionic currents, and make the patch-clamp technique ideally suited to characterizing these membrane transporters. The more complex conformational changes that accompany ion transport across the membrane by pumps, and the much lower rates of ion flux, present challenges in applying patch-clamping to directly measure such fluxes, although this has been achieved in certain modified pump-channels. We have also illustrated how the different configurations of the patch-clamp recording technique have identified and characterized the extended permeation pathway of ligand-gated ion channels. This chapter provides an overview of technical approaches to correctly apply patch-clamp recordings in the quantification of the selective ion permeation of these ligand-gated ion channels, with implications for the study of the diverse range of other channels. We describe how careful application of the patch-clamp technique, recognizing potential areas of inaccuracy, and applying appropriate voltage corrections, is critical for the precise quantification of ion channel permeation and kinetic properties. When combined with protein mutagenesis, ion channel structures, biophysical analysis, and a generous serving of patience, ingenuity, and experimental application, patch-clamping has helped elucidate the functional and molecular mechanisms of ion channels and pumps.

ACKNOWLEDGMENTS

We acknowledge the National Health and Medical Research Council of Australia and UNSW Faculty Research Grants that have supported some of our work described earlier on ion-selectivity in GlyRs.

REFERENCES

1. D.C. Gadsby, Ion channels versus ion pumps: The principal difference, in principle, *Nat. Rev. Mol. Cell. Biol.* 10 (2009) 344–352.

2. E. Neher, B. Sakmann Single-channel currents recorded from membrane of denervated frog muscle fibres, *Nature* 260 (1976) 799–802.

3. O.P. Hamill, A. Marty, E. Neher, B. Sakmann, F.J. Sigworth, Improved patch-clamp techniques for high-resolution current recording from cells and cell-free membrane patches, *Pflugers Arch.* 391 (1981) 85–100.

4. A.L. Hodgkin, A. Huxley, Action potentials recorded from inside a nerve fibre, *Nature* 144 (1939) 710–711.

5. A.L. Hodgkin, A.F. Huxley, A quantitative description of membrane current and its application to conduction and excitation in nerve, *J. Physiol.* 117 (1952) 500–544.

6. P.G. Kostyuk, O.A. Krishtal, V.I. Pidoplichko, Effect of internal fluoride and phosphate on membrane currents during intracellular dialysis of nerve cells, *Nature* 257 (1975) 691–693.

7. P.G. Kostyuk, O.A. Krishtal, Separation of sodium and calcium currents in the somatic membrane of mollusk neurons, *J. Physiol.* 270 (1977) 545–568.

8. P.G. Kostyuk, O.A. Krishtal, V.I. Pidoplichko, N.S. Veselovsky, Ionic currents in the neuroblastoma cell membrane, *Neuroscience* 3 (1978) 327–332.

9. K.S. Lee, N. Akaike, A.M. Brown, Trypsin inhibits the action of tetrodotoxin on neurones, *Nature* 265 (1977) 751–753.

10. K.S. Lee, N. Akaike, A.M. Brown, Properties of internally perfused, voltage-clamped, isolated nerve cell bodies, *J. Gen. Physiol.* 71 (1978) 489–507.

11. K.S. Lee, N. Akaike, A.M. Brown, The suction pipette method for internal perfusion and voltage-clamp of small excitable cells, *J. Neurosci. Methods* 2 (1980) 51–78.

12. R. Horn, J. Patlak, Single channel currents from excised patches of muscle membrane, *Proc. Natl. Acad. Sci. U. S. A.* 77 (1980) 6930–6934.

13. A.J. Moorhouse, A. Keramidas, A. Zaykin, P.R. Schofield, P.H. Barry, Single channel analysis of conductance and rectification in cation-selective, mutant glycine receptor channels, *J. Gen. Physiol.* 119 (2002) 411–425.

14. K. Imoto, C. Busch, B. Sakmann, M. Mishina, T. Konno, J. Nakai, H. Bujo, Y. Mori, K. Fukuda, S. Numa, Rings of negatively charged amino acids determine the acetylcholine receptor channel conductance, *Nature* 335 (1988) 645–648.

15. J.A. Peters, M.A. Cooper, J.E. Carland, M.R. Livesey, T.G. Hales, J.J. Lambert, Novel structural determinants of single channel conductance and ion selectivity in 5-hydroxytryptamine type 3 and nicotinic acetylcholine receptors, *J. Physiol.* 588 (2010) 587–595.

16. D.C. Dawson, Permeability and conductance of ion channels, in: S.G. Schulz, T.E. Andreoli, A.M. Brown, D.M. Fambrough, J.F. Hoffman and M.J. Welsh (Eds.), *Molecular Biology of Membrane Transport Disorders*, Plenum Press, New York, 1996, pp. 87–110.

17. N.L. Lassignal, A.R. Martin, Effect of acetylcholine on postjunctional membrane permeability in eel electroplaque, *J. Gen. Physiol.* 70 (1977) 23–36.

18. B. Hille, *Ion Channels of Excitable Membranes*, third ed., Sinauer Associates, Inc., Sunderland, MA, 2001.

19. T.M. Lewis, P.R. Schofield, A.M. McClellan, Kinetic determinants of agonist action at the recombinant human glycine receptor, *J. Physiol.* 549 (2003) 361–374.

20. B. Katz, R. Miledi, Further observations on acetylcholine noise, *Nat. New Biol.* 232 (1971) 124–126.

21. B. Katz, R. Miledi, The statistical nature of the acetylcholine potential and its molecular components, *J. Physiol.* 224 (1972) 665–699.

22. C.R. Anderson, C.F. Stevens, Voltage clamp analysis of acetylcholine produced end-plate current fluctuations at frog neuromuscular junction, *J. Physiol.* 235 (1973) 655–691.

23. D. Colquhoun, A.G. Hawkes, Relaxation and fluctuations of membrane currents that flow through drug-operated channels, *Proc. R. Soc. Lond. B Biol. Sci.* 199 (1977) 231–262.

24. E. Neher, C.F. Stevens, Conductance fluctuations and ionic pores in membranes, *Ann. Rev. Biophys. Bioeng.* 6 (1977) 345–381.

25. P.T.A Gray, Analysis of whole cell currents to estimate the kinetics and amplitude of underlying unitary events: Relaxation and 'noise' analysis, in: D.C. Ogden (Ed.), *Microelectrode Techniques, The Plymouth Workshop Handbook*, second ed., Company of Biologists, Cambridge, MA, 1994, pp. 189–207.

26. D. Colquhoun, A.G. Hawkes, The principles of the stochastic interpretation of ion-channel mechanisms, in: B. Sakmann and E. Neher (Eds.), *Single Channel Recording*, second ed., Plenum Press, New York, 1995, pp.397–482.

27. F.J. Sigworth, The variance of sodium current fluctuations at the node of Ranvier, *J. Physiol.* 307 (1980) 97–129.

28. S.F. Traynelis, R.A. Silver, S.G. Cull-Candy, Estimated conductance of glutamate receptor channels activated during EPSCs at the cerebellar mossy fibre-granule cell synapse, *Neuron* 11 (1993) 279–289.

29. S.B. Hladky, D.A. Haydon, Discreteness of conductance change in biomolecular lipid membranes in the presence of certain antibiotics, *Nature* 225 (1970) 451–453.

30. S.B. Hladky, D.A. Haydon, Ion movements in gramicidin channels, *Curr. Top. Membr. Transp.* 21 (1985) 327–372.

31. D.C. Ogden (Ed.), *Microelectrode Techniques: The Plymouth Workshop Handbook*, second ed., The Company of Biologists, Plymouth, 1994.

32. W. Walz (Ed.), *Patch-Clamp Analysis: Advanced Techniques*, second ed., Humana Press, Totowa, NJ, 2007.

33. P.H. Barry, Ionic permeation mechanisms in epithelia: Biionic potentials, dilution potentials, conductances and streaming potentials, *Methods Enzymol.* 171 (1989) 678–715.

34. R. Lape, D. Colquhoun, L. Sivilotti, On the nature of partial agonism in the nicotinic receptor superfamily, *Nature* 454 (2008) 722–728

35. Y. Kim, A.C. Wong, J.M. Power, S.F. Tadros, M. Klugmann, A.J. Moorhouse, P.P. Bertrand, G.D. Housley, Alternative splicing of the TRPC3 ion channel calmodulin/IP3 receptor-binding domain in the hindbrain enhances cation flux, *J. Neurosci.* 32 (2012) 11414–11423.

36. U.B. Kaupp, R. Seifert, Cyclic nucleotide-gated ion channels, *Physiol. Rev.* 82 (2002) 769–824.

37. B.S. Pallotta, K.L. Magleby, J.N. Barrett, Single channel recordings of Ca^{2+}-activated K^+ currents in rat muscle cell culture, *Nature* 293 (1981) 471–474.

38. D. Colquhoun, Practical analysis of single channel records, in: D.C. Ogden (Ed.), *Microelectrode Techniques, The Plymouth Workshop Handbook*, second ed., Company of Biologists, Cambridge, MA, 1994, pp. 101–139.

39. D. Colquhoun, F.J. Sigworth, Fitting and statistical analysis of single-channel records, in: B. Sakmann and E. Neher (Eds.), *Single Channel Recording*, second ed., Plenum Press, New York, 1995, pp. 483–587.

40. S.H. Chung, J.B. Moore, L. Xia, L.S. Premkumar, P.W. Gage, Characterization of single channel currents using digital signal processing techniques based on Hidden Markov Models, *Philos. Trans. R. Soc. Lond. B Biol. Sci.* 329 (1990) 265–285.

41. D.R. Fredkin, J.A. Rice, Maximum likelihood estimation and identification directly from single-channel recordings, *Proc. R. Soc. Lond. B Biol. Sci.* 249 (1992) 125–132.

42. C. Nicolai, F. Sachs, Solving ion channel kinetics with the QuB software. *Biophys. Rev. Lett.* 8 (2013) 1–21.

43. J. del Castillo, B. Katz, Interaction at end-plate receptors between different choline derivatives, *Proc. R. Soc. Lond. B Biol. Sci.* 146 (1957) 369–381.

44. D. Colquhoun, A.G. Hawkes, On the stochastic properties of bursts of single ion channel openings and of clusters of bursts, *Philos. Trans. R. Soc. Lond. B Biol. Sci.* 300 (1982) 1–59.

45. P. Labarca, J.A. Rice, D.R. Fredkin, M. Montal, Kinetic analysis of channel gating. Application to the cholinergic receptor channel and the chloride channel from *Torpedo californica*, *Biophys. J.* 47 (1985) 469–478.

46. D. Colquhoun, A.G. Hawkes, A note on correlations in single ion channel records, *Proc. R. Soc. Lond. B Biol. Sci.* 230 (1987) 15–52.

47. D. Colquhoun, A.G. Hawkes, K. Srodzinski, Joint distributions of apparent open times and shut times of single ion channels and the maximum likelihood fitting of mechanisms, *Proc. R. Soc. Lond. A Math. Phys. Sci.* 354, (1996) 2555–2590.

48. D. Colquhoun, C. Hatton, A.G. Hawkes, The quality of maximum likelihood estimation of ion channel rate constants, *J. Physiol.* 547 (2003) 699–728.

49. F. Qin, A. Auerbach, F. Sachs, Estimating single-channel kinetic parameters from idealized patch clamp data containing missed events, *Biophys. J.* 70 (1996) 264–280.

50. F. Qin, A. Auerbach, F. Sachs, Maximum likelihood estimation of aggregated Markov processes, *Proc. R. Soc. Lond. B Biol. Sci.* 264 (1997) 375–383.

51. S. Schorge, S. Elenes, D. Colquhoun, Maximum likelihood fitting of single channel NMDA activity with a mechanism composed of independent dimers of subunits, *J. Physiol.* 569 (2005) 395–418.

52. P. Purohit, A. Auerbach, Loop C and the mechanism of acetylcholine receptor-channel gating, *J. Gen. Physiol.* 141 (2013) 467–478.

53. P.H. Barry, J.M. Diamond, Junction potentials, electrode potentials, and other problems in interpreting electrical properties of membranes, *J. Membr. Biol.* 3 (1970) 93–122.

54. P.H. Barry, T.M. Lewis, A.J. Moorhouse, An optimized 3 M KCl salt-bridge technique used to measure and validate theoretical junction potential values in patch-clamping and electrophysiology, *Eur. Biophys. J.* 42 (2013) 631–646.

55. D.E. Goldman, Potential, impedance and rectification in membranes, *J. Gen. Physiol.* 27 (1943) 37–60.

56. A.L. Hodgkin, B. Katz, The effects of sodium ions on the electrical activity of the giant axon of the squid, *J. Physiol.* 108 (1949) 37–77.

57. P.H. Barry, The reliability of relative anion-cation permeabilities deduced from reversal (dilution) potential measurements in ion channel studies, *Cell Biochem. Biophys.* 46 (2006) 143–154.

58. M. Pusch, E. Neher, Rates of diffusional exchange between small cells and a measuring patch pipette, *Pflugers Arch.* 411 (1988) 204–211.

59. C. Oliva, I.S. Cohen, R.T. Mathias, Calculation of time constants for intracellular diffusion in whole cell patch-clamp configuration, *Biophys. J.* 54 (1988) 791–799.

60. A. Marty, E. Neher, Tight-seal whole-cell recording, in: B. Sakmann and E. Neher (Eds.), *Single-Channel Recordings*, second ed., Plenum Press, New York, 1995, 31–52.

61. J.A.H. Verheugen, D. Fricker, R. Miles, Noninvasive measurements of the membrane potential and GABergic action in hippocampal interneurons, *J. Neurosci.* 19 (1999) 2546–2555.

62. S. Sugiharto, J.E. Carland, T.M. Lewis, A.J. Moorhouse, P.H. Barry, External divalent cations increase anion-cation permeability ratio in glycine receptor channels, *Pflugers Arch.* 460 (2010) 131–152.

63. A.J. Moorhouse, P. Jacques, P.H. Barry, P.R. Schofield, The startle disease mutation Q266H, in the second transmembrane domain of the human glycine receptor, impairs channel gating, *Mol. Pharmacol.* 55 (1999) 386–395.

64. N. Akaike, N. Inomata, N. Tokutomi, Contribution of chloride shifts to the faded of γ-aminobutyric acid-gated currents in frog dorsal root ganglion cells, *J. Physiol.* 391 (1987) 219–234.

65. P.H. Barry, J.W. Lynch, Topical review. Liquid junction potentials and small cell effects in patch-clamp analysis, *J. Membr. Biol.* 121 (1991) 101–117.

66. P.H. Barry, JPCalc—A software package for calculating liquid junction potential corrections in patch-clamp, intracellular, epithelial and bilayer measurements and for correcting junction potential measurements, *J. Neurosci. Methods* 51 (1994) 107–116.

67. E. Neher, Correction for liquid junction potentials in patch-clamp experiments, *Methods Enzymol.* 207 (1992) 123–131.

68. E. Neher, Voltage offsets in patch-clamp experiments, in: B. Sakmann and E. Neher (Eds.), *Single-Channel Recording*, second ed., Plenum Press, New York, 1995, pp. 147–153.

69. P.H. Barry, S. Sugiharto, T.M. Lewis, A.J. Moorhouse, Further analysis of counterion permeation through anion-selective glycine receptor channels, *Channels* 4 (2010) 142–149.

70. S. Sugiharto, T.M. Lewis, A.J. Moorhouse, P.R. Schofield, P.H. Barry, Anion-cation permeability correlates with hydrated counterion size in glycine receptor channels, *Biophys. J.* 95 (2008) 4698–4715.

71. A. Keramidas, A.J. Moorhouse, C.R. French, P.R. Schofield, P.H. Barry, M2 pore mutations convert the glycine receptor channel from being anion- to cation-selective, *Biophys. J.* 78 (2000) 247–259.

72. A. Keramidas, A.J. Moorhouse, K.D. Pierce, P.R. Schofield, P.H. Barry, Cation-selective mutations in the M2 domain of the inhibitory glycine receptor channel reveal determinants of ion charge selectivity, *J. Gen. Physiol.* 119 (2002) 393–410.

73. N. Unwin, Refined structure of the nicotinic acetylcholine receptor at 4 angstrom resolution, *J. Mol. Biol.* 346 (2005) 967–989.

74. J.A. Peters, J.E. Carland, M.A. Cooper, M.R. Livesey, T.Z. Deeb, T.G. Hales, J.J. Lambert, Novel structural determinants of single-channel conductance in nicotinic acetylcholine and 5-hydroxytryptamine type-3 receptors, *Biochem. Soc. Trans.* 34 (2006) 882–886.

75. A. Miyazawa, Y. Fujiyoshi, N. Unwin, Structure and gating mechanism of the acetylcholine receptor pore, *Nature* 423 (2003) 949–955.

76. S.B. Hansen, H.L. Wang, P. Taylor, S.M. Sine, An ion selectivity filter in the extracellular domain of Cys-loop receptors reveals determinants for ion conductance, *J. Biol. Chem.* 283 (2008) 36066–36070.

77. M.R. Livesey, M.A. Cooper, J.J. Lambert, J.A. Peters, Rings of charge within the extracellular vestibule influence ion permeation of the 5-HT3A receptor, *J. Biol. Chem.* 286 (2011) 16008–16017.

78. M. Moroni, J.O. Mayer, C. Lahmann, L.G. Sivilotti, In glycine and GABA(A) channels, different subunits contribute asymmetrically to channel conductance via residues in the extracellular domain, *J. Biol. Chem.* 286 (2011) 13414–13422.

79. M. Brams, E.A. Gay, J.C. Sáez, A. Guskov, R. van Elk, R.C. van der Schors, S. Peigneur, J. Tytgat, S.V. Strelkov, A.B. Smit, J.L. Yakel, C. Ulens, Crystal structures of a cysteine-modified mutant in loop D of acetylcholine-binding protein, *J. Biol. Chem.* 286 (2011) 4420–4428.

80. S.P. Kelley, J.I. Dunlop, E.F. Kirkness, J.J. Lambert, J.A. Peters, A cytoplasmic region determines single-channel conductance in 5-HT3 receptors, *Nature* 424 (2003) 321–324.

81. T.G. Hales, J.I. Dunlop, T.Z. Deeb, J.E. Carland, S.P. Kelley, J.J. Lambert, J.A. Peters, Common determinants of single channel conductance within the large cytoplasmic loop of 5-hydroxytryptamine type 3 and alpha4beta2 nictonic acetylcholine receptors, *J. Biol. Chem.* 281 (2006) 8062–8071.

82. J.E. Carland, M.A. Cooper, S. Sugiharto, H.-J. Jeong, T.M. Lewis, P.H. Barry, J.A. Peters, J.J. Lambert, A.J. Moorhouse, Characterisation of the effects of charged residues in the intracellular loop on ion permeation in α1 glycine receptor-channels, *J. Biol. Chem.* 284 (2009) 2023–2030.

83. G.D. Cymes, C. Grosman, The unanticipated complexity of the selectivity-filter glutamates of nicotinic receptors, *Nat. Chem. Biol.* 8 (2012) 975–981.

84. D. Langosch, B. Laube, N. Rundstrom, V. Schmieden, J. Bormann, H. Betz, Decreased agonist affinity and chloride conductance of mutant glycine receptors associated with human hereditary hyperekplexia, *EMBO J.* 13 (1994) 4223–4228.

85. D. Langosch, A. Herbold, V. Schmieden, J. Borman, J. Kirsch, Importance of Arg-219 for correct biogenesis of alpha 1 homooligomeric glycine receptors, *FEBS Lett.* 336 (1993) 540–544.

86. B. Saul, T. Kuner, D. Sobetzko, W. Brune, F. Hanefeld, H.M. Meinck, C.M. Becker, Novel GLRA1 missense mutation (P250T) in dominant hyperekplexia defines an intracellular determinant of glycine receptor channel gating, *J. Neurosci.* 19 (1999) 869–877.

87. A. Keramidas, A.J. Moorhouse, P.R. Schofield, P.H. Barry, Ligand-gated ion channels: Mechanisms underlying ion selectivity, *Prog. Biophys. Mol. Biol.* 86 (2004) 161–204.

88. R.E. Hibbs, E. Gouaux, Principles of activation and permeation in an anion-selective Cys-loop receptor, *Nature* 474 (2011) 54–60.

89. L. Sauguet, F. Poitevin, S. Mrail, C. Van Renterghem, G. Moraga-Cid, L. Malherbe, A.W. Thompson, P. Koehl, P.J. Corringer, M. Baaden, M. Delarue, Structural basis for ion permeation mechanism in pentameric ligand-gated ion channels, *EMBO J.* 32 (2013) 728–741.

90. G.D. Cymes, C. Grosman, Tunable pKa values and the basis of opposite charge selectivities in nicotinic-type receptors, *Nature* 474 (2011) 526–530.

91. D.C. Gadsby, A. Takeuchi, P. Artigas, N. Reyes, Review. Peering into an ATPase ion pump with single-channel recordings, *Philos. Trans. R. Soc. Lond. B Biol. Sci.* 364 (2009) 229–238.

92. J.K. Hirsch, C.H. Wu, Palytoxin-induced single channel currents from the sodium pump synthesized by in vitro expression, *Toxicon* 35 (1997) 169–176.

93. N. Reyes, D.C. Gadsby, Ion permeation through the Na$^+$,K$^+$-ATPase, *Nature* 443 (2006) 470–474.

94. A. Takeuchi, N. Reyes, P. Artigas, D.C. Gadsby, Visualizing the mapped ion pathway through the Na,K-ATPase pump, *Channels* 3 (2009) 383–386.

4

PROBING CONFORMATIONAL TRANSITIONS OF MEMBRANE PROTEINS WITH VOLTAGE CLAMP FLUOROMETRY (VCF)

THOMAS FRIEDRICH

Institute of Chemistry, Technical University of Berlin, Berlin, Germany

4.1 INTRODUCTION

This chapter describes the technique of voltage clamp fluorometry (VCF), which has been widely applied to investigate conformational transitions in, for example, voltage-gated cation channels and electrogenic ion transporters. Crucial to the technique is the introduction of cysteine residues at specific locations within the amino acid sequence of a membrane protein by site-directed mutagenesis. Due to the high specificity achievable with sulfhydryl-reactive compounds, such as by maleimide-conjugation, fluorescence dyes can be covalently linked to predefined sites within the protein sequence. Fluorescence labeling is performed upon expression of the cysteine-mutated protein, for example, in *Xenopus* oocytes, and the covalent attachment of fluorescence dyes by sulfhydryl-specific coupling chemistry at strategically located positions potentially creates sensor constructs that report highly localized structural rearrangements in real time, only limited in speed by the response time of the voltage clamp electronics (typically 1 ms down to several hundreds of microseconds). Conformational transitions, either induced by biochemical stimulation or by voltage pulses during two-electrode voltage clamping, entail changes in the microenvironment of the attached fluorophore that eventually result in changes in the

Pumps, Channels, and Transporters: Methods of Functional Analysis, First Edition. Edited by Ronald J. Clarke and Mohammed A. A. Khalid.

fluorescence intensity, reflecting the evolution of conformational changes in time. The simultaneous recording of fluorescence signals and current responses in an electrophysiological experiment allows one to correlate structural changes with specific ion transport steps. The technique even allows for the investigation of net electroneutrally operating ion transporters such as the H^+,K^+-ATPase, because the only functional prerequisites are that electrogenic—hence voltage-dependent—partial reaction steps exist in the reaction cycle of the protein of interest, and that these can be isolated by appropriate choice of experimental conditions.

The field of biophysical research on membrane-embedded transporter proteins has seen a breathtaking increase in the number of available high-resolution structures, setting the stage for understanding atomic-scale properties such as substrate recognition, energy transduction, kinetics, and specificity. Despite the tremendous value of crystallographic data for our concepts of membrane transport, in the face of the sheer beauty and the accuracy of the structures on the low Ångström scale, one can easily overlook some important limitations that are inherent to these—still— structural models. Membrane protein crystallization is still a challenging task, and there is no full guarantee that the arrangement of the molecules within the crystals is truly representative of the native structure within an intact membrane environment. Furthermore, the conditions chosen to stabilize a certain reaction intermediate might necessitate the use of nonnative chemical compounds, which are at best structural analogs of the native compounds. Because protein function relies on dynamics, whereas structure resolution requires highly static arrangements, the observed intermediate structures might represent the product of dead-end processes that will never be reached during dynamic function. To bridge between static structures and dynamic function, powerful computational modeling techniques have been developed (see Chapter 15) that have dramatically increased in importance, because they provide the means to rationalize and validate structural data from a functional perspective. However, the complexity of proteins still imposes severe restrictions regarding the time scale that can be covered by molecular modeling, because even microsecond simulations require huge computational resources. On the side of the experimentalist, there is an equally high demand for means to detect the structural dynamics with sufficient site selectivity to obtain molecular information, which is independent from or complementary to the more global conclusions that can be drawn from biochemical studies of substrate transport or catalysis, respectively, or measuring ion transport as membrane currents by electrophysiology. One of these complementary techniques is voltage clamp fluorometry.

4.2   DESCRIPTION OF THE VCF TECHNIQUE

4.2.1   Generation of Single-Cysteine Reporter Constructs, Expression in *Xenopus laevis* Oocytes, Site-Directed Fluorescence Labeling

One of the most widely used expression systems is the oocyte of the African clawed toad *X. laevis*, which enables the application of powerful electrophysiological techniques such as the two-electrode voltage clamp or the patch clamp method (see Chapters

3 and 5). Upon heterologous expression of a membrane protein in *Xenopus* oocytes, only the extracellular face of the membrane is accessible for chemical labeling (unless the "vaseline gap cut-open oocyte" variant [1, 2] or the patch clamp technique on small [3] or giant [4] patches of oocyte membranes is used on those cells). Therefore, it is necessary to introduce single-cysteine residues by site-directed mutagenesis into extra-cellularly oriented loop regions, which can be identified on the basis of transmembrane topology models obtained from other biochemical assays or from structural models, if the latter are available at least from similar or homologous proteins so that structural homology modeling is feasible. In the absence of more detailed knowledge, a single-cysteine scanning mutagenesis strategy of the extracellular loop region has to be per-formed to identify potential reporter sites. Most promising candidates are positions located at the interface between the extracellular, aqueous space and the hydrophobic plasma membrane. Because many fluorescent dyes are sensitive to (among many other parameters) the hydrophobicity or polarity of the environment, large changes in the fluorescence properties can be expected if the fluorophore is moved from a solvent-exposed position to a location within the lipid or protein environment or *vice versa*. Of course, tests concerning the effect of the introduction of single-cysteine mutation changes on the functional properties of the membrane protein of interest need to be car-ried out, and, later, further tests to show whether or not functional changes occur upon covalent attachment of the cysteine-reactive dye, as exemplarily stated in Refs. 5, 6.

In vitro cRNA transcription from the mutated cDNA construct, cRNA injection, and protein expression in *Xenopus* oocytes proceeds according to established proto-cols [6–8]. Upon expression of the protein, cysteine-specific site-directed fluores-cence labeling (SDFL) is performed by incubating the oocytes for several minutes in the dark, preferably on ice, in low micromolar concentrations of a cysteine-reactive dye, such as tetramethylrhodamine-6-maleimide—TMRM, fluorescein-maleimide—FM, or Oregon Green maleimide—OrGrM, in a common buffer solution (e.g., ORI, ND96). This is followed by several extensive washes in dye-free buffer. The optimal labeling conditions (incubation time, dye concentration) need to be determined for each experimental scheme to achieve robust, specific, and low non-specific labeling or uptake of the dye into cells. Subsequently, the cells are transferred into a suitable recording chamber for two-electrode voltage clamping, which is mounted on the stage of a microscope that permits the fluorescence of labeled oocytes to be mea-sured in parallel to current recording during voltage clamping. Although many labo-ratories have refined and expanded the technical toolkit of VCF, for example, by application of the cut-open oocyte technique, patch fluorometry, and even single-molecule approaches (see Chapter 11), this chapter only deals with the most fundamental experimental schemes of VCF and the practical aspects for its imple-mentation. For details of the more refined approaches, interested readers are referred to Refs. 9, 10, or the recent review on patch fluorometry by Kusch and Zifarelli [11].

4.2.2   VCF Instrumentation

The typical components of a VCF recording system are depicted in Figure 4.1. Upon expression and after fluorescence labeling, the oocyte is transferred into a voltage clamp recording chamber (e.g., RC-10, Warner Instruments) mounted on the stage of

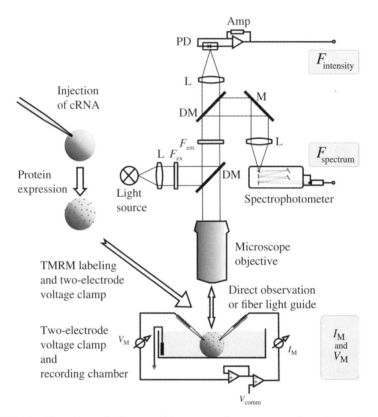

FIGURE 4.1 Experimental scheme and instrumental components of a voltage clamp fluorometry setup using the *Xenopus* oocyte expression system. Expression of a membrane protein with an exposed cysteine is achieved by cRNA injection, which is followed by SDFL with cysteine-reactive dyes (e.g., TMRM, FM, OrGrM). Labeled cells are transferred into the recording chamber of a two-electrode voltage clamp setup mounted on the stage of an upright fluorescence microscope. Light used for fluorescence excitation and fluorescence emission is coupled to or from the probe either directly via the microscope objective or via an optical fiber. Fluorescence emission is detected either by a photodiode, photomultiplier tube, or spectrometer. Signals that can be recorded in parallel are membrane voltage and current, fluorescence intensity or spectrum. By using appropriate polarization filters, also fluorescence polarization or anisotropy can be determined. Amp, amplifier; DM, dichroic mirror; F_{em}, emission filter; F_{ex}, excitation filter; I_M, membrane current; L, lens; PD, photodiode; V_{comm}, command voltage; V_M, membrane voltage.

an upright fluorescence microscope (e.g., Zeiss Axioskop II FS). Inverted microscopes have also been employed, also for confocal imaging [12], but because the oocyte is found sitting on the glass bottom of the chamber, this configuration imposes restrictions regarding efficient solution exchange if, for example, the investigated transporter needs to be activated by chemical effectors. In the upright configuration, fluorescence measurements are performed by a long-distance, high numerical aperture water immersion objective (e.g., Zeiss Achroplan 40x, NA 0.8). Alternatively, a fiber optic with high numerical aperture can be used. Fluorescence excitation is generally performed with a tungsten halogen lamp or a mercury arc lamp source of the microscope together with a filter set (excitation filter, dichroic mirror, and emission filter mounted in a filter cube) that is appropriate for the fluorescent dye. The fluorescence intensity can be measured by focusing the emitted light on a photodiode (e.g., PIN-020A, United Detector Technologies) that is connected to a fast, low-noise, high-gain amplifier of the photocurrent. For the latter, patch clamp amplifiers (e.g., Axopatch 1B, Molecular Devices) have been used [5]. Other options are the DLPCA-200 amplifier (Femto Messtechnik GmbH, Berlin) or photomultiplier tubes (e.g., Hamamatsu PMT 6356, selected for low dark current). For more refined systematic studies, fluorescence emission during VCF has also been analyzed with a spectrograph (e.g., Multispec 257i, Oriel Instruments) [5]. Also, excitation with polarized light and detection through a polarizing filter oriented in parallel or perpendicular to the polarization plane has been performed, such as for the measurement of fluorescence anisotropy of the site-specifically attached fluorophore [13, 14]. When using fiber optics in this scheme, it has to be ensured that the fibers maintain light polarization [13].

4.2.3 Technical Precautions and Controls

A major concern during the development of the VCF technique was whether (i) extracellularly exposed cysteines that are already present within the native structure of the overexpressed protein of interest or (ii) more generally, all the potentially exposed cysteines of all endogenously expressed membrane proteins present in the host cell interfere with site-specific cysteine labeling by creating a substantial fluorescence background or even by contributing to fluorescence changes induced by voltage jumps.

The first study in which the SDFL/VCF technique was described was carried out by Manuzzu et al. [12], and addressed the questions of whether and how conformational transitions within the voltage sensor of the voltage-gated *Shaker* K+ channel correlate with channel gating or, more precisely, with the kinetics of gating charge movement. Here, several amino acid positions preceding or within the fourth transmembrane segment S4 (the main constituent of the voltage sensor domain of the channel, see Fig. 4.2a) were mutated to cysteines. In this fundamentally influential study, two strategies were employed to reduce the effects of nonspecific fluorescence labeling: (i) generation of a "backbone" construct, in which all extracellularly exposed cysteines in the native sequence were replaced (e.g., by alanine or serine) resulting in an extracellular "cysteine-removed" control channel and (ii) preblocking of the oocytes some hours before the SDFL/VCF experiment. For the preblocking step, after cRNA injection and expression incubation for 3–4 days at 12°C, oocytes were placed for 1 h at 12°C in a solution containing a cysteine-reactive nonfluorescent compound (tetraglycine maleimide—TGM) to block sulfhydryl groups in all momentarily present membrane proteins. Subsequently, cells were stored for another 12–14 h at 21°C to permit further expression of channels in the plasma membrane

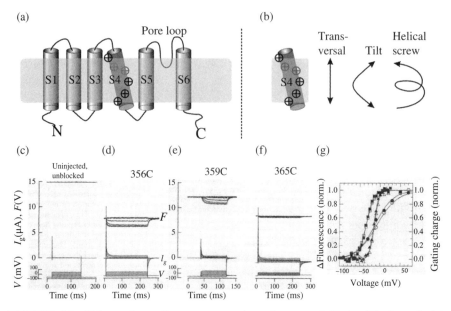

FIGURE 4.2 Voltage clamp fluorometry on the voltage-gated *Shaker* K+ channel. (a) Topology scheme for one subunit of a tetrameric, voltage-gated K+ channel. Transmembrane segments S5 and S6 line the channel pore, while the intervening pore loop carries the selectivity filter. Segments S1–S4 form the voltage sensor domain, from which S4 carries a series of evenly-spaced positively charged amino acids that contribute a major portion to the total gating charge. (b) Upon a positive voltage stimulus, a rearrangement of the voltage sensor including the S4 segment occurs that leads to the outward movement of positive charge (giving rise to the so-called gating current), which is followed by opening of the channel pore. To bring about outward movement of gating charge, the S4 segment may perform a transversal, a tilt, or a helical screw movement, or a combination thereof. (c–f) Voltage-dependent fluorescence changes of TMRM-labeled control oocytes (c) or oocytes expressing mutants M356C (d), A359C (e), or R365C (f). Simultaneous fluorescence (F, top) and gating current (I_g, middle) measurements upon voltage steps (V, bottom), with background fluorescence of cysteine-removed control channels subtracted from fluorescence records are shown. (g) Correspondence between the normalized fluorescence-voltage (F–V, filled symbols) or gating charge-voltage curves (Q–V, open symbols) from oocytes expressing mutants M356C (triangles), A359C (squares), and R365C (circles). Panels C–G were reproduced from Ref. 12 with permission from the American Association for the Advancement of Science (Science).

prior to SDFL and VCF experiments. Manuzzu et al. (using 30 min labeling with 5 µM TMRM on ice) reported that the constructs with reporter cysteines introduced into the voltage-sensing S4 segment exhibited a 2.5- to 5-fold increased labeling by TMRM compared to uninjected control oocytes or oocytes expressing the cysteine-removed control channel. The effect of preblocking was pronounced, at least on uninjected control oocytes, for which about an 80% reduction in fluorescence intensity was found upon preblocking [12]. Thus, both preblocking with nonfluorescent cysteine-reactive compounds and creation of a background construct without extracellularly accessible cysteines may significantly enhance the level of fluorescence labeling, which is specific for the introduced reporter cysteine. However, background labeling might be highly variable between individual batches of cells or laboratories: a later study by Cha and Bezanilla [13], using 40 min labeling with 5 µM TMRM at 18°C without preblocking, reported a more than 10-fold higher fluorescence intensity of oocytes expressing similar *Shaker* channel reporter cysteine constructs compared to labeled noninjected cells, suggesting that SDFL can create a robust and specific fluorescence signal, if the achievable expression level of the membrane transporter of interest is high. To give an example for a "robust" expression level: the first *Shaker* K⁺ channel constructs used for SDFL/VCF were made against the background of the "inactivation-removed" channel (deletion of the cytoplasmic N-terminus responsible for N-type inactivation) carrying the W434F mutation that renders the channel nonconductive but leaves gating (current) relaxations unaffected. The expression level for such nonconductive channels can be extremely high; from the time integrals of the recorded gating currents—typically up to the order of 10 µA over 10 ms—a total amount of gating charge during a voltage step can be up to 100 nC, which is equivalent to about 10^{12} elementary charges and about 10^{11} individual channels at the surface of the oocyte, assuming that each of the four channel subunits carries about 2.5 elementary gating charges [15]. Similar numbers for surface expression have been reported from pre-steady-state currents of the voltage-sensitive ion pump Na⁺,K⁺-ATPase [6].

4.3 PERSPECTIVES FROM EARLY MEASUREMENTS ON VOLTAGE-GATED K⁺ CHANNELS

4.3.1 Early Results Obtained with VCF on Voltage-Gated K⁺ Channels

Importantly, Mannuzzu et al. [12] as well as Cha and Bezanilla [5] observed characteristic decreases in the fluorescence of the TMRM chromophore attached to several *Shaker* K⁺ channel cysteine mutants in response to depolarizing voltage pulses, which move the voltage-sensing segment from a more or less buried (i.e., less quenched) inactivated position into a more solvent-exposed (i.e., quenched) environment in the activated state of the voltage sensor. As can be seen from Figure 4.2g, the voltage dependence of the normalized change in fluorescence amplitude upon voltage steps correlated well with the normalized integrated charge measured from gating currents recorded in parallel to the fluorescence changes, but not with the voltage dependence of ionic currents mediated by the channels, indicating that the site-specific fluorescence signal represents the graded movement of the voltage sensor from the inactivated into the activated state [12].

In addition to the correlation between gating charge movement and the voltage dependence of steady-state fluorescence amplitudes (i.e., long after a voltage step), Cha and Bezanilla [5] also investigated the temporal correlation between gating charge movement and the fluorescence signal of different chromophores that were site-specifically attached to cysteines within the extracellularly exposed S3–S4 and

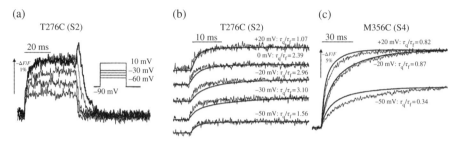

FIGURE 4.3 Temporal correlation of fluorescence changes of *Shaker* K⁺ channel reporter constructs carrying cysteine mutations in the extracellular regions of the S2 and S4 segments. (a) Fluorescence signals for mutant T276C (S2) labeled with TMRM upon pulses from −90 mV to −60 (bottom trace), −50, −40, −30, and +10 mV (thick trace). Upward movement of the traces indicates a decrease in fluorescence, as indicated by the scale arrow. (b) Superposition of fluorescence F (thin, jagged trace) and integrated gating charge Q (thick, smooth trace) during a 40 ms pulse at five different potentials measured from mutant T276C. (c) Superposition of F (thin, jagged trace) and Q (thick, smooth trace) during a 150 ms pulse for three potentials measured from the mutant M356C. For both constructs, Q and F were fitted with one or the sum of two exponential functions, and the parameters of the fit were used to calculate the ratio of the average time constant of gating charge displacement to the average time constant of fluorescence (τ_q/τ_f) at each potential. Reproduced from Ref. 5, with permission from Cell Press (Neuron).

S1–S2 loops of the *Shaker* K⁺ channel [5]. TMRM labeling of the T276C (S2) mutant also resulted in fluorescence decrease upon depolarizing voltage jumps, from −90 mV to more positive potentials, with a single-exponential time course (see Fig. 4.3a). However, the voltage dependence of steady-state fluorescence amplitudes did not correlate with the total gating charge movement and fluorescence amplitudes decreased again at potentials above −20 mV [5]. The latter indicates that the fluorophore encounters a maximally quenched environment at around −20 mV and less quenching at more positive or more negative values. Upon return from, for example, +10 mV back to the initial holding potential (−90 mV), the fluorescence signals exhibited a transient decrease in fluorescence ("blip") corresponding to the passage of the fluorophore through the maximally quenched environment followed by a single-exponential increase in fluorescence (see Fig. 4.3a). Importantly, this indicates that the extremely localized environmental changes of site-specifically attached chromophores might be much more complex than the simplified two-state scheme of a channel changing its global conformation from an inactivated to an activated state. Furthermore, comparison of the kinetics of fluorescence changes of the TMRM label bound to T276C (S2) and M356C (S4) showed that the T276C fluorescence kinetics is faster and the M356C kinetics is slower than the time integral of the gating charge (see Fig. 4.3b and c). Although the properties of the signals from the T276C and M356C mutants differed from those of the "cysteine-removed" control channel (already due to the cysteine mutations themselves), it was concluded that the fluorescence at S2 resembles the parameters of a rather fast gating charge component Q_1; whereas the fluorescence label at S4 reports kinetics and steady-state parameters of

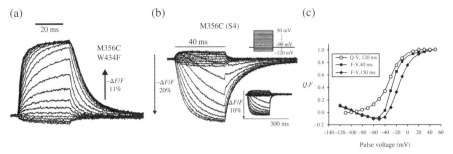

FIGURE 4.4   Voltage pulse-induced fluorescence changes recorded from the nonconducting *Shaker* K⁺ channel mutant W434F carrying the reporter cysteine mutation M356C in the extracellular region of the S4 segment upon labeling with TMRM (a) [13], OrGrM (b), or FM (inset in b) [5]. All fluorescence signals are responses to voltage pulses from −90 mV to test potentials between −120 and +50 mV (10 mV increments). Test potential durations were 40 ms for the TMRM- (a) and OrGrM-labeled channel (b), and 150 ms for the FM-labeled channel (inset in b). In (a), an upward deflection of the trace indicates a decrease in fluorescence, in (b), downward deflection denotes an increase in fluorescence with magnitudes (percent of total amplitude) indicated by the respective scale arrows. (c) Normalized integrated gating charge–voltage (Q–V) curve (200 ms integration time, open circles) and fluorescence–voltage (F-V) curves at 40 ms (closed diamonds) and 150 ms (closed squares) after onset of the test pulse measured for mutant M356C labeled with OrGrM. Panel A was reproduced from Ref. 13 with permission from the Society of General Physiologists and panels B and C from Ref. 5 with permission from Cell Press (Neuron).

the slower Q_2 component of the total gating charge [5], in accordance with previous studies [16, 17]. It is important to point out that such a good average agreement between the fluorescence parameters of TMRM attached to M356C and gating charge displacement is not necessarily expected, and rather should be considered as a lucky coincidence. As pointed out by Mannuzzu and coworkers [12], gating charge displacement is measured electrically as a voltage-dependent variable, that is, the weighted average of the position of all gating charges relative to the membrane electric field, whereas fluorescence is an optical measurement with, *inter alia*, quantum yield as a voltage-dependent variable [5].

Additional information about the nature of the underlying conformational (or environmental) changes can be obtained by conjugating different fluorescence dyes to the same amino acid position. In fact, the fluorescence responses of TMRM, FM, or OrGrM conjugated to the *Shaker* M356C mutant evoked by the same set of voltage pulses are strikingly different [5, 13]. As depicted in Figure 4.4a, the steady-state fluorescence of TMRM bound to M356C decreases in a monotonic fashion when the membrane voltage is changed from hyperpolarizing to depolarizing potentials, in good agreement with the voltage dependence of total gating charge displacement (see also Fig. 4.2g). However, steady-state fluorescence amplitudes with OrGrM (see Fig. 4.4b) or FM (see Fig. 4.4b, inset) decreased slightly only between −120 and −50 mV, but increased sharply at more depolarized potentials, with both different voltage dependence and kinetics from that of gating charge displacement. Moreover,

OrGrM labeling significantly slowed the kinetics of gating charge displacement compared to the unlabeled channel construct, an effect not seen with TMRM or FM [5]. The similarity between the voltage-dependent fluorescence amplitudes with OrGrM and FM were taken as evidence against a pH effect within the local environment, because the quantum yield of fluorescein changes in the physiological range with a pK_a of 6.4, whereas Oregon Green has a pK_a of 4.7. The differences to the TMRM responses indicate that the FM/OrGrM labels encounter not only a change between two but between three different microenvironments with distinct quantum yields, which can be attributed to electrostatic effects due to the difference in net charge (0 in case of TMRM and −2 in case of OrGrM) [5].

4.3.2   Probing the Environmental Changes: Fluorescence Spectra, Anisotropy, and the Effects of Quenchers

Because the Stokes shift (difference between the wavelength of peak absorption and peak emission) of a fluorescent probe is highly dependent on the polarity of the probe's microenvironment [18], investigation of the spectra of dyes used for SDFL/VCF can yield valuable information about the nature of the environmental changes sensed by the chromophore. In the fundamental studies of Mannuzzu et al. [12] as well as Cha and Bezanilla [5, 13], fluorescence spectra of the TMRM dye were measured in a variety of solvents with largely different polarity or quenching properties (see Fig. 4.5a and b). It was shown that the peaks of the fluorescence spectra measured on TMRM-labeled M356C or A359C *Shaker* K+ channels were close to the spectral peak in water and did not shift significantly upon changes to different membrane potentials [5], which is consistent with a graded aqueous exposure of the two residues. When solvent access was further probed with externally applied iodide, a highly efficient collisional quencher, the effects of iodide quenching at different holding potentials were less pronounced for M356C than for A359C [5]. These findings are in agreement with the report of Mannuzzu et al. [12], who observed that the specific fluorescence changes—measured as the difference in fluorescence upon a saturating voltage jump compared to the fluorescence background—were augmented by iodide about 3.6-fold for mutant A359C compared to an approximate 1.7-fold increase observed for M356C. Thus, both sites are solvent accessible, but the more outwardly located residue 356 is either sterically hindered from interaction with iodide or it lies within an aqueous crevice into which iodide cannot penetrate as well as for the 359 position. The fact that the emission peak is not shifted compared to water in the presence of iodide, whereas the Stokes shift is small in alcohols of increasing chain length and large in the aromatic solvent phenol, argues against the notion that the chromophore is moved from a purely hydrophobic to a more hydrophilic environment upon activating voltage pulses [5], but rather supports the effects of differential collisional quenching. Another difference in the access of quenchers was reported by Cha and Bezanilla [13], who compared the voltage-dependent effects of D_2O or iodide, which are less strong and stronger quenchers than water, respectively. In the case of the M356C mutant, similar quenching ratios (ratio of the relative changes in fluorescence amplitudes in the presence and absence of quencher at −90

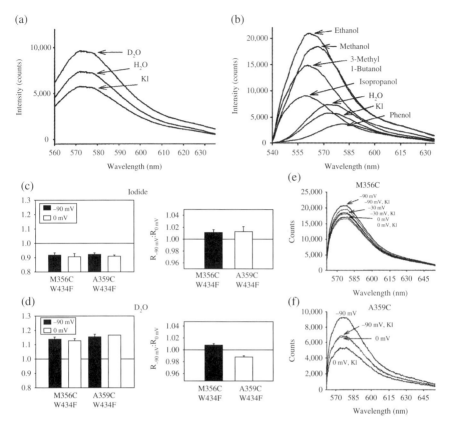

FIGURE 4.5   Utilization of spectral information for VCF: Effects of solvents and quenchers. (a) Spectral characteristics of 5 μM TMRM were measured in 120 mM *N*-methyl-*d*-glucamine (NMG)-MES with D_2O (D_2O), 120 mM NMG-MES (H_2O), and 50 mM KI + 70 mM NMG-MES (KI) [13]. (b) Spectra of 5 μM TMRM were measured in different solvents, as indicated (H_2O: 120 mM NMG-MES; KI: 50 mM KI + 70 mM NMG-MES) [5]. (c and d) Effects of collisional quenchers on TMRM fluorescence near the S4 segment upon differential solvent exposure induced by membrane voltage [13]. (c and d—left panels) The intensity ratios for *Shaker* K⁺ channel mutants M356C/W434F and A359C/W434F were determined from the fluorescence intensity after the application of 50 mM KI + 70 mM NMG-MES and the fluorescence intensity before iodide application (c), or from the fluorescence intensity after the application 120 mM NMG-MES in D_2O and the fluorescence intensity before D_2O application (d). Values below 1.0 indicate a decrease in fluorescence after application of a quencher. Fluorescence intensities were measured at either −90 mV (black columns) or 0 mV (white columns). (c and d—right panels) Ratio of the iodide intensity (c) or the D_2O intensity (d), each from the intensity ratios at −90 mV (R_{-90mV}) to the corresponding intensity ratios at 0 mV (R_{0mV}) for M356C/W434F and A359C/W434F. Values larger than 1.0 indicate that the intensity ratio at −90 mV is larger than the intensity ratio at 0 mV [13]. (e) Fluorescence spectra obtained from a TMRM-labeled oocyte expressing mutant M356C at three different holding potentials before and after the addition of 50 mM extracellular KI. From top to bottom: −90 mV; −90 mV, KI; −30 mV; −30 mV, KI; 0 mV; 0 mV, KI [5]. (f) Spectra from a TMRM-labeled oocyte expressing mutant A359C at holding potentials of −90 and 0 mV before and after the addition of 50 mM extracellular KI [5]. Panels (a, c, and d) were reproduced from Ref. 13 with permission from the Society of General Physiologists, and panels (b, e, and f) from Ref. 5 with permission from and Cell Press (Neuron).

and 0 mV) were found for both quenchers; whereas for A359C, D_2O invoked a much smaller quenching ratio than iodide. From these results, it was again concluded that iodide had better access to A359C in the activated (depolarized) state but less access in the inactivated (hyperpolarized) state, maybe due to a more negatively charged environment that would expel the iodide anions in the latter state. In contrast, the water-filled cavity for M356C, as seen from the larger quenching ratio in D_2O, might even be larger than that of A359C in the inactivated (hyperpolarized) state [13].

A comparable, in the first instance counter-intuitive pattern, was also observed when fluorescence anisotropy at various sites was compared at different membrane potentials. For TMRM-labeled M356C at −90 mV, the anisotropy was smaller than at 0 mV; whereas anisotropy increased for A359C with more negative voltages, suggesting a differential rotational flexibility that follows the pattern of D_2O access. This suggests that it is not a single parameter that constrains the fluorescence properties of the attached TMRM dye within the proteinaceous vestibule around the S4 segment [13].

4.4 VCF APPLIED TO P-TYPE ATPases

4.4.1 Structural and Functional Aspects of Na⁺, K⁺- and H⁺,K⁺-ATPase

Members of the large superfamily of P-type ATPases are found in all the kingdoms of life and share distinct structural and functional features. They transport (or flip) their substrates across membranes by utilizing the free energy of ATP hydrolysis. The P2C subgroup of P-type ATPases, to which the Na⁺,K⁺-ATPase and H⁺, K⁺-ATPase belong, is distinct, because these enzymes perform K⁺ inward transport while transporting Na⁺ or H⁺ ions out of the cell. In addition to the large catalytic α-subunit, their minimal functional unit comprises a heavily glycosylated β-subunit with a single transmembrane domain. The β-subunit plays an important role in folding and targeting of the holoprotein, but the isoform composition (four α- and three β-subunit isoforms are known in humans) also influences functional properties [19]. The Na⁺,K⁺-ATPase is a net electrogenic transporter translocating 3Na⁺ ions out of and 2K⁺ ions into the cell for each ATP molecule split, and the enzyme can well be considered as one of the most exhaustively studied ion transporters investigated by electrophysiology. In contrast, its sister protein, the H⁺,K⁺-ATPase from gastric mucosa, cannot be studied by standard electrophysiology owing to its net electro-neutral transport of 2H⁺ out versus 2K⁺ into the cell. From the viewpoint of bioenergetics, the H⁺,K⁺-ATPase is particularly interesting, because *in vivo* it generates the largest concentration gradient known in living systems. Whereas the pH in the stomach lumen can fall below 1, the pH in gastric parietal cells remains neutral, resulting in a concentration gradient of six orders of magnitude. Such a large gradient cannot be achieved by a 2H⁺:2K⁺:1 ATP stoichiometry, but requires a stoichiometry change to 1H⁺:1K⁺:1 ATP, which still today is largely hypothetical rather than firmly proven.

The reaction cycle of P-type ATPases is generally expressed in terms of the Post-Albers scheme [20, 21], which is briefly described here for Na⁺, K⁺- and H⁺,

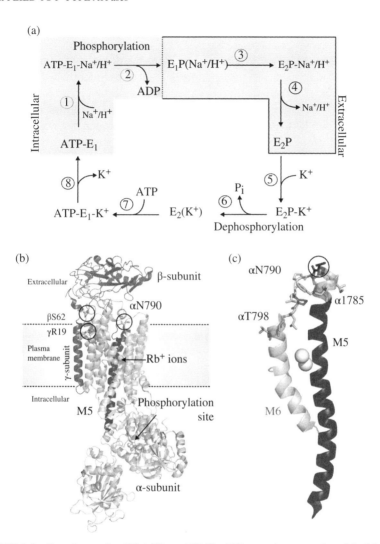

FIGURE 4.6 Reaction cycle of Na^+, K^+- and H^+,K^+-ATPase and structural model of the Na^+, K^+-ATPase. (a) Within the Post-Albers reaction scheme (see text for a detailed description), the partial reaction sequence, which can be influenced by membrane voltage in the absence of extracellular K^+, is underlain in gray. The black box indicates the voltage-dependent partial reaction steps which the Na^+,K^+-ATPase undergoes upon voltage steps in the presence of extracellular Na^+ and $[K^+]_o = 0$. (b) Structural model of the shark rectal gland Na^+,K^+-ATPase in the E2P-like conformation (PDB accession number 2ZXE, [22]) with two Rb^+ ions in the cation-binding pocket (spheres). The catalytic α-subunit, the β-subunit, and the γ-subunit are depicted in different shades. The central M5 helix is indicated and three reporter sites (N790 in the α-, S62 in the β- and R19 in the γ-subunit) are shown in ball-and-stick representation (circles). Also denoted is the phosphate analog MgF_4^{2-}, which is bound to the phosphorylation site and was used for structure stabilization. (c) Structural detail of the M5–M6 region of the Na^+,K^+-ATPase with amino acids in the interconnecting loop shown in ball-and-stick representation.

K+-ATPase (see Fig. 4.6a) to clarify the steps that can be investigated by electrophysiology or VCF. The enzymes can assume two major conformational states (E1 and E2) that differ in substrate affinities and accessibility of the binding sites. During the cycle, a phosphorylated intermediate is reversibly formed at a critical aspartate within a characteristic DKTG motif that defines the whole superfamily. Upon intracellular binding of 3Na+ (2H+) ions to the E1 conformation (step 1), a phosphointermediate with occluded Na+/H+ ions (E1P(Na+/H+)) is formed (step 2). After a conformational change to E2P (step 3), Na+/H+ ions dissociate to the extracellular space (step 4). Subsequently, K+ ions bind from the extracellular side (step 5) and become occluded. This stimulates dephosphorylation (step 6), followed by a conformational change from E2 to E1 (step 7), upon which K+ ions are released intracellularly (step 8). The gray box and the black frame indicate the reaction sequences which can be studied by voltage pulses for the H+, K+- and Na+,K+-ATPase, respectively. The general assumption for the H+,K+-ATPase, unless other evidence is available, is that intra- as well as extracellular H+ release/reverse binding and the E1P ↔ E2P conformational change are supposed to be voltage dependent (gray box in Fig. 4.6a), whereas for the Na+,K+-ATPase, there is general agreement that reverse binding of extracellular Na+ ions is the major electrogenic step (black frame in Fig. 4.6a).

Already the first high-resolution crystal structures of a P-type ATPase family member, the sarco-endoplasmic reticulum Ca^{2+}-ATPase (SERCA), which have been published from 2000 onwards in various states [23, 24], provided impressive insight into the large conformational changes that occur during the Post-Albers cycle. Because structural homology modeling became feasible for many P-type ATPases, these studies stimulated refined structure-guided mutagenesis approaches to address the molecular details of ion selectivity, transport, and catalysis. The crystal structure shows that the approximately 70Å-long α-helix M5 is a pivotal structural element (see Fig. 4.6b and c). It comprises the transmembrane segment TM5, contributes several residues to the cation-binding pocket at the center of the plasma membrane, and extends far into the phosphorylation- or P-domain of the protein. The M5 helix was, thus, hypothesized to mediate energy transduction between ATP hydrolysis and cation transport [22, 25].

4.4.2   The N790C Sensor Construct of Sheep Na+,K+-ATPase α1-Subunit

To probe the involvement of the M5 segment in conformational changes during the transport cycle, cysteine-scanning mutagenesis was performed on the whole extracellular loop between TM segments 5 and 6 (Ile785 to Thr798) of the sheep Na+,K+-ATPase α1-subunit to identify potential reporter sites for SDFL/VCF [6]. To discriminate the activity of the heterologously expressed Na+,K+-ATPase from that of the endogenous Na+ pump of *Xenopus* oocytes, the backbone construct carried mutations Q111R and N122D, which confer reduced sensitivity against the Na+,K+-ATPase-specific inhibitor ouabain (IC$_{50}$ in the millimolar range compared to micromolar in most mammalian Na+ pumps [26]). Furthermore, the two native extracellular cysteines (C911, C964) were mutated to alanines to suppress nonspecific labeling at other sites [27]. From the 11 cysteine mutants in the TM5-6 loop, only the N790C construct gave rise to functionally correlated fluorescence changes. According to the crystal structures of the Na⁺,K⁺-ATPase in the E2P-like conformation resolved later [22, 25], N790 resides at the extracellular end of helix M5 and is the outermost located amino acid of the whole TM5-6 loop (see Fig. 4.6c). Upon labeling the afore-mentioned N790C mutant of sheep Na⁺,K⁺-ATPase α1 subunit with TMRM, distinct increases in fluorescence intensity were observed upon changes of the extracellular medium from a Na⁺-rich, K⁺-free solution to K⁺-containing buffers [6]. Under the conditions of voltage clamp experiments in *Xenopus* oocytes (intracellular presence of saturating Na⁺ and ATP concentrations), the enzyme accumulates in the E2P con-formation when K⁺ is absent extracellularly because of dramatically slowed dephos-phorylation. Addition of extracellular K⁺ induces a redistribution of reaction cycle intermediates in a dose-dependent fashion toward E1-like states, which apparently gives rise to the observed fluorescence increase [6].

4.4.2.1 *Probing Voltage-Dependent Conformational Changes of Na⁺,K⁺-ATPase*

To more specifically probe whether the fluorescence change is correlated with the E2 ↔ E1 state redistribution, voltage jumps were applied in the presence of a Na⁺-rich, K⁺-free medium. Under these conditions, the Na⁺ pump is limited to carry out the reaction steps within the black box in Figure 4.6a, the enzyme mediates electro-genic Na⁺/Na⁺ exchange, and it is commonly accepted that the major electrogenic event is extracellular release or reverse binding of Na⁺, most likely related to the Na⁺ ion at the third, unique Na⁺ binding site [28, 29]. Upon jumps to hyperpolarizing (negative) membrane potentials, extracellular Na⁺ ions are driven through a narrow access channel to the binding sites, thus promoting Na⁺ occlusion and subsequent conversion to E1P, which results in negative, monoexponentially decaying transient currents (dark gray and black curve in Fig. 4.7b). Conversely, jumps from negative to depolarizing (positive) potentials drive Na⁺ ions out of the ion well, thus promoting further E1P → E2P conversion and deocclusion/release of Na⁺ ions, giving rise to positive transient currents (light gray curve in Fig. 4.7b). The voltage dependence of the integrated charge from these transient currents follows a Boltzmann-type function of the following form

$$Q(V) = Q_{max} + \frac{Q_{max} - Q_{min}}{\left(1 + \exp\left(\frac{z_q \cdot F}{R \cdot T}(V - V_{0.5})\right)\right)} \tag{4.1}$$

Here, Q_{min} and Q_{max} are the saturating values of translocated charge, $V_{0.5}$ is the mid-point potential at which the distribution is half-maximal, z_q is the slope factor or the equivalent charge (fraction of charge displaced across the transmembrane field), F the Faraday constant, R the molar gas constant, T the temperature (in K), and V the membrane potential. The resulting $Q–V$ curve is shown in Figure 4.7c, and the satu-rating levels of this curve indicate maximal accumulation of E1P (hyperpolarization) or E2P (depolarization). The reciprocal time constants obtained from the transient currents are constant at membrane potentials above about −50 mV and increase sharply at more negative voltages. Compared to reciprocal time constants of the wild-type or the extracellular cysteine-less enzyme at above −50 mV (each around 80 s⁻¹),

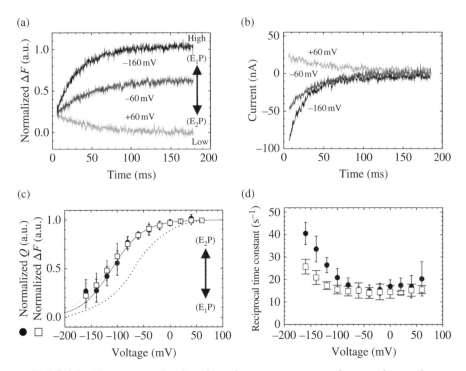

FIGURE 4.7 Fluorescence signals and transient currents measured upon voltage pulses on oocytes expressing the sheep Na^+,K^+-ATPase α_1-subunit carrying mutation N790C on the extracellular end of TM5. (a) Normalized fluorescence changes and (b) ouabain-sensitive difference or transient currents measured at $[K^+]_o = 0$ induced by voltage pulses from 0 mV to the indicated values. (c) Voltage dependence of steady-state fluorescence values (open symbols) and translocated charge (integrals of transient currents; filled symbols) from experiments as in (a) and (b). Fits of a Boltzmann-type function (Eq. 4.1/4.2) to the data sets are superimposed (parameters for F–V curve: $V_{0.5} = -117 \pm 1.4$ mV, $z_q = 0.76 \pm 0.2$; for Q–V curve: $V_{0.5} = -110 \pm 9$ mV, $z_q = 0.85 \pm 0.12$). The dotted line shows the Q–V distribution for the corresponding wild-type Na^+,K^+-ATPase ($V_{0.5} = -71$ mV, $z_q = 0.81$). (d) Voltage dependence of reciprocal time constants from fluorescence (open symbols) and current (filled symbols). Reproduced from Ref. 6 with permission from the National Academy of Sciences of the United States of America (PNAS).

the transient currents of the N790C mutant were slower (about $18 \, s^{-1}$), but the characteristic increase at hyperpolarizing potentials was preserved (see Fig. 4.7d) [6]. Also the $V_{0.5}$ value was negatively shifted from about −72 to −110 mV, whereas the slope factor z_q of about 0.8 remained unchanged. These effects indicate that the N790C mutation causes a slowing down of the $E1P \rightarrow E2P$ conformational change that is rate-limiting for extracellular Na^+ release, whereas the rate constant for reverse binding of Na^+ ions is apparently unaffected.

The fluorescence changes of the TMRM-labeled N790C construct induced by voltage jumps at $[K^+]_{ext} = 0$ (see Fig. 4.7a) followed the pattern observed from

K+ exchange experiments: Jumps to negative potentials induced fluorescence increases, consistent with a shift toward E1P, whereas a decrease in fluorescence occurred at positive potentials. Both transient currents and fluorescence changes vanished in the presence of the Na+ pump-specific inhibitor ouabain, indicating that the observed signals are specific for the Na+,K+-ATPase [6]. Moreover, the rate constants above −50 mV agreed quantitatively, and the voltage dependence of the steady-state fluorescence amplitudes (F–V curve) completely overlapped with the Q–V curve (see Fig. 4.7c), so that fluorescence also follows a Boltzmann-type function:

$$F(V) = F_{max} + \frac{F_{max} - F_{min}}{\left(1 + \exp\left(\frac{z_q \cdot F}{R \cdot T}(V - V_{0.5})\right)\right)} \tag{4.2}$$

Here, F_{min} and F_{max} are the values of the steady-state fluorescence amplitudes replacing the values for the translocated charge. A notable difference was observed at negative potentials, because the fluorescence changes were slower than the kinetics of charge movement (see Fig. 4.7d). This discrepancy can be explained by the fact that, in this voltage range, the fluorescence label senses the conformational transition (E2P → E1P), whereas the transient currents report Na+ reuptake through the extracellular ion well, a reaction step preceding the conformational change. Thus, the N790C sensor construct faithfully reports the distribution between E1(P) ↔ E2(P) conformational states and the fluorescence increase at negative potentials indicates that the dye is moved from a polar, solvent-exposed environment into a more hydrophobic, buried position. This is in agreement with a movement of the M5 segment relative to the other transmembrane helices during the E1P ↔ E2P conformational change, as proposed from biochemical experiments [30] and elicited by crystallographic data [22, 25, 31–36].

Because the N790C sensor construct provides an absolute measure of the E1P/E2P distribution, VCF could also be applied under conditions of Na+/K+ turnover transport. Recording of pre-steady-state currents of Na+,K+-ATPase in the presence of extracellular K+ is difficult because K+ enables the redistribution over all reaction cycle intermediates, thus activating additional potentially electrogenic partial reactions such as binding/occlusion of extracellular K+ [37] or the intracellular binding of Na+ [38]. Transient currents are difficult to resolve because the kinetics of the current signals is accelerated and the amplitudes decline drastically with increasing [K+], so that only few studies succeeded in reporting transient currents under Na+/K+ exchange conditions [39, 40], or by using the more slowly transported Tl+ as a K+ analog [41]. These limitations can be partially overcome with VCF. Figure 4.8a(A) shows fluorescence responses to voltage pulses at $[K^+]_{ext} = 0$, the same conditions as for Figure 4.7b. Voltage jumps from −80 mV to positive potentials elicited only small fluorescence decreases, but negative pulses induced strong voltage-dependent fluorescence increases. With 1 mM extracellular K+ (see Fig. 4.8a(B)), fluorescence signals at −80 mV started from an already increased value. With positive potentials, the fluorescence decreased significantly, whereas negative voltage jumps induced fluorescence increases, and onset of saturation could be seen in both extreme voltage

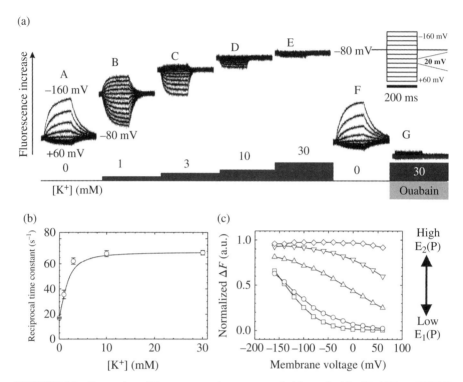

FIGURE 4.8 Properties of fluorescence changes recorded from the Na$^+$,K$^+$-ATPase N790C reporter construct in response to voltage pulses at different extracellular [K$^+$]. (a, A–F) Fluorescence signals measured continuously on a single cell in response to the voltage protocol depicted as inset (upper right); the actual [K$^+$] are indicated by the bottom scheme. (a, G) shows fluorescence signals upon the same voltage protocol in the presence of 30 mM K$^+$ and 10 mM ouabain, which blocks the Na$^+$ pump in E2P. (b) [K$^+$] dependence of reciprocal time constants from fluorescence signals induced by voltage pulses from all test potentials back to −80 mV. (c) Voltage dependence of steady-state fluorescence at different [K$^+$]: 0 mM (squares), 0.3 mM (circles), 1 mM (triangles up), 3 mM (triangles down), 10 mM (diamonds). Reproduced from Ref. 6 with permission from the National Academy of Sciences of the United States of America (PNAS).

ranges. With successively increasing extracellular [K$^+$] (see Fig. 4.8a(C and D)), stationary fluorescence at −80 mV increased further, and negative voltage pulses induced successively declining fluorescence increases, whereas positive potentials led to comparatively large fluorescence decreases. At 30 mM extracellular K$^+$, the fluorescence at −80 mV reaches a maximum, and voltage pulses no longer affected the fluorescence level (see Fig. 4.8a(E)). Repetition of the voltage protocol after removal of K$^+$ (see Fig. 4.8a(F)) induces signals equivalent to those of Figure 4.8a(A), showing reversibility of the system. Importantly, addition of ouabain at 30 mM K$^+$ decreases the fluorescence to a minimum (see Fig. 4.8a(G)), even slightly lower than obtained in the absence of K$^+$ at depolarizing voltages, indicating that the ouabain-blocked pump is arrested in E2P irrespective of the applied voltage or [K$^+$]$_{ext}$ [6].

The voltage-dependent changes of the fluorescence signals with increasing $[K^+]$ are twofold: first, the apparent rate constant of the voltage-dependent relaxation at 0 mV increases from 20 to 70 s^{-1} at saturating K^+ concentrations with an apparent K_m of about 1 mM (see Fig. 4.8b), which is close to the value that can be inferred from the K^+ dependence of stationary currents. In contrast to K^+-free conditions, in which the major electrogenic event occurs during extracellular Na^+ release or reuptake, running the full Post-Albers cycle in the presence of K^+ speeds up the total observed relaxation rate and also includes charge-translocating steps within the K^+ branch. Second, the midpoint potentials ($V_{0.5}$) of the F–V curves are positively shifted and the slope factors (equivalent charges z_q) decrease from about 0.8 at $[K^+]_{ext} = 0$ to about 0.45 at 1 mM K^+ (see Fig. 4.8c) [6]. Because z_q represents the fractional charge or, equivalently, the fraction of the transmembrane field, which a unitary charge passes during a charge-translocating event, the latter can be interpreted as apparently decreased voltage sensitivity. However, considering that z_q under these (turnover) conditions is a composite of several charge-translocating steps occurring in opposite directions, it can be estimated that the difference from the aforementioned values of about 0.35 represents the z_q value for the activation of the K^+ branch of the Post-Albers cycle, in accordance with the reported lower electrogenicity of K^+ transport steps. This value is rather a lower limit because K^+ concentrations above 3 mM induce such a large shift of the F–V curve that Boltzmann parameters cannot be determined (see Fig. 4.8c). At $[K^+] = 10$ mM, the absence of detectable fluorescence changes indicates that the conformational distribution, which is maximally poised toward E1, can no longer be shifted by voltage pulses ranging from -160 to $+60$ mV. In contrast to inhibitor experiments, however, the pump is fully active under these conditions.

4.4.2.2 The Influence of Intracellular Na$^+$ Concentrations

The previous experiments on the Na^+,K^+-ATPase were carried out under saturating intracellular $[ATP]$ and $[Na^+]$, the latter requiring a procedure for elevating the intracellular $[Na^+]$ termed "Na^+ loading" [42]. Heterologous expression of the Na^+ pump in *Xenopus* oocytes usually results in such low intracellular $[Na^+]$ that no transient or stationary pump currents can be measured without Na^+ loading. Thus, instead of loading the cells with saturating $[Na^+]$, controlled loading to defined Na^+ concentrations would enable the effects of this intracellular ligand on voltage dependence and the E1P/E2P distribution to be studied. Controlled loading of Na^+ into the cells can be achieved by coexpression of the epithelial, amiloride-sensitive Na^+ channel ENaC [43]. During expression, ENaC is inhibited by adding amiloride to the incubation buffers. Withdrawal of amiloride during an electrophysiological experiment permits the passive influx of Na^+ ions into the cell, driven by a concentration gradient or (even augmented) by the application of negative membrane potentials. After such a loading step, the achieved $[Na^+]_i$ can be measured from the reversal potential of the ENaC-mediated current using the Nernst equation. Though sounding simple in theory, this is a tedious experiment because it requires the injection of mixtures of no less than five cRNAs (three for the ENaC subunits α, β, and γ, and two for Na^+,K^+-ATPase α- and β-subunit) and the achievement of a balanced, robust expression level for both membrane proteins. Geys and coworkers undertook the challenge [44] and investigated the influence of $[Na^+]_i$ on the E1P$\leftrightarrow$E2P

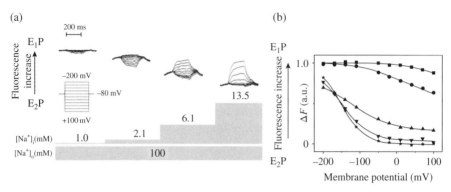

FIGURE 4.9 Dependence of the conformational distribution of Na^+,K^+-ATPase on $[Na^+]_i$. (a) Fluorescence traces in response to voltage steps at $[K^+]_o = 0$ and increasing $[Na^+]_i$ as indicated, the latter were controlled by coexpression of the epithelial Na^+ channel ENaC (see text for details). Voltage steps were performed from $-80\,mV$ to the test potentials indicated, in increments of $30\,mV$ (see inset). (b) Voltage dependence of the steady-state fluorescence amplitudes from experiments as in (a) and the related fits of a Boltzmann-type function to the data sets. The symbols relate to different $[Na^+]_i$: $1.0\,mM$ (squares), $2.1\,mM$ (circles), $6.6\,mM$ (triangles up), $12.4\,mM$ (triangles down), and $13.5\,mM$ (stars). Reprinted from Ref. 44 with permission from Cell Press (Biophysical Journal).

distribution using the TMRM-labeled N790C mutant as described above. It was shown that under electrogenic Na^+/Na^+ exchange conditions (high $[Na^+]_o$, $[K^+]_o = 0$) at $[Na^+]_i \leq 1$ mM, hardly any fluorescence changes could be induced by voltage pulses, whereas high $[Na^+]_i$ ($13.5\,mM$) elicited voltage-dependent fluorescence responses as observed previously with Na^+-loaded oocytes (see Fig. 4.9a, compare to Fig. 4.7a). At an intermediate $[Na^+]_i$ of about $2\,mM$, which is close to the apparent affinity for intracellular Na^+ reported from biochemical experiments [45], the difference to the behavior observed at high Na^+ conditions is that hardly any fluorescence changes occur upon negative voltage pulses, whereas substantial fluorescence decreases are observed at positive potentials (see Fig. 4.9a). This indicates that the pump, which at low $[Na^+]_i$ mainly accumulates in E1-like states (as inferred by the high fluorescence level), undergoes a conversion to E2P at positive voltages under these conditions [44]. This corroborates the notion that intracellular Na^+ binding is also at least weakly voltage dependent, in line with previous studies [38]. This is supported by the reduced slope factor (equivalent charge) z_q of about 0.4 at limiting $[Na^+]_i$ obtained from the F–V curves (see Fig. 4.9b) [45], which is much smaller than the value of 0.8 measured from Q–V or F–V curves under saturating $[Na^+]_i$, in which the electrogenicity is exclusively dominated by extracellular release or reverse binding of Na^+.

4.4.3   The Rat Gastric H^+,K^+-ATPase S806C Sensor Construct

From an electrophysiologist's perspective, investigation of the gastric H^+, K^+-ATPase is hampered due to its overall electroneutral outward/inward transport of equal numbers of H^+/K^+ ions, respectively, so that no stationary pump currents can be recorded. Nevertheless, the reaction cycle contains at least two electrogenic partial reactions, because transient currents could be recorded upon ATP concentration jumps in the absence of K^+ with the BLM technique [46, 47], which consequently implies that an electrogenic step must exist during the K^+ transport branch to neutralize the overall electrogenicity. However, the majority of biochemical experiments and the BLM technique (see Chapter 2) unfortunately do not preserve compartmentalization, which is a severe limitation for an enzyme that produces the largest known concentration gradients. Thus, to gain insight into the inner workings of the H^+ pump under conditions that are at least similar to those in its physiological context, measurements on intact cells are required, and here, the VCF method fully pays off.

Starting from the observation that the N790C mutant of sheep Na^+,K^+-ATPase α1-subunit was an excellent VCF reporter construct, a cysteine mutation was introduced at the homologous position (extracellular end of TM5) of rat gastric H^+,K^+-ATPase resulting in mutant S806C, which—upon SDFL with TMRM—showed remarkably robust fluorescence change signals upon voltage jumps in the absence of extracellular K^+, similar to the previous case of the Na^+ pump [7, 48].

4.4.3.1 *Voltage-Dependent Conformational Shifts of the H^+,K^+-ATPase Sensor Construct S806C During the H^+ Transport Branch* As shown in Figure 4.10, voltage jumps to negative potentials (at $[K^+]_o = 0$, pH 7.4, and 100 mM TMA^+ as major—nontransported—monovalent cation in the extracellular solution) caused fluorescence increases, whereas pulses to positive potentials induced fluorescence decreases. Similar to the interpretation of VCF signals recorded on the Na^+ pump, it was inferred that negative potentials favor a shift of the conformational state toward E1P, whereas the E2P conformation is favored at positive potentials. Specificity of these fluorescence changes was shown by the fact that the signals vanished completely in the presence of the H^+ pump inhibitor SCH28080, which acts as a K^+ congener and blocks the pump in the E2P state [50]. Voltage-dependent steady-state fluorescence also followed a Boltzmann-type function (Eq. 4.2), but the slope factor of about 0.45 was much smaller than for the Na^+,K^+-ATPase. Furthermore, the reciprocal time constants from monoexponential fits to the fluorescence traces also showed an increase at negative potentials, in line with the notion that some step correlated with inward movement of cations (H^+) through high-field access channel(s) determines electrogenicity. But, compared to the N790C mutant (or wild-type) Na^+,K^+-ATPase, the values of about $3\,s^{-1}$ at positive potentials were slower by a factor of about 7 (or 16, respectively) [7, 49, 51–53]. Both the smaller z_q and the drastically slower kinetics would reduce the expected amplitudes of these transient currents, which explains why it has never been possible to observe pre-steady-state currents that could correspond to the transient currents resulting from extracellular Na^+ release/reverse binding steps of Na^+,K^+-ATPase. If similar plasma membrane expression levels as in the case of Na^+,K^+-ATPase could be achieved, whereby the latter gives rise to transient charge movement of up to 10 nC (e.g., [54]), with z_q of about 0.8 and time constants of 100 ms ($10\,s^{-1}$ rate constant) this would result in transient current amplitudes of 0.1 μA. Such small currents are difficult to discriminate from

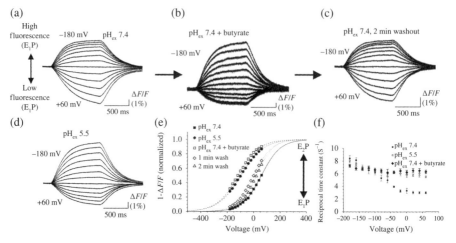

FIGURE 4.10   Effects of extra-/intracellular pH on the E1P/E2P distribution of gastric H⁺,K⁺-ATPase. (a–c) H⁺,K⁺-ATPase carrying mutation S806C on TM5 of the α-subunit serves as reporter construct for VCF upon labeling with TMRM. Fluorescence responses induced by voltage pulses from $-80\,mV$ to the test potentials indicated (20 mV increments) at $[K^+]_o = 0$ and $[Na^+]_o = 90$ mM. Recordings originated from a single oocyte, first exposed to pH_o 7.4 (a), then after 1 min in presence of 40 mM Na-butyrate at pH_o 7.4 (b), and after 2 min washout of butyrate at pH_o 7.4 (c). (d) Fluorescence responses from a different cell at pH_o 5.5. (e) Voltage dependence of normalized steady-state fluorescence amplitudes from the recordings as in (a–d). Fits of a Boltzmann-type function are superimposed to each data set, with fluorescence amplitudes normalized to saturation values from the fits. (f) Reciprocal time constants from fits of a monoexponential function to fluorescence signals at $[K^+]_o = 0$. Data obtained at pH_o 7.4 in the presence of 40 mM butyrate (diamonds) are compared to those in butyrate-free solutions at pH_o 5.5 (open squares) and at pH_o 7.4 without Na-butyrate (filled squares). Reproduced from Ref. 49 in accordance with the terms of Creative Commons Attribution (CC BY) license.

current baseline fluctuations, which almost inevitably occur in TEVC measurements on oocytes in the extreme voltage ranges, such as below $-150\,mV$ and above $+50\,mV$, that would need to be covered here.

4.4.3.2   An Intra- or Extracellular Access Channel of the Proton Pump?   If the electrogenicity of H⁺-translocating steps of the H⁺,K⁺-ATPase arises from extracellular H⁺ release or reverse binding, similar to extracellular Na⁺ interacting with the Na⁺ pump, it should be possible to shift the E1P/E2P distribution toward E1P by extracellular acidification. However, it must be kept in mind that the H⁺ pump releases protons against an extracellular pH of below 1 in the stomach lumen, so that the apparent extracellular H⁺ affinity must be of the order of 1 M or larger. By means of an extracellular high-field access channel with a z_q of 1, the effective H⁺ concentration at the binding site(s) deep within the access channel could be elevated by membrane voltage according to the formula

$$[H^+]_{eff}(V) = [H^+]_{ex,(V=0)} \cdot e^{\frac{z_q \cdot F \cdot \Delta V}{R \cdot T}} \tag{4.3}$$

Hence, a membrane potential difference (ΔV) of more than $-400\,$mV would be required to elevate the effective [H$^+$] at the binding site above 1 M at neutral extracellular pH, and even a pH of 5.5 would still require a ΔV of more than $-320\,$mV. These considerations indicate that even a deep extracellular access channel would never be quantitatively sufficient within the range of transmembrane potentials and [H$^+$] differences that can be achieved in live-cell measurements. It was therefore surprising to see that an extracellular acidification from pH 7.4 to 5.5 led to a significant shift of the voltage-dependent distribution of fluorescence amplitudes (compare Fig. 4.10a and d). However, in contrast to expectations for an extracellular access channel, the distribution was shifted in the direction of E2P (compare solid symbols in Fig. 4.10e). This, at first sight puzzling, observation could be traced to a change of the intracellular pH that is induced by extracellular acidification, which occurs due to the nonzero permeability of biological membranes toward protons. Evidence for this suggestion was obtained by controlled intracellular acidification that can be achieved in a procedure termed "acid loading" with the help of weak organic acids. With a pK_a of about 4.8, weak organic acids like butyric acid prevail in the deprotonated, charged form at neutral pH, but the still substantial amount of the protonated, neutral species can rather freely penetrate through membranes. Inside the cell, however, the weak organic acid dissociates according to its pK_a and releases protons, which can shift the intracellular pH if the influx of the weak acid is strong enough to overcome the intracellular H$^+$ buffering capacity. It was shown that 40 mM butyric acid at an extracellular pH of 7.4 induces an intracellular pH shift of about 0.5 units [49, 55]. When VCF signals were recorded on the H$^+$,K$^+$-ATPase S806C mutant under the aforementioned conditions, the signals and the corresponding F–V curve nearly completely overlapped with the signals measured previously at extracellular pH 5.5 (compare Fig. 4.10b, d). After washout of butyric acid, the signals measured previously at pH 7.4 were restored after some time (see Fig. 4.10c). This is consistent with the idea that changes in the extracellular pH by two orders of magnitude have no effect on the E1P/E2P distribution (regardless of the existence of an extracellular access channel), whereas the intracellular H$^+$ binding site(s) must have an apparent pK_a in the neutral range, so that the number of H$^+$,K$^+$-ATPase molecules that are able to enter the reaction sequence underlain in gray in Figure 4.6a is strongly affected by the concomitant change in the intracellular pH. This indicates that some intracellular H$^+$ interaction event must be voltage dependent. More precisely, as inferred from the increase of rate constants with hyperpolarization in Figure 4.10f, the voltage dependence must be dominated by an intracellular H$^+$ unbinding step. The effect of butyrate treatment or extracellular acidification on the internal pH (see Fig. 4.10f) showed that in the forward direction only the (otherwise voltage-independent) E1P $\rightarrow$ E2P, but not the backward reaction rate was accelerated, in line with a rapid, voltage-independent pre-equilibrium for intracellular H$^+$ binding to the pump that is rate-limiting for the E1P $\rightarrow$ E2P reaction.

4.4.3.3 *Effects of Extracellular Ligands: K$^+$ and Na$^+$* Upon addition of K$^+$ as extracellular ligand to stimulate H$^+$/K$^+$ exchange transport, the VCF signals of H$^+$, K$^+$-ATPase followed the behavior of the Na$^+$ pump sister enzyme (see Fig. 4.11a and

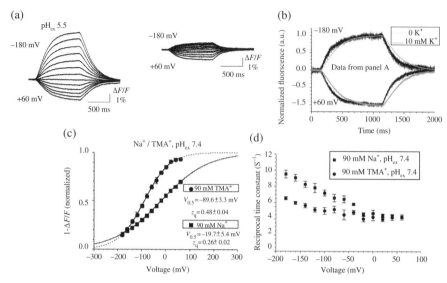

FIGURE 4.11 Effects of extracellular Na^+ or K^+ on the E1P/E2P distribution of gastric H^+,K^+-ATPase. (a) Fluorescence signals induced by voltage step protocols as in Figure 4.10 measured from TMRM-labeled oocytes expressing H^+,K^+-ATPase reporter construct S806C at $[K^+]_o = 0$ (left traces) or in solution containing 5 mM K^+ (right traces) at pH_o 5.5. (b) Comparison of the time course of the fluorescence signals from panel (a) in response to voltage pulses to -180 mV and $+60$ mV in the absence of K^+ (gray) and in the presence of 5 mM K^+ (black) after amplitude normalization. (c) Voltage dependence of steady-state fluorescence amplitudes at pH_o 7.4 and $[K^+]_o = 0$ in the presence of 90 mM extracellular Na^+ (squares) compared to Na^+-free conditions (circles, Na^+ replacement by 90 mM TMA^+). Superimposed are fits of a Boltzmann-type function (Eq. 4.2) to the data with fit parameters, as indicated. The fluorescence amplitudes were normalized to the saturation values from the fits. (d) Reciprocal time constants from fits of a monoexponential function to fluorescence signals in Na^+-free and in 90 mM Na^+-containing solutions at pH_o 7.4. Reproduced from Ref. 49 in accordance with the terms of Creative Commons Attribution (CC BY) license.

b): the fluorescence level increased and shifted toward E1(P), the maximal fluorescence changes, which could be obtained upon changes to extreme positive and negative potentials, faded, and the kinetics of the redistribution of states accelerated slightly from about 7 to $10 s^{-1}$ at $[K^+] = 5$ mM (compare the amplitude-normalized signals in Fig. 4.11b). Notably, the value of 5 mM is comparable to the concentration of half-maximal activation of cation transport inferred from Rb^+ uptake studies by AAS (see Chapter 13 and [7]). Thus, as an estimate, a maximal turnover rate of about $20 s^{-1}$ at room temperature can be inferred, which together with the measured activation energy of 95 kJ mol^{-1} (see Chapter 13 and [49]) yields a transport rate of $150 s^{-1}$ at body temperature.

Another initially counter-intuitive effect was observed upon exchange of the major extracellular cation from TMA^+ to Na^+. Neither cation species is transported by the gastric H^+ pump, but the kinetics and voltage dependence of the E1P $\leftrightarrow$ E2P redistribution changed considerably upon exchange of TMA$^+$ for Na$^+$ (see Fig. 4.11c), suggesting a distinct effect of Na$^+$ ions on the function of the pump. Unlike the case of increasing the extracellular H$^+$ concentration, the addition of Na$^+$ selectively elevated the apparent rate constant of the E2P $\rightarrow$ E1P backward reaction, as can be seen from the increase of the reciprocal rate constants at hyperpolarizing potentials in Figure 4.11d. Thus, extracellular Na$^+$, though not transported by the H$^+$ pump, stimulates E2P $\rightarrow$ E1P in a voltage-dependent, hence electrogenic, manner from the extracellular side, in line with previous findings [56]. However, despite the addition of another electrogenic step to the reaction sequence carried out by the enzyme, the slope factor z_q of the F–V distribution is decreased from about 0.45 to 0.25 (see Fig. 4.11c). Mathematical modeling of the reaction sequence underlain in gray in Figure 4.6a based on the assumption that cation interaction with the H$^+$ pump proceeds through an intra- as well as extracellular access channel could rationalize these findings [49]. These simulations showed that starting from the assumption of an intracellular access channel with $z_q = 0.5$ in the Na$^+$-free case, in which electrogenicity of extracellular cation interaction has no influence, the inclusion of the extracellular access channel with a z_q of about 0.2 by Na$^+$ addition leads to the effect that the apparently observed z_q value decreases from 0.5 (Na$^+$-free case) to about 0.3 (with Na$^+$). These considerations highlight the fact that the electrogenicity parameters measured in voltage jump relaxation experiments are merely the lumped sum of contributions from several electrogenic partial reactions, which may well be significantly lower than the z_q values of individual partial reactions [49].

4.4.4 Probing Intramolecular Distances by Double Labeling and FRET

During subsequent VCF studies on the Na$^+$,K$^+$-ATPase, conformation-sensitive reporter positions could also be identified at the extracellular interface of the β- [14] and the γ-subunit [57] of the enzyme. Especially the S62C mutant of the β-subunit turned out to perform very similarly to the αN790C construct described earlier (see Fig. 4.12b) [14]. In principle, the availability of several positions for SDFL enables multiple labeling by different cysteine-reactive dye compounds so that intramolecular distance determinations by fluorescence resonance energy transfer (FRET) may become possible. The term FRET describes the radiation-less transfer of excitation energy between a pair of fluorophores, for which the fluorescence spectrum of the "donor" molecule must overlap with the absorption spectrum of the "acceptor" molecule. Furthermore, the mutual orientation of the transition dipole moments is important, as is the distance between the chromophores, because the efficiency for such a dipole–dipole-type interaction decreases with an R^{-6} dependence according to

$$E = \frac{1}{\left(1 + \left(R / R_0\right)^6\right)} \qquad (4.4)$$

Here, the term R_0 denotes the "Förster radius," a characteristic distance for each donor–acceptor pair, which can—in principle—be calculated using Theodor Förster's theory [59]. Though the FRET signature is easy to conceive, compared to the

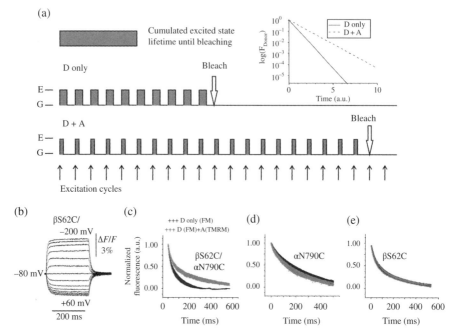

FIGURE 4.12   Indirect measurement of FRET efficiency from donor lifetime during photo-bleaching. (a) Principle of the procedure: The propensity to undergo irreversible bleaching depends on the cumulated dwell time in the excited state (E, ground state—G) so that each chromophore can undergo a certain average number of excitation cycles (arrows) before bleaching occurs ("D only"). In the presence of an appropriate acceptor ("D+A") allowing for FRET, the dwell time of donor molecules in the excited state per excitation cycle is reduced and the donor chromophore can undergo more excitation cycles before bleaching occurs. (b) Fluorescence responses of the TMRM-labeled Na^+,K^+-ATPase reporter construct carrying mutation S62C on the extracellular end of the β-subunit's transmembrane segment in response to voltage pulses from $-80\,mV$ to potentials as indicated (in $20\,mV$ steps) at $[K^+]_o = 0$ [14]. (c) FRET determination from donor (fluorescein) photobleaching. Na^+,K^+-ATPase with two cysteine reporter mutations (N790C on the α- and S62C on the β-subunit) was labeled either with FM ("D only," black symbols) or with a 1:4 mixture of FM/TMRM ("D+A," gray symbols). (d and e) Control experiments of fluorescence donor bleaching with Na^+,K^+-ATPase comprising only one cysteine reporter mutation (either N790C in the α- (d) or S62C in the β-subunit (e)) measured upon labeling with FM only (black symbols), or with a 1:4 mixture of FM/TMRM (gray symbols) [58]. For the two labeling situations in (c–e), bleaching of the fluorescein emission intensity was induced and monitored with the same excitation intensity. Panel b was reproduced from Ref. 14 with permission from the Society of General Physiologists, panels c–e were reproduced from Ref. 58 with permission from the American Society for Biochemistry and Molecular Biology (ASBMB, Journal of Biological Chemistry).

donor-only situation, the fluorescence of the donor must decrease in the presence of an appropriate acceptor and the acceptor intensity must rise concomitantly, meaning that intensity-based FRET measurements require meticulous controls [60] due to the substantial spectral overlap between excitation/emission spectra. An alternative approach, using a different signature of FRET, has been employed by Glauner and coworkers for determining intramolecular distances in the case of the aforementioned Shaker K+ channel [61]. In this case, the effect of the presence or absence of an acceptor molecule is determined based on irreversible donor bleaching (see Fig. 4.12a). Each fluorescence molecule has a certain propensity to undergo, for example, triplet conversion and subsequent bleaching by reactive oxygen species upon continuous excitation. The probability of undergoing bleaching depends on the accumulated dwell time in the excited state, so that each chromophore can pass through an average number of excitation cycles before irreversible bleaching occurs. In the presence of an appropriate acceptor chromophore within the interaction distance, the possibility of transferring excitation energy by FRET reduces the dwell time in the excited state per excitation cycle so that a donor molecule can undergo more reaction cycles before irreversible bleaching occurs, which results in a slower bleaching rate, as exemplified in Figure 4.12a (inset). From the bleaching time constants measured in the "D only" and in the "D+A" situation (τ_D and τ_{DA}, respectively), the distance R between donor and acceptor can be calculated based on the identity:

$$E = \frac{1}{\left(1 + \left(R / R_0\right)^6\right)} = 1 - \frac{\tau_{DA}}{\tau_D} \tag{4.5}$$

It is important to keep the excitation intensity identical in the "D-only" and the "D+A" situation to ensure that the average excitation rate is the same.

In the case of multiple labeling positions, the sites to which individual donor (D) or acceptor (A) chromophores bind cannot be predetermined, so that a 1:1 D/A mixture (if the bimolecular reaction rates of the two maleimide-conjugated dyes to the two cysteines available for labeling are identical) leads to 25% of molecules labeled D:D, 50% labeled D:A (or A:D), and 25% labeled A:A. Thus, in the ideal situation, 50% of the transporter molecules exhibit the desired D:A labeling scheme, and the D:D-labeled molecules create a background that is not affected by FRET, which needs to be taken into account. With even more labeling positions available, such as in the case of the tetrameric K+ channels, the combinatorial variety is even more complex, and in this case, a variation of the D:A dye mixture used for labeling in favor of A is advantageous, because by this measure, the fraction of molecules containing only one D and several A molecules can be optimized (e.g., D:A ratio 1:4, see Ref. 61). Using this procedure, it was possible to determine distances and even distance changes between labeling positions that occur during the activation of the *Shaker* K+ channel by voltage jumps, thus gaining evidence for a helical twist movement of the S4 voltage-sensing segment upon activation [61].

Using a similar experimental scheme, Dempski and coworkers used mixtures of FM and TMRM to achieve double-labeling of the Na+,K+-ATPase carrying mutations N790C in the α- and S62C in the β-subunit. From the bleaching rates measured in the "D-only" and the "D+A" labeling situation (about 35 s vs. 75 s, respectively, see Fig. 4.12c), a D–A distance of about 53Å could be determined from Equation 4.5, which agreed well with the distance between the transmembrane helix of the β-subunit and the TM5 segment of the α-subunit from crystallographic data [58]. As a control, double labeling schemes on Na$^+$,K$^+$-ATPase constructs carrying only a single cysteine (either αN790C, or βS62C) did not show differences in donor bleaching between "D-only" and "D+A" labeling scheme [58].

4.5   CONCLUSIONS AND PERSPECTIVES

VCF expands the electrophysiological toolbox to investigate molecular processes in membrane proteins that are not directly or indirectly connected to charge transloca-tion across the membrane. The technique is particularly advantageous for the study of membrane transporters with net electroneutral activity such as gastric H$^+$,K$^+$-ATPase. Due to the high sensitivity of fluorescence dyes to environmental changes in their immediate vicinity, highly localized conformational rearrangements occurring on a subnanometer scale can be investigated. Systematic investigations of the fluo-rescence properties, including fluorescence spectra and anisotropy, in the presence of different quenchers, or the use of different dyes for labeling, can provide more details about the nature of the underlying conformational changes monitored during SDFL/VCF. Multiple labeling schemes can even exploit the molecular ruler provided by Förster energy transfer to probe intramolecular distances and investigate dynamic changes that occur during the reaction cycle of membrane transporters [61, 62]. Of some concern, at least if the *Xenopus* oocyte expression system is used, is the achieve-ment of a robust expression level for the membrane protein under study. If plasma membrane expression is limited, care should be taken to remove other extracellularly accessible cysteines in the protein under study, or to apply preblocking schemes to suppress the background arising from cysteines in native membrane proteins. To identify possible SDFL reporter sites, cysteine-scanning mutagenesis approaches spanning extracellular loop regions and adjacent segments should be carried out, ideally based on authoritative topology models or structural data. Each identified reporter position will be rewarding for the study of atomic-scale molecular rear-rangements that occur in response to voltage pulses, the addition of intra- or extracel-lular ligands, toxins or inhibitors or regulatory processes, thus revealing the maximum of information to be drawn from electrophysiological studies of membrane trans-porter proteins in their natural membrane environment.

ACKNOWLEDGMENTS

The author is grateful to Prof. Ernst Bamberg (Max-Planck-Institute of Biophysics, Frankfurt, Germany) for his generous support during establishing the VCF method on P-type ATPases, Dr. Sven Geibel, Dr. Dirk Zimmermann, Dr. Giovanni Zifarelli, Dr. Neslihan N. Tavraz, Dr. Katharina L. Dürr, and Prof. Robert E. Dempski for their efforts in refining the technique and making it work, and Prof. Kazuhiro Abe (Nagoya University, Japan), Prof. Klaus Fendler and Dr. Klaus Hartung (Max-Planck-Institute of Biophysics, Frankfurt, Germany) for encouraging discussions about the kinetics and mechanism of P-type pumps. Part of the work was funded by the German Research Foundation DFG (Cluster of Excellence "Unifying Concepts in Catalysis").

REFERENCES

1. S. Kaneko, A. Akaike, M. Satoh, Cut-open recording techniques, *Methods Enzymol.* 293 (1998) 319–331.

2. E. Stefani, F. Bezanilla, Cut-open oocyte voltage-clamp technique, *Methods Enzymol.* 293 (1998) 300–318.

3. C.A. Wagner, B. Friedrich, I. Setiawan, F. Lang, S. Broer, The use of *Xenopus laevis* oocytes for the functional characterization of heterologously expressed membrane proteins, *Cell. Physiol. Biochem.* 10 (2000) 1–12.

4. D.W. Hilgemann, The giant membrane patch, in: B. Sakmann and E. Neher (Eds.), *Single-Channel Recording*, Plenum Publishing Corporation, New York, 1995, pp. 307–327.

5. A. Cha, F. Bezanilla, Characterizing voltage-dependent conformational changes in the *Shaker* K^+ channel with fluorescence, *Neuron* 19 (1997) 1127–1140.

6. S. Geibel, J.H. Kaplan, E. Bamberg, T. Friedrich, Conformational dynamics of the Na^+/K^+-ATPase probed by voltage clamp fluorometry, *Proc. Natl. Acad. Sci. U. S. A.* 100 (2003) 964–969.

7. K.L. Dürr, N.N. Tavraz, D. Zimmermann, E. Bamberg, T. Friedrich, Characterization of Na, K-ATPase and H,K-ATPase enzymes with glycosylation-deficient beta-subunit variants by voltage-clamp fluorometry in *Xenopus* oocytes, *Biochemistry* 47 (2008) 4288–4297.

8. R. Richards, R.E. Dempski, Examining the conformational dynamics of membrane proteins in situ with site-directed fluorescence labeling, *J. Vis. Exp.* (2011) e2627.

9. C.S. Gandhi, R. Olcese, The voltage-clamp fluorometry technique, *Methods Mol. Biol.* 491 (2008) 213–231.

10. M.W. Rudokas, Z. Varga, A.R. Schubert, A.B. Asaro, J.R. Silva, The *Xenopus* oocyte cut-open vaseline gap voltage-clamp technique with fluorometry, J. Vis. Exp. (2014) e51040.

11. J. Kusch, G. Zifarelli, Patch-clamp fluorometry: Electrophysiology meets fluorescence, *Biophys. J.* 106 (2014) 1250–1257.

12. L.M. Mannuzzu, M.M. Moronne, E.Y. Isacoff, Direct physical measure of conformational rearrangement underlying potassium channel gating, *Science* 271 (1996) 213–216.

13. A. Cha, F. Bezanilla, Structural implications of fluorescence quenching in the *Shaker* K^+ channel, *J. Gen. Physiol.* 112 (1998) 391–408.

14. R.E. Dempski, T. Friedrich, E. Bamberg, The beta subunit of the Na^+/K^+-ATPase follows the conformational state of the holoenzyme, *J. Gen. Physiol.* 125 (2005) 505–520.

15. M.M. White, F. Bezanilla, Activation of squid axon K^+ channels. Ionic and gating current studies, *J. Gen. Physiol.* 85 (1985) 539–554.

16. F. Bezanilla, E. Perozo, E. Stefani, Gating of Shaker K^+ channels: II. The components of gating currents and a model of channel activation, *Biophys. J.* 66 (1994) 1011–1021.

17. E. Stefani, L. Toro, E. Perozo, F. Bezanilla, Gating of Shaker K^+ channels: I. Ionic and gating currents, *Biophys. J.* 66 (1994) 996–1010.

18. J.R. Lakowicz, *Principles of Fluorescence Spectroscopy*, third ed., Springer, New York, 2006.

19. G. Crambert, U. Hasler, A.T. Beggah, C. Yu, N.N. Modyanov, J.D. Horisberger, L. Lelievre, K. Geering, Transport and pharmacological properties of nine different human Na, K-ATPase isozymes, *J. Biol. Chem.* 275 (2000) 1976–1986.

20. R.W. Albers, Biochemical aspects of active transport, *Annu. Rev. Biolchem.* 36 (1967) 727–756.

21. R.L. Post, Hegyvary, C., Kume, S., Activation by adenosine triphosphate in the phosphorylation kinetics of sodium and potassium transporting adenosine triphosphatase, *J. Biol. Chem.* 247 (1972) 6530–6540.

22. T. Shinoda, H. Ogawa, F. Cornelius, C. Toyoshima, Crystal structure of the sodium-potassium pump at 2.4 Å resolution, *Nature* 459 (2009) 446–450.

23. C. Toyoshima, M. Nakasako, H. Nomura, H. Ogawa, Crystal structure of the calcium pump of sarcoplasmic reticulum at 2.6 Å resolution, *Nature* 405 (2000) 647–655.

24. C. Toyoshima, H. Nomura, Structural changes in the calcium pump accompanying the dissociation of calcium, *Nature* 418 (2002) 605–611.

25. J.P. Morth, B.P. Pedersen, M.S. Toustrup-Jensen, T.L. Sørensen, J. Petersen, J.P. Andersen, B. Vilsen, P. Nissen, Crystal structure of the sodium-potassium pump, *Nature* 450 (2007) 1043–1049.

26. E.M. Price, J.B. Lingrel, Structure-function relationships in the Na,K-ATPase alpha subunit: Site-directed mutagenesis of glutamine-111 to arginine and asparagine-122 to aspartic acid generates a ouabain-resistant enzyme, *Biochemistry* 27 (1988) 8400–8408.

27. Y.K. Hu, J.H. Kaplan, Site-directed chemical labeling of extracellular loops in a membrane protein. The topology of the Na,K-ATPase alpha-subunit, *J. Biol. Chem.* 275 (2000) 19185–19191.

28. N. Vedovato, D.C. Gadsby, The two C-terminal tyrosines stabilize occluded Na/K pump conformations containing Na or K ions, *J. Gen. Physiol.* 136 (2010) 63–82.

29. M. Holmgren, J. Wagg, F. Bezanilla, R.F. Rakowski, P. De Weer, D.C. Gadsby, Three distinct and sequential steps in the release of sodium ions by the Na$^+$/K$^+$-ATPase, *Nature* 403 (2000) 898–901.

30. S. Lutsenko, R. Anderko, J.H. Kaplan, Membrane disposition of the M5-M6 hairpin of Na$^+$,K$^+$-ATPase alpha subunit is ligand dependent, *Proc. Natl. Acad. Sci. U. S. A.* 92 (1995) 7936–7940.

31. C. Olesen, M. Picard, A.M. Winther, C. Gyrup, J.P. Morth, C. Oxvig, J.V. Møller, P. Nissen, The structural basis of calcium transport by the calcium pump, *Nature* 450 (2007) 1036–1042.

32. C. Olesen, T.L. Sorensen, R.C. Nielsen, J.V. Møller, P. Nissen, Dephosphorylation of the calcium pump coupled to counterion occlusion, *Science* 306 (2004) 2251–2255.

33. B.P. Pedersen, M.J. Buch-Pedersen, J.P. Morth, M.G. Palmgren, P. Nissen, Crystal structure of the plasma membrane proton pump, *Nature* 450 (2007) 1111–1114.

34. T.L. Sørensen, J.V. Møller, P. Nissen, Phosphoryl transfer and calcium ion occlusion in the calcium pump, *Science* 304 (2004) 1672–1675.

35. C. Toyoshima, M. Nakasako, H. Nomura, H. Ogawa, Crystal structure of the calcium pump of sarcoplasmic reticulum at 2.6 Å resolution, *Nature* 405 (2000) 647–655.

36. M. Nyblom, H. Poulsen, P. Gourdon, L. Reinhard, M. Andersson, E. Lindahl, N. Fedosova, P. Nissen, Crystal structure of Na$^+$,K$^+$-ATPase in the Na$^+$-bound state, *Science* 342 (2013) 123–127.

37. R.F. Rakowski, L.A. Vasilets, J. LaTona, W. Schwarz, A negative slope in the current-voltage relationship of the Na^+/K^+ pump in *Xenopus* oocytes produced by reduction of external $[K^+]$, *J. Membr. Biol.* 121 (1991) 177–187.

38. J. Pintschovius, K. Fendler, E. Bamberg, Charge translocation by the Na^+/K^+-ATPase investigated on solid supported membranes: Cytoplasmic cation binding and release, *Biophys. J.* 76 (1999) 827–836.

39. A. Bahinski, M. Nakao, D.C. Gadsby, Potassium translocation by the Na^+/K^+ pump is voltage insensitive, *Proc. Natl. Acad. Sci. U. S. A.* 85 (1988) 3412–3416.

40. D.C. Gadsby, M. Nakao, A. Bahinski, Voltage dependence of transient and steady-state Na/K pump currents in myocytes, *Mol. Cell. Biochem.* 89 (1989) 141–146.

41. R.D. Peluffo, J.R. Berlin, Electrogenic K^+ transport by the Na^+-K^+ pump in rat cardiac ventricular myocytes, *J. Physiol.* 501 (1997) 33–40.

42. R.F. Rakowski, Charge movement by the Na/K pump in *Xenopus* oocytes, *J. Gen. Physiol.* 101 (1993) 117–144.

43. J.D. Horisberger, S. Kharoubi-Hess, Functional differences between alpha subunit isoforms of the rat Na,K-ATPase expressed in *Xenopus* oocytes, *J. Physiol.* 539 (2002) 669–680.

44. S.A. Geys, E. Bamberg, R.E. Dempski, Ligand-dependent effects on the conformational equilibrium of the Na^+,K^+-ATPase as monitored by voltage clamp fluorometry, *Biophys. J.* 96 (2009) 4561–4570.

45. M. Holmgren, R.F. Rakowski, Pre-steady-state transient currents mediated by the Na/K pump in internally perfused *Xenopus* oocytes, *Biophys. J.* 66 (1994) 912–922.

46. M. Stengelin, K. Fendler, E. Bamberg, Kinetics of transient pump currents generated by the (H,K)-ATPase after an ATP concentration jump, *J. Membr. Biol.* 132 (1993) 211–227.

47. H.T. van der Hijden, E. Grell, J.J. de Pont, E. Bamberg, Demonstration of the electrogenicity of proton translocation during the phosphorylation step in gastric H^+,K^+-ATPase, *J. Membr. Biol.* 114 (1990) 245–256.

48. S. Geibel, D. Zimmermann, G. Zifarelli, A. Becker, J.B. Koenderink, Y.K. Hu, J.H. Kaplan, T. Friedrich, E. Bamberg, Conformational dynamics of Na^+/K^+- and H^+/K^+-ATPase probed by voltage clamp fluorometry, *Ann. N. Y. Acad. Sci.* 986 (2003) 31–38.

49. K.L. Dürr, N.N. Tavraz, T. Friedrich, Control of gastric H,K-ATPase activity by cations, voltage and intracellular pH analyzed by voltage clamp fluorometry in *Xenopus* oocytes, *PLoS One* 7 (2012) e33645.

50. B. Wallmark, C. Briving, J. Fryklund, K. Munson, R. Jackson, J. Mendlein, E. Rabon, G. Sachs, Inhibition of gastric H^+,K^+-ATPase and acid secretion by SCH 28080, a substituted pyridyl(1,2a)imidazole, *J. Biol. Chem.* 262 (1987) 2077–2084.

51. K.L. Dürr, K. Abe, N.N. Tavraz, T. Friedrich, E_2P state stabilization by the N-terminal tail of the H,K-ATPase beta-subunit is critical for efficient proton pumping under in vivo conditions, *J. Biol. Chem.* 284 (2009) 20147–20154.

52. K.L. Dürr, I. Seuffert, T. Friedrich, Deceleration of the E_1P-E_2P transition and ion transport by mutation of potentially salt bridge-forming residues Lys-791 and Glu-820 in gastric H^+/K^+-ATPase, *J. Biol. Chem.* 285 (2010) 39366–39379.

53. K.L. Dürr, N.N. Tavraz, R.E. Dempski, E. Bamberg, T. Friedrich, Functional significance of E2 state stabilization by specific alpha/beta-subunit interactions of Na,K- and H,K-ATPase, *J. Biol. Chem.* 284 (2009) 3842–3854.

54. S. Meier, N.N. Tavraz, K.L. Dürr, T. Friedrich, Hyperpolarization-activated inward leakage currents caused by deletion or mutation of carboxy-terminal tyrosines of the Na^+/K^+-ATPase α subunit, *J. Gen. Physiol.* 135 (2010) 115–134.

55. G. Nagel, D. Ollig, M. Fuhrmann, S. Kateriya, A.M. Musti, E. Bamberg, P. Hegemann, Channelrhodopsin-1: A light-gated proton channel in green algae, *Science* 296 (2002) 2395–2398.

56. C. Polvani, G. Sachs, R. Blostein, Sodium ions as substitutes for protons in the gastric H,K-ATPase, *J. Biol. Chem.* 264 (1989) 17854–17859.

57. R.E. Dempski, J. Lustig, T. Friedrich, E. Bamberg, Structural arrangement and conformational dynamics of the gamma subunit of the Na^+/K^+-ATPase, *Biochemistry* 47 (2008) 257–266.

58. R.E. Dempski, K. Hartung, T. Friedrich, E. Bamberg, Fluorometric measurements of intermolecular distances between the alpha- and beta-subunits of the Na^+/K^+-ATPase, *J. Biol. Chem.* 281 (2006) 36338–36346.

59. T. Förster, Zwischenmolekulare Energiewanderung und Fluoreszenz, *Ann. Phys.* 6 (1948) 55–75.

60. G.W. Gordon, G. Berry, X.H. Liang, B. Levine, B. Herman, Quantitative fluorescence resonance energy transfer measurements using fluorescence microscopy, *Biophys. J.* 74 (1998) 2702–2713.

61. K.S. Glauner, L.M. Mannuzzu, C.S. Gandhi, E.Y. Isacoff, Spectroscopic mapping of voltage sensor movement in the *Shaker* potassium channel, *Nature* 402 (1999) 813–817.

62. A. Cha, G.E. Snyder, P.R. Selvin, F. Bezanilla, Atomic scale movement of the voltage-sensing region in a potassium channel measured via spectroscopy, *Nature* 402 (1999) 809–813.

5

PATCH CLAMP ANALYSIS OF TRANSPORTERS VIA PRE-STEADY-STATE KINETIC METHODS

CHRISTOF GREWER

Department of Chemistry, Binghamton University, Binghamton, NY, USA

5.1 INTRODUCTION

Transporters are transmembrane proteins that catalyze the passive or active transport of inorganic ions and/or organic solutes across biological membranes (reviewed, e.g., in Ref. 1). Traditionally, transmembrane solute flux has been experimentally assessed using radioactive, isotope-labeled solutes as the transported species (see Chapter 13). Such radiotracer flux methods have the advantage of exquisite sensitivity, being able to detect solute transport catalyzed by transporters with very low turnover rate. They are also experimentally straightforward, as long as the transported substrate is available in a radiolabeled form. However, a disadvantage of radiotracer flux methods is the low time resolution. Typically, uptake and analysis takes seconds to minutes to perform (see e.g., [2]). Thus, the transport measurements are performed at steady state, because many transporters have turnovers in the subsecond time scale. Many elementary reaction steps in the reaction cycle of membrane transport proteins occur on a millisecond timescale, even much faster than the steady-state turnover. Therefore, information on these elementary reaction steps cannot easily be obtained using radiotracer flux methods.

Pumps, Channels, and Transporters: Methods of Functional Analysis, First Edition. Edited by Ronald J. Clarke and Mohammed A. A. Khalid.
© 2015 John Wiley & Sons, Inc. Published 2015 by John Wiley & Sons, Inc.

To directly analyze elementary transporter reaction steps before the steady state is reached, pre-steady-state methods have been developed that have pushed the time resolution down into the millisecond to submillisecond range [3, 4]. Here, it is critical to perturb an initial steady state by applying a stimulus to the system. Often this is done by applying concentration jumps [5, 6], but step-changes in the trans-membrane potential have also proven useful [4, 7], in particular for electrogenic transporters, or electroneutral transporters with electrogenic partial reactions. Once the initial steady state is perturbed, relaxation occurs to a new steady state, typically associated with a redistribution of the relative occupancy of transporter states and/or conformational change of the transport protein [8–10]. This relaxation can be followed in time by using a variety of analytical methods. Because fluorescence detection (see e.g., [7, 11]) is described in Chapters 4, 7, 10, and 11, this chapter focuses on the detection of relaxation processes using electrical current as the readout. Patch clamping of transporter-expressing cells is often used to determine time-resolved transport current [12, 13]. Obviously, this is only possible for proteins that transport charged species, or that have charged domains undergoing rearrangement in the transmembrane electric field during the relaxation process (e.g., voltage-dependent potassium channels).

The analysis of the pre-steady-state kinetics of membrane transport proteins has generated a wealth of information on turnover rates of transporters, both active and passive; detailed reaction mechanism; elementary reaction steps within the transport cycle, such as substrate/ion binding/dissociation process and conformation changes; charge distribution within the transporter; and structure–function relationships, by implementing pre-steady-state methods in the presence of mutant or modified transporters [5, 12, 14–17].

Pre-steady-state kinetic methods have been applied to a large number of transporters, including neurotransmitter transporters [5, 12], sugar transporters [4, 18], phosphate transporters [19], and amino acid transporters [20, 21]. This chapter highlights key insights from two case studies: the electrogenic, Na^+-driven secondary-active transporter for the neurotransmitter glutamate (excitatory amino acid transporter, EAAT [22]) and the electroneutral Na^+-dependent amino acid exchanger (alanine serine cysteine transporter, ASCT2 [23]).

5.2   PATCH CLAMP ANALYSIS OF SECONDARY-ACTIVE TRANSPORTER FUNCTION

5.2.1   Patch Clamp Methods

The patch clamp method was initially developed by Neher and Sakmann for studying ion channels embedded in the membrane of living cells [24] (see Chapter 3 for more details). Despite several order of magnitude lower turnover rates of transporters, k_T, as compared to ion channels, the method has found widespread adoption in transporter research, in particular for transporters and exchangers that express in large numbers on the cell surface, and which are not limited by low turnover rates [25, 26]. While current clamping can be used to study the influence of the function of expressed electrogenic transporters on the resting potential of the membrane, the voltage clamp method (a practical treatment can be found in Ref. 27) allows the more direct determination of electrogenic transport current and is, thus, mostly used for functional studies. In this method, the membrane potential is held constant (clamped), and the current generated by electrogenic transport and/or capacitive charge movement is recorded by using a highly sensitive patch clamp amplifier. Currents in the range of pA can be relatively easily measured using voltage clamping of mammalian cells.

Under voltage clamp conditions, transport current, I_T, is directly proportional to the number of active transporters in the cell membrane, N_T, as well as the turnover rate of the transport cycle, k_T [1, 28]:

$$I_T = N_T z_T e_o k_T \tag{5.1}$$

In Equation 5.1, z_T is the number of charges transferred in one transport cycle, and e_o is the elementary charge. Equation 5.1 directly demonstrates that transport current is maximized by large transporter numbers under observation. For practical purposes, 10^5–10^6 transporters probably represent the minimum required to detect significant currents that are measurable with the currently used patch clamp equipment.

Transporter activity can be studied by patch clamp using several different configurations [28]. One of the most common is the whole-cell recording configuration, in which the whole cell surface is voltage-clamped and contributing to the current signal (see Fig. 5.1a). In addition, recordings can be achieved from membrane patches, which have been excised from the cell surface (see Fig. 5.1c and d). These

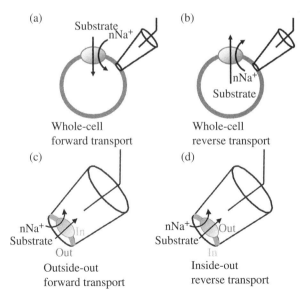

FIGURE 5.1   Illustrations of several whole-cell recording (a and b) and outside-out (c) / inside-out patch configurations (d) to measure forward and reverse transport activity.

patches represent only a fraction of the surface area of the whole cell. The inside-out and outside-out configurations are discussed later. The cell-attached mode, which is powerful for single-channel recording [24] (see Chapter 3), is not practical for studies involving transporters, and is, thus, not discussed here. The common recording configurations for transporter applications are illustrated in Figure 5.1.

5.2.2 Whole-Cell Recording

In whole-cell current recording, all of the transporters expressed in the cell membrane of a mammalian cell are contributing to the current signal [29, 30]. Its exquisite sensitivity makes the whole-cell configuration a powerful tool for biophysical studies on membrane transporters. To attain the whole-cell configuration, a glass micropipette with a diameter of 1–2 µm is first attached to the cell membrane in a cell-attached mode. Next, the membrane under the patch is ruptured, either by suction or a voltage pulse, allowing electrical access of the whole interior of the cell through the recording pipette. In addition to electrical access, the interior of the cell equilibrates with the solution in the recording pipette on a minute time scale [29]. Therefore, substrates, ions, or second messengers can be perfused into the cytosol allowing tight control over the properties of the solutions on both sides of the membrane [15]. This is an advantage of the whole-cell recording mode that cannot easily be achieved with other methods. In two-electrode voltage-clamped *Xenopus* oocytes, for example, control of the intracellular solution, while achievable [31], is much more difficult. The inclusion of transported substrate in the intracellular solution, for example, allows determination of properties under reverse transport (see Fig. 5.1b), trans-inhibition studies, as well as reversal potential measurements to determine stoichiometry of substrate/ion coupling of transport. Finally, inhibitors can be perfused through the pipette solution, to test for sidedness of the inhibition process.

Critical aspects for the success of a whole-cell recording experiment are the turnover number of the transporter, as well as the number of transporters in the cell membrane, according to Equation 5.1. Electrogenic transporters with turnover rates greater than $10 \, s^{-1}$ with reasonable expression densities should generate detectable currents in whole-cell recording.

5.2.3 Recording from Excised Patches

Inside-out membrane patches: The inside-out patch clamp technique is based on the exposure of the cytoplasmic-facing side of the membrane to the bath solution (see Fig. 5.1d). Solution changes can then be performed by using regular flow-based perfusion methods. This experimental setup allows the determination of transport kinetics in the reverse transport direction [32, 33], as well as the measurement of affinities, if there are intracellular-binding sites for substrates and/or ions.

Typically, the surface area of inside-out patches is in the range of 2–3 µm². Considering the closest packing density of highly expressing transporters, only up to 20,000–30,000 transporters will be expressed in the excised patch area. Therefore, transport currents in inside-out patches are usually small, close to the limit of detection. This problem can be somewhat overcome by averaging over many recording trials, or by increasing the size of the aperture of the recording pipette, in what has been named the giant patch method [34]. This method has been particularly successful when excising patches from *Xenopus* oocytes, but can also be used with mammalian cells. Inside-out patches have been used for studying Ca^{2+} activation and activation by intracellular substrates of secondary-active transporters [35].

Outside-out membrane patches: Starting with the whole-cell configuration, the pipette is slowly pulled away from the cell. When the contact with the cell is eventually lost, the resulting membrane fragment still attached to the tip of the pipette typically reseals, resulting in an outside-out patch [24], in which the outside of the cell membrane is exposed to the bath solution (see Fig. 5.1c). Compared to whole-cell recording, the outside-out patch method has the advantage of a smaller surface area, which allows more rapid solution exchange. The resulting higher time resolution has been exploited for studying kinetics of secondary-active transporters [5, 36]. However, the disadvantage is the smaller number of transporters in the patch. Therefore, the signal-to-noise ratio may be a problem when using outside-out patches and averaging over many signals may be required.

5.3 PERTURBATION METHODS

Substrate transport by passive transporters (with the exception of ion channels) as well as active transporters is generally described by an alternating access scheme [37], as illustrated for a hypothetical Na^+-coupled substrate transporter in Figure 5.2. This scheme is based on two major conformational states, in which the substrate-binding site faces either the extracellular side or the intracellular side of the membrane. Transitions between these major configurations often require additional substrate/ion binding/dissociation steps (see Fig. 5.2) and/or passage through intermediate conformations. Thus, the whole transport cycle can be composed of many elementary

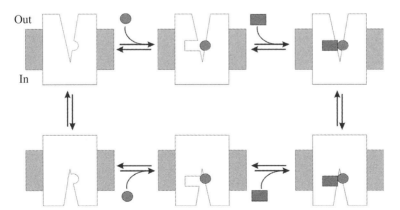

FIGURE 5.2 Simplified alternating access transport mechanism of a hypothetical Na^+ cotransporter (sphere, Na^+; rectangle, organic substrate).

reactions steps. When transporter function is assayed at steady state, the differentiation of these elementary steps is not straightforward. A solution to these difficulties is to perturb a pre-existing steady state, in which the population of transporter states does not vary with time, resulting in the relaxation to a new steady state (see e.g., [4, 10, 20, 38]). Kinetic data associated with this relaxation is typically information rich, allowing isolation of elementary reaction steps along the time axis. The interpretation of relaxation kinetics of transporters is described in Section 5.4.

How can perturbations of a steady state be achieved? The major methods that are in use at present are (i) rapid steps in the concentration of substrates/ions and (ii) steps in the transmembrane potential (useful for electrogenic partial reactions). Temperature jumps can theoretically also be used, although this approach has been less common in the transporter field [39].

5.3.1 Concentration Jumps

Rapid solution exchange and flow methods: Combined with patch clamping of mammalian cells, concentration jumps are a powerful tool to study transporter kinetics [3, 5]. Typically, solutions are applied to the cells (in whole-cell or excised patch configuration) using rapid solution exchange (flow-based method). A typical flow setup used in our laboratory is shown in Figure 5.3a. Transporter-expressing cells are selected using green fluorescent protein (GFP) coexpression as a marker (see Fig. 5.3b), and then suspended at the tip of the recording electrode and positioned in front of the flow tube (see Fig. 5.1a). This approach is somewhat analogous to the stopped-flow method (see Chapter 7), except that one reactant (the transporter in the cell membrane) is stationary. Mammalian cells, with a typical diameter in the $10 \mu m$ range, allow fast solution exchange times in the 20–100 ms range. Time resolution can be further improved by exchanging solutions at the surface of outside-out patches, for which time resolutions in the submillisecond time scale have been achieved, by

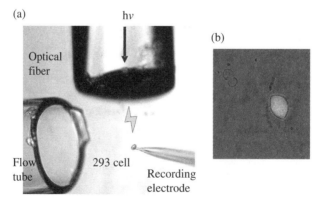

FIGURE 5.3 (a) Typical combined flow/laser photolysis setup. A HEK293 cell is shown suspended at the tip of the whole-cell recording electrode. (b) Selection of cells expressing transporters by coexpression of green fluorescent protein (GFP).

using piezo-based solution switching devices [5, 14]. The small size of outside-out patches $(1-5\,\mu m^2)$ presents a limitation for achieving a large number of transporters under observation. Therefore, signal-to-noise ratio is a significant problem, and averaging has to be employed to obtain currents that can be kinetically evaluated.

Using outside-out patches, substrates can not only be applied rapidly to the transporter, but they can also be removed after an initial steady state is established [40, 41]. Pre-steady-state kinetic analysis of current responses after substrate removal can reveal a wealth of information on transporter reaction steps that are not easily accessible with other methods. In addition, kinetic parameters of substrate dissociation may be obtained, if the dissociation is rate limiting for the current deactivation process [42, 43]. Such rate limitation can be observed for transporter-binding ligands or inhibitors with high affinity. Despite the advantages of this method, a word of caution is necessary. It is generally easier to achieve high time resolution upon application of a substrate, by using supersaturating concentrations [44], than upon its removal. If a supersaturating concentration is applied, for example, at 1000-fold the K_m of the substrate, a saturating concentration of 10-fold K_m is achieved when the solution exchange is only 1% complete. Thus, the time resolution of the method is significantly enhanced. On the other hand, if a substrate at a concentration of 1000-fold K_m is rapidly removed, it will only reach subsaturating concentration when the solution exchange is 99.9% complete. Therefore, kinetic information from substrate removal experiments is more difficult to interpret and may, more often than not, reflect limitations of the solution exchange method, rather than the intrinsic kinetics of the transporter system under observation.

Rapid concentration jumps through photolysis of caged compounds: Rapid solution exchange may not be fast enough to detect rapid reactions of membrane transporters. Such limitations in time resolution can be overcome by generating concentration jumps through photolysis of caged substrates, as previously demonstrated for ion channels [45, 46]. In this method, the transported substrate is covalently protected (caged) with a group that renders it biologically inactive [47] (see Fig. 5.4a). Often the caging group is attached to a functionally relevant moiety that is critical for binding. Carboxylic acid groups have been, for example, extensively exploited as targets for caging [48–50]. The α-carboxy-*o*-nitrobenzyl caging group (CNB) has proven very useful for protecting carboxylic acids [51] (see Fig. 5.4a). A typical experimental setup for delivering the laser light to the cell that is used in our laboratory is shown in Figure 5.3a, in which the laser light is coupled into an optical fiber, which is positioned in close proximity to the cell under observation.

The advantage of the method is that the transporters in the cell membrane can be equilibrated with the caged substrate by using slow flow methods. Once equilibration is complete, the free substrate is released by application of a brief pulse of laser light resulting in a rapid concentration jump [3, 15, 26]. The time constant of the liberation of the compound is only limited by the intrinsic kinetics of the photochemical reaction, which for many caged compounds is in the sub-millisecond range [20, 52]. Thus, problems with slow mixing because of unstirred layers on the cell surface are avoided, allowing time resolution in the sub-microsecond range for some caged compounds. A typical time-resolved transport current response of an electrogenic

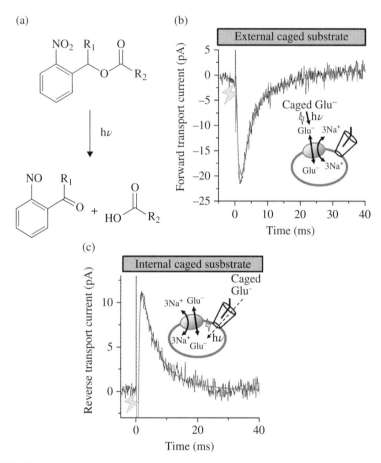

FIGURE 5.4 (a) Mechanism of photorelease of carboxylic acids through photolysis of the *o*-nitrobenzyl-protected esters. The substituent R_1 can affect the efficiency and rate of photolysis. (b) Typical forward transport current trace in response to photolysis of externally applied CNB-caged glutamate at time 0 from an EAAC1-expressing cell (see inset for recording configuration). (c) Typical reverse transport current trace in response to photolysis of CNB-caged glutamate introduced to the cytosol through the recording pipette at time 0 from an EAAC1-expressing cell (see inset for recording configuration). Experiments were performed in Na$^+$/glutamate exchange mode (no steady-state turnover).

transporter to photolytic substrate application is shown in Figure 5.4b (glutamate transporter), demonstrating several phases of current rise and decay that can be distinguished at high time resolution [53].

A second advantage of caged compounds is that they allow rapid application of the transported substrate to the intracellular side of the membrane. In the whole-cell recording configuration, for example, caged compound can be introduced into the cytoplasm by diffusion through the recording pipette. Once the caged compound in the cytoplasm has equilibrated with the pipette solution, which typically takes on the order of 2–5 min, depending on the diffusion coefficient of the compound and the access resistance of the pipette, release of the active substrate is initiated with a pulse of laser light [16]. The principle of the method is illustrated in Figure 5.4c, showing also a representative trace of reverse transport current recorded using this approach from the glutamate transporter, EAAC1. This approach combines the advantages of the whole-cell configuration (maximized signal by recording from all transporters in the membrane) with high time resolution, which cannot be achieved by using internal perfusion of the pipette [54].

5.3.2   Voltage Jumps

Another way to perturb an existing steady-state distribution of transporter states is to apply changes in the membrane potential. Typically, the voltage is changed in a step-wise fashion (voltage jump). Once the new voltage is established, the rate constants associated with voltage-dependent processes in the transport cycle are altered, according to the following equations [1]:

$$k_f(V_m) = k_f(0)\exp\left(\frac{z_Q F V_m}{2RT}\right)$$ (5.2)

$$k_b(V_m) = k_b(0)\exp\left(\frac{-z_Q F V_m}{2RT}\right)$$ (5.3)

In these equations, V_m is the membrane potential; k_f and k_b are the rate constants associated with the forward and backward reaction of the reversible, electrogenic process; z_Q is the valence of the reaction; and F, R, and T have their usual meaning. $k_f(0)$ and $k_b(0)$ are the rate constants at a membrane potential of 0 mV. If two states of the transporter reversibly convert into each other (see Fig. 5.5a and b), the location of the equilibrium between the two states depends on V_m and can, thus, be altered by step-changes in the voltage [1]. A representative current response of a glutamate transporter to a voltage jump is shown in Figure 5.5c.

   Similar to the concentration jump approaches, time resolution can be an issue when applying voltage jumps. Limitations are posed by the time necessary to charge the capacitance of the membrane, C_m, across the access or series resistance, R_a, with the time constant given as:

$$\tau = C_m R_a$$ (5.4)

For typical mammalian cells, the membrane capacitance is in the 20 pF range. With an access resistance of 5–10 MΩ, the resulting time constant is in the 100–200 μs range [27]. Therefore, problems with time resolution should only be observed with very fast transporter systems. If time resolution is a problem, techniques such as series resistance compensation can be employed in the whole-cell recording configuration, based on an estimation of the access resistance and C_m. Series resistance compensation is a feedback method that can result in rapid current oscillations that can render the recording unusable. Therefore, this method has to be applied with caution.

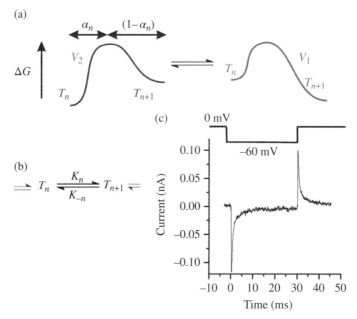

FIGURE 5.5   (a) Principle of the perturbation of voltage-dependent transition between two transporter states by the applied membrane potential, V. (b) Voltage-dependent transition between two states T_n and T_{n+1} in the transport cycle. (c) Relaxation of transport current in response to a square voltage jump from 0 to −60 mV (voltage protocol shown at top of trace) for the glutamate transporter EAAC1 in exchange mode. Adapted from Ref. 28 with permission from Annual Reviews.

Care must also be taken to differentiate specific signals from unspecific signals. This can often be achieved by applying subtraction protocols in the absence and presence of specific inhibitors that suppress the specific charge movement [21, 55].

5.4   EVALUATION AND INTERPRETATION OF PRE-STEADY-STATE KINETIC DATA

At steady state, the kinetic equations describing cyclic transport mechanisms can be usually simplified resulting in a Michaelis–Menten-type description of the kinetics. This yields an apparent K_m, describing apparent substrate/ion affinity, and a v_{max}, the maximum transport rate. If transport current is measured, then v_{max} can be substituted by I_{max}, the maximum current at saturating substrate concentration. In contrast, the interpretation of pre-steady-state kinetic data is not trivial, as it depends on the exact transport mechanism and the relative rate constants of the individual reaction steps in the transport cycle. Typically, evaluation of the data involves comparing the experimental results with predictions from kinetic transport mechanisms. Parameters that can be compared include relaxation rate constants and amplitudes of kinetic components of the relaxation process [15, 56–59].

5.4.1 Integrating Rate Equations that Describe Mechanistic Transport Models

Predictions of the behavior of cyclic transport models can be made by integrating the rate equations associated with the respective kinetic model (e.g., the six-state model shown in Fig. 5.2). Integration yields information on the expected time dependence of the transport current before the steady state is reached. Depending on the complexity of the transport mechanism, rate equations can be integrated using either analytical or numerical methods.

The analytical methods have the advantage of yielding direct mathematical relationships between observed rate constants (experimental parameter) and the model input parameters, such as the transmembrane potential, substrate/ion concentration, or rate constants of reaction steps associated with the kinetic model. Because substrate concentration, for example, can be varied, this approach allows the direct comparison and/or fit of predicted and measured kinetic quantities. The disadvantage of analytical integration is that it is typically only practical for reaction schemes with no more than two independent, reversible reaction steps before the resulting equations become too large to practically handle. Because typical cyclic transport mechanisms based on the alternating access model for a symporter, for example, consist of at least six reversible reaction steps, evaluation by analytical integration has to rely on simplification of the rate equations, for example, by using the pre-equilibrium assumption [3, 26], or by combining consecutive reactions that are not rate limiting [59]. Obviously, such simplifications are not always justified and have to be verified experimentally [60].

Integration can also be performed numerically. This procedure yields a numerical value of the population of each state in the transport cycle as a function of time, which can be converted into the time dependence of the current, if the valences of each elementary reaction steps are known. However, no direct information on observed rate constants is obtained. Data can be analyzed by directly fitting the calculated, time-dependent currents to the experimental results under varying conditions (i.e., transmembrane potentials, concentrations, temperature etc.).

A difficulty that can be encountered using numerical integration is that mechanistic problems can be easily overparameterized, meaning that the number of fit parameters is too large for a unique solution. In addition, kinetic parameters, which do not effectively alter the outcome of the numerical integration, are often included and varied in computations. This results in kinetic parameters that are ill defined and are not constrained by experimental data.

5.4.2 Assigning Kinetic Components to Elementary Processes in the Transport Cycle

Under conditions of steady-state forward transport, the transporter visits many states (Fig. 5.6, left panel) with the predominantly populated state being determined by the rate-limiting step in the transport cycle. While pre-steady-state kinetic analysis is useful in dissecting the reaction steps in forward transport conditions, interpretation

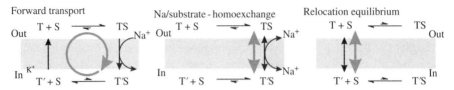

FIGURE 5.6 Restricting the transporter to visit only particular states in the transport cycle. The bold, gray arrows indicate the major transition. Reproduced from Ref. 16 with permission from the National Academy of Sciences of the United States of America.

of these pre-steady-state data based on numerical or analytical integration of rate equations is not always straightforward. The process can be aided by using experimental conditions that restrict the transporter to visit only certain states along the transport cycle, thus effectively reducing the number of kinetic constants to be determined. The principle is illustrated in Figure 5.6, with the individual cases discussed in detail later.

Substrate-binding reactions, for example, can be isolated by using low substrate concentrations resulting in apparent binding rates lower than the turnover rate of the transporter. Substrate binding becomes, thus, rate limiting [3]. The following equation is typically used to analyze the observed rate constant for binding, k_{obs}:

$$k_{obs} = k_{on}\left[\text{Substrate}\right] + k_{off} \tag{5.5}$$

Here, k_{on} and k_{off} are the intrinsic rate constants for substrate binding and dissociation, respectively. Both constants can be determined from the slope and intercept of the linear k_{obs} versus [Substrate] relationship. In the case of electrogenic binding, these rate constants are expected to become dependent on the transmembrane potential.

It should be noted that this approach is only applicable when the substrate affinity is sufficiently high to still generate a signal at low substrate concentrations that ensure rate-limiting binding. For this reason, it is typically difficult to determine rate constants of binding for cotransported ions, because their apparent affinity is often low, in the mM to 100 mM range.

A second possibility is to set ionic conditions and/or the transmembrane potential such that the transporter is restricted to certain conformational states. Examples for this strategy are the Na$^+$/glucose transporters, SGLTs [4], and Na$^+$/phosphate transporter—NaPi II [56]. Here, the pre-steady-state voltage jump experiment is carried out in the absence of the transported substrate. Thus, the transporter cannot complete the whole transport cycle, but is restricted to adopt states in which the substrate/cation sites are either not occupied at all [4, 56], or occupied by Na$^+$ ion(s) only (illustrated in Fig. 5.6, right panel). Analysis of the relaxation of the current then provides information on the rate constants and the valence of transitions of the empty transporter, as well as Na$^+$-dependent processes [8, 10]. Varying the external Na$^+$ concentration allows the differentiation between reactions of the empty or Na$^+$-bound transporter [9].

Alternatively, Na^+-driven transporters can be restricted to states in which the Na^+/substrate-binding sites are fully occupied [15, 21, 61] (Na^+/substrate exchange configuration). Here, saturating concentrations of Na^+ and substrate are applied to both sides of the membrane preventing net dissociation of both species. As noted earlier, these conditions prevent the transporter from completing a full transport cycle, but force it to shuttle back and forth between inward- and outward-facing conformations. Establishment of such an exchange equilibrium, which can be perturbed by voltage jumps [61], allows the determination of rate constants and valences of the actual substrate/ion translocation process. This strategy, which has been successfully applied to glutamate transporters [61], is illustrated in Figure 5.6, middle panel. It should be noted that perturbation analysis under exchange conditions requires that the translocation process is dependent on the membrane potential, and, thus, electrogenic. This is probably not the case for every transporter of interest. For example, transporters of the SGLT family were proposed to have electroneutral translocation steps [62].

5.5 MECHANISTIC INSIGHT INTO TRANSPORTER FUNCTION

The ability to dissect individual reaction steps in the transport cycle by using pre-steady-state kinetic methods presents a powerful tool to obtain mechanistic information on transporter function. Current mechanistic models of transporter function are based on the alternating access hypothesis [37, 63] (see Fig. 5.2). Here, the binding sites for transported substrate(s) and/or ions are accessible to either the extracellular side (outward-facing) or the intracellular side of the membrane (inward-facing), but not to both sides at the same time. Substrate movement across the low dielectric environment of the membrane is brought about by transitions between the outward- and inward-facing configurations. Such transitions can result in redistribution of charge within the membrane, thus giving rise to capacitive currents upon perturbation [15]. Simple alternating access models include one step for the outward-facing to inward-facing transition, but often additional steps have to be considered, such as external gate closure, occlusion of the substrate and/or ions, as well as internal gate opening [64]. Electrical analysis of such events has the potential to reveal such multistep processes giving detailed insight into the dynamics of the translocation process. In addition, this analysis can be coupled with temperature-dependent experiments providing information about barrier heights associated with individual transitions [53].

5.5.1 Sequential Binding Mechanism

Translocation of the substrate across the membrane is preceded by the binding of the substrate, as well as the ions that potentially drive the transport. An important mechanistic parameter is the sequence of the substrate- and ion-binding steps. First, the possibility has to be considered that the sequence of binding steps is random [65]. Thus, the transporter would not discriminate in substrate-binding affinity, no matter whether ions are already bound or not. Second, ion/substrate binding with a defined sequence is possible [65]. In this case, substrate could bind before or after the ion(s), or a combination of the two possibilities could be found [15, 66]. Third, it is possible that the sequence of binding from the extracellular side is either the same [67] or different than the sequence of dissociation to the intracellular side [16]. This concept is analogous to the first-in-first-out versus mirror mechanisms that have been proposed for soluble enzymes in classical enzymology studies [68]. Electrophysiological analysis is useful in discriminating between such binding mechanisms. For example, the ability to isolate the substrate-binding step(s) allows the determination of the exact influence of the concentration of the cotransported ions on binding only, without having to consider the effect on overall transport.

5.5.2 Electrostatics

Contributions of stationary and/or moveable charges on the transport protein to the energetics of the transmembrane movement of the substrate and co/counter-transported ions are important to understand. While interpretation of data is not straightforward, apparent valences for partial reactions are obtained by using Boltzmann analysis of pre-steady-state charge movement [4, 61, 62, 69], as well as determination of the voltage dependence at steady state [9, 13], yielding information of the valence of the rate-limiting transition. The recent availability of representative structures of members of many transporter families in several states [64, 70–72] along the transport cycle allows the structural interpretation of the experimental data. Calculation of the theoretical valence of structural transitions through computation of the electrostatic energy as a function of applied membrane potential is a powerful approach [69, 73]. Here, the Poisson–Boltzmann equation is solved numerically for different structural configurations of the transporter embedded in an implicit bilayer, yielding valences for the transition between these configurations [21, 69, 74]. The computational setup for a representative transporter is shown in Figure 5.7a. Such calculations can predict valences for transitions, which can then be compared to experimental data, providing detailed information as to which step(s) in the transport cycle are electrogenic or electroneutral [21, 69].

5.5.3 Structure–Function Analysis

A large number of studies have investigated the function of mutated secondary-active transporters (see e.g., [75] for a review on lac permease). Analysis of the effects of site-specific mutations allows the determination of the direct impact of specific amino acid side chains in processes such as ion/substrate binding, as well as conformational changes. Traditionally, functional analysis of mutant transporters has been performed at steady-state. However, the determination of the exact reaction step(s) that is/are influenced by the mutation is not straightforward under steady-state conditions, because disruption of any step will disrupt the whole transport cycle, and, thus, function of the transporter.

When performing pre-steady-state kinetic experiments, individual reaction steps in the transport cycle can be isolated, as described in the previous section. This ability

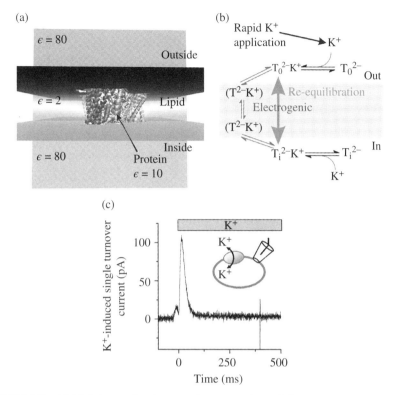

FIGURE 5.7 (a) Model setup for electrostatic computations of charge movement associated with transitions in the glutamate transport cycle. (b) Electrostatic computations predict negative charge of the empty transporter binding sites, which overcompensates for the positive charge of the transported K^+ ion. Application of K^+ to the extracellular side in the K^+ exchange mode is predicted to result in outward current due to inward movement of negative charge of the binding sites. (c) In agreement with the theoretical predictions, outward transient current is observed in the single-turnover experiment. Adapted from Ref. 69 with permission from the American Society of Biochemistry and Molecular Biology.

allows the identification of the specific step(s) that is/are affected by the mutation, because steps that are unaffected can still be observed [17, 61, 69, 76, 77]. For example, a mutation may interfere with the conformational changes associated with substrate translocation and/or binding, but does not affect ion binding. In this case, charge movement associated with, for example, the Na^+ binding steps can still be observed, while steady-state transport is inhibited. In another scenario, mutations may interfere with conformational changes of the *apo*-form of the transporter, but not the substrate-bound form. In this case, partial reactions associated with the substrate-dependent half-cycle can still be observed, while steady-state transport is, again, inhibited. Examples of both of these scenarios have been demonstrated in the literature [78, 79].

5.6 CASE STUDIES

5.6.1 Glutamate Transporter Mechanism

Plasma membrane glutamate transporters are members of the solute carrier 1 (SLC1) family [80]. They are electrogenic, Na^+-driven, secondary-active transporters, that use the free energy stored in the transmembrane concentration gradient for Na^+, as well as the transmembrane potential, to take up glutamate into the cell against its own concentration gradient [81]. The Na^+ concentration gradient is established and maintained by the sodium pump through hydrolysis of the primary energy source ATP. For mammals, five transporter subtypes have been cloned and named the excitatory amino acid transporters (EAATs) 1–5 [22, 81–83]. While the structures of the mammalian members of the family are not known, crystal structures of a highly homologous archeal aspartate transporter, GltPh, from *Pyrococcus horikoshii* were established [64, 70, 84]. The crystal structures indicate that the transporter is assembled as a homotrimer. The subunits of the trimer are thought to operate independently from each other [85]. Glutamate transporters also catalyze a chloride conductance, which is gated by Na^+ and glutamate binding to the transporter [15, 86].

Our understanding of the mechanism of glutamate transport has been facilitated by the crystal structures and biochemical analysis of function, but also by kinetic studies. Pre-steady-state analysis, in particular, has generated a wealth of information on individual reaction steps in the transport cycle, as well as the sequence of events during transport [15, 16]. From the large number of published results, just two relevant examples are highlighted here.

Electrogenic Na^+ binding: Voltage jump analysis has shown that reaction steps associated with the binding of Na^+ to the glutamate-free transporter are electrogenic. Step-changes in the transmembrane potential in the absence of glutamate but the presence of Na^+ result in sizable transient currents that decay within less than 10 ms [10]. No steady-state component was observed because the transport cycle cannot be completed without bound glutamate. The transient currents were found to be capacitive, because stepping the voltage back to the original value resulted in charge movement of identical magnitude, but opposite sign, as would be expected for charging and discharging of a capacitor [10, 15]. The charge movement was dependent on the membrane potential and the extracellular Na^+ concentration. A likely explanation for the behavior is that voltage jumps to more negative potentials facilitate Na^+ binding to its external binding site, thus generating transient inward current that subsides once the binding equilibrium has been established. Upon removal of the negative voltage stimulus, Na^+ dissociates from its binding site creating outward transient current. A potential issue with this interpretation is that the time constant of the transient current is in the millisecond range, which is too slow for a diffusion-controlled Na^+ binding process at $[Na^+]$ in the 100 mM range. Therefore, it is rather likely that the time constant of the process is determined by a conformational change that is associated with the Na^+ binding process.

K^+-induced relocation of the glutamate-free carrier: Pre-steady-state kinetic analysis has also been used to obtain information about the K^+-induced relocation of the

glutamate transporter. This reaction restores the glutamate-binding site to face the extracellular side, after intracellular glutamate dissociation is complete. It requires intracellular K^+ association [87]. After relocation, K^+ dissociates to the extracellular side and a new transport cycle can start. Based on results from electrophysiological measurements at steady state, it was hypothesized that K^+-induced relocation is associated with outward movement of negative charge [3, 11]. To test this hypothesis, K^+ was applied to the transporter in the presence of saturating concentrations of intracellular K^+ using rapid solution exchange (principle of the method shown in Fig. 5.7b). After extracellular K^+ binding, the K^+-exchange equilibrium reequilibrates resulting in charge movement, if the process is electrogenic (see Fig. 5.7b). The experimental results support this hypothesis, showing transient outward current upon K^+ application [69] (see Fig. 5.7c). This result directly shows that negative charge of the transporter-binding sites overcompensates the positive charge of the bound K^+, leading to an overall negatively charged transporter-K^+ complex [69] (see Fig. 5.7b).

5.6.2 Electrogenic Charge Movements Associated with the Electroneutral Amino Acid Exchanger ASCT2

Another member of the SLC1 family with high sequence similarity with the glutamate transporters is the alanine serine cysteine transporter, ASCT2 [88, 89]. ASCT2, together with ASCT1 [90], constitutes transport activity of the classically defined system ASC. ASCT2 is specific for small, neutral, hydrophilic amino acids. In contrast to the EAATs, ASCT2 does not catalyze net transport of substrate, but rather acts as an exchanger [21, 89, 91]. Thus, net uptake of amino acid requires efflux of a different amino acid. Amino acid exchange by ASCT2 requires the presence of Na^+, but may not be driven by the transmembrane $[Na^+]$ gradient [21, 89].

Based on the 1:1 stoichiometry of internal/external amino acid exchange, it was proposed that the exchange process is electroneutral [92]. However, flux measurements at steady state indicated a slight voltage dependence of transport [93].

Pre-steady-state kinetic analysis in combination with electrophysiology is particularly useful for studying obligate exchange, because, by nature of the exchange process, no steady-state current is associated with amino acid flux across the membrane. Consistent with this expectation, ASCT2 did not generate steady-state current when alanine, a transported substrate, was applied to the extracellular side of the membrane [91]. However, when alanine was rapidly released from a caged precursor, methoxyindoline (MNI) caged alanine, a transient inward current was observed, which is consistent with inward movement of positive charge [20, 21]. Because alanine is neutral, the charge movement is, most likely, caused by the inward movement of the bound Na^+ ion(s). These results provided the first direct evidence for electrogenicity of the amino acid exchange reaction catalyzed by ASCT2. This pre-steady-state experiment requires that the intracellular solution contains saturating alanine and Na^+ concentrations, to ensure that the substrate-binding site is exposed to the extracellular side, before alanine release. It is a "single turnover" experiment, because net inward movement of the amino acid can only occur until the translocation equilibrium is established [20, 21].

Based on flux studies, determination of the [Na$^+$] dependence of exchange, as well as rapid kinetic experiments, the mechanism of amino acid exchange by ASCT2 was derived as shown in Figure 5.8e. In contrast to the EAATs, Na$^+$ binding to the empty transporter occurs with high (sub mM) apparent affinity [89, 91]. Thus, it is likely that Na$^+$ does not dissociate during amino acid exchange and rather plays a modulatory role instead of actively driving substrate transport. The time course of the decay of alanine-induced transient currents is in the range of 1–2 ms. This result indicates that equilibration of the amino acid translocation step(s) is fast, providing limits for the turnover rate of the transporter, which had not been determined previously [21].

If amino acid exchange is electrogenic, then it should be possible to perturb the exchange equilibrium by step-changes of the membrane potential. Consistent with this expectation, voltage jumps resulted in transient currents that were capacitive in nature (see Fig. 5.8a–c). The voltage dependence of the charge movement displayed Boltzmann-like behavior, consistent with voltage-dependent redistribution of the electrogenic amino acid exchange equilibrium [21] (see Fig. 5.8d). This result confirmed the observations from the single-turnover concentration jump experiments.

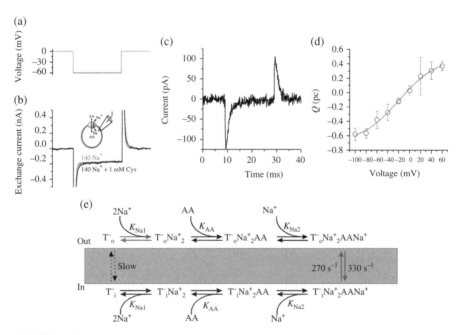

FIGURE 5.8 (a and b) Original current recording of an ASCT2-expressing cell in response to a voltage jump (a) in exchange mode (see symbol in inset). (c) Subtraction protocol (140 mM Na$^+$/1 mM Cys – 140 mM Na$^+$ in (a)) reveals the cysteine-specific component of the current. (d) Voltage dependence of the cysteine-specific charge movement. The solid line represents a fit of the Boltzmann equation to the data. (e) Predicted mechanism of amino acid (AA) exchange by ASCT2. Adapted from Ref. 21 with permission from Rockefeller University Press.

The voltage dependence of the transient current also allowed the direct determination of the valence of the exchange reaction through analysis with the Boltzmann equation. This valence was estimated as 0.7 [21] (see Fig. 5.8d). It should be noted that the recent development of competitive inhibitors should make voltage jump analysis of ASCT2 function more straightforward, as charge movements can be isolated through their specific inhibition [21, 94, 95].

A predicted exchange mechanism for amino acid transport by ASCT2, based on both steady-state analysis and pre-steady-state analysis is shown in Figure 5.8e. This mechanism includes sequential binding of the modulatory Na^+ ions, as well as the amino acid substrate, and electrogenic amino acid transport across the membrane. In contrast to the homologous glutamate transporters, relocation of the empty transporter is slow, thus prohibiting amino acid inward transport without exchange.

5.7 CONCLUSIONS

Pre-steady-state kinetic analysis of transporter function is a valuable tool to gain insight into the mechanistic details of coupled and uncoupled substrate and ion movement across the membrane. It is also powerful for structure–function studies in combination with site-directed mutagenesis and modification, because it allows the determination of the detailed mechanism of the impact of mutations on function, that is, at which reaction steps do the mutations impose their effects. This makes it easier to avoid misinterpretation of mutagenesis results by providing additional means to identify indirect effects. The most important challenge in obtaining results is related to the time resolution of the available perturbation methods. Often fast, individual reaction steps in transporters take place on a millisecond to submillisecond time scale. Thus, adequate perturbation methods have to be used to avoid artifacts due to limitation of kinetics through slow mixing from unstirred layers, or other experimental factors that impede time resolution. Applying concentration jumps using caged compounds has been particularly successful to avoid mixing artifacts and push the time resolution into the sub-10 µs regime. It is expected that pre-steady-state kinetic analysis will continue to provide critical mechanistic information on transporters the dynamics of which have not yet been studied on a rapid time scale.

REFERENCES

1. P. Läuger, *Electrogenic Ion Pumps*, Sinauer Associates Inc., Sunderland, MA, 1991, pp. 61–91.
2. S.M. Jarvis, J.R. Hammond, A.R. Paterson, A.S. Clanachan, Species differences in nucleoside transport. A study of uridine transport and nitrobenzylthioinosine binding by mammalian erythrocytes, *Biochem. J.* 208 (1982) 83–88.
3. C. Grewer, N. Watzke, M. Wiessner, T. Rauen, Glutamate translocation of the neuronal glutamate transporter EAAC1 occurs within milliseconds, *Proc. Natl. Acad. Sci. U. S. A.* 97 (2000) 9706–9711.

4. D.D. Loo, A. Hazama, S. Supplisson, E. Turk, E.M. Wright, Relaxation kinetics of the Na$^+$/glucose cotransporter, *Proc. Natl. Acad. Sci. U. S. A.* 90 (1993) 5767–5771.

5. J.I. Wadiche, M.P. Kavanaugh, Macroscopic and microscopic properties of a cloned glutamate transporter/chloride channel, *J. Neurosci.* 18 (1998) 7650–7661.

6. N. Watzke, T. Rauen, E. Bamberg, C. Grewer, On the mechanism of proton transport by the neuronal excitatory amino acid carrier 1, *J. Gen. Physiol.* 116 (2000) 609–622.

7. D.D. Loo, B.A. Hirayama, E.M. Gallardo, J.T. Lam, E. Turk, E.M. Wright, Conformational changes couple Na$^+$ and glucose transport, *Proc. Natl. Acad. Sci. U. S. A.* 95 (1998) 7789–7794.

8. C.C. Lu, D.W. Hilgemann, GAT1 (GABA:Na$^+$:Cl$^-$) cotransport function. Kinetic studies in giant *Xenopus* oocyte membrane patches, *J. Gen. Physiol.* 114 (1999) 445–457.

9. S. Mager, J. Naeve, M. Quick, C. Labarca, N. Davidson, H.A. Lester, Steady states, charge movements, and rates for a cloned GABA transporter expressed in *Xenopus* oocytes, *Neuron* 10 (1993) 177–188.

10. J.I. Wadiche, J.L. Arriza, S.G. Amara, M.P. Kavanaugh, Kinetics of a human glutamate transporter, *Neuron* 14 (1995) 1019–1027.

11. H.P. Larsson, A.V. Tzingounis, H.P. Koch, M.P. Kavanaugh, Fluorometric measurements of conformational changes in glutamate transporters, *Proc. Natl. Acad. Sci. U. S. A.* 101 (2004) 3951–3956.

12. A. Bicho, C. Grewer, Rapid substrate-induced charge movements of the GABA transporter GAT1, *Biophys. J.* 89 (2005) 211–231.

13. C.C. Lu, D.W. Hilgemann, GAT1 (GABA:Na$^+$:Cl$^-$) cotransport function. Steady state studies in giant *Xenopus* oocyte membrane patches, *J. Gen. Physiol.* 114 (1999) 429–444.

14. D.E. Bergles, A.V. Tzingounis, C.E. Jahr, Comparison of coupled and uncoupled currents during glutamate uptake by GLT-1 transporters, *J. Neurosci.* 22 (2002) 10153–10162.

15. N. Watzke, E. Bamberg, C. Grewer, Early intermediates in the transport cycle of the neuronal excitatory amino acid carrier EAAC1, *J. Gen. Physiol.* 117 (2001) 547–562.

16. Z. Zhang, Z. Tao, A. Gameiro, S. Barcelona, S. Braams, T. Rauen, C. Grewer, Transport direction determines the kinetics of substrate transport by the glutamate transporter EAAC1, *Proc. Natl. Acad. Sci. U. S. A.* 104 (2007) 18025–18030.

17. Z. Tao, Z. Zhang, C. Grewer, Neutralization of the aspartic acid residue Asp-367, but not Asp-454, inhibits binding of Na$^+$ to the glutamate-free form and cycling of the glutamate transporter EAAC1, *J. Biol. Chem.* 281 (2006) 10263–10272.

18. A. Hazama, D.D. Loo, E.M. Wright, Presteady-state currents of the rabbit Na$^+$/glucose cotransporter (SGLT1), *J. Membr. Biol.* 155 (1997) 175–186.

19. I.C. Forster, K. Kohler, J. Biber, H. Murer, Forging the link between structure and function of electrogenic cotransporters: the renal type IIa Na$^+$/P$_i$ cotransporter as a case study, *Prog. Biophys. Mol. Biol.* 80 (2002) 69–108.

20. Z. Zhang, G. Papageorgiou, J.E. Corrie, C. Grewer, Pre-steady-state currents in neutral amino acid transporters induced by photolysis of a new caged alanine derivative, *Biochemistry* 46 (2007) 3872–3880.

21. C.B. Zander, T. Albers, C. Grewer, Voltage-dependent processes in the electroneutral amino acid exchanger ASCT2, *J. Gen. Physiol.* 141 (2013) 659–672.

22. Y. Kanai, M.A. Hediger, Primary structure and functional characterization of a high-affinity glutamate transporter, *Nature* 360 (1992) 467–471.

23. S. Broer, N. Brookes, Transfer of glutamine between astrocytes and neurons, *J. Neurochem.* 77 (2001) 705–719.

24. O.P. Hamill, A. Marty, E. Neher, B. Sakmann, F.J. Sigworth, Improved patch-clamp techniques for high-resolution current recording from cells and cell-free membrane patches, *Pflugers Arch.* 391 (1981) 85–100.

25. M. Kappl, K. Hartung, Rapid charge translocation by the cardiac Na^+-Ca^{2+} exchanger after a Ca^{2+} concentration jump, *Biophys. J.* 71 (1996) 2473–2485.

26. C. Grewer, S.A. Madani Mobarekeh, N. Watzke, T. Rauen, K. Schaper, Substrate translocation kinetics of excitatory amino acid carrier 1 probed with laser-pulse photolysis of a new photolabile precursor of D-aspartic acid, *Biochemistry* 40 (2001) 232–240.

27. R. Sherman-Gold (Ed.), *The Axon Guide*, Axon Instruments, Inc., Foster City, CA, 1993.

28. C. Grewer, A. Gameiro, T. Mager, K. Fendler, Electrophysiological characterization of membrane transport proteins, *Annu. Rev. Biophys.* 42 (2013) 95–120.

29. P. Jauch, O.H. Petersen, P. Läuger, Electrogenic properties of the sodium-alanine cotransporter in pancreatic acinar cells: I. Tight-seal whole-cell recordings, *J. Membr. Biol.* 94 (1986) 99–115.

30. P. Jauch, P. Läuger, Electrogenic properties of the sodium-alanine cotransporter in pancreatic acinar cells: II. Comparison with transport models, *J. Membr. Biol.* 94 (1986) 117–127.

31. X.Z. Chen, M.J. Coady, J.Y. Lapointe, Fast voltage clamp discloses a new component of presteady-state currents from the Na^+-glucose cotransporter, *Biophys. J.* 71 (1996) 2544–2552.

32. G.A. Sauer, G. Nagel, H. Koepsell, E. Bamberg, K. Hartung, Voltage and substrate dependence of the inverse transport mode of the rabbit Na^+/glucose cotransporter (SGLT1), *FEBS Lett.* 469 (2000) 98–100.

33. N. Watzke, C. Grewer, The anion conductance of the glutamate transporter EAAC1 depends on the direction of glutamate transport, *FEBS Lett.* 503 (2001) 121–125.

34. D.W. Hilgemann, C.C. Lu, Giant membrane patches: Improvements and applications, *Methods Enzymol.* 293 (1998) 267–280.

35. D.W. Hilgemann, A. Collins, S. Matsuoka, Steady-state and dynamic properties of cardiac sodium-calcium exchange. Secondary modulation by cytoplasmic calcium and ATP, *J. Gen. Physiol.* 100 (1992) 933–961.

36. T.S. Otis, C.E. Jahr, Anion currents and predicted glutamate flux through a neuronal glutamate transporter, *J. Neurosci.* 18 (1998) 7099–7110.

37. O. Jardetzky, Simple allosteric model for membrane pumps, *Nature* 211 (1966) 969–970.

38. J.B. Koenderink, S. Geibel, E. Grabsch, J.J. De Pont, E. Bamberg, T. Friedrich, Electrophysiological analysis of the mutated Na,K-ATPase cation binding pocket, *J. Biol. Chem.* 278 (2003) 51213–51222.

39. J. Wang, M.A. El-Sayed, Temperature jump-induced secondary structural change of the membrane protein bacteriorhodopsin in the premelting temperature region: A nanosecond time-resolved Fourier transform infrared study, *Biophys. J.* 76 (1999) 2777–2783.

40. T.S. Otis, M.P. Kavanaugh, Isolation of current components and partial reaction cycles in the glial glutamate transporter EAAT2, *J. Neurosci.* 20 (2000) 2749–2757.

41. K. Erreger, C. Grewer, J.A. Javitch, A. Galli, Currents in response to rapid concentration jumps of amphetamine uncover novel aspects of human dopamine transporter function, *J. Neurosci.* 28 (2008) 976–989.

42. R. Callender, A. Gameiro, A. Pinto, C. De Micheli, C. Grewer, Mechanism of inhibition of the glutamate transporter EAAC1 by the conformationally constrained glutamate analogue (+)-HIP-B, *Biochemistry* 51 (2012) 5486–5495.

43. G.P. Leary, D.C. Holley, E.F. Stone, B.R. Lyda, L.V. Kalachev, M.P. Kavanaugh, The central cavity in trimeric glutamate transporters restricts ligand diffusion, *Proc. Natl. Acad. Sci. U. S. A.* 108 (2011) 14980–14985.

44. R.S. Brett, J.P. Dilger, P.R. Adams, B. Lancaster, A method for the rapid exchange of solutions bathing excised membrane patches, *Biophys. J.* 50 (1986) 987–992.

45. C. Grewer, G.P. Hess, On the mechanism of inhibition of the nicotinic acetylcholine receptor by the anticonvulsant MK-801 investigated by laser-pulse photolysis in the microsecond-to-millisecond time region, *Biochemistry* 38 (1999) 7837–7846.

46. G.P. Hess, C. Grewer, Development and application of caged ligands for neurotransmitter receptors in transient kinetic and neuronal circuit mapping studies, *Methods Enzymol.* 291 (1998) 443–473.

47. S.R. Adams, R.Y. Tsien, Controlling cell chemistry with caged compounds, *Annu. Rev. Physiol.* 55 (1993) 755–784.

48. L. Niu, R. Wieboldt, D. Ramesh, B.K. Carpenter, G.P. Hess, Synthesis and characterization of a caged receptor ligand suitable for chemical kinetic investigations of the glycine receptor in the 3-microseconds time domain, *Biochemistry* 35 (1996) 8136–8142.

49. R. Wieboldt, D. Ramesh, B.K. Carpenter, G.P. Hess, Synthesis and photochemistry of photolabile derivatives of gamma-aminobutyric acid for chemical kinetic investigations of the gamma-aminobutyric acid receptor in the millisecond time region, *Biochemistry* 33 (1994) 1526–1533.

50. D. Ramesh, R. Wieboldt, L. Niu, B.K. Carpenter, G.P. Hess, Photolysis of a protecting group for the carboxyl function of neurotransmitters within 3 microseconds and with product quantum yield of 0.2, *Proc. Natl. Acad. Sci. U. S. A.*, 90 (1993) 11074–11078.

51. R. Wieboldt, K.R. Gee, L. Niu, D. Ramesh, B.K. Carpenter, G.P. Hess, Photolabile precursors of glutamate: Synthesis, photochemical properties, and activation of glutamate receptors on a microsecond time scale, *Proc. Natl. Acad. Sci. U. S. A.* 91 (1994) 8752–8756.

52. W. Maier, J.E. Corrie, G. Papageorgiou, B. Laube, C. Grewer, Comparative analysis of inhibitory effects of caged ligands for the NMDA receptor, *J. Neurosci. Methods* 142 (2005) 1–9.

53. C. Mim, Z. Tao, C. Grewer, Two conformational changes are associated with glutamate translocation by the glutamate transporter EAAC1, *Biochemistry* 46 (2007) 9007–9018.

54. A.M. Brown, K.S. Lee, T. Powell, Voltage clamp and internal perfusion of single rat heart muscle cells, *J. Physiol.* 318 (1981) 455–477.

55. L. Parent, S. Supplisson, D.D. Loo, E.M. Wright, Electrogenic properties of the cloned Na$^+$/glucose cotransporter: I. Voltage-clamp studies, *J. Membr. Biol.* 125 (1992) 49–62.

56. G. Lambert, I.C. Forster, G. Stange, J. Biber, H. Murer, Properties of the mutant Ser-460-Cys implicate this site in a functionally important region of the type IIa Na$^+$/P$_i$ cotransporter protein, *J. Gen. Physiol.* 114 (1999) 637–652.

57. D.W. Hilgemann, C.C. Lu, GAT1 (GABA:Na$^+$:Cl$^-$) cotransport function. Database reconstruction with an alternating access model, *J. Gen. Physiol.* 114 (1999) 459–475.

58. C. Mim, P. Balani, T. Rauen, C. Grewer, The glutamate transporter subtypes EAAT4 and EAATs 1-3 transport glutamate with dramatically different kinetics and voltage dependence but share a common uptake mechanism, *J. Gen. Physiol.* 126 (2005) 571–589.

59. A. Gameiro, S. Braams, T. Rauen, C. Grewer, The discovery of slowness: Low-capacity transport and slow anion channel gating by the glutamate transporter EAAT5, *Biophys. J.* 100 (2011) 2623–2632.

60. W. Maier, R. Schemm, C. Grewer, B. Laube, Disruption of interdomain interactions in the glutamate binding pocket affects differentially agonist affinity and efficacy of *N*-methyl-D-aspartate receptor activation, *J. Biol. Chem.* 282 (2007) 1863–1872.

61. J. Mwaura, Z. Tao, H. James, T. Albers, A. Schwartz, C. Grewer, Protonation state of a conserved acidic amino acid involved in Na^+ binding to the glutamate transporter EAAC1, *ACS Chem. Neurosci.* 12 (2012) 1073–1083.

62. B. Mackenzie, D.D. Loo, M. Panayotova-Heiermann, E.M. Wright, Biophysical characteristics of the pig kidney Na^+/glucose cotransporter SGLT2 reveal a common mechanism for SGLT1 and SGLT2, *J. Biol. Chem.* 271 (1996) 32678–32683.

63. T.J. Crisman, S. Qu, B.I. Kanner, L.R. Forrest, Inward-facing conformation of glutamate transporters as revealed by their inverted-topology structural repeats, *Proc. Natl. Acad. Sci. U. S. A.* 106 (2009) 20752–20757.

64. D. Yernool, O. Boudker, Y. Jin, E. Gouaux, Structure of a glutamate transporter homologue from *Pyrococcus horikoshii*, *Nature* 431 (2004) 811–818.

65. R.J. Turner, Kinetic analysis of a family of cotransport models, *Biochim. Biophys. Acta-Biomembr.* 649 (1981) 269–280.

66. Y. Kanai, S. Nussberger, M.F. Romero, W.F. Boron, S.C. Hebert, M.A. Hediger, Electrogenic properties of the epithelial and neuronal high affinity glutamate transporter, *J. Biol. Chem.* 270 (1995) 16561–16568.

67. U. Hopfer, R. Groseclose, The mechanism of Na^+-dependent D-glucose transport, *J. Biol. Chem.* 255 (1980) 4453–4462.

68. I.H. Segel, *Enzyme Kinetics: Behavior and Analysis of Rapid Equilibrium and Steady-State Enzyme Systems*, Wiley-Interscience, New York, 1993.

69. C. Grewer, Z. Zhang, J. Mwaura, T. Albers, A. Schwartz, A. Gameiro, Charge compensation mechanism of a Na^+-coupled, secondary active glutamate transporter, *J. Biol. Chem.* 287 (2012) 26921–26931.

70. O. Boudker, R.M. Ryan, D. Yernool, K. Shimamoto, E. Gouaux, Coupling substrate and ion binding to extracellular gate of a sodium-dependent aspartate transporter, *Nature* 445 (2007) 387–393.

71. J. Abramson, I. Smirnova, V. Kasho, G. Verner, H.R. Kaback, S. Iwata, Structure and mechanism of the lactose permease of *Escherichia coli*, *Science* 301 (2003) 610–615.

72. A. Yamashita, S.K. Singh, T. Kawate, Y. Jin, E. Gouaux, Crystal structure of a bacterial homologue of Na^+/Cl^--dependent neurotransmitter transporters, *Nature* 437 (2005) 215–223.

73. K.M. Callenberg, O.P. Choudhary, G.L. de Forest, D.W. Gohara, N.A. Baker, M. Grabe, APBSmem: A graphical interface for electrostatic calculations at the membrane, *PLoS One* 5 (2010) e12722.

74. M. Grabe, H. Lecar, Y.N. Jan, L.Y. Jan, A quantitative assessment of models for voltage-dependent gating of ion channels, *Proc. Natl. Acad. Sci. U. S. A.* 101 (2004) 17640–17645.

75. H.R. Kaback, I. Smirnova, V. Kasho, Y. Nie, Y. Zhou, The alternating access transport mechanism in LacY, *J. Membr. Biol.* 239 (2011) 85–93.

76. Z. Zhang, A. Gameiro, C. Grewer, Highly conserved asparagine 82 controls the interaction of Na⁺ with the sodium-coupled neutral amino acid transporter SNAT2, *J. Biol. Chem.* 283 (2008) 12284–12292.

77. Z. Tao, C. Grewer, Cooperation of the conserved aspartate 439 and bound amino acid substrate is important for high-affinity Na⁺ binding to the glutamate transporter EAAC1, *J. Gen. Physiol.* 129 (2007) 331–344.

78. Y. Bismuth, M.P. Kavanaugh, B.I. Kanner, Tyrosine 140 of the gamma-aminobutyric acid transporter GAT-1 plays a critical role in neurotransmitter recognition, *J. Biol. Chem.* 272 (1997) 16096–16102.

79. G. Pines, Y. Zhang, B.I. Kanner, Glutamate 404 is involved in the substrate discrimination of GLT-1, a (Na⁺ + K⁺)-coupled glutamate transporter from rat brain, *J. Biol. Chem.* 270 (1995) 17093–17097.

80. Y. Kanai, B. Clemencon, A. Simonin, M. Leuenberger, M. Lochner, M. Weisstanner, M.A. Hediger, The SLC1 high-affinity glutamate and neutral amino acid transporter family, *Mol. Aspects Med.* 34 (2013) 108–120.

81. G. Pines, N.C. Danbolt, M. Bjoras, Y. Zhang, A. Bendahan, L. Eide, H. Koepsell, J. Storm Mathisen, E. Seeberg, B.I. Kanner, Cloning and expression of a rat brain L glutamate transporter, *Nature* 360 (1992) 464–467.

82. J.L. Arriza, S. Eliasof, M.P. Kavanaugh, S.G. Amara, Excitatory amino acid transporter 5, a retinal glutamate transporter coupled to a chloride conductance, *Proc. Natl. Acad. Sci. U. S. A.* 94 (1997) 4155–4160.

83. W.A. Fairman, R.J. Vandenberg, J.L. Arriza, M.P. Kavanaugh, S.G. Amara, An excitatory amino-acid transporter with properties of a ligand-gated chloride channel, *Nature* 375 (1995) 599–603.

84. N. Reyes, C. Ginter, O. Boudker, Transport mechanism of a bacterial homologue of glutamate transporters, *Nature* 462 (2009) 880–885.

85. C. Grewer, P. Balani, C. Weidenfeller, T. Bartusel, Z. Tao, T. Rauen, Individual subunits of the glutamate transporter EAAC1 homotrimer function independently of each other, *Biochemistry* 44 (2005) 11913–11923.

86. J.I. Wadiche, S.G. Amara, M.P. Kavanaugh, Ion fluxes associated with excitatory amino acid transport, *Neuron* 15 (1995) 721–728.

87. B.I. Kanner, A. Bendahan, Binding order of substrates to the sodium and potassium ion coupled L-glutamic acid transporter from rat brain, *Biochemistry* 21 (1982) 6327–6330.

88. A. Broer, N. Brookes, V. Ganapathy, K.S. Dimmer, C.A. Wagner, F. Lang, S. Broer, The astroglial ASCT2 amino acid transporter as a mediator of glutamine efflux, *J. Neurochem.* 73 (1999) 2184–2194.

89. A. Broer, C. Wagner, F. Lang, S. Broer, Neutral amino acid transporter ASCT2 displays substrate-induced Na⁺ exchange and a substrate-gated anion conductance, *Biochem. J.* 346 (2000) 705–710.

90. J.L. Arriza, M.P. Kavanaugh, W.A. Fairman, Y.N. Wu, G.H. Murdoch, R.A. North, S.G. Amara, Cloning and expression of a human neutral amino acid transporter with structural similarity to the glutamate transporter gene family, *J. Biol. Chem.* 268 (1993) 15329–15332.

91. C. Grewer, E. Grabsch, New inhibitors for the neutral amino acid transporter ASCT2 reveal its Na⁺-dependent anion leak, *J. Physiol.* 557 (2004) 747–759.

92. M. Valdeolmillos, J. Garcia-Sancho, B. Herreros, Differential effects of transmembrane potential on two Na⁺-dependent transport systems for neutral amino acids, *Biochim. Biophys. Acta-Biomembr.* 858 (1986) 181–187.

93. O. Bussolati, P.C. Laris, B.M. Rotoli, V. Dall'Asta, G.C. Gazzola, Transport system ASC for neutral amino acids. An electroneutral sodium/amino acid cotransport sensitive to the membrane potential, *J. Biol. Chem.* 267 (1992) 8330–8335.

94. T. Albers, W. Marsiglia, T. Thomas, A. Gameiro, C. Grewer, Defining substrate and blocker activity of alanine-serine-cysteine transporter 2 (ASCT2) ligands with novel serine analogs, *Mol. Pharmacol.* 81 (2012) 356–365.

95. C.S. Esslinger, K.A. Cybulski, J.F. Rhoderick, N_γ-Aryl glutamine analogues as probes of the ASCT2 neutral amino acid transporter binding site, *Bioorg. Med. Chem.* 13 (2005) 1111–1118.

6

RECORDING OF PUMP AND TRANSPORTER ACTIVITY USING SOLID-SUPPORTED MEMBRANES (SSM-BASED ELECTROPHYSIOLOGY)

FRANCESCO TADINI-BUONINSEGNI[1] AND KLAUS FENDLER[2]

[1] *Department of Chemistry "Ugo Schiff", University of Florence, Sesto Fiorentino, Italy*
[2] *Department of Biophysical Chemistry, Max Planck Institute of Biophysics, Frankfurt/Main, Germany*

6.1 INTRODUCTION

Electrical techniques play a central role among the various methods for functional characterization of membrane transport proteins. They have been used for ion channels, ion pumps, and secondary active transporters to monitor the transport of charged substrates, to identify electrogenic partial reactions, and to determine kinetic parameters of the transporter, such as rate constants and substrate affinities.

Standard electrophysiological methods, that is, voltage-clamp and patch-clamp techniques, are especially useful in the analysis of ion channels [1]. They have provided valuable information about the function of ion channels in a variety of cell types under physiological and pathological conditions (see Chapter 3). Also for eukaryotic transporters, these techniques have been proven useful [2]. However, apart from a few rare exceptions, bacterial transporters cannot be investigated using voltage-clamp or patch-clamp methods because of the small size of bacteria and because bacterial transporters are difficult to express in mammalian cells. Why are

Pumps, Channels, and Transporters: Methods of Functional Analysis, First Edition. Edited by Ronald J. Clarke and Mohammed A. A. Khalid.
© 2015 John Wiley & Sons, Inc. Published 2015 by John Wiley & Sons, Inc.

bacterial transporters of importance? Bacterial homologues of mammalian transporters have been widely used for protein crystallization and structure determination, as they can be conveniently prepared and purified in large amounts. However, structural information needs complementation with powerful functional techniques for an insightful interpretation. Indeed, based on the combination of structural and functional analyses of bacterial transporters, principles of the mechanism of transporters in general are emerging.

A similar limitation is encountered in the case of eukaryotic transporters from intracellular compartments, like the sarcoplasmic reticulum or parietal cell vesicles. They cannot be easily investigated using standard electrophysiology because of the small size of these cytoplasmic structures. In these cases, electrophysiology based on solid-supported membranes (SSM-based electrophysiology) can be useful [3]. In fact, SSM-based electrophysiology has been used successfully for the investigation of charge translocation processes in several ion pumps, for example, P-type ATPases [4] and a number of secondary active transporters [2, 5, 6], and for the analysis of ion channels [7]. The technique allows direct measurement of charge displacements yielding information about movement of charged substrates within the membrane transporter as well as about electrogenic conformational transitions associated with the transport process. SSM-based electrophysiology is also attractive for screening applications in drug discovery because of its robustness and its potential for automation [8, 9].

The SSM represents a convenient model system for a lipid bilayer membrane with the advantage of being mechanically so stable that solutions can be rapidly exchanged at the surface [10]. In particular, the SSM consists of a hybrid alkanethiol/phospholipid bilayer supported by a gold electrode. Proteoliposomes, membrane vesicles, or membrane fragments containing the transport protein of interest are adsorbed to an SSM and are activated using a fast solution exchange technique [3, 4]. By rapidly changing from a solution containing no substrate for the protein to one that contains a substrate, the protein can be activated and an electrical current can be detected, which provides information about the displaced charge and temporal development of the transport process.

Adsorption of proteoliposomes or membrane fragments to an SSM allows a variety of transport proteins to be immobilized on the SSM surface in a simple spontaneous process and has the advantage of providing an aqueous environment on both sides of the membrane for the transport protein. This experimental approach is much easier and more effective than direct incorporation of the protein in the planar membrane, which requires complicated reconstitution procedures, leading to a superior signal-to-noise ratio and time resolution of the electrical measurement.

6.2   THE INSTRUMENT

An SSM experiment consists of three steps. First, the SSM is formed by painting a small amount of lipid solution on an octadecanethiol-functionalized gold surface on the sensor chip. Then a protein sample, proteoliposomes, membrane vesicles, or membrane fragments containing the protein of interest are immobilized on the SSM surface. Finally, a rapid substrate concentration jump activates the transport protein and the charge displacement concomitant with its transport activity is recorded by capacitive coupling to the capacitance of the SSM. The procedures employed for individual instruments are slightly different. In the following, we describe an instrument built and used in our laboratories. The sensor chip consists of a microstructured gold electrode on a BOROFLOAT glass substrate with a circular active area of approximately $0.8 \, mm^2$. The chips are pretreated with a 1 mM octadecanethiol solution in ethanol or propanol leading to the formation of a hydrophobic octadecanethiol monolayer on gold. Then, lipid is deposited by application of a small amount of lipid solution ($15 \, mg \, ml^{-1}$ diphytanoyl-phosphatidylcholine in decane) for the formation of the SSM. Finally, the sensor chip is mounted into the cuvette. The lipid monolayer forms spontaneously on top of the alkanethiol film when the electrode is rinsed with buffer in the cuvette. After the formation of the SSM, its capacitance and conductance are monitored until they become constant after approximately 90 min. The capacitance and the conductance of the SSM have characteristic values ranging from 0.2 to $0.5 \, \mu F \, cm^{-2}$ and from 20 to $100 \, nS \, cm^{-2}$, respectively.

Before use, the protein samples, that is, suspensions of proteoliposomes, membrane vesicles, or membrane fragments, are thawed on ice and briefly sonicated with a tip sonicator or ultrasonic bath sonicator. $30 \, \mu l$ of the suspension are applied to the SSM using a standard pipette via the exit bore of the cuvette. The membranes are allowed to adsorb to the SSM for 1 h at room temperature.

6.2.1   Rapid Solution Exchange Cuvette

Figure 6.1 shows a schematic of the cuvette used for SSM-based electrophysiology (left panel) and the SSM with adsorbed proteoliposomes or membrane fragments (right panel). The SSM sensor is mounted in a flow-through cuvette. Using an electromagnetic valve, the inflowing solution is rapidly switched to a solution

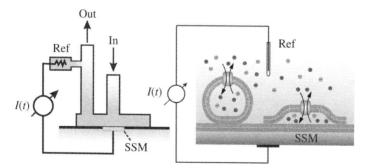

FIGURE 6.1   The SSM and the flow-through cuvette. The left panel shows the flow pathway in the cuvette allowing rapid solution exchange at the surface of the SSM. An expanded view of the SSM with adsorbed proteoliposomes or membrane fragments containing the transport protein is shown on the right panel. The gold layer of the SSM and a reference electrode (ref) serve as measuring electrodes and are connected to the current measuring circuitry $I(t)$.

containing the transporter substrate. The ensuing transporter current is measured using an amplifier connected to the SSM and the reference electrode (for details, see Ref. 3).

A rapid solution exchange at the surface of the SSM is desirable for two reasons: (i) The current detection principle of capacitive coupling leaves a relatively short time window of less than 200 ms for recording of the transporter current [3]. Longer solution exchange times lead to smaller signals and a reduced signal-to-noise ratio. (ii) One of the advantages of SSM-based electrophysiology is the possibility of rapid activation of the transporter, allowing isolation of partial reactions in the transport cycle (see [11, 12]). Therefore, the cuvette has been optimized for a rapid solution exchange. Rapidly switching valves and a short distance between the terminal valve and the SSM are critical components in the design of a fast solution exchange [12]. Small dimensions of the SSM are also helpful. Using an SSM area of 0.8 mm^2, a solution exchange time constant of 2 ms has been demonstrated, allowing the detection of a rapid charge displacement in the melibiose permease (MelB) [12]. Higher sensitivity at the expense of time resolution may be obtained with larger SSM areas.

6.2.2 Setup and Flow Protocols

The cuvette and the entire fluid pathway, including solution containers and valves, are mounted in a Faraday cage. The solution containers are pressurized with compressed air or nitrogen gas. Pressures of 0.2–1.0 bar are used in the experiments. An example of a laboratory constructed instrument is shown in Figure 6.2. Solution flow is controlled via the valves V_1 and the terminal valve V_2. The current generated at the SSM is measured via a reference electrode positioned at the solution outlet and an external current amplifier. A function generator allows assessment of the electrical properties, that is, capacitance and conductance, of the SSM. Valve control and current recording are accomplished via a computer equipped with a commercial A/D converter/digital output device and appropriate software. For technical details, see Refs. 3, 13.

The setup shown in Figure 6.2 is equipped for a single solution exchange protocol, that is, two different solutions are sequentially conducted through the cuvette: the nonactivating (NA) solution followed by the activating (A) solution containing the substrate of the transporter immobilized on the SSM. Exchanging from NA to A solution produces a substrate concentration jump at the SSM initiating transport and concomitant charge displacement. Finally, the activating solution is again replaced by nonactivating solution, so that before and in between experiments, the SSM is incubated with nonactivating solution. More complicated flow protocols are easy to establish by adding more solution containers and valves [3]. For example, a double solution exchange experiment employs an additional resting solution, which is conducted through the cuvette after the NA/A/NA solution exchange. Thus, in between experiments, the SSM is incubated in resting solution, allowing the establishment and/or maintenance of ion gradients. An example is given in Section 6.5.3.

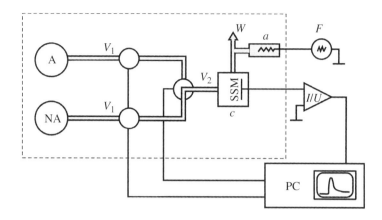

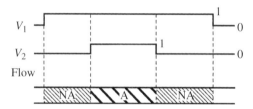

FIGURE 6.2 Flow configuration, valve timing, and flow protocol for a single solution exchange configuration. The entire fluid pathway is mounted in a Faraday cage (dashed line). The current is measured with reference to an Ag/AgCl electrode (*a*) and amplified by a current/voltage converter (*I/U*). A function generator (*F*) is used to measure the electrical properties of the SSM. Valves are controlled, and current is recorded by computer (PC). After passing through the cuvette, the solutions are directed to a waste container (*W*). The two pressurized containers with nonactivating (NA) and activating (A) solutions are connected to the cuvette (*c*) via two-way isolation valves V_1 and a three-way isolation valve V_2. Opening of valves V_1 initiates flow of NA. Subsequently, when V_2 is switched on, A is directed to the cuvette. Adapted from Ref. 3 with permission from Elsevier.

6.2.3 Protein Preparations

SSM-based electrophysiology has been employed for the functional characterization of various membrane transport proteins. The target proteins can be analyzed in a wide range of membrane systems:

1. Membrane vesicles or enriched membrane fragments from native tissue like pig or rabbit kidney, rabbit skeletal muscle, pig gastric mucosa, and eel electric organ
2. Membrane vesicles with overexpressed protein from recombinant expression in bacteria and mammalian cell lines such as CHO, HEK, and COS-1
3. Recombinantly produced purified protein reconstituted into proteoliposomes

For the adsorption of protein-containing membranes on the SSM surface, the protein preparation can be diluted with the resting or nonactivating buffer solution used for the formation of the SSM. The dilution factor has to be optimized empirically for each protein preparation.

A proteoliposome suspension suitable for the SSM measurement has a typical lipid concentration of approximately $10\,mg\,ml^{-1}$ and a lipid-to-protein ratio of approximately 5–10 (w/w). Membranes from native tissue or from bacterial or mammalian cells are usually prepared at a total protein concentration of approximately 2–$10\,mg\,ml^{-1}$. As a general rule, it is favorable to use between 10 and $30\,\mu g$ total protein for the adsorption on one SSM sensor.

For long-term storage, the protein preparations can be flash frozen in a resting buffer and stored in liquid nitrogen or in a freezer at $-80°C$ for months.

6.2.4 Commercial Instruments

Commercial instruments for SSM-based electrophysiology are now available (Nanion Technologies, Munich, Germany). These instruments are based on the surface electrogenic event reader (SURFE^{2}R) technology, which was originally established and developed by IonGate Biosciences (Frankfurt am Main, Germany). A description of the SURFE^{2}R technology can be found in Refs. 8, 9.

The SURFE^{2}R N1 instrument is a single-channel, semiautomated analysis device that is well suited for basic research in academia and industry. The instrument allows recording from one protein preparation at a time using a specially designed SSM sensor, which consists of a circular gold electrode of 3 mm diameter on a glass base. The SURFE^{2}R instruments use an open liquid injection geometry in contrast to the closed flow-through cuvette described previously (see Fig. 6.1). The device is equipped with an autosampler for liquid handling, and the solution exchange protocol is controlled by computer.

The SURFE^{2}R N96 instrument is a fully automated measurement system designed specifically for applications in transporter-based drug discovery, where higher throughput and lower reagent consumption are required (see Fig. 6.3). The instrument operates with SSM sensor plates in the standard 96-well format. A movable fluidic unit is used for liquid handling: It collects solutions for each measurement and injects them on the sensor surface in individual wells of the 96-well plate, thus generating a rapid solution exchange. Approximately 2000 measurements can be carried out per day with the SURFE^{2}R N96 instrument.

6.3 MEASUREMENT PROCEDURES, DATA ANALYSIS, AND INTERPRETATION

6.3.1 Current Measurement, Signal Analysis, and Reconstruction of Pump Currents

Current signals are recorded using a specific solution exchange protocol as discussed earlier. As an example, we show the current recorded from the NhaA Na^+/H^+ exchanger from *Helicobacter pylori* reconstituted in proteoliposomes and adsorbed

(a)

(b)

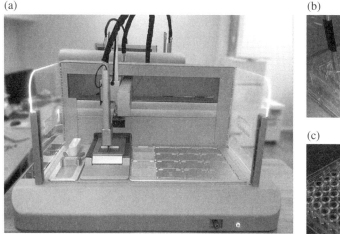

(c)

FIGURE 6.3 (a) TheSURFE²R N96 instrument. The movable fluidic unit is positioned above the 96 well plate. It is equipped with an eight channel fluid injection head. (b) Fluid injection head and 96 well plate. Each of the eight channels uses two separate tubes for simultaneous fluid injection and withdrawal. It collects solutions for each measurement and injects them on the electrode surface into individual wells of the 96-well sensor plate. (c) SSM sensors are arranged in 8 rows and 12 columns to match the standard microtiter plate format (96 wells).

to the SSM (see Fig. 6.4). Transport is initiated by addition of 100 mM Na^+ in the activating (A) solution. This protein transports 2 H^+ for 1 Na^+. Therefore, upon Na^+ addition, a negative transient current is observed (indicated by an arrow at t ~0.55 s) corresponding to the transport of positive charge out of the liposomes. When Na^+ is removed ($t = 1.0$ s), a current in the opposite direction is observed representing the inverse process. Only the current upon addition of the substrate (Na^+) is used for the analysis.

The adsorbed membrane fragments, membrane vesicles, or proteoliposomes and the underlying planar bilayer form a compound membrane. This system and the respective current detection principle (called capacitive coupling) merit some special consideration. The electrical behavior of the compound membrane can be described and analyzed by an equivalent circuit (see Fig. 6.5) [14, 15]. Note that the equivalent circuit and consequently the electrical behavior of the system is essentially the same whether membrane fragments or membrane vesicles or liposomes are used. When activated, the transporter generates an electrical current into the liposome, which charges the liposomal membrane and subsequently decreases the transporter current via the voltage dependence of the transporter activity. Therefore, currents measured in the capacitively coupled system are always transient currents and are strictly correlated with the transporter-generated current by the electrical components of the compound membrane system. The interpretation of the transient currents is not trivial [3]. However, in many cases, the peak current is a reliable

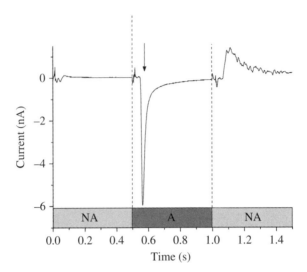

FIGURE 6.4 Transient current during rapid solution exchange on an SSM using the single solution exchange protocol. The displayed current is from the NhaA Na$^+$/H$^+$ exchanger from *H. pylori* reconstituted into proteoliposomes. The negative transient current indicated by the arrow corresponds to the exchange activity of NhaA initiated by a 100 mM Na$^+$ concentration jump. Buffer composition: 100 mM KP$_i$ (pH 8.5) plus 300 mM KCl in the nonactivating (NA) and 200 mM KCl + 100 mM NaCl in the activating (A) solution. Adapted from Ref. 3 with permission from Elsevier.

estimate of the steady-state transport activity of the transporter [16, 17]. But pre-steady-state components of the transporter current can also be detected and analyzed in the capacitively coupled system [11, 12, 18, 19]. The most straightforward approach to the interpretation of the transient currents is a reconstruction of the transporter current by numerically processing the measured current on the basis of circuit analysis as described in the following.

The current generated by the transporter $I_p(t, V_p)$ is distorted by capacitive coupling via the network of the compound membrane formed by the SSM and the adsorbed liposome (V_p is the voltage across the liposome). The resulting measured current is $I(t)$. The compound membrane can be described by an equivalent circuit (see Fig. 6.5) where C_p and C_m are the capacitances of the liposome, and that of its contact area with the SSM and G_p is the conductance of the liposome [15, 20]. Using the parameters of the equivalent circuit, the original current of the transporter can be reconstructed according to an algorithm suggested by Läuger and coworkers [20]. However, this algorithm neglects the voltage dependence of the transporter activity. During extended activity, the transporter builds up a voltage, V_p, across the liposomal membrane. In the following, we present a modification of the algorithm for reconstruction of the original current of the voltage-dependent transporter.

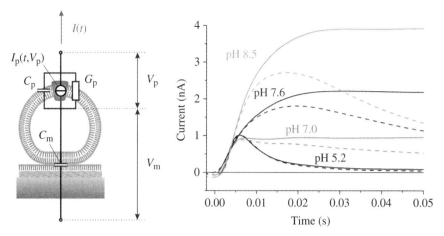

FIGURE 6.5 Liposome adsorbed to the SSM and the equivalent circuit describing the electrical properties of the compound membrane formed by the liposome and the underlying SSM. The transporter generates the pump current $I_p(t, V_p)$ that charges the liposome membrane to the voltage V_p. Capacitive coupling via the SSM–liposome contact region C_m yields the measured current $I(t)$. The right panel shows the measured currents (dashed lines) and the reconstructed transporter currents (solid lines) obtained at different pH by the method described in the text in the case of the H⁺/lactose symporter LacY.

A linear voltage dependence of the transporter activity is assumed, yielding a current

$$I_p(t, V) = I_p^T(t) + I_p^\infty(V)$$

Here, the current is dissected in a voltage-independent transient current $I_p^T(t)$ and a voltage-dependent steady-state current $I_p^\infty(V)$. The rationale behind this is that the relaxation into a steady state represented by the transient current consists of a single turnover, during which so little charge is translocated that the activity of the trans-porter is hardly affected. During the steady-state activity, on the other hand, a voltage is built up across the liposomal membrane, which reduces the current. For simplicity, we limit the analysis to a time range where the voltage dependence of the transporter activity is approximately linear [14]:

$$I_p^\infty(V) = I_p^\infty(0)\left(1 - \frac{|V_p|}{V^*}\right)$$

$I_p^\infty(0)$ is the steady-state current at zero voltage, and V^* is the reversal potential (see Fig. 6.5). Under these conditions, the reconstructed transporter current at zero voltage is given by [20]

$$I_p(t, 0) = \left(1 + \frac{C_p}{C_m}\right)\left\{I(t) + (k_0 + k_\infty)\int_0^t I(t)\,dt\right\}$$

With

$$k_0 = \frac{G_p}{C_m + C_p}$$

$$k_\infty = \frac{I_p^\infty(0)}{V^*\left(C_m + C_p\right)}$$

This is completely analogous to the equation given by Ref. 20 if $1/\tau_1$ in Ref. 20 is replaced by $k_0 + k_\infty$. This is the so-called reciprocal system time constant [14, 15] that in the case of a voltage-sensitive transporter depends, as expected, on the charging current $I_p^\infty(0)$. The reciprocal system time constant is introduced into the measured current [15] as an additional component and can, therefore, be determined from the experimental current traces. This allows reconstruction of the transporter current $I_p(t,0)$ using the earlier equation.

Figure 6.5 shows the reconstructed currents generated by the lactose permease from *Escherichia coli* (LacY) using the procedure described previously. These currents are shown on a reduced timescale (<50 ms) because the reconstruction algorithm fails at larger times. The reconstructed currents clearly demonstrate the increasingly lower steady-state turnover of LacY when decreasing the pH. At pH 5.2, the current decays nearly to zero after an initial transient charge displacement.

6.3.2   Voltage Measurement

As an alternative to current measurement described in the previous section, voltage signals generated by an electrogenic transport protein adsorbed to a SSM can be recorded under open circuit conditions, that is, using an amplifier of virtually infinite impedance. This method requires the almost complete suppression of leakage currents from the measuring circuit in order to detect changes in the electric potential during the time period of seconds.

In some early studies on the Na^+,K^+-ATPase [20, 21] and the light-driven proton pump bacteriorhodopsin [22] on black lipid membranes, the kinetics of these ion pumps was investigated by voltage measurements. More recently, measurements of current and voltage signals were combined to investigate the ion transport mechanism of the sarco(endo)plasmic reticulum Ca^{2+}-ATPase (SERCA) and the Na^+,K^+-ATPase on an SSM [23]. In these experiments, native vesicles incorporating SERCA or purified membrane fragments containing the Na^+,K^+-ATPase were adsorbed to the SSM and activated by concentration jumps of different substrates, that is, ATP, Ca^{2+}, or Na^+ ions. The corresponding current and voltage signals were recorded and attributed to electrogenic events in the reaction cycles of the two enzymes.

From the experimental point of view, voltage signals were measured by connecting the reference Ag/AgCl electrode and working gold electrode (SSM) to a voltage amplifier, with an input resistance of about $10^{14}\Omega$.

Figure 6.6 shows the current and voltage signals generated by SERCA-containing vesicles adsorbed to an SSM and subjected to ATP activation. Both the ATP-induced current and voltage are suppressed in the presence of thapsigargin (solid lines in Fig. 6.6), a highly specific and potent SERCA inhibitor, thereby demonstrating that the current and corresponding voltage signals are associated with pump activity and related to electrogenic Ca^{2+} displacement by the enzyme.

It is worth noticing that, under the condition of ideal capacitive coupling, the current $I(t)$ and the time derivative of the voltage $V(t)$ are identical, apart from a scaling factor $-AC_f$, where A is the total area of the SSM and C_f is the specific capacitance of the SSM [20], that is,

$$I(t) \approx -AC_f \frac{dV}{dt}$$

This expectation was actually tested for the signals reported in Figure 6.6 [23]. It was shown that the time derivative of $V(t)$ (computed from the voltage signal of Fig. 6.6b, dotted line) and the experimental current $I(t)$ (Fig. 6.6a, dotted line) recorded from the same SSM superimpose very well, in agreement with ideal capacitive coupling conditions.

Voltage measurements are particularly useful for the analysis of slow charge displacements that yield very small amplitudes in current recordings. In fact, it was demonstrated [23] that if the solution flow rate is reduced to slow down the SERCA response, the ATP-induced current peak is hardly measurable (see inset of Fig. 6.6a), while the corresponding voltage signal can be detected with higher accuracy thanks to a much more favorable signal-to-noise ratio (see inset of Fig. 6.6b). The slow response of the SERCA pump induced by flow rate reduction is evident from the broader shape of the current signal (see inset of Fig. 6.6a) and from the slow rise of the voltage signal, which reaches its maximum value within 2 s after protein activation (see inset of Fig. 6.6b).

The results discussed previously indicate that voltage measurements are preferable for the investigation of slow processes that generate very small current signals, for example, for the analysis of low turnover transporters.

6.3.3 Solution Exchange Artifacts

In current and voltage measurements, control experiments are usually carried out with the protein-free SSM or even better with empty (protein-free) liposomes in order to exclude electrical artifacts that can be generated by fast solution exchange at the SSM surface [10]. In fact, the replacement of one buffer solution by another buffer of different composition may yield signal artifacts because of binding of solutes to the lipid surface of the SSM [24]. These electrical artifacts, which are independent of the presence of protein, are difficult to circumvent and can only be minimized by the appropriate choice of solutes.

As a general rule, special care should be taken to keep nonactivating and activating solutions as similar as possible. Small differences in ionic strength, pH, and

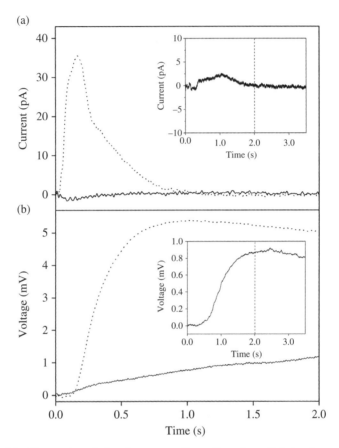

FIGURE 6.6 Current (panel a) and corresponding voltage (panel b) signals recorded from SERCA following a 100 μM ATP concentration jump in the presence of 10 μM free Ca²⁺, at normal flow rate (180 μl s⁻¹; dotted lines), at reduced flow rate (40 μl s⁻¹, inset), and in the presence of 100 nM thapsigargin (solid lines). Adapted with permission from Ref. 23. Copyright (2009), American Chemical Society.

temperature may generate current artifacts of comparable magnitude and shape as the protein-induced signals [3].

A few simple guidelines are now provided in order to minimize these electrical artifacts. (i) The basic buffer (solution with buffer and background salts but without activating substrate) for the activating solution and nonactivating solution must be the same. This means that the basic buffer is prepared in one batch and then divided into two aliquots. The activating substrate is added to one aliquot. (ii) A compensating inert compound, which is known not to affect the protein, is added to the nonactivating solution at the same concentration as the substrate. For instance, if transport by the Na⁺/H⁺ exchanger NhaA is initiated with 100 mM NaCl in the activating solution, 100 mM KCl should be used in the nonactivating solution as an inert compensation. (iii) A high ionic strength is recommended to reduce electrical artifacts. For example, in Na^+ concentration jumps on NhaA or Ca^{2+} concentration jumps on SERCA, a high concentration of KCl (100–200 mM) in the basic buffer can reduce artifacts due to binding of cations to the SSM surface. (iv) Before protein samples are adsorbed to the SSM, the solution exchange protocols should be tested on the protein-free SSM. Such preliminary control is valuable to check for the presence of artifacts and their magnitude and polarity. (v) It is preferable to use activating compounds that have a high water solubility. Lipophilic and amphiphilic compounds may generate large artifacts if used at high concentrations.

6.4 P-TYPE ATPases INVESTIGATED BY SSM-BASED ELECTROPHYSIOLOGY

P-type ATPases are a large and varied family of membrane transporters that are of fundamental importance in virtually all living organisms. These membrane proteins couple the energy provided by ATP hydrolysis to the transport of various ions. One of their prominent tasks is to generate and maintain crucial electrochemical potential gradients across biological membranes [25, 26]. A specific feature of P-type ATPases is the formation of a phosphorylated intermediate state during their enzymatic cycle.

Charge transfer in P-type ATPases was investigated by SSM-based electrophysiology in order to obtain insight into the ion transport mechanism (for a review, see Ref. 4). In particular, SSM-based electrophysiology allowed the identification of electrogenic steps and the assignment of rate constants to partial reactions in the enzymatic cycle of these membrane transporters.

P-type ATPases can be conveniently investigated on an SSM using a wide range of preparations [4]: membranes from native tissues (sarcoplasmic reticulum Ca^{2+}-ATPase from rabbit skeletal muscle and H^+,K^+-ATPase from pig gastric mucosa), purified membrane fragments (Na^+,K^+-ATPase from pig or rabbit kidney), reconstituted proteoliposomes (Na^+,K^+-ATPase from shark rectal gland), and heterologous expression in mammalian cells (human Cu^+-ATPases ATP7A and ATP7B in microsomes from COS-1 cells).

In the following section, we focus our attention on how the SSM-based electrophysiology has contributed to unraveling the transport mechanism of two prominent members of the P-type ATPase family, that is, the sarcoplasmic reticulum Ca^{2+}-ATPase and the human Cu^+-ATPase (ATP7A and ATP7B).

6.4.1 Sarcoplasmic Reticulum Ca^{2+}-ATPase

SERCA (sarco(endo)plasmic reticulum Ca^{2+}-ATPase) is a mammalian membrane-bound protein sustaining Ca^{2+} transport and involved in cell Ca^{2+} signaling and homeostasis [27]. This enzyme hydrolyzes one molecule of ATP in order to transport two Ca^{2+} ions against their electrochemical potential gradient from the cytoplasm to the lumen of sarcoplasmic reticulum (SR) in muscle cells. The SERCA transport activity plays an essential role in lowering cytosolic calcium concentration as required for muscle relaxation. It is now well established that the Ca^{2+} pump also countertransports two to three protons and that the transport cycle is electrogenic. General information on SERCA catalytic function and molecular structure is given in several reviews [28–32].

SR vesicles containing SERCA are obtained by extraction from fast-twitch skeletal muscle, as described in Eletr and Inesi [33]. The SR vesicles contain a high quantity of Ca^{2+}-ATPase (SERCA1 isoform), which accounts for approximately 50% of the total protein and which reaches a density in this membrane of about $30,000\,\mu m^{-2}$ [34], providing a very useful experimental system for functional and structural studies.

SERCA1 is a 996 amino acid protein [35] comprising ten transmembrane helical segments and a cytosolic headpiece including three distinct domains (A, N, and P) that are directly involved in catalytic activity. The ATPase cycle begins with high-affinity binding of two Ca^{2+} ions derived from the cytosolic medium, followed by ATP utilization and formation of a phosphoenzyme intermediate. The free energy derived from ATP is used by the phosphoenzyme for a conformational transition that favors translocation and release of the bound Ca^{2+} against its concentration gradient. Ca^{2+} ions are delivered to the intravesicular lumen in exchange for lumenal protons, which are translocated across the membrane to the cytosolic side during subsequent enzyme dephosphorylation. Hydrolytic cleavage of the phosphoenzyme is the final step, which allows the enzyme to undergo a new transport cycle.

Studies on structural and mechanistic properties of SERCA are numerous, and this enzyme is one of the best investigated ion pumps. High-resolution crystal structures of SERCA have been obtained in various conformational states (or their analogs), almost covering the entire reaction cycle, and are described in detailed reviews [31, 32, 36, 37].

SSM-based electrophysiology has provided useful information on the ion transport mechanism of SERCA, in particular the SERCA1a isoform [19, 38, 39]. The SSM electrode is well suited for the adsorption of membrane-bound SERCA1a and its activation by Ca^{2+} or ATP concentration jumps (see Fig. 6.7). Current signals induced by a $10\,\mu M$ Ca^{2+} jump in the absence of ATP and a $100\,\mu M$ ATP jump in the presence of $10\,\mu M$ Ca^{2+} are shown in Figure 6.8. The charge obtained by numerical integration of each transient current is attributed to sequential electrogenic events [19], corresponding to Ca^{2+} binding to the ATPase transport sites (solid line in Fig. 6.8), and then to vectorial translocation and release of bound Ca^{2+} into the lumen of the vesicle after phosphoenzyme formation by utilization of ATP (dotted line in Fig. 6.8).

Exchange of Ca^{2+} with H^+ upon vectorial translocation is a specific feature of SERCA [39, 40], facilitating lumenal Ca^{2+} release [37, 41]. It was demonstrated that the stoichiometry of Ca^{2+}/H^+ countertransport is approximately 1/1 when the lumenal and medium pH is near neutrality [41]. It is shown in the inset of Figure 6.8 that the current produced by the ATP concentration jump at neutral pH (solid line) decreases significantly if the ATP jump is performed at alkaline pH (dashed line). This result indicates that when the lack of H^+ limits H^+/Ca^{2+}exchange (i.e., at alkaline pH), vectorial translocation of bound Ca^{2+} is prevented, even though K^+ is present in high concentration and may neutralize acidic residues at alkaline pH.

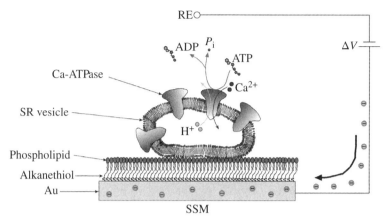

FIGURE 6.7 Schematic diagram of a SR vesicle adsorbed to a SSM and subjected to an ATP concentration jump (not drawn to scale). RE is the Ag/AgCl reference electrode. If the ATP jump induces net charge displacement, a transient current $I(t)$ is measured along the external circuit (the spheres are electrons). For simplicity, only four Ca^{2+}-ATPase molecules are shown in the vesicle. Reprinted from Ref. 4 with permission from Elsevier.

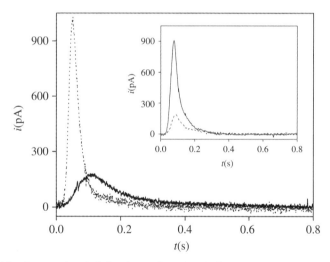

FIGURE 6.8 Current signals following a $10\,\mu M$ free Ca^{2+} concentration jump in the absence of ATP (solid line) and a $100\,\mu M$ ATP concentration jump in the presence of $10\,\mu M$ free Ca^{2+} (dotted line). The inset shows transient currents obtained after $100\,\mu M$ ATP concentration jumps on SERCA preincubated with $10\,\mu M$ free Ca^{2+} and $100\,mM$ KCl at pH 7 (solid line) and 8 (dashed line). Adapted from Ref. 4 with permission from Elsevier.

This suggests a requirement for specific H^+ binding at the Ca^{2+} transport sites in order to obtain Ca^{2+} release. Furthermore, cation/H^+ exchange at the transport sites following phosphoenzyme formation occurs in SERCA, but does not occur in copper-ATPases [39].

SSM-based electrophysiology is conveniently employed to study not only native SERCA1a but also the recombinant wild-type and mutant enzyme [19, 42], which can be obtained by heterologous expression in microsomes of COS-1 cells. In particular, the effects of site-directed mutations on current produced by Ca^{2+} and ATP concentration jumps have been investigated. The effect of Asp-351 mutation in recombinant SERCA was analyzed. Asp-351 is the residue receiving the ATP γ-phosphate at the catalytic site in the cytosolic P-domain of SERCA to form the aspartyl phosphorylated intermediate. Its mutation prevents phosphoenzyme formation and therefore produces catalytic inactivation of the ATPase, even though Ca^{2+} binding is retained. Accordingly, it was found that the D351N mutant yields a transient current upon a Ca^{2+} jump, but no current signal upon an ATP jump in the presence of Ca^{2+}. The study of recombinant mutant SERCA has provided useful information for a detailed characterization of the enzyme's transport cycle, especially concerning the Ca^{2+} binding mechanism, Ca^{2+}/H^+ exchange, and competitive Mg^{2+} binding.

SSM-based electrophysiology is well suited for the analysis of drug interactions with membrane transporters. In this respect, the effects of various inhibitors with different affinities for SERCA were investigated using SSM-based electrophysiology [43–45]. A molecular mechanism was proposed to explain the effect of each inhibitor, and the reaction step and/or intermediate of the pump cycle affected by the inhibitor was identified. The interaction of the new drug istaroxime with SERCA2a, that is, the isoform found in the cardiac muscle, deserves special mention [46, 47]. In this study, it was shown that istaroxime enhances SERCA2a activity, Ca^{2+} uptake, and Ca^{2+}-dependent charge movements into SR vesicles from healthy or failing hearts of dogs. Although not directly demonstrated, the most probable explanation of these activities is that istaroxime acts by displacing the regulatory protein phospholamban (PLN) from the SERCA2a/PLN complex, thereby removing the inhibitory action of PLN on this complex.

Finally, it is worth mentioning a very recent study on the adsorption of SERCA-containing vesicles on a negatively charged SSM, that is, a gold-supported octadecanethiol/phosphatidylserine bilayer. With the aim of fabricating an SSM with higher affinity to vesicles with incorporated membrane transporters, the adsorption process of vesicles containing either native or recombinant SERCA was characterized by combining surface plasmon resonance (SPR) and current measurements [48]. It was found that for both native and recombinant SERCA, the current measured on the serine-based SSM is about three times higher than that recorded on the conventional SSM. SPR measurements demonstrated that the higher current amplitude detected on the negatively charged SSM is correlated with a greater quantity of vesicles adsorbed on the serine-based SSM in the presence of Ca^{2+} and Mg^{2+}. It is suggested that Ca^{2+} and Mg^{2+} ions may accumulate at the negatively charged SSM surface, thus contributing to a more stable vesicle–SSM interaction. In other words, Ca^{2+} and Mg^{2+} coordinated to serine headgroups of the negatively charged SSM may promote a further interaction with the external surface of the membrane vesicles. The enhanced adsorption of membrane vesicles on the serine-based SSM may be useful to study membrane preparations with a low concentration of transport protein generating small current signals, as in the case of various recombinantly expressed proteins.

6.4.2 Human Cu⁺-ATPases ATP7A and ATP7B

The copper ATPases ATP7A and ATP7B are included in the P_{1B}-ATPase subfamily, which is selective for transition metal ions. Both proteins normally reside in the trans-Golgi network (TGN) of the cell for the metallation of copper-dependent enzymes of the secretory pathway [49]. ATP7A and ATP7B catalyze the ATP-dependent copper transport across the membrane into the lumen of the TGN via a process that involves formation of a transient phosphoenzyme intermediate [49, 50]. In addition to this predominant biosynthetic role, ATP7A and ATP7B also participate in the export of excess copper from the cells [49]. These two functions are associated with distinct intracellular localization of the transporters. At normal copper concentrations, the localization in the TGN reflects their role in the delivery of copper to cuproenzymes. In contrast, at high copper concentrations, Cu⁺-ATPases respond to hormone release and other signaling pathways, by sequestering copper into vesicles that subsequently fuse with the plasma membrane releasing copper into the extracellular milieu [49].

ATP7A and ATP7B share 56% sequence identity and are 160–170 kDa proteins with eight transmembrane segments (as compared to 10 in SERCA), including a copper binding site involved in catalytic activation and transport, and a cytosolic headpiece including the N, P, and A domains, which are common to other P-type ATPases [51]. A specific feature of mammalian Cu⁺-ATPases is an N–metal binding domain with six additional copper binding sites, which may play a regulatory role.

Various physiological processes depend on adequate and timely copper transport mediated by these proteins, for example, development, neurological function, intestinal copper absorption, and kidney function [52]. The physiological importance of ATP7A/B in humans is illustrated by the deleterious consequences of copper-ATPase inactivation on cell metabolism. In fact, inactivation of either ATP7A or ATP7B is associated with severe metabolic disorders, known as Menkes (ATP7A) and Wilson (ATP7B) diseases.

Due to the very low native abundance of ATP7A and ATP7B, biochemical characterization requires recombinant protein obtained by heterologous expression in insect or mammalian cell lines (see references in Ref. 53). In particular, Inesi and coworkers recently reported heterologous expression of ATP7A and ATP7B in COS-1 cells infected with adenovirus vector and functional characterization of membrane-bound copper-ATPase obtained with the microsomal fraction of infected cells [54, 55].

Radioactive assays of copper transport with ATP7A and ATP7B are rendered difficult by unfavorable characteristics (very short half-life and high gamma emission) of the ⁶⁴Cu radioactive isotope. Even though ⁶⁴Cu has been used in steady-state copper transport assays [56], it would be very difficult to employ ⁶⁴Cu in direct measurements of copper movements within the short timescale of a single catalytic cycle. However, experimental demonstration of copper movements across ATP7A and ATP7B was obtained by measurements of charge transfer following an ATP concentration jump on COS-1 microsomes containing recombinant Cu⁺-ATPase

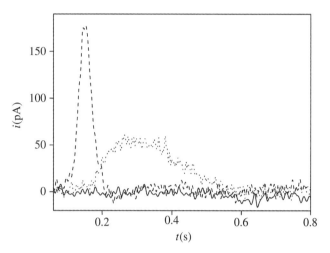

FIGURE 6.9 Transient currents induced by 100 μM ATP concentration jumps on SERCA in the presence of 10 μM free Ca^{2+} (dashed line) and on ATP7B in the presence of 5 μM $CuCl_2$ (dotted line) or 1 mM bathocuproinedisulfonate (solid line). Adapted from Ref. 57 with permission from Elsevier.

adsorbed to an SSM [39, 57]. It is shown in Figure 6.9 that, in the presence of 5 μM $CuCl_2$ (and 10 mM dithiothreitol to reduce Cu^{2+} to Cu^+), a 100 μM ATP jump on ATP7B induces a transient current (dotted line), which is not observed in the presence of bathocuproinedisulfonate, a specific copper chelator (solid line). The observed charge movement is attributed to vectorial translocation of bound copper, and it is related to the formation of a phosphoenzyme intermediate by the utilization of ATP [57]. It is of interest to compare the different time frames of charge movements generated by ATP addition to the recombinant ATP7B (in the presence of 5 μM $CuCl_2$, dotted line) or to the recombinant SERCA (in the presence of 10 μM free Ca^{2+}, dashed line). Analysis of the time frames shows that the decay time constant of the ATP7B signal is 140 ± 4 ms, which is about one order of magnitude larger than that obtained with SERCA, that is, 25 ± 0.3 ms. This result indicates that the charge transfer kinetics sustained by ATP7B are significantly slower than that of SERCA, even though the general catalytic mechanism retains similar features.

Another important difference between ATP7A/B and SERCA was observed in ATP concentration jump experiments at different pH values. In fact, contrary to SERCA, ATP-dependent charge transfer by ATP7A or ATP7B is not affected by raising the pH [39], thereby indicating that Cu^+/H^+ exchange at the intramembrane transport site does not play a role in copper release. As suggested by structural studies of *Legionella pneumophila* copper-ATPase (LpCopA) [58, 59], ATP7A and ATP7B may therefore operate without countertransport, which has never been demonstrated for a P_{1B}-ATPase.

Finally, SSM-based electrophysiology was very recently used to investigate the mechanism of interaction of platinum-based anticancer drugs with human Cu^+-ATPases (ATP7A and ATP7B) [60]. Pt anticancer drugs, for example, cisplatin, carboplatin, and oxaliplatin, are extensively employed as chemotherapeutic agents against various types of tumors. Unfortunately, malignant cells soon become resistant to this treatment, thus considerably limiting the cure rate. Experimental evidence indicated that Cu^+-ATPases can mediate resistance to Pt drugs [61, 62]. However, the mechanism by which Cu^+-ATPases mediate resistance to such drugs is not fully understood.

To gain information on the interaction with, and translocation of, Pt drugs by human Cu^+-ATPases, ATP-dependent charge movements were measured on COS-1 microsomes, enriched with recombinant ATP7A, ATP7B, or some selected mutants, adsorbed to a SSM [60]. Current measurements demonstrate that Pt drugs (cisplatin and oxaliplatin) activate Cu^+-ATPases and undergo ATP and phosphoenzyme-dependent translocation with a mechanism identical to that of copper. NMR spectroscopy shows a slow (hours) formation rate of stable Pt drug chelate adducts with the CXXC Cu-binding motif within a recombinant domain of the ATP7A N-terminal extension. These results suggest that translocation of Pt drugs by Cu^+-ATPases at short incubation times and sequestration of these drugs in the N-terminal extension at long incubation times may both contribute to drug resistance *in vivo*.

6.5 SECONDARY ACTIVE TRANSPORTERS

Secondary active transporters in the plasma membrane are of fundamental importance for the cell. To name only a few functions, they catalyze the uptake of nutrients, export toxic compounds, translocate macromolecules, regulate cell turgor, and create ion gradients essential for the function of other membrane proteins. Functional assays for secondary active transporters are notoriously difficult. They not only require a quantitative assessment of substrate transport but also the establishment of a "driving" electrochemical gradient. This is only possible in a compartmentalized system such as a cell, membrane vesicle, or proteoliposome. SSM-based electrophysiology provides such a compartmentalized system allowing application of a "driving" substrate gradient and the analysis of transport.

In the following, we present three cases of secondary active transporters that exploit the particular advantages of SSM-based electrophysiology. SSM-based electrophysiology is a convenient tool for the analysis of bacterial transporters as shown in the first two examples, a bacterial antiporter and a bacterial cotransporter. An additional advantage of SSM-based electrophysiology is its rugged design and potential for automation, which make it suitable for drug screening. This is demonstrated using a mammalian glutamate transporter, which is relevant for pharmacological drug development.

6.5.1   Antiport: Assessing the Forward and Reverse Modes of the NhaA Na$^+$/H$^+$ Exchanger of *E. coli*

The Na$^+$/H$^+$ exchanger NhaA is essential for the Na$^+$ and H$^+$ homeostasis in *E. coli* [63]. Its homologues are widely spread in enterobacteria and found even in humans [64]. NhaA transports 2 H$^+$ ions in exchange for 1 Na$^+$ ion across the bacterial membrane. Compared to other transporters, its transport capacity is extremely high. Turnover numbers of up to $1000\,s^{-1}$ have been reported [65].

In general, secondary active transporters are believed to function according to an alternative access model [66]: Transport is brought about by providing alternative access of a common binding site to both sides of the membrane. Conformational transitions are most probably involved in substrate transport of NhaA although their exact nature has yet to be determined. Secondary active transporters have been shown to be highly reversible, a property that has far-reaching consequences as, for example, in ischemia, where reversal of the astrocytic glutamate transporter is considered a major cause of neuronal damage [67].

SSM-based electrophysiology [3] allows the measurement of NhaA activity at well-defined solution composition on both sides of the membrane. These data allowed the proposal of a transport mechanism based on competing Na$^+$ and H$^+$ binding to a common transport site and the development of a kinetic model quantitatively explaining the experimental results [11]. Forward and reverse transport directions were investigated at zero membrane potential using preparations with inside out (ISO)- and right side out (RSO)-oriented transporters with Na$^+$ or H$^+$ gradients as the driving force. Examples of these measurements are given in Figure 6.10. Because of the capacitive coupling detection principle, the monitored currents are transient. However, the peak currents were demonstrated to represent a reliable estimate of the steady-state transport activity of NhaA [16].

The forward mode is the physiological transport direction of NhaA in the bacterial membrane, where an inside-directed proton-motive force drives Na$^+$ export (see Fig. 6.10 left panel). On the SSM, the forward mode can be studied in ISO membrane vesicles that are prepared using a French press protocol [11]. A Na$^+$ concentration jump generates a Na$^+$ gradient driving Na$^+$ to the periplasmic side (inner side of the vesicle) and H$^+$ to the cytoplasmic side (outer side of the vesicle). A negative current (ΔNa) is generated because 2 H$^+$ are transported per Na$^+$. Alternatively, RSO proteoliposomes can be used together with a driving H$^+$ gradient applied by a pH jump to low pH outside. A positive current is observed (ΔpH) in this case.

In the reverse mode of NhaA, 1 Na$^+$ ion is transported to the cytoplasm of the bacterial cell per 2 H$^+$ ions (see Fig. 6.10 right panel). This can be realized on the SSM by using RSO proteoliposomes and applying a Na$^+$ concentration jump (ΔNa). Likewise, the reverse mode can be observed by applying a pH jump to high pH, which creates an outside directed H$^+$ gradient (ΔpH).

Using SSM-based electrophysiology, the forward and reverse modes of NhaA can be readily monitored by applying the appropriate preparations and flow protocols. For NhaA, a study of the substrate dependencies in both transport modes yielded

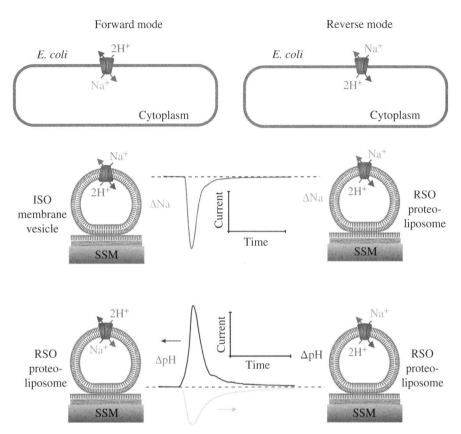

FIGURE 6.10 Transport modes of NhaA in the bacterial cell and under experimental conditions. The left column shows the "forward" transport mode of NhaA corresponding to its physiological transport direction in *Escherichia coli*; the right column shows the reverse transport mode. In the upper panel, a graphical representation of both transport modes in the bacterial cell is given. The lower two panels show the experimental preparations used for the investigation of the two transport modes. Na⁺ gradient (ΔNa)- or pH gradient (ΔpH)-driven transport activity of NhaA and the corresponding transient currents observed on the SSM are schematically depicted. Note the different polarity of the currents at different conditions. Typical currents were 1 nA for ISO membrane vesicles and 10 nA for RSO proteoliposomes. The timescale in the figure corresponds to approximately 10 ms. Reproduced from Ref. 11 with permission from the American Society for Biochemistry and Molecular Biology.

virtually identical kinetic parameters. In particular, the same pH profile was obtained and K_m values for Na⁺ were identical within error limits [11]. Furthermore, forward and reverse mode data were analyzed using a kinetic model based on a common binding site for H⁺ and Na⁺. Forward and reverse modes were characterized by the same kinetic parameters, leading to the conclusion that this transporter is highly symmetrical [11].

6.5.2   Cotransport: A Sugar-Induced Electrogenic Partial Reaction in the Lactose Permease LacY of *E. coli*

The lactose permease of *E. coli* (LacY) is a galactopyranoside/H^+symporter that belongs to the major facilitator superfamily of membrane transport proteins [68, 69]. LacY uses the proton-motive force at the bacterial plasma membrane for the uptake of lactose under physiological conditions. One molecule of lactose is transported per proton. LacY [70] and other secondary active transporters most probably function according to the alternating access model. Sugar transport proceeds as shown in the kinetic model in Figure 6.11a: The proton binds first to the outward-facing transporter, C_{out}, and then the sugar. Then a conformational transition to the inward-facing transporter, C_{in}, via a putative occluded state, $C_{occluded}$, allows cytoplasmic release of first the sugar and then the proton. In the last step, the carrier reorients to C_{out}.

For a detailed investigation of the alternating access transport mechanism, resolution of the partial reactions in the reaction cycle is essential. This requires a method with sufficient time resolution such as SSM-based electrophysiology, which allows concentration jumps as rapid as 2 ms [12]. Such a fast solution exchange is used to detect rapid charge displacements in the reaction cycle of LacY and to determine rate constants [71]. An example of this type of analysis is given in Figure 6.11.

It has been proposed that the rate-limiting step of LacY sugar transport in the absence of a proton-motive force is deprotonation [72]. Therefore, at sufficiently low pH, steady-state transport ceases as cytoplasmic deprotonation becomes slow. Interestingly, even at pH 5.2, a transient charge displacement remains, which has been assigned to a reaction preceding deprotonation, a sugar-induced electrogenic reaction with low electrogenicity [71]. This rapid charge displacement is analyzed in Figure 6.11.

Figure 6.11b shows transient currents recorded at pH 5.2 and different sugar concentrations. The time course of the transient current can be analyzed to determine the rate constant of the reaction associated with the observed charge displacement. For this purpose, the time resolution of the SSM setup has to be taken into account, which is mainly determined by the speed of the solution exchange at the SSM surface. The time resolution is represented by the transfer function of the setup (dashed line in the figure), which was determined as described [12, 71]. The observed relaxation rate constant, k_{obs}, at different sugar concentrations was determined using a least-square deconvolution algorithm and is plotted in Figure 6.11c.

The hyperbolic dependence of k_{obs} on the sugar concentration is expected for a two-step reversible binding mechanism where the first step is electroneutral and the second electrogenic:

$$E + S \overset{K_D}{\leftrightarrow} ES \underset{k^-}{\overset{k^+}{\leftrightarrow}} ES^*$$

Assuming rapid substrate binding, the values for the forward ($k^+ = 181 \pm 14\,s^{-1}$) and reverse ($k = 33 \pm 5\,s^{-1}$) rate constants and the dissociation constant ($K_D = 23 \pm 6\,mM$) can be determined from the data [71].

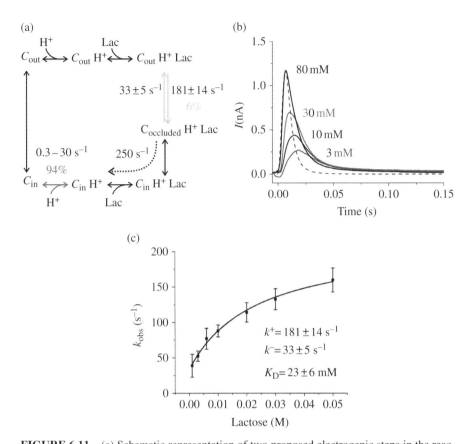

FIGURE 6.11 (a) Schematic representation of two proposed electrogenic steps in the reaction cycle of LacY. Inward-facing, outward-facing, and occluded conformations are marked with subscripts in, out, and occluded, respectively. Addition of sugar leads to rapid sugar binding (C_{out}H$^+$Lac) followed by formation of the occluded state ($C_{occluded}$H$^+$Lac) and is accompanied by a minor electrogenic reaction (6%) with forward and reverse rate constants of 181 s^{-1} and 33 s^{-1}, respectively. Deocclusion and release of the sugar are electroneutral and proceed with an estimated rate constant of approximately 250 s^{-1}. H$^+$ release on the cytoplasmic side (94%) is the major electrogenic step in the overall transport cycle and characterized by a pH-dependent rate constant (0.3–30 s^{-1}). (b) Transient currents measured on wild-type LacY proteoliposomes after different lactose concentration jumps as indicated in the figure at pH 5.2. Nonactivating solutions contained 100 mM of glucose; activating solutions a concentration of x mM lactose plus 100 − x mM glucose to maintain a constant sugar concentration. Solutions were prepared in 100 mM potassium phosphate buffer at pH 5.2 plus 1 mM DTT. All traces were recorded on the same sensor. $t=0$ corresponds to the time when activating solution reaches the sensor surface. The transfer function of the experiment representing the time resolution is shown as a dashed line. (c) Rate constants determined from the transient currents at different lactose concentrations at pH 5.2. From the transient currents, the rate constants were calculated with an iterative least-square deconvolution algorithm [71]. Average rate constants and SE from three to four data sets at a given lactose concentration are given. Adapted from Ref. 71 with permission from the American Chemical Society.

The kinetic parameters determined from the experiment are summarized in Figure 6.11a. Sugar binding triggers a rapid reaction. Because evidence is accumulating for a sugar-bound occluded state in LacY [73], we tentatively assign this reaction to the formation of the occluded state $C_{occluded}H^+Lac$. Integration of the transient current and comparison with the transport currents at maximum turnover show that the occlusion reaction is only weakly electrogenic (only 6% of the total charge translocation in one turnover) [17]. Additional experiments [71] allowed an estimation of the rate constants of the following reactions in the cycle such as the deocclusion step and the rate-limiting proton release step (see Fig. 6.11a). In conclusion, SSM-based electrophysiology allowed the delineation of electrogenic steps and the determination of rate constants during the complete substrate transfer process of LacY.

6.5.3 The Glutamate Transporter EAAC1: A Robust Electrophysiological Assay with High Information Content

The neuronal glutamate transporter excitatory amino acid carrier 1 (EAAC1), present in inhibitory GABAergic neurons [74] and excitatory glutamatergic synapses [75], is responsible for the reuptake of glutamate, the most important excitatory neurotransmitter in the central nervous system. Impaired function of the corresponding human glutamate transporter, excitatory amino acid transporter 3 (EAAT3), has been proposed to be involved in neuronal diseases such as Parkinson's disease, Alzheimer's disease, and epilepsy [76, 77] and is thus of interest for physiological and pharmacological research. Here, we show an analysis of the murine EAAC1, which shares about 90% sequence identity with its human homologue.

During its transport cycle, EAAC1 takes up three Na^+ ions, one L-glutamate (L-Glu) ion, and one proton from the exterior of the cell and countertransports one K^+ ion from the interior. The mechanism of the transporter is fully reversible [78], which is critical in special pathological situations, for example, stroke, due to the degeneration of the ATP levels under these conditions [79, 80]. The members of the glutamate transporter family share the feature of a channel-like anion conductance [81]. Two different conductance modes can be distinguished: a substrate-independent [82] and a substrate-gated anion conductance [83, 84]. Both phenomena have in common that Na^+ has to be bound to the protein.

The neuronal glutamate transporter EAAC1 is a complex multisubstrate enzyme. Its analysis requires a sophisticated assay technique that allows the assessment of its different transport modes. SSM-based electrophysiology is an alternative technique for the investigation of transporters. It is robust, does not require skilled personnel, and is well suited for automation. An example is given in Figure 6.12 where the murine EAAC1 expressed in Chinese hamster ovary (CHO) cells is selectively studied under various conditions. A high information content assay was established, allowing not only the analysis of different active and passive transport modes of the protein but also the assessment of partial reactions of the reaction cycle [85].

Figure 6.12 shows current traces recorded with EAAC1 together with the corresponding substrate environment (right column in the figure). The spike on

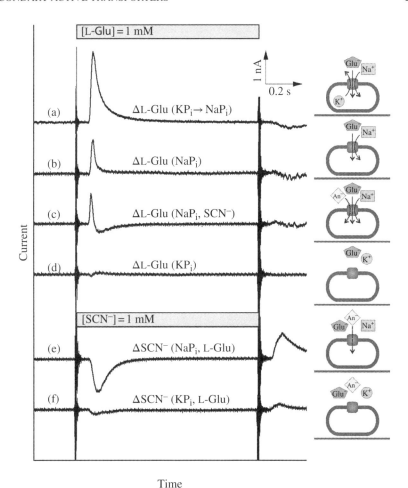

FIGURE 6.12 Different transport modes of the EAAC1 (120 mM KP_i buffer, pH 7.4). Pictograms on the right show the corresponding transporter situation. (a–d) Signal traces induced by 1 mM L-Glu concentration jumps. (a) Double solution exchange with KP_i buffer in resting solution and NaP_i buffer in nonactivating and activating solutions, (b) single solution exchange in NaP_i buffer solutions, (c) single solution exchange experiment in NaP_i buffer and 1 mM of the permeating anion SCN^-, and (d) single solution exchange experiment in KP_i buffer solutions. (e and f) Signal traces induced by 1 mM SCN^- concentration jumps with 1 mM L-Glu in the background in single solution exchange experiments: (e) experiment in NaP_i buffer, (f) experiment in KP_i buffer. Adapted from Ref. 85 with permission from Elsevier.

the left side of the current trace marks the time point of switching of the valves to the activating solution. The activating solution reaches the SSM approximately 100 ms later, and a transient current is recorded representing the transport activity of the glutamate transporter. A second spike on the right side of the trace represents the valve-switching artifact when reverting back to the nonactivating solution. The bar above the current traces shows the presence of the substrates: L-glutamate (L-Glu in a–d) and thiocyanate (SCN⁻ in e and f). All traces are recorded with a single solution exchange protocol except trace a, where a double solution exchange protocol [3] was used to establish an outward-directed K⁺ gradient.

The figure shows two fundamentally different transport activities of EAAC1: the tightly coupled L-Glu/Na⁺/K⁺ transport activity (a–d) and a channel-like activity, the anion conductance, with SCN⁻ as the permeable anion (e and f). The largest signal is observed when all coupling ions are present: Na⁺ and H⁺ outside and K⁺ inside the vesicles. Under these conditions, steady-state transport is observed after the L-Glu concentration jump as indicated by the large amplitude and slow decay of the current transient (a). Rapidly decaying currents of smaller amplitude are recorded when only a partial reaction is possible because K⁺ inside is missing (b and c). The slow negative current component in the presence of SCN⁻ (c) is interpreted as a compensating flow of negative charges (SCN⁻) into the vesicles. Finally, trace d demonstrates the requirement for the cotransported Na⁺ ion. The anion conductivity of the EAAC1 glutamate transporter is tested in trace e. A negative transient current is observed after an SCN⁻ concentration jump. This activity is strictly dependent on the presence of Na⁺ ions (f). In conclusion, the SSM-based assay presented in Figure 6.12 allowed the analysis of three different transport modes of EAAC1 by selectively controlling solutions inside and outside the membrane vesicles: steady-state turnover, L-Glu-induced L-Glu and Na⁺ internalization, and anion conductance.

6.6 CONCLUSIONS

SSM-based electrophysiology has been employed for a direct measurement of charge movements of a variety of electrogenic membrane transporters. More than 30 different transporters from mammalian and bacterial origin have so far been investigated using this technique, among them ATPases, cotransporters, exchangers, uniporters, and ion channels. Using the high time resolution of the technique, rate constants of transport steps can be determined, and electrogenic partial reactions can be identified.

While this technique is well suited for the investigation of fundamental questions concerning the transport mechanism of membrane transporters, the SSM sensor combined with robotized instrumentation contains potential for industrial applications. This is expected to become a promising platform technology for drug screening and development.

ACKNOWLEDGMENTS

F.T.-B. acknowledges the financial support from the Italian Ministry of University and Research and EnteCassa di Risparmio di Firenze. K.F. acknowledges the financial support by the Deutsche Forschungsgemeinschaft (SFB 807). We also thank Andrea Brüggemann (Nanion Technologies) for providing the components of Figure 6.3.

REFERENCES

1. O.P. Hamill, A. Marty, E. Neher, B. Sakmann, F.J. Sigworth, Improved patch-clamp techniques for high-resolution current recording from cells and cell-free membrane patches, *Pflugers Arch.* 391 (1981) 85–100.

2. C. Grewer, A. Gameiro, T. Mager, K. Fendler, Electrophysiological characterization of membrane transport proteins, *Annu. Rev. Biophys.* 42 (2013) 95–120.

3. P. Schulz, J.J. Garcia-Celma, K. Fendler, SSM-based electrophysiology, *Methods* 46 (2008) 97–103.

4. F. Tadini-Buoninsegni, G. Bartolommei, M.R. Moncelli, K. Fendler, Charge transfer in P-type ATPases investigated on planar membranes, *Arch. Biochem. Biophys.* 476 (2008) 75–86.

5. C. Ganea, K. Fendler, Bacterial transporters: Charge translocation and mechanism, *Biochim. Biophys. Acta-Bioenerg.* 1787 (2009) 706–713.

6. J. Garcia-Celma, A. Szydelko, R. Dutzler, Functional characterization of a ClC transporter by solid-supported membrane electrophysiology, *J. Gen. Physiol.* 141 (2013) 479–491.

7. P. Schulz, B. Dueck, A. Mourot, L. Hatahet, K. Fendler, Measuring ion channels on solid supported membranes, *Biophys. J.* 97 (2009) 388–396.

8. B. Kelety, K. Diekert, J. Tobien, N. Watzke, W. Dörner, P. Obrdlik, K. Fendler, Transporter assays using solid supported membranes: A novel screening platform for drug discovery, *Assay Drug Dev. Technol.* 4 (2006) 575–582.

9. S. Geibel, N. Flores-Herr, T. Licher, H. Vollert, Establishment of cell-free electrophysiology for ion transporters: Application for pharmacological profiling, *J. Biomol. Screen.* 11 (2006) 262–268.

10. J. Pintschovius, K. Fendler, Charge translocation by the Na^+/K^+-ATPase investigated on solid supported membranes: Rapid solution exchange with a new technique, *Biophys. J.* 76 (1999) 814–826.

11. T. Mager, A. Rimon, E. Padan, K. Fendler, Transport mechanism and pH regulation of the Na^+/H^+ antiporter NhaA from *Escherichia coli*: An electrophysiological study, *J. Biol. Chem.* 286 (2011) 23570–23581.

12. J.J. Garcia-Celma, B. Dueck, M. Stein, M. Schlueter, K. Meyer-Lipp, G. Leblanc, K. Fendler, Rapid activation of the melibiose permease MelB immobilized on a solid-supported membrane, *Langmuir* 24 (2008) 8119–8126.

13. A. Bazzone, W.S. Costa, M. Braner, O. Calinescu, L. Hatahet, K. Fendler, Introduction to solid supported membrane based electrophysiology, *J. Vis. Exp.* (2013) e50230.

14. E. Bamberg, H.J. Apell, N.A. Dencher, W. Sperling, H. Stieve, P. Läuger, Photocurrents generated by bacteriorhodopsin on planar bilayer membranes, *Biophys. Struct. Mech.* 5 (1979) 277–292.

15. K. Fendler, S. Jaruschewski, A. Hobbs, W. Albers, J.P. Froehlich, Pre-steady-state charge translocation in NaK-ATPase from eel electric organ, *J. Gen. Physiol.* 102 (1993) 631–666.

16. D. Zuber, R. Krause, M. Venturi, E. Padan, E. Bamberg, K. Fendler, Kinetics of charge translocation in the passive downhill uptake mode of the Na$^+$/H$^+$ antiporter NhaA of *Escherichia coli*, *Biochim. Biophys. Acta-Bioenerg.* 1709 (2005) 240–250.

17. J.J. Garcia-Celma, I.N. Smirnova, H.R. Kaback, K. Fendler, Electrophysiological characterization of LacY, *Proc. Natl. Acad. Sci. U. S. A.* 106 (2009) 7373–7378.

18. J. Pintschovius, K. Fendler, E. Bamberg, Charge translocation by the Na$^+$/K$^+$-ATPase investigated on solid supported membranes: Cytoplasmic cation binding and release, *Biophys. J.* 76 (1999) 827–836.

19. F. Tadini-Buoninsegni, G. Bartolommei, M.R. Moncelli, R. Guidelli, G. Inesi, Pre-steady state electrogenic events of Ca^{2+}/H$^+$ exchange and transport by the Ca^{2+}-ATPase, *J. Biol. Chem.* 281 (2006) 37720–37727.

20. R. Borlinghaus, H.J. Apell, P. Läuger, Fast charge translocations associated with partial reactions of the Na,K-pump: I. Current and voltage transients after photochemical release of ATP, *J. Membr. Biol.* 97 (1987) 161–178.

21. R. Borlinghaus, H.J. Apell, Current transients generated by the Na$^+$/K$^+$-ATPase after an ATP concentration jump: Dependence on sodium and ATP concentration, *Biochim. Biophys. Acta-Biomembr.* 939 (1988) 197–206.

22. M. Holz, M. Lindau, M.P. Heyn, Distributed kinetics of the charge movements in bacteriorhodopsin: Evidence for conformational substates, *Biophys. J.* 53 (1988) 623–633.

23. G. Bartolommei, M.R. Moncelli, G. Rispoli, B. Kelety, F. Tadini-Buoninsegni, Electrogenic ion pumps investigated on a solid supported membrane: Comparison of current and voltage measurements, *Langmuir* 25 (2009) 10925–10931.

24. J.J. Garcia-Celma, L. Hatahet, W. Kunz, K. Fendler, Specific anion and cation binding to lipid membranes investigated on a solid supported membrane, *Langmuir* 23 (2007) 10074–10080.

25. W. Kühlbrandt, Biology, structure and mechanism of P-type ATPases, *Nat. Rev. Mol. Cell Biol.* 5 (2004) 282–295.

26. M. Bublitz, J.P. Morth, P. Nissen, P-type ATPases at a glance, *J. Cell Sci.* 124 (2011) 2515–2519.

27. M. Brini, E. Carafoli, Calcium pumps in health and disease, *Physiol. Rev.* 89 (2009) 1341–1378.

28. L. de Meis, A.L. Vianna, Energy interconversion by the Ca^{2+}-dependent ATPase of the sarcoplasmic reticulum, *Annu. Rev. Biochem.* 48 (1979) 275–292.

29. G. Inesi, C. Sumbilla, M.E. Kirtley, Relationships of molecular structure and function in Ca^{2+} transport ATPase, *Physiol. Rev.* 70 (1990) 749–760.

30. J.P. Andersen, B. Vilsen, Structure-function relationships of cation translocation by Ca^{2+}- and Na$^+$, K$^+$-ATPases studied by site-directed mutagenesis, *FEBS Lett.* 359 (1995) 101–106.

31. C. Toyoshima, Structural aspects of ion pumping by Ca^{2+}-ATPase of sarcoplasmic reticulum, *Arch. Biochem. Biophys.* 476 (2008) 3–11.

32. J.V. Møller, C. Olesen, A.M. Winther, P. Nissen, The sarcoplasmic Ca^{2+}-ATPase: Design of a perfect chemi-osmotic pump, *Q. Rev. Biophys.* 43 (2010) 501–566.

33. S. Eletr, G. Inesi, Phospholipid orientation in sarcoplasmic membranes: Spin-label ESR and proton MNR studies, *Biochim. Biophys. Acta-Biomembr.* 282 (1972) 174–179.

34. C. Franzini-Armstrong, D.G. Ferguson, Density and disposition of Ca^{2+}-ATPase in sarcoplasmic reticulum membrane as determined by shadowing techniques, *Biophys. J.* 48 (1985) 607–615.

35. D.H. MacLennan, C.J. Brandl, B. Korczak, N.M. Green, Amino-acid sequence of a Ca^{2+}-Mg^{2+}-dependent ATPase from rabbit muscle sarcoplasmic reticulum, deduced from its complementary DNA sequence, *Nature* 316 (1985) 696–700.

36. C. Toyoshima, F. Cornelius, New crystal structures of PII-type ATPases: Excitement continues, *Curr. Opin. Struct. Biol.* 23 (2013) 507–514.

37. M. Bublitz, M. Musgaard, H. Poulsen, L. Thøgersen, C. Olesen, B. Schiøtt, J.P. Morth, J.V. Møller, P. Nissen, Ion pathways in the sarcoplasmic reticulum Ca^{2+}-ATPase, *J. Biol. Chem.* 288 (2013) 10759–10765.

38. F. Tadini-Buoninsegni, G. Bartolommei, M.R. Moncelli, G. Inesi, R. Guidelli, Time-resolved charge translocation by sarcoplasmic reticulum Ca-ATPase measured on a solid supported membrane, *Biophys. J.* 86 (2004) 3671–3686.

39. D. Lewis, R. Pilankatta, G. Inesi, G. Bartolommei, M.R. Moncelli, F. Tadini-Buoninsegni, Distinctive features of catalytic and transport mechanisms in mammalian sarco-endoplasmic reticulum Ca^{2+} ATPase (SERCA) and Cu^+ (ATP7A/B) ATPases, *J. Biol. Chem.* 287 (2012) 32717–32727.

40. G. Inesi, F. Tadini-Buoninsegni, Ca^{2+}/H^+ exchange, lumenal Ca^{2+} release and Ca^{2+}/ATP coupling ratios in the sarcoplasmic reticulum ATPase, *J. Cell Commun. Signal.* 8 (2014) 5–11.

41. X. Yu, L. Hao, G. Inesi, A pK change of acidic residues contributes to cation countertransport in the Ca-ATPase of sarcoplasmic reticulum. Role of H^+ in Ca^{2+}-ATPase countertransport, *J. Biol. Chem.* 269 (1994) 16656–16661.

42. Y. Liu, R. Pilankatta, D. Lewis, G. Inesi, F. Tadini-Buoninsegni, G. Bartolommei, M.R. Moncelli, High-yield heterologous expression of wild type and mutant Ca^{2+} ATPase: Characterization of Ca^{2+} binding sites by charge transfer, *J. Mol. Biol.* 391 (2009) 858–871.

43. G. Bartolommei, F. Tadini-Buoninsegni, S. Hua, M.R. Moncelli, G. Inesi, R. Guidelli, Clotrimazole inhibits the Ca^{2+}-ATPase (SERCA) by interfering with Ca^{2+} binding and favoring the E_2 conformation, *J. Biol. Chem.* 281 (2006) 9547–9551.

44. G. Bartolommei, F. Tadini-Buoninsegni, M.R. Moncelli, S. Gemma, C. Camodeca, S. Butini, G. Campiani, D. Lewis, G. Inesi, The Ca^{2+}-ATPase (SERCA1) is inhibited by 4-aminoquinoline derivatives through interference with catalytic activation by Ca^{2+}, whereas the ATPase E_2 state remains functional, *J. Biol. Chem.* 286 (2011) 38383–38389.

45. F. Tadini-Buoninsegni, G. Bartolommei, M.R. Moncelli, D.M. Tal, D. Lewis, G. Inesi, Effects of high-affinity inhibitors on partial reactions, charge movements, and conformational states of the Ca^{2+} transport ATPase (sarco-endoplasmic reticulum Ca^{2+} ATPase), *Mol. Pharmacol.* 73 (2008) 1134–1140.

46. M. Ferrandi, P. Barassi, F. Tadini-Buoninsegni, G. Bartolommei, I. Molinari, M.G. Tripodi, C. Reina, M.R. Moncelli, G. Bianchi, P. Ferrari, Istaroxime stimulates SERCA2a and accelerates calcium cycling in heart failure by relieving phospholamban inhibition, *Br. J. Pharmacol.* 169 (2013) 1849–1861.

47. C.L. Huang, SERCA2a stimulation by istaroxime: A novel mechanism of action with translational implications, *Br. J. Pharmacol.* 170 (2013) 486–488.

48. A. Sacconi, M.R. Moncelli, G. Margheri, F. Tadini-Buoninsegni, Enhanced adsorption of Ca-ATPase containing vesicles on a negatively charged solid-supported-membrane for the investigation of membrane transporters, *Langmuir* 29 (2013) 13883–13889.

49. S. Lutsenko, N.L. Barnes, M.Y. Bartee, O.Y. Dmitriev, Function and regulation of human copper-transporting ATPases, *Physiol. Rev.* 87 (2007) 1011–1046.

50. J.M. Argüello, M. González-Guerrero, D. Raimunda, Bacterial transition metal P1B-ATPases: Transport mechanism and roles in virulence, *Biochemistry* 50 (2011) 9940–9949.

51. P. Gourdon, O. Sitsel, J.L. Karlsen, L.B. Møller, P. Nissen, Structural models of the human copper P-type ATPases ATP7A and ATP7B, *Biol. Chem.* 393 (2012) 205–216.

52. S. La Fontaine, M.L. Ackland, J.F. Mercer, Mammalian copper-transporting P-type ATPases, ATP7A and ATP7B: Emerging roles, *Int. J. Biochem. Cell Biol.* 42 (2010) 206–209.

53. G. Inesi, Calcium and copper transport ATPases: Analogies and diversities in transduction and signaling mechanisms, *J. Cell Commun. Signal.* 5 (2011) 227–237.

54. R. Pilankatta, D. Lewis, C.M. Adams, G. Inesi, High yield heterologous expression of wild-type and mutant Cu^+-ATPase (ATP7B, Wilson disease protein) for functional characterization of catalytic activity and serine residues undergoing copper-dependent phosphorylation, *J. Biol. Chem.* 284 (2009) 21307–21316.

55. Y. Liu, R. Pilankatta, Y. Hatori, D. Lewis, G. Inesi, Comparative features of copper ATPases ATP7A and ATP7B heterologously expressed in COS-1 cells, *Biochemistry* 49 (2010) 10006–10012.

56. Y.H. Hung, M.J. Layton, I. Voskoboinik, J.F. Mercer, J. Camakaris, Purification and membrane reconstitution of catalytically active Menkes copper-transporting P-type ATPase (MNK; ATP7A), *Biochem. J.* 401 (2007) 569–579.

57. F. Tadini-Buoninsegni, G. Bartolommei, M.R. Moncelli, R. Pilankatta, D. Lewis, G. Inesi, ATP dependent charge movement in ATP7B Cu^+-ATPase is demonstrated by pre-steady state electrical measurements, *FEBS Lett.* 584 (2010) 4619–4622.

58. P. Gourdon, X.Y. Liu, T. Skjørringe, J.P. Morth, L.B. Møller, B.P. Pedersen, P. Nissen, Crystal structure of a copper-transporting PIB-type ATPase, *Nature* 475 (2011) 59–64.

59. M. Andersson, D. Mattle, O. Sitsel, T. Klymchuk, A.M. Nielsen, L.B. Møller, S.H. White, P. Nissen, P. Gourdon, Copper-transporting P-type ATPases use a unique ion-release pathway, *Nat. Struct. Mol. Biol.* 21 (2014) 43–48.

60. F. Tadini-Buoninsegni, G. Bartolommei, M.R. Moncelli, G. Inesi, A. Galliani, M. Sinisi, M. Losacco, G. Natile, F. Arnesano, Translocation of platinum anticancer drugs by human copper ATPases ATP7A and ATP7B, *Angew. Chem. Int. Ed. Engl.* 53 (2014) 1297–1301.

61. R. Safaei, P.L. Adams, M.H. Maktabi, R.A. Mathews, S.B. Howell, The CXXC motifs in the metal binding domains are required for ATP7B to mediate resistance to cisplatin, *J. Inorg. Biochem.* 110 (2012) 8–17.

62. N.V. Dolgova, D. Olson, S. Lutsenko, O.Y. Dmitriev, The soluble metal-binding domain of the copper transporter ATP7B binds and detoxifies cisplatin, *Biochem. J.* 419 (2009) 51–56.

63. E. Padan, E. Bibi, I. Masahiro, T.A. Krulwich, Alkaline pH homeostasis in bacteria: New insights, *Biochim. Biophys. Acta-Biomembr.* 1717 (2005) 67–88.

64. C.L. Brett, M. Donowitz, R. Rao, Evolutionary origins of eukaryotic sodium/proton exchangers, *Am. J. Physiol. Cell Physiol.* 288 (2005) C223–C239.

65. D. Taglicht, E. Padan, S. Schuldiner, Overproduction and purification of a functional Na^+/H^+ antiporter coded by nhaA (ant) from *Escherichia coli*, *J. Biol. Chem.* 266 (1991) 11289–11294.

66. O. Jardetzky, Simple allosteric model for membrane pumps, *Nature* 211 (1966) 969–970.

67. Y. Seki, P.J. Feustel, R.W. Keller, Jr., B.I. Tranmer, H.K. Kimelberg, Inhibition of ischemia-induced glutamate release in rat striatum by dihydrokinate and an anion channel blocker, *Stroke* 30 (1999) 433–440.

68. M.H. Saier, Jr., J.T. Beatty, A. Goffeau, K.T. Harley, W.H. Heijne, S.C. Huang, D.L. Jack, P.S. Jahn, K. Lew, J. Liu, S.S. Pao, I.T. Paulsen, T.T. Tseng, P.S. Virk, The major facilitator superfamily, *J. Mol. Microbiol. Biotechnol.* 1 (1999) 257–279.

69. M.H. Saier, Jr., Families of transmembrane sugar transport proteins, *Mol. Microbiol.* 35 (2000) 699–710.

70. H.R. Kaback, Structure and mechanism of the lactose permease, *C. R. Biol.* 328 (2005) 557–567.

71. J.J. Garcia-Celma, J. Ploch, I. Smirnova, H.R. Kaback, K. Fendler, Delineating electrogenic reactions during lactose/H⁺ symport, *Biochemistry* 49 (2010) 6115–6121.

72. H.R. Kaback, M. Sahin-Toth, A.B. Weinglass, The kamikaze approach to membrane transport, *Nat. Rev. Mol. Cell Biol.* 2 (2001) 610–620.

73. H. Kumar, V. Kasho, I. Smirnova, J.S. Finer-Moore, H.R. Kaback, R.M. Stroud, Structure of sugar-bound LacY, *Proc. Natl. Acad. Sci. U. S. A.* 111 (2014) 1784–1788.

74. D. Yernool, O. Boudker, Y. Jin, E. Gouaux, Structure of a glutamate transporter homologue from *Pyrococcus horikoshii*, *Nature* 431 (2004) 811–818.

75. Y. He, P.R. Hof, W.G. Janssen, J.D. Rothstein, J.H. Morrison, Differential synaptic localization of GluR2 and EAAC1 in the macaque monkey entorhinal cortex: A postembedding immunogold study, *Neurosci. Lett.* 311 (2001) 161–164.

76. M. Palmada, J.J. Centelles, Excitatory amino acid neurotransmission. Pathways for metabolism, storage and reuptake of glutamate in brain, *Front. Biosci.* 3 (1998) d701–d718.

77. J.P. Sepkuty, A.S. Cohen, C. Eccles, A. Rafiq, K. Behar, R. Ganel, D.A. Coulter, J.D. Rothstein, A neuronal glutamate transporter contributes to neurotransmitter GABA synthesis and epilepsy, *J. Neurosci.* 22 (2002) 6372–6379.

78. M. Szatkowski, B. Barbour, D. Attwell, Non-vesicular release of glutamate from glial cells by reversed electrogenic glutamate uptake, *Nature* 348 (1990) 443–446.

79. Y. Kanai, M.A. Hediger, The glutamate/neutral amino acid transporter family SLC1: Molecular, physiological and pharmacological aspects, *Pflugers Arch.* 447 (2004) 469–479.

80. C. Grewer, A. Gameiro, Z. Zhang, Z. Tao, S. Braams, T. Rauen, Glutamate forward and reverse transport: From molecular mechanism to transporter-mediated release after ischemia, *IUBMB Life* 60(2008) 609–619.

81. S. Eliasof, C.E. Jahr, Retinal glial cell glutamate transporter is coupled to an anionic conductance, *Proc. Natl. Acad. Sci. U. S. A.* 93 (1996) 4153–4158.

82. T.S. Otis, C.E. Jahr, Anion currents and predicted glutamate flux through a neuronal glutamate transporter, *J. Neurosci.* 18 (1998) 7099–7110.

83. S. Eliasof, F. Werblin, Characterization of the glutamate transporter in retinal cones of the tiger salamander, *J. Neurosci.* 13 (1993) 402–411.

84. W.A. Fairman, R.J. Vandenberg, J.L. Arriza, M.P. Kavanaugh, S.G. Amara, An excitatory amino-acid transporter with properties of a ligand-gated chloride channel, *Nature* 375 (1995) 599–603.

85. R. Krause, N. Watzke, B. Kelety, W. Dörner, K. Fendler, An automatic electrophysiological assay for the neuronal glutamate transporter mEAAC1, *J. Neurosci. Methods* 177 (2009) 131–141.

7

STOPPED-FLOW FLUORIMETRY USING VOLTAGE-SENSITIVE FLUORESCENT MEMBRANE PROBES

RONALD J. CLARKE[1] AND MOHAMMED A. A. KHALID[2]

[1] *School of Chemistry, University of Sydney, Sydney, New South Wales, Australia*
[2] *Department of Chemistry, Faculty of Applied Medical Sciences, University of Taif, Turabah, Saudi Arabia*

7.1 INTRODUCTION

Conceptually, stopped flow is probably the simplest of all rapid-reaction techniques. Its use is not specific to pumps, channels, and transporters, but in combination with voltage-sensitive fluorescent membrane probes, it has yielded much valuable information on their kinetics and mechanisms. Stopped-flow involves the rapid mixing of two reactant solutions, each simultaneously delivered into an observation chamber after the flow of the reactants has ceased. The detection of the course of the reaction of interest can be via UV/visible spectrophotometry or conductometry, but in this chapter, we will concentrate on fluorometric detection. The time resolution that can be achieved is determined by the time required to mix the two reactants, which, using specially designed mixing jets, can be as short as 1–2 ms. There is a huge range of biochemical and chemical reactions that occur over the millisecond-to-second timescale. Therefore, apart from being the most conceptually simple, rapid-reaction technique, stopped-flow is also the most versatile and most widely used. In fact, in a recent review, Olsen and Gutfreund [1] expressed the opinion that stopped-flow, together with the related quenched-flow technique, has probably resulted in more

Pumps, Channels, and Transporters: Methods of Functional Analysis, First Edition. Edited by Ronald J. Clarke and Mohammed A. A. Khalid.
© 2015 John Wiley & Sons, Inc. Published 2015 by John Wiley & Sons, Inc.

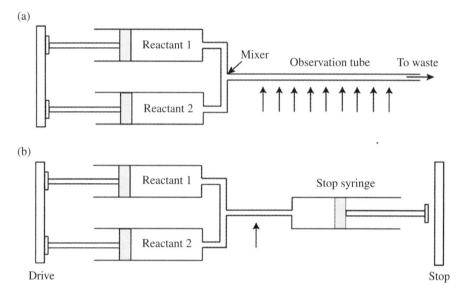

FIGURE 7.1   Basic design of the flow cells of the continuous-flow (a) and stopped-flow (b) techniques. Adapted from Ref. 2 by permission of Oxford University Press.

major contributions to enzymology, molecular biology, and many areas of chemistry than some developments that have been rewarded with the Nobel Prize.

The history of the stopped-flow technique is that it evolved out of economic necessity from a previously developed technique: continuous-flow. The continuous-flow technique involves mixing two reactants, which then flow with a constant velocity down a long tube. If the flow velocity is constant and accurately known, then each distance down the tube past the mixing point corresponds to a particular reaction time (see Fig. 7.1a). By making measurements of the extent of reaction at different distances along the tube, a complete reaction profile can be constructed. This method was first developed by Rutherford [3] in 1897 while working with J. J. Thomson at the Cavendish Laboratory in Cambridge for measuring the rate of recombination of gaseous ions produced by exposure to X-ray radiation. The same concept was taken up in the 1920s by Hartridge and Roughton [4] from the Physiological Laboratory, Cambridge, for the use on solution-phase reactions, in particular the reactions of ligands with hemoglobin and myoglobin. Although excellent time resolution can be achieved using the continuous-flow technique, its major drawback in studying biochemical reactions is that flow must be maintained for considerable lengths of time and hence large amounts of reactants are required. This may not have been a major issue for Hartridge and Roughton if they had sufficient PhD students willing to offer their blood for hemoglobin purification or a nearby abattoir, but for research on more precious proteins, it could well be prohibitive. To reduce the amount of material consumed, in 1936, Millikan [5], a collaborator of Hartridge and Roughton's, devised a miniaturized continuous-flow apparatus, but for many biochemical reactions, the expenditure of biological reactants was still too great and the technique has never become widely used.

   To reduce the amount of material required further, Britton Chance [6–10], working in 1940 for a time at Millikan and Roughton's laboratory in Cambridge before returning to the University of Pennsylvania, proposed and developed the stopped-flow technique. In this technique, after a continuous-flow period, the flow is abruptly stopped and the course of the reaction is monitored within the observation chamber. Although the time resolution achievable is not as good as in the continuous-flow method, stopped-flow significantly reduces the expenditure of expensive chemicals or biochemicals. It also allows the complete time course of a reaction to be followed in a single measurement, rather than having to take a series of measurements at different positions as is the case with continuous-flow. Stopped-flow has since become a hugely successful technique. Once commercial instruments became available in the 1960s, the use of stopped-flow was no longer restricted to a small number of specialized laboratories. Nowadays, stopped-flow instruments are in most well-equipped biochemistry or chemistry departments around the world.

7.2 BASICS OF THE STOPPED-FLOW TECHNIQUE

7.2.1 Flow Cell Design

The method has undergone many technical improvements since Britton Chance's first stopped-flow instruments. However, the most fundamental has probably been the introduction in the mid-1950s of the stopping syringe by Quentin Gibson [11–13], then at the University of Sheffield, but who collaborated extensively on the mechanisms of globin proteins with Roughton, the pioneer of continuous-flow. The most important aspect of any flow method is complete mixing, which is only possible if the flow is turbulent, not laminar. Gibson realized this necessitated very abrupt stopping for the stopped-flow method. To achieve this, he introduced a stopping syringe at the outlet from the observation chamber (see Fig. 7.1b). During the continuous-flow period of a stopped-flow measurement, Gibson's stopping syringe gradually fills until its plunger hits a mechanical stop and flow suddenly ceases. Gibson was unable to find a British manufacturer for his newly designed stopped-flow because the companies he approached were all of the opinion that the demand wouldn't be great enough [1]. The first commercial stopped-flow instrument, based on Gibson's design, was produced in 1965 by the US company Durrum Inc. (a predecessor of the present Dionex), based in Palo Alto, CA. Now, companies in the United Kingdom (TgK Scientific, Applied Photophysics), the United States (Update Instruments, KinTek Corporation, On-Line Instrument Systems), and France (Bio-Logic Science Instruments) are all producing stopped-flow instruments. Although they all have their differences, the principles of their flow cells are fundamentally based on Gibson's design.

7.2.2 Rapid Data Acquisition

Another crucial aspect for any stopped-flow instrument, and for any other rapid-reaction technique (e.g., temperature jump or voltage jump), is rapid data acquisition. It doesn't matter how short the mixing time of a stopped-flow instrument is if the

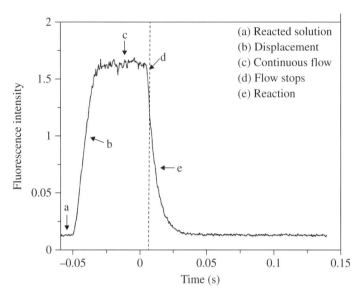

FIGURE 7.2 A typical stopped-flow fluorescence trace. Reproduced from Ref. 2 by permission of Oxford University Press.

data can't be captured at a sufficiently high rate to allow the calculation of rate constants for the system under study. In the absorbance- or fluorescence-based mode of detection, a photomultiplier is used to detect the change in light intensity caused by the reaction. The current output of the photomultiplier is then converted into a voltage by measuring the potential difference that the current produces across a known load resistor. Up until the 1980s, the transient changes in the voltage signal were commonly recorded by inputting the photomultiplier output into the voltage port of an analogue storage oscilloscope. Analysis of the data was then carried out by photographing the kinetic traces captured on the screen of the oscilloscope using a Polaroid camera, manually digitizing the data from the photographs and fitting to one or more exponential time functions to obtain an observed rate constant.

As with virtually any other instrument-based experimental technique, stopped flow has benefitted greatly by the revolution in electronic and computer technology. Storage oscilloscopes are now a thing of the past. In modern stopped-flow instruments, the analogue voltage signal coming from the photomultiplier is input directly into a computer, which converts it into a digital signal via an analogue-to-digital converter board. The experimental traces can then be directly averaged and fitted at the computer, reducing the time necessary for data analysis immensely.

An example of a typical stopped-flow fluorescence trace is shown in Figure 7.2. In this example, we have arbitrarily chosen a reaction that causes a decrease in fluorescence, but the principles are the same for a reaction causing an increase in fluorescence. In stopped-flow measurements, it is common practice to sometimes record the fluorescence level before the actual experiment starts. This is known as a "pre-trigger." In any stopped-flow experiment, it is necessary first to prime the flow circuit by flushing through with small volumes of the two reactants so that any solutions remaining from an old experiment or from cleaning are removed. Therefore, an experiment starts with reacted solution already in the observation chamber. This corresponds to a low level of fluorescence in the experiment shown in Figure 7.2. Once the actual experiment starts and the drive syringes are pushing reactant solution into the observation chamber, the reacted solution is displaced and the fluorescence rises due to the high fluorescence of the unreacted solution in the example. After a short time, the rate at which fresh high fluorescent reactant enters the observation chamber via the drive syringes equals the rate at which the fluorescence decreases by the two solutions reacting with one another, and the fluorescence level reaches a plateau region. This is the continuous-flow region of the measurement. Once the stopping syringe is completely filled and the flow suddenly stops, no more fresh solution enters the observation chamber, and the fluorescence decays due to the reaction. This is the stopped-flow region, which is used for data fitting to obtain the observed rate constant of the reaction.

7.2.3   Dead Time

One of the most important considerations for any stopped-flow instrument is what's known as the "dead time." In Figure 7.1b, it can be seen that the point of observation is not at the point where the two solutions first meet. To obtain reliable measurements of rate constants, it's important that the two reactants be completely mixed before observing the reaction. Therefore, the observation chamber of the flow circuit must be positioned a certain distance further along the flow system. This means that solution reaching the observation chamber has already been reacting for a short period of time before observation commences. The term dead time is used because no data is captured during this initial part of the reaction. Of course, the magnitude of the dead time depends on the flow velocity, but typically for research-grade commercial stopped-flow instruments, the value is around 1–2 ms.

For very fast reactions, it could be the case that the entire reaction is over during the dead time. In this case, stopped-flow measurements would just yield a horizontal flat line, that is, a constant fluorescence level, and to obtain kinetic information on the system, one would have to resort to a different experimental technique with a higher time resolution, for example, a chemical relaxation method such as temperature jump. However, fast reactions that involve a very large change in fluorescence can still be measured using stopped flow if a sufficiently large change in fluorescence can still be observed following the dead time. This is demonstrated in Figure 7.3. As reactions become faster, more and more of the total amplitude of the fluorescence change is lost during the dead time. However, for any exponential decay or rise of the fluorescence due to a first order or pseudo first-order reaction, the same time constant or observed rate constant characterizes the entire curve. Therefore, even if only the tail of the reaction can be captured, this could be enough to analyze the data and determine the rate constant.

From the discussion just presented, it would seem that the shorter the dead-time, the better. However, if the dead time is too short, then it could be the case that the solutions are still incompletely mixed when they enter the observation cell. Then,

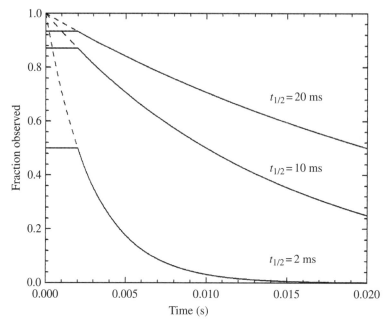

FIGURE 7.3 Stopped-flow fluorescence traces showing the increasing proportion of the total amplitude lost during the dead time as the speed of the reaction increases. The dead time was assumed to be 2 ms. For first-order reactions with half-lives of 20 ms, 10 ms, and 2 ms, the percentages of the reaction amplitude lost in the dead time are 6.7%, 13%, and 50%, respectively. Used with permission from TgK Scientific Ltd., Bradford on Avon, United Kingdom.

as described previously, the rate constant values obtained from the data analysis would actually be underestimates of the true values. Therefore, the value of the dead time of an instrument is a compromise. It must be sufficiently long to ensure complete mixing before observation starts but sufficiently short that not too much of the reaction is lost.

Experimental systems for measuring the dead time of a stopped-flow instrument in the fluorescence and absorbance modes and for testing the efficiency of mixing are described by Eccleston et al. [2].

7.3   COVALENT VERSUS NONCOVALENT FLUORESCENCE LABELING

As described previously, rapid-reaction flow methods in the solution phase were first developed to study the interaction of globin proteins such as hemoglobin and myoglobin with small ligands. These proteins contain intrinsic chromophores, porphyrins, whose UV/visible absorbance spectra are sensitive to ligand binding. Therefore, it is obvious that the first stopped-flow and continuous-flow measurements were carried out using the absorbance mode of detection. However, absorbance is a relative measurement. It requires the determination of both the intensity of the incident and the transmitted light. In contrast, fluorescence is an absolute measurement. It just requires the measurement of the total light intensity emitted by a sample on its irradiation with exciting light. Therefore, fluorescence is inherently a much more sensitive mode of detection than absorbance. Just as stopped-flow arose out of continuous-flow driven by a desire to limit the expenditure of precious protein material, fluorescence-based stopped-flow is a logical choice over absorbance-based stopped-flow if one wishes to reduce protein consumption even further, which most researchers do.

7.3.1   Intrinsic Fluorescence

Most proteins contain one or more fluorescent tryptophan residues. Therefore, there is no need to label the protein if a reaction which one is interested in involves a sufficiently large change in tryptophan fluorescence. This is the most desirable situation because any extrinsic probe has the potential to influence protein activity in some way. By observing tryptophan fluorescence, one can study the protein in its native state. Fluorescence-based stopped-flow measurements are, therefore, most commonly performed using excitation in the ultraviolet region, because tryptophan can be selectively excited (over phenylalanine and tyrosine) in the wavelength range 295–305 nm [14]. The wavelength maximum of the fluorescence emission spectrum of tryptophan is sensitive to the polarity and the hydrogen-bonding capacity of its surrounding solvent. Therefore, protein conformational changes that alter the exposure of tryptophan to aqueous surroundings can be expected to show significant changes in tryptophan fluorescence intensity. For this reason, stopped-flow studies based on tryptophan fluorescence are extensively used in protein folding investigations, a topic of great current interest.

However, not all proteins containing tryptophan exhibit large changes in their fluorescence when they undergo conformational changes. It could be the case that some tryptophan residues are so deeply buried within the protein matrix that a conformational change doesn't significantly alter the exposure of the residue to water or the polarity of its surroundings. An example of the use of tryptophan fluorescence to probe the mechanism of an ion pump is the study of Karlish and Yates [15] who used stopped-flow to investigate the kinetics of conversion of membrane-bound Na^+,K^+-ATPase between its Na^+-selective E1 and K^+-selective E2 conformational states. However, the amplitudes of the fluorescence changes they observed were only around 2% at most, which made analysis difficult, particularly at high ATP concentrations when the reaction becomes faster. Therefore, in many cases, it is necessary to use extrinsic fluorescent probes to improve the signal-to-noise ratio or to obtain additional information not obtainable using tryptophan fluorescence.

Apart from the amplitude of the signal, other important considerations when choosing whether to use intrinsic protein fluorescence or an extrinsic probe are the wavelengths of excitation and emission. If one uses UV irradiation, the potential problem of components in the reaction mixture other than the target fluorophore absorbing either the exciting light or the emitted fluorescence, and thereby quenching the fluorescence signal, is greater than if one uses longer wavelength visible light. This problem was a limiting factor in the study of Karlish and Yates [15] in which the overlap between the absorbance spectrum of ATP and the emission spectrum of tryptophan prevented measurements being carried out up to a saturating ATP concentration.

7.3.2   Covalently Bound Extrinsic Fluorescent Probes

There are a large number of fluorescent molecules that can be used to covalently label proteins and probe protein activity. For a more comprehensive coverage of the range of probes available, we suggest the reader consult other sources [14, 16]. Here, we concentrate on some basic principles and selected examples to illustrate the potential advantages and disadvantages of different probes.

One of the most commonly used fluorescent labels is fluorescein. A linking agent is needed in order to attach this probe to an amino acid residue of a protein. Three of the most widely used linkers are iodoacetamide, maleimide, and isothiocyanates. Iodoacetamides and maleimides attack sulfhydryl groups and hence target cysteine residues of the protein. Isothiocyanates attack amines and hence target basic amino acid residues such as lysine and arginine. Because of the problem of small tryptophan fluorescence amplitudes and quenching via ATP described previously, Karlish [17] turned to labeling of the Na^+,K^+-ATPase by fluorescein-5′-isothiocyanate (FITC), which labels lysine-501 of the protein [18]. Faller et al. [19] also used FITC to label the H^+,K^+-ATPase, a P-type ATPase related to the Na^+,K^+-ATPase, which is responsible for the acidification of the stomach. Both Karlish [17] and Faller et al. [19] were able to use the FITC-labeled enzymes to follow via stopped-flow the kinetics of conformational changes of these ion pumps when they underwent conformational changes. However, a major disadvantage of FITC labeling for kinetic studies is that the probe blocks the ATP binding [20]. Therefore, no kinetic studies of reactions activated by ATP can be carried out.

Labeling of the Na^+,K^+-ATPase by fluorescein, but using the iodoacetamide linker rather than isothiocyanate, was first introduced by Kapakos and Steinberg [21]. They found that the fluorescein label (5-IAF) showed significant changes in its fluorescence when the protein interacted with the ligands Na^+, K^+, and ATP, but, in contrast to FITC, the ATPase activity was unaffected. Subsequently, Steinberg and Karlish [22] were able to use 5-IAF-labeled enzyme to study the kinetics of the $E2 \rightarrow E1$ conformational transition via stopped-flow and show that the rate of the transition reached a saturating level at high ATP concentrations. This study would have been impossible using FITC because of blockage of the ATP site by the label. These studies demonstrate the importance of the site of labeling for any kinetic investigations of membrane protein mechanisms employing covalent fluorescent labeling.

Other covalent fluorescent labels that have been used to study reactions of the Na^+,K^+-ATPase include N-[p-(2-benzimidazoyl)phenyl]maleimide (BIPM) [23] and tetramethylrhodamine-6-maleimide (TMRM) [24]. Similar to 5-IAF, BIPM was found to have no effect on either Na^+,K^+-ATPase or H^+,K^+-ATPase activity and has been used in stopped-flow studies of both of these enzymes [23, 25]. The TMRM labeling [24] was carried out using mutations to the native Na^+,K^+-ATPase so that only a single cysteine residue was available for reaction with TMRM. This form of site-directed fluorescent labeling was pioneered in the laboratories of Isacoff [26] and Bezanilla [27] for use on the Shaker K$^+$ channel. Although this is a powerful method for localizing amino acid residues involved in conformational changes of membrane proteins, a disadvantage is that both the mutations to the proteins and the TMRM labeling can modify the rate constants relative to the native enzyme [24]. Site-directed TMRM labeling and its use in combination with the voltage clamp technique is the subject of Chapter 4.

7.3.3   Noncovalently Bound Extrinsic Fluorescent Probes

Noncovalently bound extrinsic fluorescent probes have the advantage that they don't chemically modify the protein under investigation. However, they can still modify protein function. Therefore, one needs to test carefully if this is the case and possibly take any inhibition into account in data analysis or experimental design.

A noncovalent probe that has been extensively used together with stopped-flow to study reactions of the Na$^+$,K$^+$-ATPase is eosin Y (often simply referred to as eosin). Eosin was first used in this application by Skou and Esmann, who initially attempted to covalently attach it to the enzyme via a maleimide linkage [28]. However, in the course of their studies, they found that even noncovalently bound eosin responded to the addition of Na$^+$ or K$^+$ to the enzyme with a fluorescence change [29]. By comparing the interaction of the enzyme with ATP in the presence of eosin, they were able to conclude that eosin binds directly in the enzyme's ATP binding site. Unfortunately, the blockage of the ATP binding site by eosin precludes its use in any kinetic studies of partial reactions of the enzyme involving ATP hydrolysis. However, Skou and Esmann [30] used it via stopped-flow to study the rates of transition between the E1 and E2 conformational states of the enzyme, which is associated with a change in affinity of the enzyme for eosin, just as there is for ATP. By stopped-flow measurements, in which he mixed an enzyme–eosin complex with ADP or ATP, Esmann [31] was also able to determine rate constants for nucleotide displacement from the enzyme.

Perhaps the most useful noncovalently bound fluorescent probes for studying the kinetics of ion pumps are voltage-sensitive probes. All electrogenic ion pumps, such as the Na$^+$,K$^+$-ATPase, transport net charge across the membrane. The current they generate thus builds up a transmembrane electrical potential difference across the membrane, which can be sensed by some voltage-sensitive dyes. The conformational changes they undergo during ion transport can also cause local changes in electric field strength within the membrane. These can be sensed by other voltage-sensitive dyes. Because all voltage-sensitive dyes respond to electrical field strength changes within or across membranes, they can indirectly react to the activity of pumps, channels, or transporters. There is no need for them to interact directly with any membrane protein to be able to kinetically follow its activity. However, this doesn't exclude the possibility that some voltage-sensitive dyes may also interact directly with membrane proteins as well as with the surrounding lipid phase. The remaining sections of this chapter are devoted to voltage-sensitive dyes exclusively.

7.4 CLASSES OF VOLTAGE-SENSITIVE DYES

Voltage-sensitive dyes are commonly divided into two classes based on their response times to rapid changes in the transmembrane voltage: fast dyes and slow dyes. Fast dyes have response times of less than milliseconds. Dyes belonging to this class include styrylpyridinium and annellated hemicyanine dyes, merocyanine dyes, and 3-hydroxychromone dyes. Slow dyes have response times greater than milliseconds. Dyes belonging to this class include cationic carbocyanine and rhodamine dyes and anionic oxonol dyes. The response times of the dyes determine the type of information that can be gained on the mechanisms of pumps, channels, or transporters.

Readers wishing to gain more detailed information on the range of dyes available, their response mechanisms, and their photostability and phototoxicity are referred to a recent review [32]. Here, we concentrate predominantly on aspects related to their application in determining the mechanism of ion pumps.

7.4.1 Slow Dyes

Slow dyes (see Fig. 7.4) respond via a movement of dye across the entire membrane. When the transmembrane potential difference changes, slow dyes redistribute themselves across the membrane. For example, if the cytoplasm becomes more positive relative to the extracellular fluid, anionic dyes accumulate in the cytoplasm. Similarly, if the cytoplasm becomes more negative relative to the extracellular fluid, cationic dyes accumulate in the cytoplasm. Any accumulation of dye in the cytoplasm also perturbs the partitioning of dye between the aqueous and membrane phases. Under most circumstances, the total intracellular volume of a suspension of cells is much smaller than the total volume of the extracellular fluid. Therefore, an accumulation of dye in the cytoplasm is also expected to result in an increase in dye bound to the membrane. If the fluorescence of dye bound to the membrane and in an aqueous environment is different, the dye responds to the change in transmembrane potential with a fluorescence change.

Because slow dyes respond to the transmembrane potential difference, they can only be used to study pumps, channels, or transporters in intact cells, intact organelles, or reconstituted into synthetic lipid vesicles. The greater than millisecond response time of slow dyes is due to the requirement of their mechanism that they move across the entire membrane. Many individual partial reactions in the mechanisms of pumps, channels, or transporters are much faster than this. Therefore, although slow dyes are useful in detecting changes in transmembrane potential due to changes in the activity of ion-transporting membrane proteins, their application in kinetic studies of such proteins is limited.

It is, however, possible to use slow dyes to follow the steady-state activity of electrogenic ion pumps. When reconstituted into lipid vesicles, the Na^+,K^+-ATPase causes a change in transmembrane potential because it pumps 3 Na^+ ions in one direction and 2 K^+ ions in the other direction across the membrane, that is, there is a net transport of one positive charge per enzyme turnover per ATP molecule hydrolyzed. Because the mechanisms of ion pumps must involve substantial

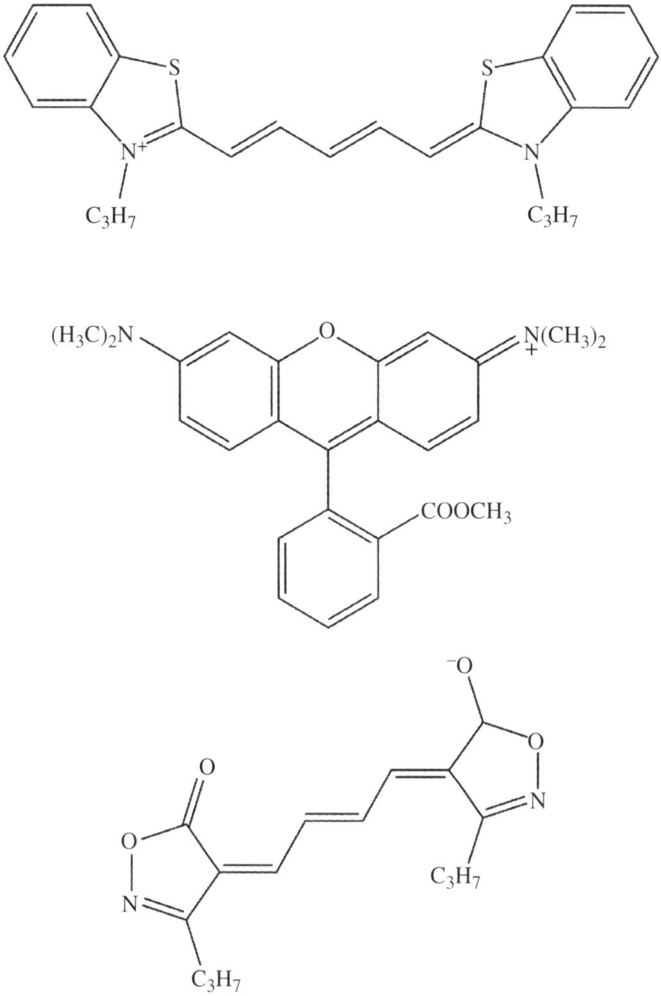

FIGURE 7.4 Classes of slow voltage-sensitive fluorescent dyes. From above, examples of voltage-sensitive carbocyanine, rhodamine and oxonol dyes are shown, respectively. Reprinted from Ref. 32 with kind permission from Springer Science and Business Media.

conformational changes in order to transport ions against a concentration gradient, the ion fluxes they produce across membranes are generally much lower than those of ion channels. Maximum turnovers in the range $10–100\,s^{-1}$ at normal physiological temperatures are typical. A stopped-flow investigation [33] of the interaction of the slow dye oxonol VI with lipid vesicles showed that it responds to changes in transmembrane potential in less than a second. This is sufficiently fast that it could be used in a fluorimeter to kinetically follow the build-up of the transmembrane potential by the Na^+,K^+-ATPase in lipid vesicles [34, 35]. Much more detailed kinetic information can, however, be obtained using fast dyes. We therefore now turn our attention to them.

7.4.2   Fast Dyes

Fast dyes (see Fig. 7.5) respond to changes in intramembrane electric field strength via a reorientation of the dye molecule within the membrane and/or a redistribution of the dye's electrons within its chromophore. The amplitudes of their fluorescence responses are in general significantly smaller than those of slow dyes, but for kinetic studies, this disadvantage is far outweighed by the advantage of their much faster response time.

For research on the kinetics of ion pumps, by far, the most widely used fast dyes are those based on the aminostyrylpyridinium chromophore, in particular the dye RH421 (see Fig. 7.5). The development of this chemical class of dyes arose out of quantum mechanical calculations by Loew et al. [36], which led to the prediction that the aminostyrylpyridinium chromophore should be ideally suited to respond to changes in membrane potential via an electrochromic mechanism. Electrochromism can be defined as an electric-field-induced wavelength shift in the UV/visible absorbance spectrum or fluorescence excitation spectrum of a molecule. For such a wavelength shift to occur, the molecule must undergo a large charge shift on excitation, that is, the electron distribution must be very different in the excited state compared to the ground state. This is indeed the case for the aminostyrylpyridinium chromophore (see Fig. 7.6). The reason for the electrochromic wavelength shift is that any local electric field within the membrane causes different degrees of stabilization or destabilization of the ground and excited states of membrane-bound dye. Thus, the energy gap between the ground and excited states and the wavelength of maximum absorbance, λ_{max}, are electric-field-dependent. Because this mechanism only involves a redistribution of the probe's electrons, the response is expected to be very fast, that is, on the femtosecond timescale. If one uses fixed wavelengths of excitation and emission, the electrochromic shift in the excitation spectrum would cause either an increase or decrease in the observed fluorescence intensity depending on whether the excitation wavelength used is on the low- or high-wavelength side of λ_{max}.

After theoretically predicting the potential of the aminostyrylpyridinium chromophore, Loew and collaborators went ahead and synthesized a range of dyes based on the chromophore and successfully demonstrated their electrical responses in both model membrane systems and intact cells [37–41]. Probably, the most widely used and most well-known dye originating in Loew's laboratory at the State University of New York at Binghamton is now di-8-ANEPPS, which contains a naphthyl derivative of the aminostyrylpyridinium chromophore. The name is an acronym of its chemical name.

Another important laboratory that contributed to the development of fast response dyes is that of Grinvald [42, 43] at the Weizmann Institute in Rehovot, Israel. The dyes developed there are known as RH dyes, but they are also based on the same aminostyrylpyridinium chromophore used in Loew's laboratory. The RH designation derives from the name of Grinvald's collaborator Rina Hildesheim, who actually synthesized the dyes. The most well-known dye originating in their laboratory is RH421 (see Fig. 7.5).

Although the ANEPPS and RH dyes were synthesized with an electrochromic mechanism in mind, testing on both model membrane systems and cell preparations

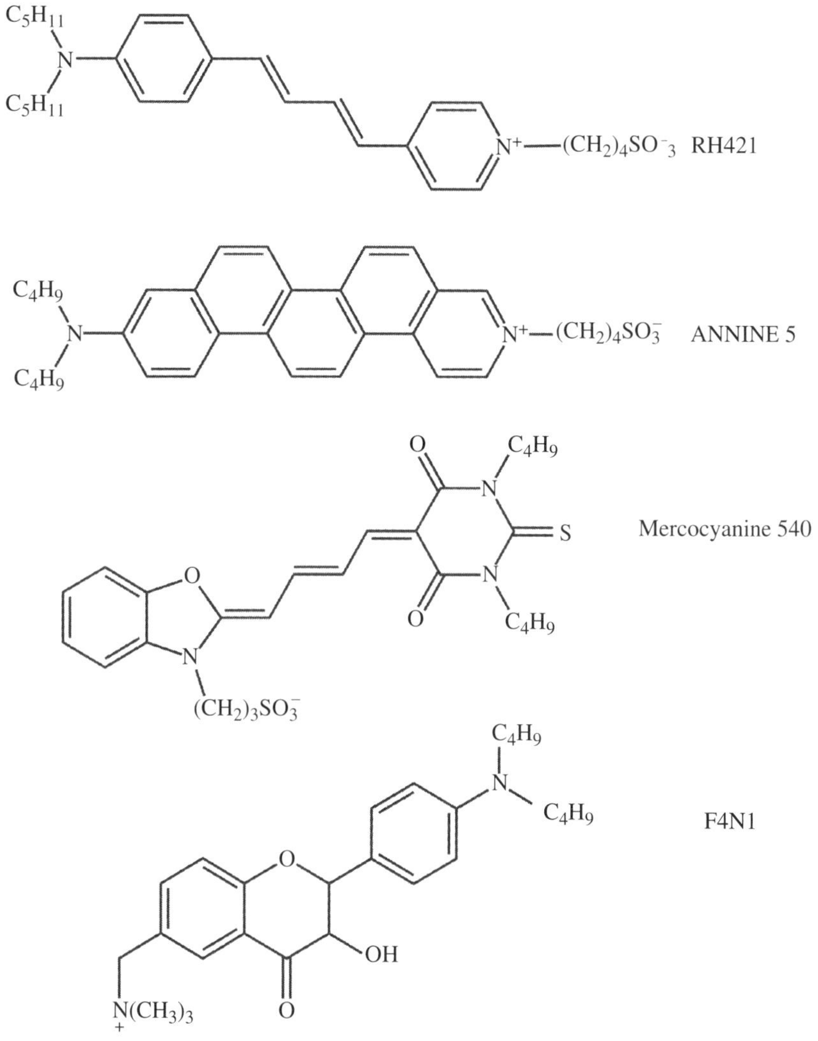

FIGURE 7.5 Classes of fast voltage-sensitive fluorescent dyes. From above, examples of voltage-sensitive styrylpyridinium, annellated hemicyanine, merocyanine and 3-hydroxychromone dyes are shown, respectively. Reprinted from Ref. 32 with kind permission from Springer Science and Business Media.

suggests [39, 40, 44, 45] that for many of them, their electrical responses are not purely electrochromic. For example, Fluhler et al. [40] stated that "The impressive sensitivity of RH421 in neuroblastoma cells is clearly too large to be attributable to electrochromism." The same conclusion was reached from studies on model membrane systems and Na^+, K^+-ATPase-containing membrane fragments [44, 45]. In these systems, evidence for a reorientation/solvatochromic contribution to the overall response of RH421 was found. From measurements in chloroform, RH421 was

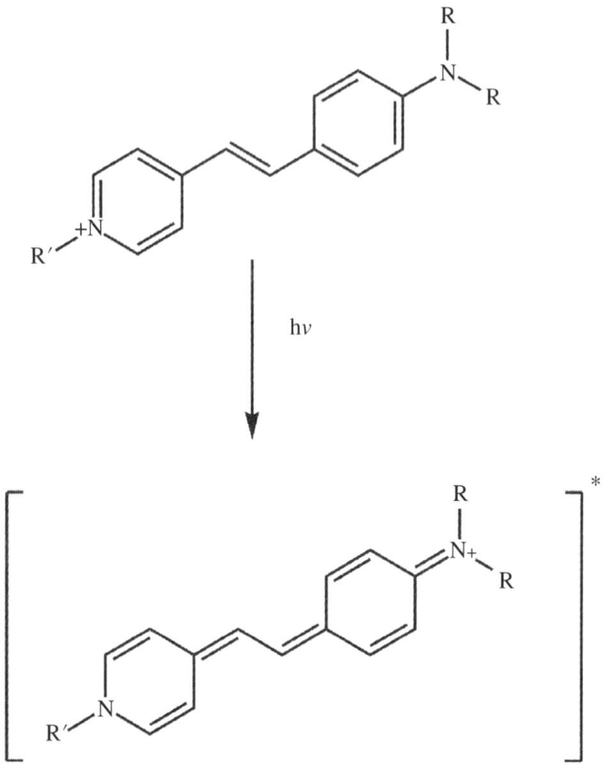

FIGURE 7.6 Electron redistribution occurring during excitation of the aminostyrylpyridinium chromophore. Reprinted from Ref. 32 with kind permission from Springer Science and Business Media.

found [44] to have a dipole moment of 12 (±2) Debye. Any change in the transmembrane electrical potential would cause a change in electric field strength within the membrane, which would result in the reorientation of the RH421 dipole within the membrane. Lipid membranes are known to possess a sharp polarity gradient on going from the lipid headgroup region into the hydrocarbon interior of the membrane. Because of their amphiphilic structure, the dyes are thought to reside precisely in this region of sharply changing polarity. Therefore, any small reorientation of the dye population in the membrane brought about by a change in the transmembrane electric field could significantly change the average local polarity the dye molecules experience. This would cause a solvatochromic shift in their absorbance and fluorescence excitation spectra, because a change in polarity would differentially stabilize or destabilize the ground and excited states of the dye, just as a transmembrane electric field does. Furthermore, if the dye reorients so that the long axis of its chromophore becomes more perpendicular to the membrane surface, this would strengthen its interaction with the transmembrane field and yield a larger electrochromic shift. Thus, electrochromism and solvatochromism could work together to increase the overall response of a dye to the transmembrane electrical potential.

Apart from the chromophore, other important structural aspects of the RH and ANEPPS dyes are that they each possess a negatively charged sulfonate group attached via an alkyl chain linker to the pyridinium ring and two hydrocarbon chains attached to the anilino nitrogen (see, e.g., RH421 in Fig. 7.5). These groups have two important purposes. Firstly, the hydrophilic sulfonate group acts as an anchor fixing that end of the molecule at the interface of the membrane with the adjacent aqueous medium. The hydrophobic hydrocarbon chains, on the other hand, insert deeply into the hydrophobic interior of the membrane. Thus, together, the sulfonate group and the hydrophobic chains tend to orient the dye molecules so that the major component of their chromophore is aligned perpendicular to the membrane surface, which strengthens the interaction with the transmembrane electric field.

The other purpose of the sulfonate group is to prevent the dye from diffusing across the membrane [38, 40]. Measurements on synthetic bilayers have shown [38] that the direction of the fluorescence change on application of a transmembrane electric field reverses if dye is added to the inside of the bilayer rather than the outside. This is because the polarity of the field is opposite for dye in the internal and external leaflets of the bilayer. Thus, an inside positive electrical potential destabilizes the positive charge on the pyridinium nitrogen of dye in the inner leaflet, whereas, if dye molecules are in the outer leaflet, positive charges on their pyridinium nitrogens are stabilized. If dye were present in equal concentrations in both leaflets, their electrochromic responses would, therefore, exactly cancel. The same applies to a reorientation/solvatochromic response mechanism. When using the dyes to detect changes in transmembrane electrical potential in cells, cell organelles, or closed vesicles, it is important, therefore, to just add the dyes to one side of the membrane. The dyes may not be completely impenetrable to the membrane, but the sulfonate group ensures that any flip-flop that does occur is very slow.

7.5   MEASUREMENT OF THE KINETICS OF THE Na⁺,K⁺-ATPase

Voltage-sensitive fast dyes have been most widely used to image the transmembrane potential and electrical activity of excitable cells, in particular neurons [46–48]. Here, we consider a more specific application, that is, fundamental research on the mechanisms of individual ion-transporting membrane proteins. RH and ANEPPS dyes have been applied in investigations of the mechanisms of a number of ion pumps, for example, the Na⁺,K⁺-ATPase [49–51], the fungal plasma membrane H⁺-ATPase [52–54], bacteriorhodopsin [55], the gastric H⁺,K⁺-ATPase [51, 56], and the sarcoplasmic reticulum Ca^{2+}-ATPase [57].

The first work on the mechanism of an ion pump using fast styrylpyridinium dyes (RH160 and RH421) was that of Klodos and Forbush [49, 50] on the Na⁺,K⁺-ATPase. They used open membrane fragments from dog kidney containing a high surface density of Na⁺,K⁺-ATPase molecules for their measurements. Because the fragments were open on all sides to the surrounding electrolyte solution, there could be no transmembrane potential initially present, and after activation of the Na⁺,K⁺-ATPase by the addition of ATP, the pumping of Na⁺ and K⁺ ions across the membrane by the protein couldn't produce any transmembrane potential. Klodos and Forbush [49, 50], therefore, concluded that the fluorescence changes they observed were due to local electric field effects within the membrane or a direct interaction with the Na^+,K^+-ATPase.

Nagel et al. [53, 54] also used an open membrane fragment system for their studies on the fungal H^+-ATPase with the dye RH160. They observed that the fluorescence responses that occurred on addition of ATP or the inhibitor vanadate were inconsistent with a purely electrochromic mechanism of the dye, that is, in accord with other studies on the mechanisms of styrylpyridinium dyes described in Section 7.4.2.

In the following sections, we will limit ourselves to one particular case study: the use of the dye RH421 in stopped-flow investigations of the mechanism of the Na^+,K^+-ATPase. In the context of this case study, we explain all the important practical considerations relevant to the use of any fast voltage-sensitive dye in kinetic studies of ion pumps. We have chosen this example because up to now, it is the most extensively studied system and the experimental approaches required to obtain the best possible data are already well established.

7.5.1   Dye Concentration

One of the simplest but most important considerations in any kinetic study using fast dyes is what dye concentration to use. Frank et al. [58] found that micromolar concentrations of RH421 inhibit the steady-state activity of the Na^+,K^+-ATPase. Subsequently, using several different time-resolved experimental techniques, Kane et al. [59] were able to localize the reaction step of the enzyme inhibited by the dye. They found that there was no effect of micromolar concentrations of RH421 on the kinetics of phosphorylation of the Na^+,K^+-ATPase and concluded that RH421 inhibits a reaction downstream from phosphorylation, possibly the E1P to E2P conformational transition of the enzyme, which is involved in deocclusion of Na^+ and its release to the extracellular medium. For future kinetic studies, the important practical point is that the RH421 concentration should be kept in the submicromolar range to avoid any dye-induced inhibition and obtain true values of any rate constants.

There are two possible origins for RH421-induced inhibition of the Na^+, K^+-ATPase. One is a direct binding of RH421 to the protein; another is indirect through dye binding to the lipid phase of the membrane. First, we will consider direct protein binding. It is known that RH421 can bind to globular proteins, for example, bovine serum albumin [60] and ribulose-1,5-bisphosphate carboxylase/oxygenase (rubisco) [58], and to polyamino acids, for example, poly(L-lysine), poly(L-arginine), and poly(L-tyrosine) [58]. On the other hand, no spectral changes have been detected when poly(L-glutamic acid), poly(L-aspartic acid), or poly(L-serine) is added to the dye [58]. The observed interactions with poly(L-lysine) and poly(L-arginine) suggest that one mode of dye binding to proteins is via interaction between the negatively charged sulfonate group of the dye and the positively charged side chains of the basic amino acid residues lysine and arginine. Schwappach et al. [60] concluded that RH421 interacts with the protein as well as with the lipid bilayer. This conclusion was based on two findings: (i) resonance energy transfer between the dye and fluorescence probes on the enzyme and (ii) the observation of biexponential fluorescence lifetime decays of RH421 bound to Na$^+$,K$^+$-ATPase-containing membrane fragments. However, more extensive fluorescence studies of the dye have shown that its photochemistry is complex, and multiexponential fluorescence decays are observed even for dye in homogeneous solvents [45, 61–63]. Therefore, the biexponential nature of the fluorescence lifetime decays reported by Schwappach et al. [60] cannot be used as evidence for direct binding to the Na$^+$,K$^+$-ATPase. Recently, using a cytoplasmic fragment of the Na$^+$,K$^+$-ATPase, the group of Kubala (unpublished results) found evidence that at micromolar concentrations, RH421 could interact directly with the ATP binding site, that is, similar to eosin. However, based on the kinetic measurements of Kane et al. [59], an inhibition of the Na$^+$,K$^+$-ATPase in membrane fragments due to a direct competition between ATP and RH421 for binding to the same site seems unlikely because this would significantly slow the kinetics of ATP phosphorylation, in contradiction to experimental observations. It's possible that in the presence of both protein and lipid membrane, the dye may bind preferentially to the lipid membrane rather than to the ATP binding site, in part perhaps because much more lipid is available for binding than ATP sites. Nevertheless, this doesn't exclude the possibility that RH421 may bind elsewhere on the Na$^+$,K$^+$-ATPase and lead to inhibition of pumping activity of the protein in membrane fragments.

The second possible cause of dye-induced Na$^+$,K$^+$-ATPase inhibition is via an effect from dye bound to the lipid phase of the membrane. For any dye to be able to detect changes in intramembrane electric field strength, it must have either a net charge or be zwitterionic, as is the case with RH421. Therefore, charges on the dyes could potentially influence the kinetics of charge movements associated with protein activity. Via electrical measurements on planar lipid bilayers, Malkov and Sokolov [64] showed that the dyes RH421, RH237, and RH160 all increase the electric field strength in the boundary layer of the membrane because of the positive charge on their pyridinium rings. The negatively charged sulfonate group presumably has no effect because it would be localized in the high dielectric constant medium of the adjacent aqueous phase. It has also been found [65] that the electric field produced by bound dye molecules in the membrane affects neighboring dye molecules and shifts their fluorescence excitation spectrum. If the intramembrane electric field strength produced by dye binding to the membrane becomes too high, it's possible that this affects the kinetics of the Na$^+$,K$^+$-ATPase. According to calculations of Malkov and Sokolov [64], the fields produced by the dyes are likely to have only a minor effect on ion conduction through the center of the protein because of screening from the intervening protein mass. There could, however, be other reactions of the Na$^+$,K$^+$-ATPase where there is a stronger effect.

Another important aspect to the question of the dye concentration to use is the effect that the dye has on its own response. Even at concentrations below 1 μM, where one would expect negligible inhibition of the Na$^+$,K$^+$-ATPase, Frank et al. [58] have found that the fluorescence response of RH421 to the addition of ATP to the enzyme decreases rapidly with increasing dye concentration (see Fig. 7.7). Of course in all experiments, the dye concentration needs to be high enough that one can detect its fluorescence, but Figure 7.7 indicates that the dye concentration should be as low as possible to achieve the maximum possible relative fluorescence change. In their

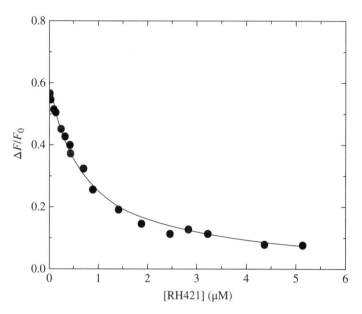

FIGURE 7.7   Decrease in the relative fluorescence change, $\Delta F/F_0$, of Na$^+$,K$^+$-ATPase-containing membrane fragments noncovalently labeled with RH421 as the concentration of the probe increases. The solid line represents a fit of the equation $\Delta F/F_0 = (\Delta F/F_0)_{max} \cdot (K/K + c)$ to the experimental data, where $(\Delta F/F_0)_{max}$ is the theoretical maximum relative fluorescence at infinite dilution of the dye, c is the total dye concentration, and K is the dye concentration at which the relative fluorescence change has dropped to half of its infinite dilution value. The values obtained from fitting were $(\Delta F/F_0)_{max} = 0.593$ (± 0.009) and $K = 0.73$ (± 0.04) µM. Reproduced from Ref. 58 with permission from Elsevier.

stopped-flow experiments in which they mixed the Na$^+$,K$^+$-ATPase-containing membrane fragments with ATP, Kane et al. [59] added 150 nM of RH421 to the protein before mixing. Because equal volumes from both drive syringes were mixed, the RH421 concentration in the stopped-flow observation cell was 75 nM after mixing. This is definitely low enough to avoid any inhibition of the protein's activity as well as to avoid any significant drop-off in the dye's response.

The reason for the drop in dye response with increasing dye concentrations isn't entirely clear at this stage. However, possible explanations have been proposed. It seems reasonable that if the dye is responding to changes in local electric field strength due to the activity of the Na$^+$,K$^+$-ATPase, then dye molecules close to a protein molecule should produce a larger response than those located further away. This explains the much lower response of RH421 when used on Na$^+$,K$^+$-ATPase reconstituted into lipid vesicles [66] in comparison to measurements on Na$^+$,K$^+$-ATPase-containing membrane fragments, which have a very high protein density of up to 10^5 pumps µm^{-2} [67]. Dye molecules located in the membrane but at a large distance from protein molecules would produce a constant background fluorescence and decrease the fluorescence response in membranes with a low protein density relative to those with a high protein density. This by itself doesn't explain the concentration-dependent drop in dye response observed in membrane fragments. However, if the dye binds preferentially in the vicinity of Na⁺,K⁺-ATPase molecules, as efficient RH421 quenching of the tryptophan fluorescence of Na⁺,K⁺-ATPase via resonance energy transfer suggests [58], then, as the dye concentration increases, the amount of dye binding at distances far from a Na⁺,K⁺-ATPase would gradually increase. Thus, the constant background RH421 fluorescence would increase and the relative fluorescence response to Na⁺,K⁺-ATPase activity would decrease.

Another reason or contributing factor to the concentration-dependent drop in dye response could lie in the electric field effect of dye molecules on one another [58], described previously in the context of Na⁺,K⁺-ATPase inhibition. Dye molecules in the membrane surrounding any other particular dye molecule would create a constant electric field strength, which would oppose any changes in local electric field strength arising from Na⁺,K⁺-ATPase activity. In effect, the dye molecules would polarize each other's electron clouds and dipoles and make it harder for any dye molecule to respond via either an electrochromic or reorientational/solvatochromic mechanism. Therefore, the largest fluorescence response would be expected at low dye concentrations where each dye molecule is isolated in the membrane, unaffected by the fields of any other dyes molecules, so that it responds completely to fields originating from the Na⁺,K⁺-ATPase alone.

7.5.2 Excitation Wavelength and Light Source

RH421 responds to a change in its local electric field strength with a shift in its fluorescence excitation spectrum (see Fig. 7.8). Therefore, if the local electric field strength within a membrane changes due to Na⁺,K⁺-ATPase activity or any other cause, there will be a certain excitation wavelength where the excitation spectra before and after the change cross. If one carried out stopped-flow measurements using this particular excitation wavelength, no change in fluorescence would be detected at all. Therefore, one needs to be careful in selecting an appropriate excitation wavelength.

In any system where the fluorescence excitation spectrum shifts, the highest relative changes in fluorescence are observed on the flanks of the spectrum, that is, on the low-wavelength (blue) edge or the high-wavelength (red) edge of the spectrum. For a number of reasons, excitation on the red edge is preferable. Lower energy red-edge excitation is less likely to cause photochemical damage to the dye. It is also less likely to be absorbed by any other components in the reaction mixture, for example, buffers, and cause unwanted background fluorescence. Long wavelength excitation also decreases the amount of light scattering, which is an important consideration when using membrane systems.

The fluorescence intensity detected in any steady-state fluorometric technique such as stopped-flow fluorimetry depends on the steady-state proportion of dye molecules in the excited state relative to the ground state, which increases with the number of photons absorbed per unit time. Therefore, one needs a lamp with a high photon flux at the desired wavelength of excitation to detect a high fluorescence intensity. This is particularly important for stopped-flow measurements, where there

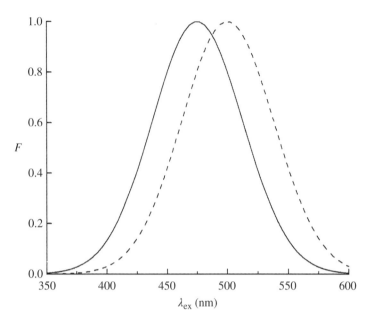

FIGURE 7.8 Shift in the normalized fluorescence excitation spectrum of RH421 expected on changing the intramembrane electric field strength. If the amount of positive charge within the membrane decreases, the fluorescence excitation spectrum of RH421 shifts to longer wavelengths, that is, from the solid curve to the dotted curve.

is a limited time over which the fluorescence is measured. In measurements of fluorescence spectra, many individual spectra can be recorded and averaged to increase the signal-to-noise ratio. However, in stopped-flow, an individual measurement is often over in less than a second. Therefore, one needs to be able to maximize the fluorescence intensity detected. This generally necessitates using an arc lamp, that is, xenon (Xe), mercury (Hg), or mercury–xenon (Hg–Xe), rather than a filament lamp such as a quartz–tungsten halogen (QTH) lamp. The output of QTH lamps is more stable than that of arc lamps, but their intensities are too low for fluorescence, except for measurements in the near infrared.

For the excitation of RH421 in stopped-flow measurements on the Na$^+$,K$^+$-ATPase, the authors have used Hg or Hg–Xe lamps. The advantage of these lamps is that they possess very intense lines corresponding to transitions in the electronic spectrum of Hg vapor. For RH421, there is a perfectly positioned line at 577 nm, which lies exactly on the red edge of the dye's excitation spectrum and is, thus, optimal for yielding large relative fluorescence changes.

7.5.3 Monochromators and Filters

The excitation wavelength can be selected by using an excitation grating monochromator. The transmission efficiency of a monochromator, however, varies with the wavelength of the transmitted light and depends on its construction. The wavelength of maximum transmission is termed the "blaze wavelength" [14]. Because the most common application of stopped-flow fluorimetry involves the measurement of tryptophan fluorescence, commercial stopped-flow instruments are generally supplied with an excitation monochromator grating with a blaze wavelength in the ultraviolet region. However, for more efficient excitation of RH421 at 577 nm, it's desirable to use a grating with a blaze wavelength in the visible range. The authors use a grating with a blaze wavelength of 500 nm. This improves the signal-to-noise ratio of RH421 fluorescence transients, but it doesn't preclude measurements at lower excitation wavelengths for other applications.

To select the fluorescence emission wavelength range, in principle, it's possible to use an emission monochromator in front of the photomultiplier used for detection. However, fluorescence light is always lost on passage through a monochromator. Therefore, to increase the amount of fluorescence light detected, the authors prefer to use a colored glass cutoff filter, that is, a Schott RG665 filter, with 50% transmission at 665 nm. This allows all of the fluorescence above 665 nm to be collected. The large wavelength difference between excitation at 577 nm and emission at ≥665 nm ensures that very little scattered exciting light reaches the photomultiplier and the vast majority of the detected light is actual fluorescence.

If one wished, in principle, one could also use a combination of glass filters or an interference filter on the excitation side rather than a monochromator. This could further increase the intensity of the 577 nm light reaching the sample and increase the emitted fluorescence further. However, increasing the exciting light intensity is a two-edged sword, because it always carries with it the danger that one could at the same time increase the likelihood of photochemical reactions such as bleaching of the dye. Therefore, apart from increasing the efficiency of excitation, it is also worthwhile to consider the efficiency of light detection.

7.5.4 Photomultiplier and Voltage Supply

A voltage needs to be applied across the dynodes of the photomultiplier to amplify the signal for the photon flux of fluorescent light hitting the photocathode of a photomultiplier to produce a measurable output. The higher the applied voltage, the higher the signal (initially a photocurrent, but converted to a photovoltage via a resistor). However, a higher applied voltage also amplifies the noise. Therefore, to obtain the best signal-to-noise ratio, it is best to use the lowest possible applied voltage and use a photomultiplier with a particularly high sensitivity in the wavelength range of detection.

For the long wavelength detection of RH421 fluorescence at wavelengths ≥665 nm, it is desirable to use a photomultiplier tube with a high red sensitivity. A good choice is the Hamamatsu multialkali side-on R928 photomultiplier tube. With the high red sensitivity, the applied photomultiplier voltage can be reduced, reducing the signal noise, and the intensity of the exciting light can be reduced, reducing the danger of photochemical damage. The R928 tube still has excellent sensitivity down into the ultraviolet region, so that its use doesn't preclude other applications.

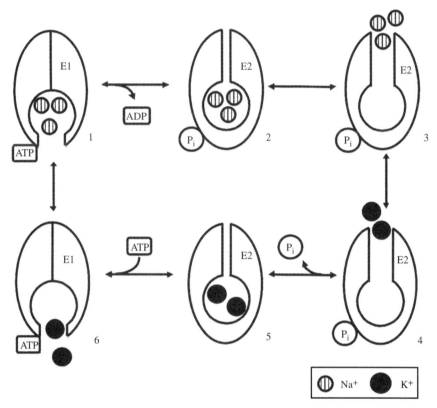

FIGURE 7.9   Albers–Post model of the Na$^+$,K$^+$-ATPase reaction cycle. The entire protein is embedded in the cell plasma membrane (not shown). The top of the protein interfaces with the extracellular medium, whereas the bottom of the protein interfaces with the cytoplasm of the cell. Thus, each cycle, in which one ATP molecule is hydrolyzed to inorganic phosphate P$_i$, involves the pumping of 3Na$^+$ ions out of the cell and 2K$^+$ ions in. Adapted from Ref. 68 with permission from Wiley.

7.5.5   Reactions Detected by RH421

The sequence of reaction steps that the Na$^+$,K$^+$-ATPase undergoes in pumping Na$^+$ and K$^+$ across the plasma membrane is described by the Albers–Post or E1–E2 cycle (see Fig. 7.9). Analogous reaction cycles but with different transported ions can be drawn for all other P-type ATPases. In principle, there are three ways in which stopped flow can be used together with RH421 to determine rate constants of reaction steps of this cycle:

a) Mixing Na$^+$,K$^+$-ATPase with ATP in the presence of Na$^+$ and Mg^{2+}, but in the absence of K$^+$. Reactions occurring: $E1(Na^+)_3 + ATP \rightarrow E1P(Na^+)_3 + ADP$ immediately followed by $E1P(Na^+)_3 \rightarrow E2P + 3Na^+$. Fluorescence of RH421 increases.

b) Mixing Na$^+$,K$^+$-ATPase with Na$^+$ with or without ATP, but in the absence of Mg^{2+} and K$^+$. Reactions occurring: $E2 \rightarrow E1(Na^+)_3$ or $E2ATP \rightarrow E1(Na^+)_3 ATP$. Fluorescence of RH421 decreases.

c) Mixing Na$^+$,K$^+$-ATPase, pre-equilibrated with Na$^+$, Mg^{2+}, and ATP, with K$^+$. Reactions occurring: $E2P + 2K^+ \rightarrow E2P(K^+)_2$ followed immediately by $E2P(K^+)_2 \rightarrow E2(K^+)_2 + P_i$. Fluorescence of RH421 decreases.

In each of these different types of experiments, the Na$^+$,K$^+$-ATPase-containing membrane fragments must be premixed with RH421, added from an ethanolic stock solution, because incorporation of dye into the membrane fragments is a relatively slow process occurring over tens of seconds [69]. As soon as the dye is added to aqueous solution, it forms large aggregates, which need to disaggregate so that the dye can insert in monomeric form into the membrane. The Na$^+$,K$^+$-ATPase-containing membrane fragments noncovalently labeled with RH421 are added to one drive syringe and the substrates with which the enzyme is to be mixed are added to the second drive syringe. RH421 could, in principle, also be added to the second drive syringe to avoid dissociation of dye from the membrane fragments. However, because the dye binds strongly to the membrane, the amount of dissociation is likely to be very small, and the fluorescence of membrane-bound dye is far higher than that of dye in aqueous solution. Therefore, in practice, it seems unnecessary to include dye in the second syringe.

Examples of fluorescence transients obtained using RH421 in the first type of experiment are shown in Figure 7.10 [70]. In this study, Na$^+$,K$^+$-ATPase-containing membrane fragments noncovalently labeled with RH421 were equilibrated in a buffer containing Na$^+$ and varying concentrations of Mg^{2+} and then rapidly mixed with ATP. K$^+$ was completely excluded from the solutions in both drive syringes. The exclusion of K$^+$ inhibits dephosphorylation of the enzyme and after mixing with ATP causes it to accumulate in the E2P state, which is associated with a high fluorescence of RH421. The transients shown here demonstrate the large amplitudes of the fluorescence changes that can be obtained, that is, relative fluorescence changes of up to 400%. The measurements were made using Na$^+$,K$^+$-ATPase-containing membrane fragments purified from shark rectal glands. In this case, the protein concentration of the preparation was particularly high, that is, 4.82 mg ml^{-1}. Smaller relative fluorescence changes are more typical, but values of more than 100% are not unusual.

The aim of the measurements shown in Figure 7.10 was to determine the enzyme's dissociation constant, K_d, for Mg^{2+} from the Mg^{2+} concentration dependence of the observed rate constants, k_{obs}, or reciprocal relaxation times, $1/\tau$, of the traces. Mg^{2+} is a necessary cofactor of ATP. Thus, phosphorylation of the enzyme by ATP doesn't proceed unless ATP is complexed with Mg^{2+}. Taking into account the competition for Mg^{2+} complexation by free ATP in solution in the data analysis, the enzyme's K_d for Mg^{2+} could be determined to be 0.069 ($\pm$0.010) mM. This value is indistinguishable from the K_d for complexation of Mg^{2+} by free ATP in solution in the absence of Na$^+$,K$^+$-ATPase of 0.071 ($\pm$0.003) mM, which was determined by isothermal titration

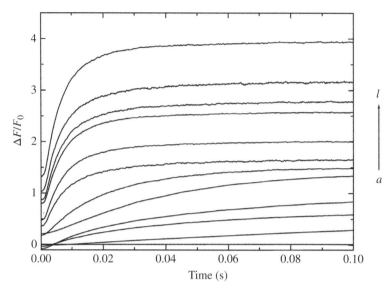

FIGURE 7.10   Stopped-flow fluorescence transients obtained on mixing Na$^+$,K$^+$-ATPase-containing membrane fragments noncovalently labeled with RH421 with ATP in the presence of 130 mM NaCl and varying concentrations of MgCl$_2$. The curves a to l represent increasing Mg^{2+} concentrations from 0 to 5.0 mM after mixing. Reprinted from Ref. 70 with permission from Elsevier.

calorimetry [71]. The similarity of these two values suggests that the enzyme itself has no significant effect on the strength of Mg^{2+} complexation and that Mg^{2+} is held within the enzyme indirectly via complexation with ATP.

This is purely one example of a stopped-flow kinetics study of the Na$^+$,K$^+$-ATPase utilizing RH421. Many others have also been published; see, for example, Refs. 50, 59, 72–84. Together with the results from other kinetic studies, sufficient information has now been accumulated to allow theoretical simulations of the entire reaction cycle of the Na$^+$,K$^+$-ATPase to be carried out under physiological conditions [85].

7.5.6   Origin of the RH421 Response

The origin of the fluorescence response of RH421 in open membrane fragments is a subject that is still the topic of current research. It is clear that RH421 responds to changes in local electric field strength, but what is the origin of the changes in local field strength?

Let us consider first the ATP mixing experiment that produces the large increases in RH421 fluorescence shown in Figure 7.10. Using chymotrypsin-modified Na$^+$, K$^+$-ATPase, which is still able to be phosphorylated but is no longer able to transport Na$^+$ ions, Stürmer et al. [86] found that the fluorescence changes of RH421 previously induced by the addition of ATP to native Na$^+$,K$^+$-ATPase were abolished, thus indicating that the phosphorylation reaction alone is insufficient to produce a fluorescence change. Furthermore, Pratap and Robinson [72] found that after treatment with oligomycin, an inhibitor that has no effect on the rate of phosphorylation and is thought to act by blocking the conversion of E1P to E2P, the rate of the fluorescence change of RH421-labeled Na$^+$,K$^+$-ATPase membrane fragments induced by mixing with ATP decreased by two orders of magnitude. In a similar fashion, Cornelius [76] found that the addition of ADP dramatically slowed the kinetics of the RH421 fluorescence change. This indicates that ADP increases the back reaction rate of a step preceding the step responsible for the fluorescence change. It is known that ADP can stimulate the dephosphorylation of the E1P state (see Fig. 7.9). All of these results are consistent with the conclusion that the RH421 fluorescence change is due to either the deocclusion of Na$^+$, which occurs simultaneously with the conformational transition $E1P(Na^+)_3 \rightarrow E2PNa^+_3$, or the subsequent release of Na$^+$ from $E2PNa^+_3$ to the external solution. The term "deocclusion" refers to an opening of the enzyme to expose ion binding sites to the adjacent aqueous solution. "Occlusion," on the other hand, is a closing of the enzyme so that ions become trapped within the protein matrix with no access to aqueous solutions on either side of the membrane.

Over the last few years, crystal structures of the Na$^+$,K$^+$-ATPase have been determined by X-ray crystallography [87–91]. Based on these structures, it is now possible to theoretically calculate the electric field strength changes in the adjacent membrane when Na$^+$ or K$^+$ ions are released to or bind from, respectively, the extracellular solution. The results of these calculations [92] indicate that it is very unlikely that RH421 directly senses the electric field strength changes that arise in the membrane due to the release or binding of Na$^+$ or K$^+$ to or from the extracellular solution. The fields are expected to be effectively screened by the intervening mass of protein. In comparison, experimental data supporting the conclusion that the dye detects occlusion or deocclusion reactions has recently been obtained [92]. A useful tool for discriminating binding and occlusion is the cation benzyltriethylammonium (BTEA), which is able to bind to the transport sites of the E2P state but which can't be occluded because of its bulky size. It has been found that the addition of K$^+$, Rb$^+$, or Cs$^+$ ions, which are all known to be capable of being occluded, to phosphorylated Na$^+$,K$^+$-ATPase-containing membrane fragments causes a significant drop in RH421 fluorescence. This is in stark contrast to BTEA, which causes only a very small increase in fluorescence. These results, therefore, imply that the large drop in RH421 fluorescence when ions interact with the E2P state and the large increase in fluorescence when the enzyme is phosphorylated in the presence of Na$^+$ ions are due to changes in the state of ion occlusion, not in the state of ion binding per se.

The mechanism by which ion occlusion might cause an RH421 fluorescence change still requires further investigation. However, it was suggested by Frank et al. [58] that a protein conformational change, such as occlusion, might cause a reorganization of the lipids surrounding the protein. This appears to be a feasible origin for the RH421 response because it is known from fluorescence measurements [93] on pure lipid vesicles that the excitation spectrum of RH421 is sensitive to lipid packing, which influences the magnitude of the lipid electrical dipole potential. The effect of lipid packing on dipole potential could be a direct consequence of the change in packing density of the lipid dipoles and associated water dipoles within the membrane/water interface but could also arise partially from a change in water penetration into the membrane, producing a change in the local dielectric constant (to which the electric field strength is inversely proportional). Therefore, any reorganization of the lipids surrounding the protein occurring as the result of a protein conformational change could well change the electric field within the membrane and hence yield an RH421 fluorescence response. Further support for such a mechanism comes from low resolution X-ray crystallographic measurements [94] on the related P-type ATPase, the sarcoplasmic reticulum Ca^{2+}-ATPase, which indicate that conformational changes of the protein are indeed accompanied by local deformations of the surrounding lipid bilayer. Further investigations into the origin of the RH421 response could, therefore, yield valuable information of fundamental scientific importance concerning the nature of lipid–protein interactions.

7.6　CONCLUSIONS

The utilization of stopped-flow fluorimetry together with voltage-sensitive fluorescent probes is a powerful combination of techniques that has already yielded much valuable information on the kinetics and mechanism of the Na^+,K^+-ATPase. However, these are not standardized research procedures that will produce immediate results of publishable quality. To obtain the best quality data, one needs to be aware of the pitfalls, for example, protein inhibition by the probe, inhibition by the probe of its own response, and the appropriate choice of excitation wavelength. To this point in time, the methods described in this chapter have only been applied to ion pump research in a limited number of laboratories. It is hoped that the tips provided here will help in extending their use and that more valuable information on a wider range of membrane proteins will be obtained. A further factor that may have limited the application of fast voltage-sensitive dyes is the lack of a solid understanding of the origin of their fluorescence response. However, rather than a disincentive, the authors feel that this should really be a stimulus for further research. The resolution of the issue of the dyes' response is likely to lead to information of broad importance to the understanding, not just of a particular ion pump, but how lipids and proteins interact with one another in membranes in general.

ACKNOWLEDGMENTS

The authors would like to acknowledge helpful correspondence with Ted King of TgK Scientific. R.J.C. received financial support from the Australian Research Council (Discovery Grants DP-12003548 and DP150103518) and from the Alexander von Humboldt Foundation.

REFERENCES

1. J.S. Olson, H. Gutfreund, Quentin Howieson Gibson 9 December 1918—16 March 2011, *Biogr. Mems. Fell. R. Soc.*, 60 (2014) 169–210.
2. J.F. Eccleston, J.P. Hutchinson, H.D. White, Stopped-flow techniques, in: S.E. Harding and B.Z. Chowdhry (Eds.), *Protein-Ligand Interactions: Structure and Spectroscopy. A Practical Approach*. Oxford University Press, Oxford, 2001, pp. 201–237.

3. E. Rutherford, The velocity and rate of recombination of the ions of gases exposed to Röntgen radiation, *Phil. Mag.* 44 (1897) 422–440.

4. H. Hartridge, F.J.W. Roughton, A method of measuring the velocity of very rapid chemical reactions, *Proc. R. Soc. Lond. A* 104 (1923) 376–394.

5. G.A. Millikan, Photoelectric methods of measuring the velocity of rapid reactions. III – A portable micro-apparatus applicable to an extended range of reactions, *Proc. R. Soc. Lond. A* 155 (1936) 277–292.

6. B. Chance, The accelerated flow method for rapid reactions, Part I. Analysis, *J. Franklin Inst.* 229 (1940) 455–476.

7. B. Chance, The accelerated flow method for rapid reactions, Part I. Analysis (continued), *J. Franklin Inst.* 229 (1940) 613–640.

8. B. Chance, The accelerated flow method for rapid reactions, Part II. Design, construction and tests, *J. Franklin Inst.* 229 (1940) 737–766.

9. B. Chance, Rapid and sensitive spectrophotometry. I. The accelerated and stopped-flow methods for the measurement of the reaction kinetics and spectra of unstable compounds in the visible region of the spectrum, *Rev. Sci. Instrum.* 22 (1951) 619–627.

10. B. Chance, V. Legallais, Rapid and sensitive spectrophotometry. II. A stopped-flow attachment for a stabilized quartz spectrophotometer, *Rev. Sci. Instrum.* 22 (1951) 627–634.

11. Q.H. Gibson, Stopped-flow apparatus for the study of rapid reactions, *Discuss. Faraday Soc.* 17 (1954) 137–139.

12. Q.H. Gibson, F.J.W. Roughton, The kinetics of dissociation of the first oxygen molecule from fully saturated oxyhaemoglobin in sheep blood solutions, *Proc. R. Soc. Lond. B* 143 (1955) 310–334.

13. Q.H. Gibson, L. Milnes, Apparatus for rapid and sensitive spectrophotometry, *Biochem. J.* 91 (1964) 161–171.

14. J.R. Lakowicz, *Principles of Fluorescence Spectroscopy*, third ed., Springer, New York, 2006, pp. 530–534, pp. 31–49, 67–72.

15. S.J.D. Karlish, D.W. Yates, Tryptophan fluorescence of (Na$^+$ + K$^+$)-ATPase as a tool for study of the enzyme mechanism, *Biochim. Biophys. Acta* 527 (1978) 115–130.

16. I. Johnson, M. Spence (Eds.), *The Molecular Probes Handbook, A Guide to Fluorescent Probes and Labeling Technologies*, eleventh ed., Life Technologies Corporation, Carlsbad, CA, 2010.

17. S.J.D. Karlish, Characterization of conformational changes in (Na,K)ATPase labeled with fluorescein at the active site, *J. Bioenerg. Biomembr.* 12 (1980) 111–136.

18. R.A. Farley, C.M. Tran, C.T. Carilli, D. Hawke, J.E. Shively, The amino acid sequence of a fluorescein-labeled peptide from the active site of (Na,K)-ATPase, *J. Biol. Chem.* 259 (1984) 9532–9535.

19. L.D. Faller, R.A. Diaz, G. Scheiner-Bobis, R. Farley, Temperature dependence of rates of conformational changes reported by fluorescein 5′-isothiocyanate modification of H$^+$,K$^+$- and Na$^+$,K$^+$-ATPases, *Biochemistry* 30 (1991) 3503–3510.

20. G. Scheiner-Bobis, J. Antonipillai, R.A. Farley, Simultaneous binding of phosphate and TNP-ADP to FITC-modified Na$^+$,K$^+$-ATPase, *Biochemistry* 32 (1993) 9592–9599.

21. J.G. Kapakos, M. Steinberg, Fluorescent labeling of (Na$^+$ + K$^+$)-ATPase by iodoacetamidofluorescein, *Biochim. Biophys. Acta* 693 (1982) 493–496.

22. M. Steinberg, S.J.D. Karlish, Studies on conformational changes in Na,K-ATPase labeled with 5-iodoacetamidofluorescein, *J. Biol. Chem.* 264 (1989) 2726–2734.

23. M. Nagai, K. Taniguchi, K. Kangawa, H. Matsuo, S. Nakamura, S. Iida, Identification of *N*-[*p*-(2-benzimidazoyl)phenyl]maleimide-modified residue participating in dynamic fluorescence changes accompanying Na⁺,K⁺-dependent ATP hydrolysis, *J. Biol. Chem.* 261 (1986) 13197–13202.

24. S. Geibel, J.H. Kaplan, E. Bamberg, T. Friedrich, Conformational dynamics of the Na⁺/ K⁺-ATPase probed by voltage clamp fluorometry, *Proc. Natl. Acad. Sci.* 100 (2003) 964–969.

25. H. Eguchi, S. Kaya, A. Shimada, Y. Ootomo, K. Nomoto, M. Kikuchi, Y. Usida, K. Taniguchi, ATP-induced dynamic fluorescence changes of a *N*-[*p*-(2-benzimidazoyl) phenyl]maleimide probe at cys241 in the α-chain of pig stomach H⁺,K⁺-ATPase, *J. Biochem.* 122 (1997) 659–665.

26. L.M. Mannuzu, M.M. Moronne, E.Y. Isacoff, Direct physiological measure of conformational rearrangement underlying potassium channel gating, *Science* 271 (1996) 213–216.

27. A. Cha, F. Bezanilla, Characterizing voltage-dependent conformational changes in the Shaker K⁺ channel with fluorescence, *Neuron* 19 (1997) 1127–1140.

28. J.C. Skou, M. Esmann, Effects of ATP and protons on the Na:K selectivity of the (Na⁺ + K⁺)-ATPase studied by ligand effects on intrinsic and extrinsic fluorescence, *Biochim. Biophys. Acta-Biomembr.* 601 (1980) 386–402.

29. J.C. Skou, M. Esmann, Eosin, a fluorescent probe of ATP binding to the (Na⁺ + K⁺)-ATPase, *Biochim. Biophys. Acta-Biomembr.* 647 (1981) 232–240.

30. J.C. Skou, M. Esmann, The effects of Na⁺ and K⁺ on the conformational transitions of (Na⁺ + K⁺)-ATPase, *Biochim. Biophys. Acta-Protein Struct. Molec. Enzym.* 746 (1983) 101–113.

31. M. Esmann, Determination of rate constants for nucleotide dissociation from Na, K-ATPase, *Biochim. Biophys. Acta-Biomembr.* 1110 (1992) 20–28.

32. R.J. Clarke, Electric field sensitive dyes, in: A.P. Demchenko (Ed.), *Advanced Fluorescence Reporters in Chemistry and Biology I: Fundamentals and Molecular Design*, Springer Series on Fluorescence 8, Springer, Heidelberg (2010) pp. 331–344.

33. R.J. Clarke, H.-J. Apell, A stopped-flow kinetic study of the interaction of potential-sensitive oxonol dyes with lipid vesicles, *Biophys. Chem.* 34 (1989) 225–237.

34. H.-J. Apell, B. Bersch, Oxonol VI as an optical indicator for membrane potentials in lipid vesicles, *Biochim. Biophys. Acta-Biomembr.* 903 (1987) 480–494.

35. R.J. Clarke, H.-J. Apell, P. Läuger, Pump current and Na⁺/K⁺ coupling ratio of Na⁺/K⁺-ATPase in reconstituted lipid vesicles, *Biochim. Biophys. Acta-Biomembr.* 981 (1989) 326–336.

36. L.M. Loew, G.W. Bonneville, J. Surow, Charge shift optical probes of membrane potential. Theory, *Biochemistry* 17 (1978) 4065–4071.

37. L.M. Loew, S. Scully, L. Simpson, A.S. Waggoner, Evidence for a charge-shift electrochromic mechanism in a probe of membrane potential, *Nature* 281 (1979) 497–499.

38. L.M. Loew, L.L. Simpson, Charge-shift probes of membrane potential. A probable electrochromic mechanism for *p*-aminostyrylpyridinium probes on a hemispherical lipid bilayer, *Biophys. J.* 34 (1981) 353–365.

39. L.M. Loew, L.B. Cohen, B.M. Salzberg, A.L. Obaid, F. Bezanilla, Charge-shift probes of membrane potential. Characterization of aminostyrylpyridinium dyes on the squid giant axon, *Biophys. J.* 47 (1985) 71–77.

40. E. Fluhler, V.G. Burnham, L.M. Loew, Spectra, membrane binding, and potentiometric responses of new charge shift probes, *Biochemistry* 24 (1985) 5749–5755.

41. L.M. Loew, L.B. Cohen, J. Dix, E.N. Fluhler, V. Montana, G. Salama, W. Jian-young, A naphthyl analog of aminostyryl pyridinium class of potentiometric membrane dyes show consistent sensitivity in a variety of tissue, cell, and model membrane preparations, *J. Membr. Biol.* 130 (1992) 1–10.

42. A. Grinvald, R. Hildesheim, I.C. Farber, L. Anglister, Improved fluorescent probes for the measurement of rapid changes in membrane potential, *Biophys. J.* 39 (1982) 301–308.

43. A. Grinvald, A. Fine, I.C. Farber, R. Hildesheim, Fluorescence monitoring of electrical responses from small neurons and their processes, *Biophys. J.* 42 (1983) 195–198.

44. R.J. Clarke, A. Zouni, J.F. Holzwarth, Voltage sensitivity of the fluorescent probe RH421 in a model membrane system, *Biophys. J.* 68 (1995) 1406–1415.

45. N.V. Visser, A. van Hoek, A.J.W.G. Visser, J. Frank, H.-J. Apell, R.J. Clarke, Time-resolved fluorescence investigations of the interaction of the voltage-sensitive probe RH421 with lipid membranes and proteins, *Biochemistry* 34 (1995) 11777–11784.

46. L.M. Loew, Potentiometric dyes: Imaging electrical activity of cell membranes, *Pure Appl. Chem.* 68 (1996) 1405–1409.

47. M. Zochowski, M. Wachowiak, C.X. Falk, L.B. Cohen, Y.-W. Lam, S. Antic, D. Zecevic, Imaging membrane potential with voltage-sensitive dyes, *Biol. Bull.* 198 (2000) 1–21.

48. A. Grinvald, R. Hildesheim, VSDI: A new era in functional imaging of cortical dynamics, *Nat. Rev. Neurosci.* 5 (2004) 874–885.

49. I. Klodos, B. Forbush III, Rapid conformational changes of the Na/K pump revealed by a fluorescent dye, *J. Gen. Physiol.* 92 (1988) 46a.

50. B. Forbush III, I. Klodos, Rate-limiting steps in Na translocation by the Na/K pump, in: J.H. Kaplan and P. DeWeer (Eds.), *The Sodium Pump: Structure, Mechanism, and Regulation*, Rockefeller University Press, New York, 1991, pp. 211–225.

51. I. Klodos, Partial reactions in Na⁺/K⁺- and H⁺/K⁺-ATPase studied with voltage-sensitive fluorescent dyes, in: E. Bamberg and W. Schoner (Eds.), *The Sodium Pump: Structure, Mechanism, Hormonal Control and Its Role in Disease*, Steinkopff Verlag, Darmstadt, 1994, pp. 517–528.

52. C.L. Slayman, I. Klodos, G. Nagel, Probing for the molecular mechanisms of active transport, in J. Dainty, M.I. de Michelis, E. Marré and F. Rasi-Caldogno (Eds.), *Plant Membrane Transport: The Current Position*, Elsevier, Amsterdam, 1989, pp. 467–472.

53. G. Nagel, E. Bashi, C.L. Slayman, Spectral tagging of reaction substrates in a sodium pump analogue: The plasma membrane H⁺-ATPase of Neurospora, in: J.H. Kaplan and P. DeWeer (Eds.), *The Sodium Pump: Recent Developments*, Rockefeller University Press, New York, 1991, pp. 493–498.

54. G. Nagel, E. Bashi, F. Firouznia, C.L. Slayman, Probing the molecular mechanism of a P-type H⁺-ATPase by means of fluorescent dyes, in: E. Quagliariello and F. Palmieri (Eds.), *Molecular Mechanisms of Transport*, Elsevier, Amsterdam, 1992, pp. 33–40.

55. J. Heberle, N.A. Dencher, Surface-bound optical probes monitor proton translocation and surface potential changes during the bacteriorhodopsin photocycle, *Proc. Natl. Acad. Sci. U. S. A.* 89 (1992) 5996–6000.

56. A. Diller, O. Vagin, G. Sachs, H.-J. Apell, Electrogenic partial reactions of the gastric H,K-ATPase, *Biophys. J.* 88 (2005) 3348–3359.

57. C. Butscher, M. Roudna, H.-J. Apell, Electrogenic partial reactions of the SR-Ca-ATPase investigated by a fluorescence method, *J. Membr. Biol.* 168 (1999) 169–181.

58. J. Frank, A. Zouni, A. van Hoek, A.J.W.G. Visser, R.J. Clarke, Interaction of the fluorescent probe RH421 with ribulose-1,5-bisphosphate carboxylase/oxygenase and with Na^+,K^+-ATPase membrane fragments, *Biochim. Biophys. Acta-Biomembr.* 1280 (1996) 51–64.

59. D.J. Kane, K. Fendler, E. Grell, E. Bamberg, K. Taniguchi, J.P. Froehlich, R.J. Clarke, Stopped-flow kinetic investigations of conformational changes of pig kidney Na^+, K^+-ATPase, *Biochemistry* 36 (1997) 13406–13420.

60. B. Schwappach, W. Gassman, P.A. Fortes, Interaction of the voltage-sensitive dye RH421 with (Na,K)-ATPase, *Biophys. J.* 59 (1991) 339a.

61. Zouni, A., R.J. Clarke, A.J.W.G. Visser, N.V. Visser, J.F. Holzwarth, Static and dynamic studies of the potential-sensitive membrane probe RH421 in dimyristoylphosphatidylcholine vesicles, *Biochim. Biophys. Acta-Biomembr.* 1153 (1993) 203–212.

62. N.V. Visser, A. van Hoek, A.J.W.G. Visser, R.J. Clarke, J.F. Holzwarth, Time-resolved polarized fluorescence of the potential-sensitive dye RH421 in organic solvents and micelles, *Chem. Phys. Lett.* 231 (1994) 551–560.

63. T.H.N. Pham, R.J. Clarke, Solvent dependence of the photochemistry of the styrylpyridinium dye RH421, *J. Phys. Chem. B* 112 (2008) 6513–6520.

64. D.Y. Malkov, V.S. Sokolov, Fluorescent styryl dyes of the RH series affect a potential drop on the membrane/solution boundary, *Biochim. Biophys. Acta-Biomembr.* 1278 (1996) 197–204.

65. R.J. Clarke, D.J. Kane, Optical detection of membrane dipole potential: Avoidance of fluidity and dye-induced effects, *Biochim. Biophys. Acta-Biomembr.* 1323 (1997) 223–239.

66. R. Bühler, W. Stürmer, H.-J. Apell, P. Läuger, Charge translocation by the Na,K-pump: I. Kinetics of local electric field changes studied by time-resolved fluorescence measurements, *J. Membr. Biol.* 121 (1991) 141–161.

67. N. Deguchi, P.L. Jørgensen, A.B. Maunsbach, Ultrastructure of the sodium pump. Comparison of thin sectioning, negative staining, and freeze-fracture of purified, membrane-bound (Na^+,K^+)-ATPase, *J. Cell Biol.* 75 (1977) 619–634.

68. G. Scheiner-Bobis, The sodium pump. Its molecular properties and mechanics of ion transport, *Eur. J. Biochem.* 269 (2002) 2424–2433.

69. A. Zouni, R.J. Clarke, J.F. Holzwarth, Kinetics of solubilization of styryl dye aggregates by lipid vesicles, *J. Phys. Chem.* 98 (1994) 1732–1738.

70. A. Pilotelle-Bunner, F. Cornelius, P. Sebban, P.W. Kuchel, R.J. Clarke, Mechanism of Mg^{2+} binding in the Na^+,K^+-ATPase, *Biophys. J.* 96 (2009) 3753–3761.

71. A. Pilotelle-Bunner, J.M. Mathews, F. Cornelius, H.-J. Apell, P. Sebban, R.J. Clarke, ATP binding equilibria of the Na^+,K^+-ATPase, *Biochemistry* 47 (2008) 13103–13114.

72. P.R. Pratap, J.D. Robinson, Rapid kinetic analyses of the Na^+/K^+-ATPase distinguish among different criteria for conformational change, *Biochim. Biophys. Acta-Biomembr.* 1151 (1993) 89–98.

73. R.J. Clarke, D.J. Kane, H.-J. Apell, M. Roudna, E. Bamberg, Kinetics of the Na^+-dependent conformational changes of rabbit kidney Na^+,K^+-ATPase, *Biophys. J.* 75 (1998) 1340–1353.

74. D.J. Kane, E. Grell, E. Bamberg, R.J. Clarke, Dephosphorylation kinetics of pig kidney Na^+,K^+-ATPase, *Biochemistry* 37 (1998) 4581–4591.

75. F. Cornelius, N.U. Fedosova, I. Klodos, E_2P phosphoforms of Na,K-ATPase. II. Interaction of substrate and cation-binding sites in Pi phosphorylation of Na,K-ATPase, *Biochemistry* 37 (1998) 16686–16696.

76. F. Cornelius, Rate determination in phosphorylation of shark rectal Na,K-ATPase by ATP: Temperature sensitivity and effects of ADP, *Biophys. J.* 77 (1999) 934–942.

77. C. Saudan, E. Schick, P. Bugnon, E. Lewitzki, P. Buet, A.E. Merbach, E. Grell, Quantifying transient conformational transitions of Na,K-ATPase: volume changes and activation volumes, in: K. Taniguchi and S. Kaya (Eds.), *Na/K-ATPases and Related ATPases*, Elsevier, Amsterdam, 2000, pp. 441–444.

78. C. Lüpfert, E. Grell, V. Pintschovius, H.-J. Apell, F. Cornelius, R. J. Clarke, Rate limitation of the Na^+,K^+-ATPase pump cycle, *Biophys. J.* 81 (2001) 2069–2081.

79. P.A. Humphrey, C. Lüpfert, H.-J. Apell, F. Cornelius, R.J. Clarke, Mechanism of the rate-determining step of the Na^+,K^+-ATPase pump cycle, *Biochemistry* 41 (2002) 9496–9507.

80. R.J. Clarke, H.-J. Apell, B.Y. Kong, Allosteric effect of ATP on Na^+,K^+-ATPase conformational kinetics, *Biochemistry* 46 (2007) 7034–7044.

81. R.J. Clarke, D.J. Kane, Two gears of pumping by the sodium pump, *Biophys. J.* 93 (2007) 4187–4196.

82. M. Khalid, G. Fouassier, H.-J. Apell, F. Cornelius, R.J. Clarke, Interaction of ATP with the phosphoenzyme of the Na^+,K^+-ATPase, *Biochemistry* 49 (2010) 1248–1258.

83. M. Khalid, F. Cornelius, R.J. Clarke, Dual mechanisms of allosteric acceleration of the Na^+,K^+-ATPase by ATP, *Biophys. J.* 98 (2010) 2290–2298.

84. S.L. Myers, F. Cornelius, H.-J. Apell, R.J. Clarke, Kinetics of K^+ occlusion by the phosphoenzyme of the Na^+,K^+-ATPase, *Biophys. J.* 100 (2011) 70–79.

85. R.J. Clarke, M. Catauro, H.H. Rasmussen, H.-J. Apell, Quantitative calculation of the role of the Na^+,K^+-ATPase in thermogenesis, *Biochim. Biophys. Acta-Bioenerg.* 1827 (2013) 1205–1212.

86. W. Stürmer, R. Bühler, H.-J. Apell, P. Läuger, Charge translocation by the Na,K-pump: II. Ion binding and release at the extracellular face, *J. Membr. Biol.* 121 (1991) 163–176.

87. J.P. Morth, B.P. Pedersen, M.S. Toustrup-Jensen, T.L.-M. Sørensen, J. Petersen, J.P. Andersen, B. Vilsen, P. Nissen, Crystal structure of the sodium-potassium pump, *Nature* 450 (2007) 1043–1049.

88. T. Shinoda, H. Ogawa, F. Cornelius, C. Toyoshima, Crystal structure of the sodium-potassium pump at 2.4 Å resolution, *Nature* 459 (2009) 446–450.

89. H. Ogawa, T. Shinoda, F. Cornelius, C. Toyoshima, Crystal structure of the sodium-potassium pump (Na^+,K^+-ATPase) with bound potassium and ouabain, *Proc. Natl. Acad. Sci.* 106 (2009) 13742–13747.

90. M. Nyblom, H. Poulsen, P. Gourdon, L. Reinhard, M. Anderson, E. Lindahl, N. Fedosova, P. Nissen, Crystal structure of Na^+,K^+-ATPase in the Na^+-bound state, *Science* 342 (2013) 123–127.

91. R. Kanai, H. Ogawa, B. Vilsen, F. Cornelius, C. Toyoshima, Crystal structure of a Na^+-bound Na^+,K^+-ATPase preceding the E1P state, *Nature* 502 (2013) 201–206.

92. L.J. Mares, A. Garcia, H.H. Rasmussen, F. Cornelius, Y.A. Mahmmoud, J.R. Berlin, B. Lev, T.W. Allen, R.J. Clarke, Identification of electric-field-dependent steps in the Na^+,K^+-pump cycle, *Biophys. J.* 107 (2014) 1352–1363.

93. R.J. Clarke, Effect of lipid structure on the dipole potential of phosphatidylcholine bilayers, *Biochim. Biophys. Acta-Biomembr* 1327 (1997) 269–278.

94. Y. Sonntag, M. Musgaard, C. Olesen, B. Schiøtt, J.V. Møller, P. Nissen, L. Thøgersen, Mutual adaptation of a membrane protein and its lipid bilayer during conformational changes, *Nat. Commun.* 2 (2011) 304.

8

NUCLEAR MAGNETIC RESONANCE SPECTROSCOPY

PHILIP W. KUCHEL

School of Molecular Bioscience, University of Sydney, Sydney, New South Wales, Australia

8.1 INTRODUCTION

In this chapter, we introduce you to the background theory and applications of nuclear magnetic resonance (NMR) spectroscopy to the measurement of fluxes of water, solutes, and ions across cell membranes. Due to the different chemical environment inside and outside cells, signals arising from NMR-active nuclei on either side of a cell membrane can be distinguished. Time-resolved changes in the intensity of these signals enable the kinetics of membrane pores, channels, and pumps to be analyzed.

All experiments described later are considered only in outline; should you decide to apply the methods to your own system of study, you will need to consult the materials and methods sections of the cited papers.

Textbooks on NMR spectroscopy abound. Some of the excellent ones are [1–8]. Those readers familiar with NMR spectroscopy may wish to skip this section and proceed to the applications in Sections 8.2–8.10. The descriptions of NMR phenomena given here are only qualitative, but they are intended to be accurate, even though the concepts are deeply rooted in quantum mechanics [1]. The subsections in this introduction should provide sufficient insight into the various NMR phenomena to be obvious how they might be used in membrane transport studies.

Pumps, Channels, and Transporters: Methods of Functional Analysis, First Edition. Edited by Ronald J. Clarke and Mohammed A. A. Khalid.
© 2015 John Wiley & Sons, Inc. Published 2015 by John Wiley & Sons, Inc.

8.1.1 Definition of NMR

NMR spectroscopy generates spectra that reflect the absorption, by matter, of electromagnetic radiation that is in the radio-frequency (RF) range. A notable feature of the NMR phenomenon is that it only takes place if the sample resides in an externally applied magnetic field. The mechanism of the absorption is based on the interaction between the magnetic component of an RF electromagnetic wave and the magnetic dipole of an atomic nucleus. By virtue of its interaction with the applied magnetic field, the otherwise degenerate energy levels of the nucleus are split (called the Zeeman effect). In the classical description of the system, the nucleus "precesses," as does a spinning top in the Earth's gravitational field, around the direction of the applied magnetic field, denoted by $\mathbf{B}_0$ (this is a vector, hence the bold letter, and the subscript zero specifies this as the externally applied magnetic field).

8.1.2 Why So Useful?

Because NMR spectroscopy measures energy absorption at wavelengths of electromagnetic radiation that are in the centimeter range, the extent of background scattering is small. The Rayleigh scattering equation indicates that the extent of nonspecific scattering of electromagnetic radiation at a given angle away from the direction of the input beam varies as the inverse fourth power of the wavelength [9]. Thus, only specific frequencies of photons are absorbed, as described by the quantum-mechanical nature of the energy transfer, so optically opaque samples such as cell suspensions can reveal their internal molecular secrets to NMR spectroscopy!

In other words, because the electromagnetic radiation used in NMR experiments is of such a long wavelength, scattering by particles in a suspension on the size scale of a micrometer or less do not scatter the radiation and hence do not interfere with the measurement. Therefore, measurements of solute concentrations can be made of optically opaque systems, which would be very difficult to do using electronic spectroscopy techniques, for example, UV–vis absorption or fluorescence, where the wavelength of the incident radiation is in the visible range.

8.1.3 Magnetic Polarization

If a sample of water, say, in a long cylindrical glass tube, is placed inside a solenoidal magnet, it becomes magnetized, with its north pole pointing in the direction of the south pole of the solenoid. We say that the sample has become magnetically polarized. This polarization is a consequence of the proton in the nucleus of each hydrogen atom possessing a magnetic dipole. In fact, a much larger contribution to the overall induced "magnetization" (magnetic dipole moment per unit volume) comes from each electron that also has a magnetic dipole (see Chapter 12). But in terms of NMR experiments, it is the nuclear magnetization with which we are concerned.

Not all atomic nuclei lead to magnetic polarization, but those that do are amenable to study by NMR spectroscopy. More is said about this below, but note that in a biochemical-NMR context, the key nuclear magnetic isotopes (nuclides) are ^{1}H, ^{13}C, ^{19}F, ^{23}Na, and ^{31}P.

The magnitude of the magnetization of the sample is accounted for by a Boltzmann distribution of nuclear magnetic dipoles between two energy levels. (For some nuclides that are termed "quadrupolar," there are more than two energy levels.) The net magnetization is a consequence of there being more nuclear dipoles in the low-energy state, pointing in the direction of $\mathbf{B}_0$, than pointing in the opposite direction in a high-energy state. Input of energy from an RF field perturbs the system away from the Boltzmann distribution at equilibrium.

8.1.4 Larmor Equation

Because a magnetic nucleus also has angular momentum (as for any spinning mass), and because it is not necessarily perfectly aligned with $\mathbf{B}_0$, it is subjected to a torque that makes it precess about the direction of $\mathbf{B}_0$. The precession frequency is called the Larmor frequency; it is this frequency at which the energy of electromagnetic radiation in the RF range is transferred to the nuclear system. The matching of the two frequencies to give energy transfer is referred to as resonance, hence the appellation, NMR.

The master equation of NMR is the Larmor equation:

$$\boldsymbol{\omega} = -\gamma \mathbf{B}_{loc} \tag{8.1}$$

where the precession frequency (a vector, so we use plain bold type to denote it) is $\boldsymbol{\omega}$ (in $\mathrm{rad\,s^{-1}}$). The direction of this vector indicates the axis around which precession takes place, and the convention used to declare which is the positive direction is based on a Cartesian coordinate system. The z-axis is chosen as the direction of the main magnetic field, along the bore of a modern super conducting solenoidal magnet. (Apologies for the sleight-of-hand with the use of $\mathbf{B}_{loc}$ and not $\mathbf{B}_0$ in Equation 8.1; this is explained below.)

The γ in Equation 8.1 is the magnetogyric ratio; it is the ratio of the angular momentum to the magnetic dipole moment of the nucleus. This proportionality constant (units $\mathrm{MHz\,T^{-1}}$) has a unique value for each nuclide, and it defines, according to Equation 8.1, the resonance frequency at a given magnetic field strength. Hence, the resonance frequency is under the control of the experimenter; change the strength of the magnetic field ($|\mathbf{B}_0|$, in units of Tesla, where the vertical bars denote the magnitude of the vector) and the resonance frequency changes in direct proportion to the changed magnetic field strength.

In a typical modern NMR spectrometer ($|\mathbf{B}_0| = 7$–$20\,\mathrm{T}$), the resonance frequencies of the nuclides are in the RF, that is, MHz, range.

8.1.5 Chemical Shift

Every element of the periodic table has at least one isotope whose nuclei are magnetic. Such a nucleus is said to have nonzero spin, and each such nuclide has its own value of the magnetogyric ratio. This makes it amenable to separate identification from other nuclides based on its NMR absorption frequency. If the story ended with the Larmor equation and the NMR-receptive nuclides, we would only be able to distinguish between some isotopes of the elements of the periodic table based on their "signature" NMR resonance frequencies. But something quite remarkable happens with atoms of a given nuclide at different locations in a particular molecule. Electrons surrounding a nucleus form a cloud that shields the applied magnetic field B_0 away from the nucleus, thus reducing the local magnetic field at the nucleus. Then, according to the Larmor equation, Equation 8.1, the resonance frequency is less than might have otherwise been the case. And, within a molecule, different atoms in different locations have subtle variations in the shielding effect of the electrons. Hence, this different chemical environment brings about a change in the resonance frequency that is referred to as the "chemical shift." Because this shielding is very subtle, and leads to changes in RF resonance frequency that are approximately a millionth fraction of the main RF frequency, it is usual to express the "chemical shift" in units of parts per million (ppm). We denote it by δ, as follows:

$$\delta\,(\text{ppm}) = \frac{v_0 - v_{\text{sample}}}{v_0} \times 10^6 \tag{8.2}$$

where v_0 is the resonance frequency (typically in the MHz range) of a particular set of nuclear spins in a reference compound and v_{sample} is that of the spins in an individual moiety in the sample. The chemical shift of nuclear spins in a particular chemical location in a molecule is independent of the magnitude of B_0. Therefore, spectra from different NMR spectrometers can be directly compared in terms of the chemical shifts of the spectral peaks (more correctly referred to as resonances).

8.1.6   Free Induction Decay

An NMR spectrum is generated from the response of the nuclear spins in a sample to an RF pulse (or a sequence of pulses), usually of a few microseconds duration (see Fig. 8.1 for a set of valuable pulse sequences). An RF pulse induces a coherence of the directions of magnetization vectors of the nuclear spins as they precess around B_0 at their Larmor frequency; this net magnetization vector induces a voltage in the receiver coil that embraces the sample, in the same way that a bar magnet induces a voltage in a coil when it is moved rapidly past it. The induced voltage decays over time; hence, the time domain signal is called a free (since the system is *free* of the RF field after the pulse) induction decay (FID). The time over which the FID approaches zero varies from one nuclide to another, but for ^{1}H atoms in molecules like glucose, it is approximately 1 s, while for ^{23}Na$^+$ in aqueous solution, it is typically approximately 50 ms. (There is more discussion on this in Section 8.1.7.)

A computer connected to the NMR spectrometer records the FID and converts it into a spectrum, which is a graph of absorption intensity as a function of frequency or alternatively, chemical shift (see Fig. 8.2). This elegant transformation from a time domain function to a frequency domain function is performed by a numerical Fourier transformation in the computer. While it was a tedious process 30 years ago, it is almost instantaneous today.

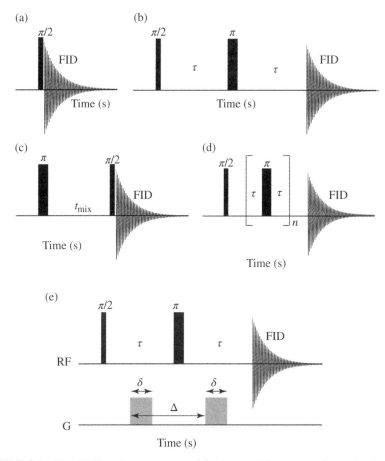

FIGURE 8.1 Five NMR pulse sequences of fundamental importance in applications like those described in this chapter; the various symbols denote key times in the sequences with $\pi/2$ and π being RF pulses of sufficient duration to nutate the magnetization through the indicated number of radians. (a) The simplest of all, "pulse and acquire"; (b) the "Hahn spin echo" in which 2τ is the "echo time"; (c) the "inversion recovery," used for measuring T_1, in which t_{mix} is typically varied over a sequence of around 12 values; (d) the "Carr–Purcell–Meiboom–Gill," used for measuring T_2, in which τ and n are varied over a range of values in making the measurement; and (e) the "pulsed field gradient spin echo" (PGSE), used for measuring the translational diffusion coefficient, D; the rectangles (green in the online version) denote magnetic field gradient pulses of duration δ and separation Δ and whose magnitude is usually varied through 12 values in the experiment.

8.1.7 Pulse Excitation

It is worth noting that the intense RF pulse of short duration excites a band of frequencies; and the effective bandwidth is inversely proportional to the duration of the pulse [1, 3, 8]. Therefore, experimentally, the power and duration of the RF pulse are chosen so that the full chemical shift domain is spanned across the spectrum.

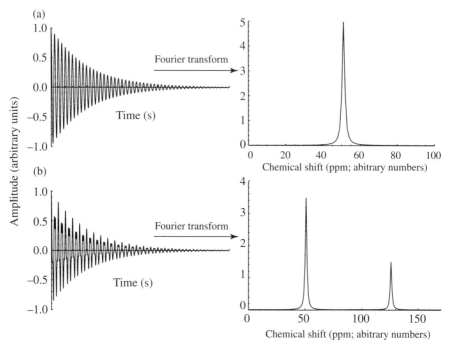

FIGURE 8.2 Free induction decays (FIDs) together with their numerical Fourier transforms (FT), simulated in *Mathematica* [10]. Note the exponential envelopes of the FIDs and the, so-called, Lorentzian line shapes in the frequency spectra. It can be shown mathematically that the FT of an exponential function of time is the Lorentzian function of frequency. (a) Single frequency present in the FID as made evident in the FT. (b) Two frequencies present in the FID. Note the extra repetitive complexity in the FID that arises from the second frequency. In real experimental FIDs, the trace is often much less regular in spite of having an exponential envelope.

To explain how an RF pulse can lead to the excitation of a large number of resonance frequencies, consider the following analogy. Imagine a table with a tuning fork sitting on it. How do we set this ringing without directly hitting it? Well, we place on the table next to it another tuning fork of the same intrinsic frequency that we have previously hit with a percussion hammer. The vibrations from the activated tuning fork are transmitted via the table to the originally nonvibrating tuning fork, which then begins to vibrate; this is called "resonance energy transfer." If there were two or more tuning forks with different intrinsic frequencies, how might we excite them all? Well, we could use a set of previously percussed tuning forks each with a frequency that matches one of those on the table. But there is another simpler strategy: "Whack" the table with a heavy hammer! This intense impulse excites a broad range of frequencies that immediately sets all the tuning forks vibrating.

The NMR counterpart of the hammer impulse is called a "hard" RF pulse; it sets the nuclear spins of different resonance frequencies "ringing." It must be intense (usually 100–1000 W) and, as stated earlier, of only a few microseconds duration.

8.1.8 Relaxation Times

The envelope of an FID can generally be described by a single decaying exponential function, the decay constant of which is called the transverse relaxation time constant, or T_2 [1, 3, 8]. The decay is a consequence of the loss of coherence (separation into different precession frequencies) between the groups of precessing nuclear spin magnetizations. Hence, the net magnetization projected onto the plane transverse to the z-axis (viz., the x,y-plane; the axis of the receiver coil lies parallel to this plane) declines. So the voltage induced in the receiver coil consequently decays. (Note that in practice T_2 is not measured from the FID but by a pulse sequence that generates an echo in the voltage induced in the receiver coil, so it is called a "spin-echo" pulse sequence; we encounter it in Section 8.8.1 and in Figure 8.1b.)

A second relaxation process takes place concomitantly with transverse relaxation: When the RF pulse is applied to a sample, the magnetic polarization undergoes reorientation, changing the Boltzmann distribution between the two energy levels. The return of the system to Boltzmann equilibrium commences as soon as energy input ceases, and it follows a time dependence that is described by a rising exponential function. The relaxation time constant in the expression is called T_1, and the process is referred to as longitudinal relaxation because it entails a return of the magnetic polarization to the direction of $\mathbf{B}_0$, the z-direction [1, 3, 8]. The mechanism underlying this relaxation is fluctuations in the magnetic field at the Larmor frequency in the molecules surrounding a nucleus—the so-called sample "lattice." Hence, T_1 is often referred to as the "spin-lattice relaxation time" or more correctly (because this does not commit to the actual mechanism) the "longitudinal relaxation time constant."

8.1.9 Splitting of Resonance Lines

The resonance lines in an NMR spectrum often contain "fine structure": For example, the peak might be split into subcomponents. The most common mechanism for this is called "scalar coupling," whereby nuclear spins in atoms that are covalently bonded to each other through electrons (from one to four bonds apart in a molecule) sense each other's magnetic dipoles via interactions with the dipoles of their shared electrons [1, 3, 8]. This supplementation or subtraction of magnetic field strength means that what was a single peak will be split into two or more peaks depending on how many adjacent magnetic nuclei there are. Note that this splitting is advantageous because it provides additional information about the chemical environment of the nucleus. However, it can also be detrimental in data analysis because it decreases peak amplitude and can create quite complicated spectra with overlapping peaks that are difficult to assign to particular atoms in a molecule or mixture of molecules.

8.1.10 Measuring Membrane Transport

Any technique used to measure transmembrane exchange (transport) of an ion, solute, or water relies on being able to identify and quantify the ions or molecules on either side of the cell membrane [11–13]. Using NMR spectroscopy to do this entails exploiting differences in one or more of the various NMR parameters, including relaxation time constants, chemical shifts, and splitting patterns, or exploiting differences in diffusion coefficient (which, as we will see in Section 8.8, can be accurately measured using NMR). These differences are brought about by transmembrane differences in paramagnetic ions that affect the relaxation times T_1 or T_2; the presence of metal chelates on one side of the membrane and not on the other, which bring about a chemical shift difference; differences in the viscosity that affect the diffusion coefficient in one compartment more than the other; and differences in pH that can affect chemical shift and scalar coupling constants. Examples of the application of all these effects are given in the following sections.

8.2 COVALENTLY-INDUCED CHEMICAL SHIFT DIFFERENCES

8.2.1 Arginine Transport

Perhaps the simplest NMR experiment for monitoring transport of a molecule across the plasma membrane of a cell is depicted in Figure 8.3a. It requires that the molecule is changed chemically once it enters the cell (see Fig. 8.3b). If the chemical transformation occurs much more rapidly than the transport process, then the rate of appearance of the product directly reports on the kinetics of the transport. Such is the case with arginine transport into the human red blood cell (RBC) [14]. These cells

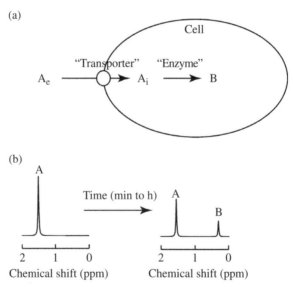

FIGURE 8.3 Intracellular covalent transformation of a solute as the basis for measuring its transport into a cell. The product of the enzymatic reaction has a different chemical shift of the "reporter groups" from that of the substrate. (a) The reaction scheme involving membrane transport and intracellular conversion to a product. (b) Stylized NMR spectra, before and at some later time after transport has taken place.

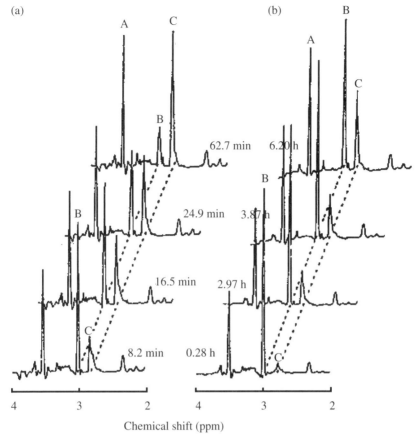

FIGURE 8.4 ^{1}H NMR spectra of arginine incubated with a hemolysate (a) and whole RBCs (b). Arginine is hydrolyzed by arginase inside the RBCs to ornithine and urea. The chemical shift of the δ-CH$_2$ group differs in the two amino acids so when arginine is hydrolyzed its resonance intensity (peak B) declines while the ornithine intensity (peak C) rises. In the hemolysate, there is no membrane impediment to the arginine so the rate of conversion to ornithine is six times faster than in the whole RBCs; hence, the rate of membrane transport was estimated. (Peak A is from both arginine and ornthine.)

contain arginase, which catalyzes with high activity the hydrolysis of arginine to ornithine and urea. The δ-CH$_2$ resonance in the ^{1}H spin-echo NMR spectrum of arginine is well resolved from that of ornithine. To explore the extent to which the overall hydrolytic reaction is dominated by membrane transport, the RBCs are ruptured by sonification or three cycles of freezing and thawing in liquid nitrogen. By comparing Figure 8.4a and b, it can be seen that the reaction has reached a greater extent after approximately 60 min in the hemolysate than the whole cells achieve after approximately 6 h, indicating a sixfold rate difference. Thus, the transport kinetics of arginine under conditions of different substrate concentration, and in the presence of effector molecules, can be studied using this facile assay method.

8.2.2   Other Examples

Human RBCs contain proteases and peptidases, so peptide transport can be measured using the same approach as depicted in Figure 8.3a. As an example, imidodipeptide transport was measured this way [15].

But beware, a stark artifact can "rear its ugly head" with this method. Cell lysis is largely unavoidable during prolonged incubation times (hours), even with robust cells like RBCs; and consequently, small amounts of the intracellular converting enzyme can be released to the suspension medium. Extracellular conversion of the transported solute is then mistakenly interpreted as being due to transport to the inside of the cells. Appropriate control experiments are needed to account for this effect in quantitative kinetic analysis of such systems.

8.3   SHIFT-REAGENT-INDUCED CHEMICAL SHIFT DIFFERENCES

8.3.1   DyPPP

The ^{23}Na NMR spectrum of Na$^+$ in a suspension of cells is a rather uninformative single resonance (called a "singlet"). However, with the introduction of a shift reagent that is suitable for aqueous solutions, it became possible to measure the Na$^+$ concentration inside and outside cells in a suspension. The first such shift reagent was dysprosium tripolyphosphate [16]. The authors showed that with millimolar concentrations of this complex, separate resonances are observed for intra- and extracellular Na$^+$ in RBC suspensions (see Fig. 8.5); so changes in the distribution of Na$^+$ can be measured over time by recording sequential ^{23}Na NMR spectra.

8.3.2   TmDTPA and TmDOTP

Myriad cation-shift reagents, based on lanthanide chelates, have been reported since the Guptas' paper in 1982 [16]. The thulium complex of the pentaacetic acid derivative of diethylamine (DTPA) is a favorable shift reagent, suitable for a range of inorganic cations, and it is effective at approximately 5 mM. The complex is shown in Figure 8.6a [17]. Another chelate that is also very useful is thulium-1,4,7, 10-tetraazacyclodecane-1,4,7,10-tetrakis(methylenephosphonate) (TmDOTP; see Section 8.10). These complexes are readily synthesized in the laboratory from the free chelator and typically the nitrate or chloride salt of the lanthanide. Keeping a watchful eye on the pH is crucial during the addition of the salt solution to the chelator, as a high pH leads rapidly to precipitation of the lanthanide oxide. It is also usual to prepare the solution so there is a slight excess of the chelator, recognizing a 1:1 stoichiometry for DTPA–lanthanide and DOTP–lanthanide, but 2:1 for tripolyphosphate.

8.3.3   Fast Cation Exchange

The two previous examples are concerned with transport on the min-to-hour time-scale. However, Figure 8.7a pertains to a system of Na$^+$ exchange that occurs on the 1-s timescale. This is an exciting prospect for NMR spectral analysis [11]. This

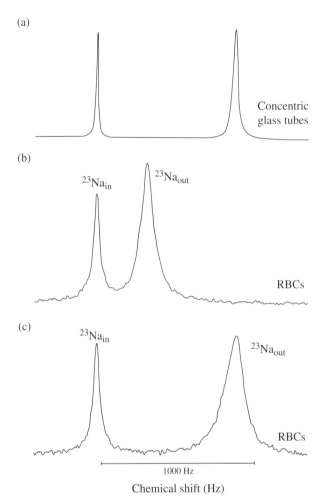

FIGURE 8.5 ^{23}Na NMR spectra showing the effect of dysprosium tripolyphosphate on the chemical shift of ^{23}Na$^+$. (a) Concentric glass NMR tubes, one of which contained 1 mM of the shift reagent in 150 mM NaCl; (b) RBCs in the saline solution with the shift reagent at 2 mM (which is membrane impermeable) on the outside of them; and (c) RBCs with the shift reagent at 5 mM in the extracellular medium. Adapted from Ref. 16 with permission from Elsevier.

method requires the inversion or saturation of the spin state (e.g., the ^{23}Na$^+$ population outside a compartment) and then detecting the perturbation of the signal from the population of ions on the other side of the membrane. The requisite selective RF irradiation is routinely available on modern NMR spectrometers, and the selective inversion experiment is akin to that achieved with the inversion recovery pulse sequence depicted in Figure 8.1c.

Figure 8.7b illustrates another useful aspect of lanthanide-chelate shift reagents: dysprosium-chelates shift the ^{23}Na$^+$ NMR resonance to a lower frequency, while thulium-chelates shift it to a higher frequency. Using this fact, a sample of two

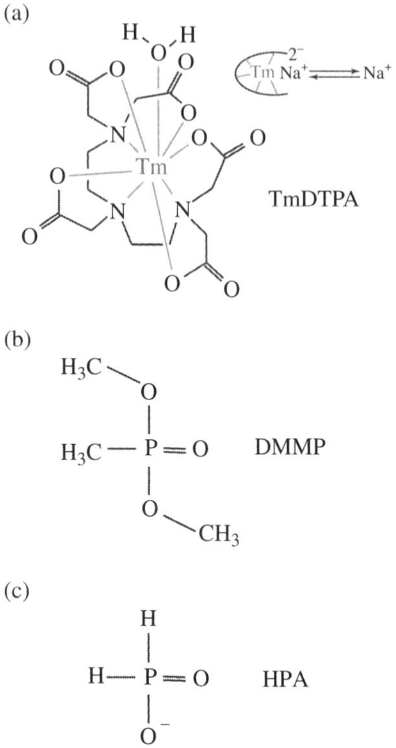

FIGURE 8.6 Three key molecules used in NMR measurements of cell function. (a) Thulium-DTPA (Tm-diethylenetriaminepentacetate), while the inset depicts the rapid exchange of Na$^+$ in the bulk medium with the thulium ion, giving rise to a changed average value of $\mathbf{B}_{loc}$. (b) Dimethyl methylphosphonate (DMMP), which shows separate chemical shifts from its populations inside and outside RBCs; this is due to the different average extent of hydrogen bonding of water to the phosphoryl oxygen atom that is brought about by the higher concentration of proteins that compete for the hydrogen bonds of water inside relative to outside. This we call the "hydrogen-bond split peak effect." (c) Hypophosphite anion (HPA); this also shows the "hydrogen-bond split peak effect" in RBC suspensions. It exchanges across the cell membrane via the Band3 protein and is used to measure the membrane potential.

different phospholipid vesicles was prepared, one containing DyPPP, while the suspension medium contained TmPPP. The exchange of ^{23}Na$^+$ was engineered to be rapid, on the subsecond timescale, by also incorporating the Na$^+$-ionophore monensin into the phospholipid vesicles when they were made. Figure 8.7b shows the ^{23}Na NMR spectrum and reveals the success of making the two-vesicle type of sample. Furthermore, by using magnetization transfer spectra, a full kinetic analysis was done on this system. The two-dimensional exchange spectrum (2D-EXSY; which we will not discuss here) showed that ^{23}Na$^+$ ions mostly exchange between members of the opposite population via the intermediate medium, while direct exchange also takes place, albeit to a much smaller extent; presumably, this exchange is between vesicles that are in close contact with each other [18].

(a)

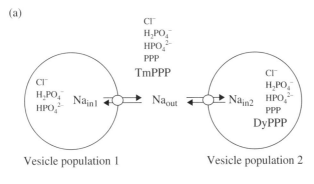

(b)

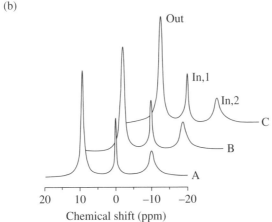

FIGURE 8.7   Two different shift reagents used with ^{23}N NMR to study the kinetics of $^{23}Na^+$ exchange between two different populations of synthetic phospholipid vesicles. The transmembrane exchange was mediated by the ionophore, monensin. (a) The reaction scheme showing the main components of the media in each compartment. (b) ^{23}Na NMR spectra acquired at different temperatures, showing the three resonances that were assigned to the respective compartments: (A) 25°C, (B) 35°C, and (C) 40°C. Adapted from Ref. 18 with permission from Elsevier.

The significance of this type of experiment lies in its use as a model system, to explore interactions (perhaps receptor–ligand mediated) between different vesicle populations, and the role effector molecules might play in these interactions.

8.4   pH-INDUCED CHEMICAL SHIFT DIFFERENCES

8.4.1   Orthophosphate

It had been shown in 1972 that the metabolism of ^{13}C-labeled glucose could be monitored with ^{13}C NMR spectroscopy of whole intact yeast cells [19]. However, it was the observation that ^{31}P NMR spectroscopy could be applied to mammalian cells (RBCs) to report on the intracellular pH and metabolism of a phosphorylated intermediate (2,3-bisphosphoglycerate) that really captured the imagination of biochemists [20]. This work was followed by the first observation of ATP and the estimation of pH in intact muscle [21].

The pH estimation in cells is worth closer scrutiny here, as it is a parameter that is often readily and usefully estimated coincidentally in the course of other more detailed biophysical studies of cells by NMR spectroscopy. The method relies on the finding that the protonation–deprotonation rate of orthophosphate in aqueous media is rapid at pH values around its pK_a of approximately 7.2. The reaction is

$$HPO_4^{2-} + H^+ \rightleftarrows H_2PO_4^-$$

Because the electronic shielding of the ^{31}P nuclei differs in the two differently protonated and charged species, they have intrinsically different chemical shifts. In fact, by recording ^{31}P NMR spectra of phosphate solutions at pH ~4 and ~9, it is seen that the values for the acid and base forms, δ_{acid} and δ_{base}, respectively, are separated by approximately 2.3 ppm.

The fast exchange implies that a ^{31}P nucleus in the orthophosphate is subjected to a local magnetic field that is the weighted average of the relative concentrations of the two species. Thus,

$$\delta_{observed} = \frac{\left[HPO_4^{2-}\right]\delta_{base} - \left[H_2PO_4^-\right]\delta_{acid}}{\left[HPO_4^{2-}\right] + \left[H_2PO_4^-\right]} \tag{8.3}$$

By adapting the Henderson–Hasselbalch equation to the two extreme chemical shifts measured in separate experiments, the estimate of pH is given by

$$pH = pK_a' + \log_{10}\left(\frac{\delta_{observed} - \delta_{acid}}{\delta_{base} - \delta_{observed}}\right) \tag{8.4}$$

Note the prime on pK_a': It is there to remind us that it is an apparent pK_a that must be used for our particular buffer or intracellular conditions because the value varies, especially with the ionic strength and presence of covalent cations.

8.4.2   Methylphosphonate

If a molecular species with an NMR-receptive nuclide has a pK_a in the pH range of either side of a cell membrane, and if it is in fast exchange between two (or more) protonation states, then its transfer across the membrane can be followed by NMR spectroscopy. Such a molecule is methylphosphonate. It is a valuable ^{31}P NMR pH probe in its own right [22, 23], and its transport can readily be tracked using sequential ^{31}P NMR spectra. This could be of use in studies of divalent anion transport in cells.

8.4.3   Triethylphosphate: ^{31}P Shift Reference

All NMR methods used for measuring pH with different nuclides require a chemical shift reference compound to be added to the sample; or in some favorable circumstances, an endogenous metabolite can be identified as having a chemical shift that is well defined, is out of the way of other resonances, and is pH insensitive. In fact, any other parameter value that depends on an accurate measurement of chemical shift requires a chemical shift reference. All such cases will not be mentioned here, but because ^{31}P NMR methods have been presented previously, we note that triethylphosphate [24] is a useful reference compound because it is water soluble, nontoxic and rapidly exchanges across cell membranes.

8.5 HYDROGEN-BOND-INDUCED CHEMICAL SHIFT DIFFERENCES

8.5.1 Phosphonates: DMMP

Another mechanism giving rise to different NMR chemical shifts that depend on a molecule's location inside or outside cells is the different average extent of hydrogen-bond formation of water to the phosphoryl–oxygen or carbonyl–oxygen or to a fluorine atom in the molecule. The difference arises because the high concentration of protein inside the cells, relative to outside, causes the average extent of hydrogen bonding of water to the probe solute to be less inside the cells. The effect, albeit subtle, is nevertheless of the order of tens of hertz (~0.1 ppm) for phosphonates, somewhat less for carbonyls, and more for ^{19}F-labeled molecules (see below).

Dimethyl methylphosphonate (DMMP; see Fig. 8.6b) is a water-soluble, neutral molecule that rapidly enters RBCs by crossing the phospholipid part of the membrane. It appears not to enter via other transporters such as those for water or glucose. It was the compound that first showed the "hydrogen-bond split peak effect" and on which the physical basis of the effect was worked out [25–28]. The separation of the ^{31}P NMR DMMP resonance into two is shown in Figure 8.8a. Interestingly, as the RBC volume is changed by adding concentrated NaCl solution to the suspension, the separation of the two DMMP resonances increases [25]. Therefore, after calibration of the split, relative to RBC volume (estimated by cell counting and centrifugal hematocrit measurement [29] where the haematocrit gives the volume percent of cells in the sample), the mean volume of the RBCs can be tracked by recording sequential ^{31}P NMR spectra. Thus, the rate of net K[+] entry into the cells in response to adding the K[+]-selective ionophore, valinomycin, is able to be measured [25].

8.5.2 HPA

Another compound that also shows the "hydrogen-bond split peak effect" is the hypophosphite anion (HPA; see Fig. 8.6c). The reaction scheme for its interaction with RBCs is shown in Figure 8.9a. It exchanges across the RBC membrane via the chloride–bicarbonate pore (also called Band3 or capnophorin). Because it is a stable monovalent anion with $pK_a \sim 1.1$, it distributes in all the water available across the RBC membrane, with a concentration ratio that reflects the membrane potential of

(a)

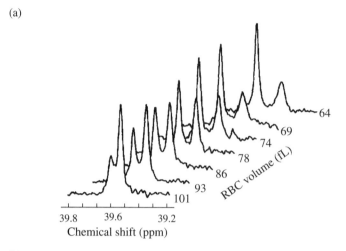

(b)

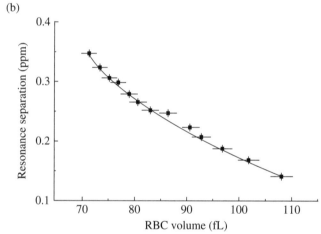

FIGURE 8.8 ³¹P NMR of DMMP in suspensions of RBC in a range of cell volumes. (a) As the cell volume was decreased by osmotic shrinkage by adding concentrated NaCl solutions, the extent of the "split peak effect" increased. (b) Graph of the magnitude of the difference in chemical shift between the resonances assigned to the intra- and extracellular populations of DMMP as a function of mean RBC volume. Adapted from Ref. 25 with permission from Elsevier.

the cell. Assuming the distribution of the anion across the membrane reaches equilibrium, the Nernst equation for membrane potential states

$$\Delta\psi = \frac{RT}{F}\ln\left(\frac{[\text{HPA}_i]}{[\text{HPA}_e]}\right) \tag{8.5}$$

where R is the universal gas constant, T is the absolute temperature, and F is the Faraday constant [12].

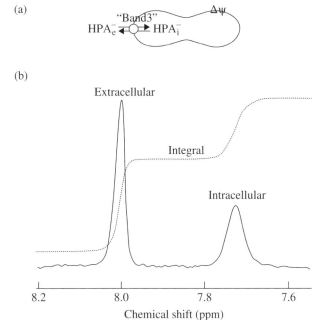

FIGURE 8.9 ^{31}P NMR measurement of RBC membrane potential. (a) Schematic representation of HPA exchange back-and-forth via Band3 (capnophorin; AE1) in RBCs; the relative concentrations on either side of the cell membrane is a measure of the membrane potential. (b) ^{31}P NMR spectrum of HPA in a suspension of human RBCs; the relative values of the integrals (areas; dotted line) is used in the Nernst equation to estimate the membrane potential, $\Delta\Psi$ [12]. Adapted from Ref. 30 with permission from Elsevier.

The volume fraction of the water in the sample that is outside the cells is simply one minus the packing density (hematocrit, i.e., the volume fraction of RBCs) of the sample; the latter is measured by using a capillary centrifuge [29]. The volume fraction of water that is inside the RBCs is based on the hematocrit and the measured mean hemoglobin content of the cells because the partial specific volume of hemoglobin is known, and it is approximately 95% of the protein in the cell. Typically, 71.4% of the volume of a human RBC of normal volume is occupied by water. Therefore, the ratio of the integrals of the ^{31}P NMR spectrum divided by the relative water volume fractions of each compartment yields the transmembrane concentration ratio [HPA$_i$]/[HPA$_e$]. Typically, in a normal sample of human RBCs, the membrane potential is about -10 mV [12].

8.5.3 Fluorides

The strongly electronegative fluorine atom of fluorine-containing molecules forms a polarization bond with hydrogen atoms on molecules like water; this is also a hydrogen bond. Once the mechanism of the split peak effect had been worked out for the phosphoryl compounds, it was surmised that the effect would be evident with

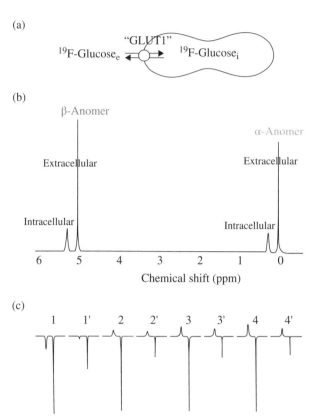

FIGURE 8.10 ^{19}F NMR analysis of the kinetics of transmembrane exchange of glucose (2-^{19}F-2-deoxy-D-glucose) in human RBCs. (a) The reaction scheme, showing that the glucose transporter in these cells is GLUT1. (b) ^{19}F NMR spectrum of ^{19}F-glucose showing the "split peak effect" that arises as a result of the hydrogen bonding of water to the fluorine atom on the modified glucose and that even this form of glucose exists in aqueous solution as a mixture of α- and β-anomers. (c) The series of 1-D exchange spectra (β-anomer section only) used to estimate the rate constant that characterizes the exchange. Adapted from Ref. 31 with permission from The Biochemical Society and Portland Press.

^{19}F-labeled glucose in ^{19}F NMR spectra. Thus, 2-^{19}F-2-deoxy-D-glucose that exchanges across the RBC membrane was shown to yield well-resolved resonances in suspensions of human RBCs (see Fig. 8.10a and b). Note that the simple aldoses and ketoses exist in aqueous solution in equilibrium mixtures of α- and β-anomers. Because the fluorine atom is small, it behaves like a hydrogen atom in terms of its contribution to molecular volume. So the substituted glucose is surmised to behave like natural glucose when it binds to and is transported by glucose transporters. Indeed, Figure 8.10b shows a greater abundance of the β-anomer over the α-anomer, but it is not the ratio 64:36 typically found with unlabeled glucose. A possible complication in using it as a probe of the natural sugars is hydrogen bonding by the fluorine atom. Fluorinated compounds might have additional binding energy if hydrogen-bond donor groups reside in the binding site of the protein.

We showed that the β-anomer is preferentially transmitted across the membrane in our work on the GLUT1 transporter in human RBCs [31]. This conclusion was based on magnetization-inversion experiments; some representative spectra from such an experiment are shown in Figure 8.10c. While the details of the analysis are not given here (through space constraints), it is worth learning how to inspect these spectra for evidence of the exchange process: The best pair of spectra in which to see this effect is those labeled 4 and 4′. Spectrum 4 is from a zero mixing (or exchange) time experiment; it shows the situation after the magnetization of the spins had been "prepared" by two RF pulses and a particular delay period such that the magnetization (resonance) of the intracellular pool is positive and that of the extracellular pool is negative, and the mixing time is taken to be 0 s. Spectrum 4′ was acquired from the setup used for spectrum 4 but after 1 s of mixing time. Because transmembrane exchange has occurred, the positive magnetization of the intracellular fluoroglucose has passed into the pool of negative magnetization of the extracellular pool; but because the magnetization of the extracellular pool is significantly more negative, it suppresses that of the intracellular pool even more. On the other hand, the extracellular peak is additionally diminished in negative intensity as a result of longitudinal relaxation.

As stated earlier, a full kinetic analysis was carried out for these spectra, and maximal velocities and binding affinities were estimated. It was shown that the α-anomer is more rapidly transported than the β-anomer [31], but the physiological significance of this is still unknown.

8.6 IONIC-ENVIRONMENT-INDUCED CHEMICAL SHIFT DIFFERENCES

8.6.1 Cs⁺ Transport

The 54 electrons in 6 energy levels that surround the $^{133}Cs^+$ nucleus exert a strong magnetic field shielding effect on it. Thus, $^{133}Cs^+$ has a chemical shift that is very sensitive to the ionic environment of aqueous media because the electron cloud is readily distorted by electric fields associated with other components of the sample. Even without any added shift reagents, the ^{133}Cs NMR spectrum from a suspension of RBCs that have the added cation shows separate intra- and extracellular resonances.

Because Cs^+ is a congener of K^+, it can be used with ^{133}Cs NMR spectral time courses in the kinetic analysis of cation exchange in cells. One example is the kinetics of Cs^+ transport into isolated hepatocytes [32].

8.7 RELAXATION TIME DIFFERENCES

8.7.1 Mn²⁺ Doping

The oldest NMR method for quantitatively measuring an aspect of cellular function was developed to study water exchange in human RBCs. It was already known at the time that the rate of exchange of water in many cell types, but especially in RBCs, is very fast [33–35].

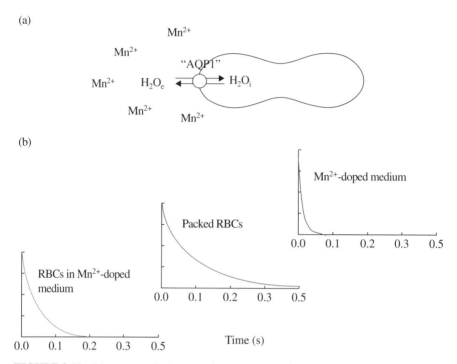

FIGURE 8.11   Manganese-doping experiment used with ^{1}H NMR spectroscopy, to measure water transport via aquaporin 1 (AQP1) in RBCs. (a) Depiction of the reaction scheme showing (largely) membrane impermeable Mn^{2+} ions outside the cells; it rapidly relaxes the magnetization of the RF-excited ^{1}H$_2$O molecules. (b) Schematic representation of the relaxation of the ^{1}H NMR signal from water in the Mn^{2+}-doped medium (upper right graph), the RBCs alone (packed cells; middle graph), and the suspension of RBCs in the Mn^{2+}-containing medium (left graph). The shape of the latter curve contains the information about the kinetics of transmembrane exchange.

The method involves "doping" the extracellular medium with a salt of the paramagnetic cation, Mn^{2+}, as depicted in Figure 8.11a. The Mn^{2+} greatly reduces the apparent transverse relaxation time constant, T_2, of the water in the extracellular medium while hardly changing that of the intracellular water (as represented in Fig. 8.11b). However, with a very high concentration of Mn^{2+}, approximately 100 mM, the water resonance is quenched almost immediately after the RF pulse has been applied to record the ^{1}H NMR spectrum. On the other hand, the water inside the cells relaxes relatively slowly, but any signal from the water that passes out of the cells is also immediately quenched. Therefore, the rate of decay of the water signal is dominated by the rate of water efflux from the cells. The mathematical analysis of the data is straightforward and uses a separate measurement of the T_2 of water in a sample of packed RBCs [36].

The method has been refined for use in a modern high magnetic field NMR spectrometer by employing the Carr–Purcell–Meiboom–Gill (CPMG) T_2-measuring RF pulse sequence (see Fig. 8.1d) [36]. It has been applied to RBCs from a large number of animals in a quest to understand the rationale for the very fast water exchange in the RBCs of most animals, especially humans, and other physically active species [37].

8.8   DIFFUSION COEFFICIENT DIFFERENCES

8.8.1   Stejskal–Tanner Plot

The NMR method for measuring translational self-diffusion coefficients, D, of water or solutes dissolved in it employs the pulsed field gradient spin-echo (PGSE) RF pulse sequence (see Fig. 8.1e). During the sequence, two magnetic field gradient pulses are applied in the $\mathbf{B}_0$ direction, of magnitude denoted by g. These are generated by a pair of solenoids that carry current in opposite directions (a so-called Maxwell pair, which generates a linearly varying magnetic field, viz., a field gradient of constant magnitude), while the DC current is from a special rapidly switchable power supply. The pulses are delivered during each of the two τ intervals of the pulse sequence. Spectra are recorded for a sequence of values of g. A plot is constructed of the natural logarithm of the intensity of the NMR resonance of the molecule of interest, relative to that acquired when the magnetic field gradient is zero (the so-called signal "attenuation"), versus the variable denoted by b^2, where $b^2 = \gamma^2 g^2 \delta^2 (\Delta - \delta/3)$. This is called a Stejskal–Tanner plot [38].

For water and solutes that undergo diffusion that is unbounded and in an isotropic medium, the lines are straight and the negative slope is the value of D, as shown in the two left-hand graphs in Figure 8.12b.

In modern practice, the PGSE pulse sequence has become more complicated in an effort to obviate the effects of convection in the sample (which gives an artifactually high estimate of D) and to suppress unwanted dominating resonances such as those from water and buffer components. However, the data analysis is the same as before, but with a form of the Stejskal–Tanner plot that uses more complicated expressions for b^2, which depend on the details of the number of RF and field gradient pulses and their relative magnitudes and times of application [39–42].

8.8.2   Andrasko's Method

The Mn-doping method of Conlon and Outhred [33] for measuring water transport in RBCs carries the risk of the divalent cations affecting the transport mechanism. In Andrasko's method [43], the PGSE pulse sequence (see Fig. 8.1e) is used to report on the diffusion of water in RBC suspensions.

The central idea is that the translational motion of water molecules outside the cells is merely "obstructed"; thus, despite cells being "in the way" of the water molecules, they could still diffuse almost endlessly away given sufficient time. But inside the cells, water motion is truly "restricted"; thus, the trajectories of translational motion fold back onto themselves as a result of the water molecules impacting the

(a)

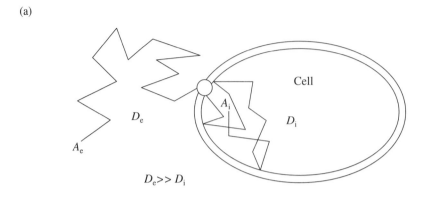

(b)

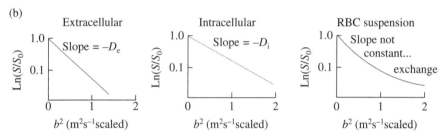

FIGURE 8.12 Transmembrane exchange of water measured with ^{1}H NMR PGSE spectroscopy, exploiting the difference in apparent diffusion coefficient in the intra- and extracellular media, D_i and D_e, respectively. (a) Depiction of the random walk of species A in the two compartments showing the relatively constrained motion inside the cell but with transmembrane exchange. (b) The diffusion-analyzing graphs (Stejskal–Tanner plots; slope $=-D$ for an isolated system) for the extracellular medium alone (left), the intracellular medium alone (middle), and when the RBCs are in a suspension showing the curved line that carries the information about the kinetics of transmembrane exchange.

inner surface of the cell membrane (see Fig. 8.12a). If there were no transmembrane exchange, the corresponding Stejskal–Tanner equation would be a superposition of two straight lines (see Fig. 8.12b, left and central graphs) and hence would be curved as in Figure 8.12b, right graph.

Because the water molecules do exchange across the cell membrane, a given molecule will have a diffusion coefficient that is the average of its Ds in the two regions, weighted by the time it spends in each region during the experiment. As a consequence, the curved nature of the plot contains information about the rate of exchange (see Fig. 8.12b, right graph). The theoretical analysis of this situation is quite subtle and was first presented by Kärger in 1969 [44, 45] as a two-site exchange theory; it is spelt out in Ref. 46.

If the apparent D values in the two compartments differ by around an order of magnitude, it is easy to estimate the exchange rate constant from the Stejskal–Tanner plot by using the two-site exchange theory described by Andrasko [43]. The estimate of the mean residence time of a water molecule in a human RBC at 25°C was 17 ms, a value greater than the one reported by Conlon and Outhred [33–35], or later by Benga et al. [36], with the Mn-doping method, but well within an order of magnitude of it.

8.9 SOME SUBTLE SPECTRAL EFFECTS

8.9.1 Scalar (J) Coupling Differences

The magnitude of the splitting of a resonance can be pH sensitive and hence used to measure pH in different compartments of biological systems. However, it has rarely been used because of the extra demand it places on high spectral resolution, as the splitting may not be resolved in cellular systems. However, an oxidized form of HPA (see Fig. 8.6c), the phosphite anion, is ideal for this purpose [47]. In isotonic solution, the divalent anion PHO_3^{2-} has a one bond proton–phosphorus scalar coupling constant of $^1J_{PH} = 568.1$ Hz that increases to 620.7 Hz for the monovalent anion $PH(OH)O_2^-$ with a measured $pK_a = 6.19$.

In a manner analogous to the Henderson–Hasselbalch equation of chemical shifts, the one for scalar couplings is

$$pH = pK_a + \log_{10}\left(\frac{J_{acid} - J_{observed}}{J_{observed} - J_{base}}\right) \qquad (8.6)$$

where J_{acid} and J_{base} are the splittings (J-coupling constants) at the pH extremes.

The ^{31}P NMR spectra of these species give well-resolved doublets for the intra- and extracellular spaces of RBC suspensions with pH difference across the cell membrane, with an error of only ±0.01 pH units in the range 5–7. The cytosolic pH in human RBCs is typically 7.1–7.2, so such variations due to metabolic changes are clearly observed.

A major drawback in those pH measuring methods that use the chemical shift (see Section 8.4.1) is the necessity for an independent chemical shift reference (see Section 8.4.3). On the other hand, scalar couplings are exquisitely sensitive to the electronic state of a molecule, and large changes are frequently observed at pH values that fall in the region of its pK_a; it is the latter condition that is often hard to satisfy for biological applications. Furthermore, scalar coupling constants are independent of the magnitude of the static magnetic field $\mathbf{B}_0$, and they do *not* require an internal reference, so their infrequent use in NMR studies of cellular systems is somewhat surprising.

8.9.2 Endogenous Magnetic Field Gradients

We now make a slight detour into the realm of magnetostatics to arrive at the definition of magnetic susceptibility. We need this term to have tangible meaning when we use it in discussing magnetic field gradients in and around cells in a suspension inside the NMR magnet.

8.9.2.1 Magnetic Induction and Magnetic Field Strength

The magnetic induction within an electromagnetic solenoid, $\mathbf{B}_0$, is different from the magnetic field strength, $\mathbf{H}_0$, due to the presence of polarizable media. The magnetic induction (or flux density; $J\,m^{-2}\,A^{-1} = T$) $\mathbf{B}_0$ that exists in a region of space inside an electromagnetic solenoid, like that used in an NMR spectrometer, is a function of the number of turns of the coil and the current in the windings. The latter defines the magnetic field strength $\mathbf{H}_0$ in units of amperes per meter ($A\,m^{-1}$). $\mathbf{B}$ is also a function of the medium or material in that space. The relationship that connects the two vectors for an empty (vacuum) solenoid is

$$\mathbf{B}_0 = \mu_0 \mathbf{H}_0 \tag{8.7}$$

where μ_0 ($4\pi \times 10^{-7}\,T\,m\,A^{-1}$, or alternatively $H\,m^{-1}$) is the magnetic permeability of free space (vacuum). When matter is present, it is magnetized (positively or negatively) and contributes $\mathbf{M}$ to the magnetic field strength, so $\mathbf{B}_0$ becomes

$$\mathbf{B}_0 = \mu_0 \left(\mathbf{H}_0 + \mathbf{M} \right) \tag{8.8}$$

When the matter behaves linearly with respect to the magnetization induced by $\mathbf{H}_0$, and it is homogeneous and isotropic, $\mathbf{M}$ is proportional to $\mathbf{H}_0$, so we define the "magnetic susceptibility" χ_m as the dimensionless ratio $\mathbf{M}/\mathbf{H}$. Hence, the magnetic permeability of the matter is

$$\mu = \frac{\mathbf{B}_0}{\mathbf{H}_0} = \mu_0 \left(1 + \chi_m \right) \tag{8.9}$$

Alternatively,

$$\mathbf{B}_0 = \mu_0 \left(1 + \chi_m \right) \mathbf{H}_0 \tag{8.10}$$

Extensive lists of the values of magnetic susceptibilities of common solvents and substances are tabulated in scientific handbooks; a subset of these that are useful in biological studies is given in Ref. 48. When χ_m is positive, the medium is said to be paramagnetic, and when negative, it is said to be diamagnetic.

8.9.2.2 Magnetic Field Gradients Across Cell Membranes and CO Treatment of RBCs

Because the magnetic susceptibility of the solutions inside and outside cells differ, the values of $\mathbf{B}_0$ inside and outside the cell are different. The physical and chemical discontinuity at the cell membrane means that there is a gradient of magnetic induction across the membrane [48–51]. With RBCs in a suspension, this adds to the chemical shift difference of the split peak effect with compounds like DMMP (see Section 8.5.1). This contribution is less than the hydrogen-bonding contribution to the chemical shift difference if the hemoglobin is in the oxy- or carbonmonoxy-(Fe(II)) states (which are diamagnetic) [28]. However, if the hemoglobin is in the met-(Fe(III)) or deoxy states, it is paramagnetic. Thus, the magnetic field gradient is large, and the NMR spectral signal-to-noise is often severely degraded (see [52]). For this reason, in most of our work on RBCs, we bubble the cell suspensions with CO for 5–15 min before NMR experiments to convert the paramagnetic form of hemoglobin to the diamagnetic form.

8.9.2.3 Exploiting Magnetic Field Gradients in Membrane Transport Studies

When the spin-echo pulse sequence (see Fig. 8.1a) is applied to a suspension of CO-treated RBCs in which the extracellular medium has been made paramagnetic with Fe(III)-ferrioxamine, then the resonances from solutes outside the cells are broadened, while inside the cells, they remain relatively sharp. Most importantly, for the same concentration of the solute, the resonance intensity is less outside than inside. We say that the "specific intensity" (resonance area per unit of concentration) is smaller outside than inside the cells.

This method for measuring membrane transport is elegant in concept, but it is very difficult to calibrate the concentration estimation. Nevertheless, it has been used to measure L-lactate [49], L-alanine [53], and choline [54] transport into human RBCs with ^{1}H spin-echo NMR spectral time courses.

8.9.3 Residual Quadrupolar (ν_Q) Coupling

Another NMR attribute that has only recently become accessible to studies of biological systems is quadrupolar coupling. Of the 102 nonradioactive NMR-receptive nuclides, 66 (64%) have spin $I > \frac{1}{2}$ and are therefore quadrupolar [55]. Such nuclei possess an electric quadrupole moment, and in an environment that is aligned in $\mathbf{B}_0$ of the NMR spectrometer, this manifests itself as a split resonance. Because the extent of splitting is much less than if the spins were in the solid state, it is referred to as a residual quadrupolar splitting or coupling (RQC). Such splitting is made obvious if molecules that contain nuclei with $I > \frac{1}{2}$ are guests in a stretchable gel such as gelatin, and the gel is held stretched. A means of doing this is to draw the gelatin in its fluid state at 40–80°C into a silicone rubber tube that is then sealed at one end with a plastic plug. The tube with the gel is inserted into a glass pipe like that of a bottomless NMR sample tube. The gelatin sets by cooling to less than approximately 25°C, and the gel can be held in a stretched state by using a thumbscrew that rests on the upper end of the glass tube.

Commonly encountered quadrupolar nuclei are ^{2}H, ^{6}Li, and ^{14}N with spin quantum number $I = 1$; ^{23}Na, ^{39}K, and ^{87}Rb (the useful K congener) with spin $= 3/2$; ^{17}O, ^{27}Al, and ^{95}Mo with $I = 5/2$; and ^{133}Cs with $I = 7/2$. Spectra of ^{2}H$_2$O and salts containing ions of some of these nuclides are shown in Ref. 56. The number of multiplet components in the relevant NMR spectrum is simply twice the spin quantum number; so stretching a gelatin gel made with ^{2}H$_2$O has two peaks in its ^{2}H NMR spectrum.

When RBCs are incorporated into liquid gelatin at 42°C using isotonic saline made with ^{2}H$_2$O as the solvent, and then cooled to 25°C, the sample forms a gel. When the gel-plus-RBCs is stretched, as described previously, the ^{2}H NMR spectrum has three peaks; the two outer peaks are from the extracellular deuterated water, and the central peak is from the intracellular deuterated water. This spectral feature occurs because the gel medium and hence the ^{2}H$_2$O molecules outside the cells are partially aligned with $\mathbf{B}_0$, whereas the cytoplasmic contents of the cells are not aligned, although overall each cell becomes stretched in the z-direction. Because the spectrum has separately identifiable populations of water on either side of the RBC membrane, the sample is amenable to magnetization transfer analysis of the exchange of the deuterated water. The fine details of the experiment and data analysis are given in Ref. 57, but it is worth noting that the estimate of the mean residence time of ^2H_2O molecules in RBCs is approximately 20 ms at 25°C; this value is close to the values given in Sections 8.7.1 and 8.8.2 determined by two other NMR methods.

Incorporating RBCs into stretched gels was the first time that residual quadrupolar splittings had been used to distinguish extra- from intracellular molecules and then used to analyze a membrane transport process [57].

8.10   A CASE STUDY: THE STOICHIOMETRIC RELATIONSHIP BETWEEN THE NUMBER OF Na⁺ IONS TRANSPORTED PER MOLECULE OF GLUCOSE CONSUMED IN HUMAN RBCs

Let us now consider an example in which $^{23}Na^+$ transport was measured with ^{23}Na NMR spectroscopy in human RBCs using the shift reagent TmDOTP. In parallel experiments, the consumption of [1-^{13}C]D-glucose was measured in a sequence of ^{13}C NMR spectra. The aim was to estimate the number of Na⁺ ions that are pumped out of the human RBC per molecule of glucose consumed; in other words, we aimed to estimate the stoichiometric ratio of Na⁺-pumped/glucose consumed [58] by human RBCs.

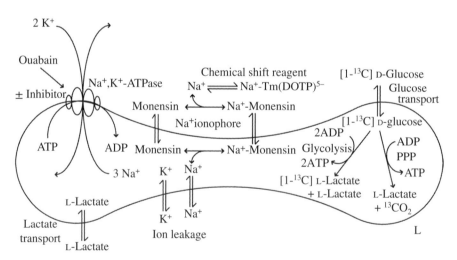

FIGURE 8.13   Estimation of the stoichiometric relationship between the number of Na⁺ ions pumped out of the human RBC per molecule of glucose consumed. Reaction scheme used for the mathematical model with which the stoichiometric ratio was estimated. The experiment involved the Na⁺-ionophore monensin, the Na⁺,K⁺-ATPase inhibitor ouabain, and the shift reagent TmDOTP to give separate ^{23}Na NMR signals from the intra- and extracellular compartments. Adapted from Ref. 58 with permission from Elsevier.

The significance of making this estimate is that only approximately 50% of the ATP turnover in the RBC had been accounted for (and this is unfortunately still the case despite what is described here!); so perhaps the "lost ATP" could be due to inefficiency of the Na^+,K^+-ATPase that pumps $3Na^+$ ions out and $2K^+$ ions into the cells, with the hydrolysis of one molecule of ATP.

The experimental design involved the enhancement of Na^+ leakage into the cells with the Na^+-selective ionophore monensin, thus increasing Na^+,K^+-ATPase activity in its known homeostatic response. In addition, the experiment was carried out without and with the well-known Na^+,K^+-ATPase inhibitor, ouabain. The overall reaction scheme is shown in Figure 8.13.

Figure 8.14 shows how the net influx of $^{23}Na^+$ was readily measured over several hours from the ^{23}Na NMR spectra, because of the two well-resolved resonances. When the Na^+,K^+-ATPase was inhibited, it no longer could compete by pumping out the Na^+ that leaked in via monensin, so the influx time course had a larger slope than

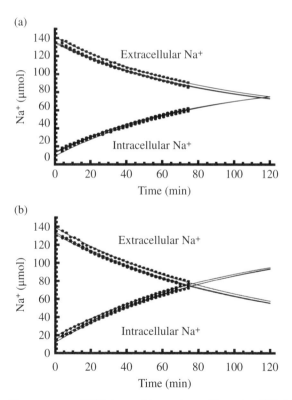

FIGURE 8.14 Time courses of $^{23}Na^+$ quantities estimated by using ^{23}Na NMR spectra of suspensions of human RBCs that were consuming [1-^{13}C]D-glucose, in the presence of the Na^+-ionophore monensin, and in the absence (a) and presence (b) of the Na^+,K^+-ATPase inhibitor, ouabain. Note the more rapid net transport in the presence of ouabain. Adapted from Ref. 58 with permission from Elsevier.

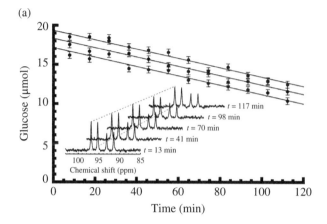

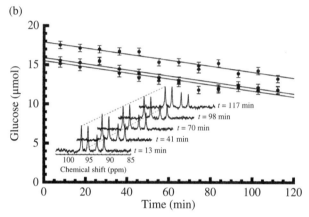

FIGURE 8.15 ^{13}C NMR time courses of [1-^{13}C]D-glucose in the experiment carried out for Figure 8.14. (a) Shows the simultaneous decline of the resonances of both anomers of [1-^{13}C]D-glucose, in the absence of the Na$^+$,K$^+$-ATPase inhibitor, ouabain. (b) Shows the time course of glucose concentration in the presence of the inhibitor. The insets show the corresponding ^{13}C NMR spectral series. Adapted from Ref. 58 with permission from Elsevier.

when the pump was working. The difference in the two slopes (with and without ouabain in Fig. 8.14a and b) gave the estimated rate of Na$^+$ pumping.

The ^{13}C NMR spectra (see Fig. 8.15a and b) show the pleasing result that when the Na$^+$,K$^+$-ATPase was inhibited with ouabain, the rate of [1-^{13}C]D-glucose consumption was decreased. This finding exemplifies the tight coupling that exists between the rate of glycolysis and ATP turnover; this has been the topic of many studies over many years, and it is a feature predicted by metabolic control analysis of computer models of human RBC glycolysis (see, e.g., [59–62]).

From measurements of the hematocrit, and hence the known volume of the cells and water in the sample, we calculated the difference in rate of glucose consumption and Na$^+$ pumping. The stoichiometry estimates were made by fitting, using Bayesian analysis, a mathematical model based on Figure 8.13 to the NMR time course data. The overall conclusion was that the ratio of Na^+ ions pumped out of the RBC per molecule of glucose consumed is close to 6:1. This is encouraging confirmation of the accepted 3:2 $Na^+:K^+$ pumping stoichiometry of the Na^+,K^+-ATPase, but as alluded to earlier, it therefore does not account for the "lost ATP" in the human RBCs. The quest to solve this "holy grail question" is ongoing [58, 63, 64].

8.11 CONCLUSIONS

A huge diversity of NMR spectroscopic methods are available for studying the exchange of water, solutes, and ions across cell membranes. These methods all use a large number of NMR-related parameters, whose values differ intrinsically between the inside and outside of cells or can be manipulated by the addition of various reagents to the suspension medium. While the list of methods given in this chapter is long, it is not exhaustive, and other methods will most probably be developed as new NMR technologies emerge, such as rapid dissolution dynamic nuclear polarization (RD-DNP) that has recently been applied to measure the (initially) zero trans exchange of ^{13}C-labeled urea in human RBCs on the second-to-minute timescale [65]. Another challenge is the extension of many of the methods described previously to solid tissues, as opposed to cell suspensions; and beyond those studies, there lies the largely unexplored realm of membrane transport *in vivo*.

ACKNOWLEDGMENTS

The work was written during the tenure of an Australian Research Council Discovery Project Grant. Dr Max Puckeridge is thanked for valuable input into the work especially in Section 8.10. Drs. Bob Chapman and Dmitry Shishmarev are thanked for valuable comments on the chapter.

REFERENCES

1. M.H. Levitt, *Spin Dynamics*, second ed., Wiley-Blackwell, Chichester, 2008, pp. 744.

2. H. Guenther, *NMR Spectroscopy: An Introduction*, John Wiley & Sons, Ltd, Chichester, 1980.

3. R.K. Harris, *Nuclear Magnetic Resonance Spectroscopy: A Physicochemical Approach*, Pitman, London, 1983.

4. R. Freeman, *A Handbook of Nuclear Magnetic Resonance*, second ed., Addison-Wesley-Longman, Harlow, 1997.

5. E. Fukushima, S.B.W. Roeder, *Experimental Pulse NMR: A Nuts and Bolts Approach*, Westview Press, Boulder, CO, 1981.

6. B. Bluemich, *Essential NMR for Scientists and Engineers*, Springer, Heidelberg, 2005.

7. P.J. Hore, J.A. Jones, S. Wimperis, *NMR: The Toolkit*, Oxford University Press, Oxford, 2000.

8. J. Keeler, *Understanding NMR Spectroscopy*, John Wiley & Sons, Ltd, Chichester, 2005.

9. C. Tanford, *Physical Chemistry of Macromolecules*, John Wiley & Sons, Inc., New York, 1966.

10. S. Wolfram, *The Mathematica Book, Version 8.0.1.0*, Wolfram Media Inc., Champaign, IL, 2011.

11. P.W. Kuchel, Spin-exchange NMR spectroscopy in studies of the kinetics of enzymes and membrane transport. *NMR Biomed.* 3 (1990) 102–119.

12. K. Kirk, NMR methods for measuring membrane transport rates, *NMR Biomed.* 3 (1990) 1–16.

13. P.W. Kuchel, K. Kirk, G.F. King, NMR methods for measuring membrane transport, in H.J. Hilderson and G.B. Ralston (Eds.), *Subcellular Biochemistry 23: Physicochemical Methods in the Study of Biomembranes*, Plenum Press, New York, 1994, pp. 247–327.

14. P.W. Kuchel, B.E. Chapman, Z.H. Endre, G.F. King, D.R. Thorburn, M.J. York, Monitoring metabolic reactions in human erythrocytes using NMR spectroscopy, *Biomed. Biochim. Acta* 43 (1984) 719–726.

15. G.F. King, P.W. Kuchel, A proton NMR study of imidodipeptide transport and hydrolysis in the human erythrocyte: Possible physiological roles for the coupled system, *Biochem. J.* 220 (1984) 553–560.

16. R.K. Gupta, P. Gupta, Direct observation of resolved resonances from intra-cellular and extracellular ^{23}Na ions in NMR studies of intact cells and tissues using dysprosium(III) tripolyphosphate as paramagnetic shift reagent, *J. Magn. Reson.* 47 (1982) 344–350.

17. M. Puckeridge, B.E. Chapman, A.D. Conigrave, P.W. Kuchel, Quantitative model of NMR chemical shifts of $^{23}Na^+$ induced by TmDOTP: Applications in studies of Na^+ transport in human erythrocytes, *J. Inorg. Biochem.* 115 (2012) 211–219.

18. A.R. Waldeck, P.W. Kuchel, ^{23}Na-NMR study of ionophore-mediated cation exchange between two populations of liposomes, *Biophys. J.* 64 (1993) 1445–1455.

19. R.T. Eakin, L.O. Morgan, C.T. Gregg, N.A. Matwiyof, ^{13}C nuclear magnetic-resonance spectroscopy of living cells and their metabolism of a specifically labeled ^{13}C substrate, *FEBS Lett.* 28 (1972) 259–264.

20. R.B. Moon, J.H. Richards, Determination of intracellular pH by ^{31}P magnetic-resonance, *J. Biol. Chem.* 248 (1973) 7276–7278.

21. D.I. Hoult, S.J.W. Busby, D.G. Gadian, G.K. Radda, R.E. Richards, P.J. Seeley, Observation of tissue metabolites using ^{31}P nuclear magnetic resonance, *Nature* 252 (1974) 285–287.

22. R.J. Labotka, Measurement of intracellular pH and deoxyhemoglobin concentration in deoxygenated erythrocytes by ^{31}P nuclear magnetic resonance. *Biochemistry* 23 (1984) 5549–5555.

23. I.M. Stewart, B.E. Chapman, K. Kirk, P.W. Kuchel, V.A. Lovric, J.E. Raftos, Intracellular pH in stored erythrocytes: Refinement and further characterization of the ^{31}P NMR methylphosphonate procedure, *Biochim. Biophys. Acta-Mol. Cell Res.* 885 (1986) 23–33.

24. K. Kirk, J.E. Raftos, P.W. Kuchel, Triethyl phosphate as an internal ^{31}P NMR reference in biological samples, *J. Magn. Reson.* 70 (1986) 484–487.

25. K. Kirk, P.W. Kuchel, Red cell volume changes monitored using a new ^{31}P NMR procedure, *J. Magn. Reson.* 62 (1985) 568–572.

26. K. Kirk, P.W. Kuchel, Physical basis of the effect of hemoglobin on the phosphorus-31 NMR chemical shifts of various phosphoryl compounds, *Biochemistry* 27 (1988) 8803–8810.

27. K. Kirk, P.W. Kuchel, Characterization of transmembrane chemical shift differences in the ^{31}P NMR spectra of various phosphoryl compounds added to erythrocyte suspensions, *Biochemistry* 27 (1988) 8795–8802.

28. K. Kirk, P.W. Kuchel, The contribution of magnetic susceptibility effects to transmembrane chemical shift differences in the ^{31}P NMR spectra of oxygenated erythrocyte suspensions, *J. Biol. Chem.* 263 (1988) 130–134.

29. J.V. Dacie, S.M. Lewis, *Practical Haematology*, fifth ed., Churchill Livingstone, Edinburgh, 1975.

30. K. Kirk, P.W. Kuchel, R.J. Labotka, Hypophosphite as a ^{31}P nuclear magnetic resonance probe of membrane potential in erythrocyte suspensions, *Biophys. J.* 54 (1988) 241–247.

31. J.R. Potts, P.W. Kuchel, Anomeric preference of fluoroglucose exchange across human red-cell membranes. ^{19}F-n.m.r. studies, *Biochem. J.* 281 (1992) 753–759.

32. R.M. Wellard, W.R. Adam, Functional hepatocyte cation compartmentation demonstrated with [133]Cs NMR. *Magn. Reson. Med.* 48 (2002) 810–818.

33. T. Conlon, R. Outhred, Water diffusion permeability of erythrocytes using an NMR technique, *Biochim. Biophys. Acta-Biomembr.* 288 (1972) 354–361.

34. R. Outhred, T. Conlon, Volume dependence of erythrocyte water diffusion permeability, *Biochim. Biophys. Acta-Biomembr.* 318 (1973) 446–450.

35. T. Conlon, R. Outhred, Temperature dependence of erythrocyte water diffusion permeability, *Biochim. Biophys. Acta-Biomembr.* 511 (1978) 408–418.

36. G. Benga, B.E. Chapman, C.H. Gallagher, D. Cooper, P.W. Kuchel, NMR studies of diffusional water permeability of red-blood-cells from macropodid marsupials (kangaroos and wallabies), *Comp. Biochem. Physiol. A Physiol.* 104 (1993) 799–803.

37. P.W. Kuchel, G. Benga, Why does the mammalian red blood cell have aquaporins? *Biosystems* 82 (2005) 189–196.

38. P.W. Kuchel, G. Pagès, K. Nagashima, S. Velan, V. Vijayaragavan, V. Nagararajan, K.H. Chuang, Stejskal–Tanner equation derived in full, *Concepts Magn. Reson. Part A* 40A (2012) 205–214.

39. W.S. Price, Pulsed-field gradient nuclear magnetic resonance as a tool for studying translational diffusion. 1. Basic theory, *Concepts Magn. Reson.* 9 (1997) 299–336.

40. W.S. Price, Pulsed-field gradient nuclear magnetic resonance as a tool for studying translational diffusion: Part II. Experimental aspects. *Concepts Magn. Reson.* 10 (1998) 197–237.

41. K.I. Momot, P.W. Kuchel, Pulsed field gradient nuclear magnetic resonance as a tool for studying drug delivery systems. *Concepts Magn. Reson. Part A* 19A (2003) 51–64.

42. K.I. Momot, P.W. Kuchel, PFG NMR diffusion experiments for complex systems, *Concepts Magn. Reson. Part A* 28A (2006) 249–269.

43. J. Andrasko, Water diffusion permeability of human erythrocytes studies by a pulsed gradient NMR technique, *Biochim. Biophys. Acta-Gen. Subj.* 428 (1976) 304–311.

44. J. Kärger, Determination of diffusion in a 2 phase system by pulsed field gradients, *Ann. Phys.* 24 (1969) 1–2.

45. J. Kärger, Influence of 2-phase diffusion on spin-echo attenuation regarding relaxation in measurements using pulsed field gradients, *Ann. Phys.* 27 (1971) 107–108.

46. A.R. Waldeck, P.W. Kuchel, A.J. Lennon, B.E. Chapman, NMR diffusion measurements to characterise membrane transport and solute binding, *Prog. Nucl. Magn. Reson. Spectrosc.* 30 (1997) 39–68.

47. T.R. Eykyn, P.W. Kuchel, Scalar couplings as pH probes in compartmentalized biological systems: ^{31}P NMR of phosphite, *Magn. Reson. Med.* 50 (2003) 693–696.

48. P.W. Kuchel, B.E. Chapman, W.A. Bubb, P.E. Hansen, C.J. Durrant, M.P. Hertzberg, Magnetic susceptibility: Solutions, emulsions, and cells, *Concepts Magn. Reson. Part A* 18A (2003) 56–71.

49. K.M. Brindle, F.F. Brown, I.D. Campbell, C. Grathwohl, P.W. Kuchel, Application of spin-echo nuclear magnetic-resonance to whole cell systems: Membrane-transport, *Biochem. J.* 180 (1979) 37–44.

50. P.W. Kuchel, B.T. Bulliman, Perturbation of homogeneous magnetic fields by isolated single and confocal spheroids. Implications for NMR spectroscopy of cells, *NMR Biomed.* 2 (1989) 151–160.

51. C.J. Durrant, M.P. Hertzberg, P.W. Kuchel, Magnetic susceptibility: Further insights into macroscopic and microscopic fields and the sphere of Lorentz, *Concepts Magn. Reson. Part A* 18A (2003) 72–95.

52. A.J. Lennon, N.R. Scott, B.E. Chapman, P.W. Kuchel, Hemoglobin affinity for 2,3-bisphosphoglycerate in solutions and intact erythrocytes: studies using pulsed-field gradient nuclear-magnetic-resonance and Monte–Carlo simulations, *Biophys. J.* 67 (1994) 2096–2109.

53. S. Brand, K. Brindle, F. Brown, I. Campbell, C. Eliot, D. Foxall, E. Jaroszkiewicz, R. Simpson, P. Styles, Studies of transport in suspensions of whole cells using NMR, *Biochem. Soc. Trans.* 9 (1981) 176P.

54. A.J. Jones, P.W. Kuchel, Measurement of choline concentration and transport in human erythrocytes by ^{1}H NMR: Comparison of normal blood and that from lithium-treated psychiatric-patients, *Clin. Chim. Acta* 104 (1980) 77–85.

55. P.W. Kuchel, Quadrupolar splitting in stretched hydrogels, *eMagRes.* 3 (2014) 171–180

56. D.J. Philp, C. Naumann, P.W. Kuchel, Relative intensities of components of quadrupolar-split multiplets in NMR spectra: Rationale for a simple rule, *Concepts Magn. Reson. Part A* 40A (2012) 90–99.

57. P.W. Kuchel, C. Naumann, $^{2}H_2O$ quadrupolar splitting used to measure water exchange in erythrocytes, *J. Magn. Reson.* 192 (2008) 48–59.

58. M. Puckeridge, B.E. Chapman, A.D. Conigrave, S.M. Grieve, G.A. Figtree, P.W. Kuchel, Stoichiometric relationship between Na$^+$ ions transported and glucose consumed in human erythrocytes: Bayesian analysis of ^{23}Na and ^{13}C NMR time course data, *Biophys. J.* 104 (2013) 1676–1684.

59. P.J. Mulquiney, W.A. Bubb, P.W. Kuchel, Model of 2,3-bisphosphoglycerate metabolism in the human erythrocyte based on detailed enzyme kinetic equations: in vivo kinetic characterization of 2,3-bisphosphoglycerate synthase/phosphatase using ^{13}C and ^{31}P NMR. *Biochem. J.* 342 (1999) 567–580.

60. P.J. Mulquiney, P.W. Kuchel, Model of 2,3-bisphosphoglycerate metabolism in the human erythrocyte based on detailed enzyme kinetic equations: Equations and parameter refinement. *Biochem. J.* 342 (1999) 581–596.

61. P.J. Mulquiney, P.W. Kuchel, Model of 2,3-bisphosphoglycerate metabolism in the human erythrocyte based on detailed enzyme kinetic equations: Computer simulation and metabolic control analysis, *Biochem. J.* 342 (1999) 597–604.

62. P.W. Kuchel, P.J. Mulquiney, Combined NMR experimental and computer-simulation study of 2,3-bisphosphoglycerate metabolism in human erythrocytes, in A. Cornish-Bowden and M.L. Cardenas (Eds.), *Technological and Medical Implications of Metabolic Control Analysis*, 74, NATO Science Series, Kluwer Academic Publishers, Boston, MA, 2000, pp 139–145.

63. M. Puckeridge, P.W. Kuchel, Membrane flickering of the human erythrocyte: Constrained random walk used with Bayesian analysis, *Eur. Biophys. J.* 43 (2014) 157–167.

64. M. Puckeridge, B.E. Chapman, A.D. Conigrave, P.W. Kuchel, Membrane flickering of the human erythrocyte: Physical and chemical effectors, *Eur. Biophys. J.* 43 (2014) 169–177.

65. G. Pagès, M. Puckeridge, L.F. Guo, Y.L. Tan, C. Jacob, M. Garland, P.W. Kuchel, Transmembrane exchange of hyperpolarized ^{13}C-urea in human erythrocytes: Subminute timescale kinetic analysis. *Biophys. J.* 105 (2013) 1956–1966.

9

TIME-RESOLVED AND SURFACE-ENHANCED INFRARED SPECTROSCOPY

JOACHIM HEBERLE

Institute of Experimental Physics, Department of Physics, Free University of Berlin, Berlin, Germany

9.1 INTRODUCTION

Infrared (IR) absorption spectroscopy has long been an established tool for chemists to identify small molecules by their characteristic vibrational "fingerprint." The IR spectrum reflects bond strengths and masses by exciting vibrational levels. Moreover, IR spectroscopy senses noncovalent forces on atoms that fine-tune vibrational frequencies and, thus, is able to probe structural properties of biological samples "through space." Akin to NMR spectroscopy, the number of bands is very large and the frequencies of the vibrations are similar. The result is a congested IR spectrum composed of overlapping bands that are often difficult to disentangle. The analysis of the amide I mode of proteins ($C=O$ stretching vibration of the peptide bond) yields secondary structure information in a way comparable to circular dichroism (CD) spectroscopy. In order to derive information on individual bands and groups in the protein, difference spectra between two functional states of a protein are recorded by applying a perturbation such as light, electron transfer, addition of a substrate, temperature, or others. The resulting difference spectrum indicates the functionally relevant constituents of the protein that are altered in the course of

Pumps, Channels, and Transporters: Methods of Functional Analysis, First Edition. Edited by Ronald J. Clarke and Mohammed A. A. Khalid.

the reaction, while the contributions from those parts of the protein that are unaffected cancel out. In the ideal case of a comprehensive band assignment, the vibrational bands reveal all structural information on the reaction, on the possible intermediates, conformational changes, and dynamics. The latter is a consequence of the inherently high temporal resolution of IR spectroscopy. Using pump–probe techniques, IR spectroscopy with time resolution in the femtosecond range is almost routinely achieved these days. The time range from nanosecond to second is tracked by transient spectroscopy. This time range is particularly relevant for conformational changes that are associated with the physiological activity of proteins and, thus, is discussed here.

The first IR study covering the fingerprint region of protein modes (1000–1800 cm^{-1}) was performed on rhodopsin in the second half of the 1970s using a conventional dispersive IR spectrophotometer [1, 2]. Subsequent application of this technique to the investigation of bacteriorhodopsin (bR) photoreactions [3] led to the identification of the C=O modes from transiently protonated amino acid side chains involved in light-driven proton transfer [4]. Parallel to these studies, Fourier transform infrared (FTIR) spectroscopy was recognized as a means to record broadband difference spectra of intermediate states in the bR photocycle [5, 6]. Starting from these pioneering investigations, a whole body of studies succeeded in resolving the reaction mechanism of the light-driven proton pump bR on the level of single amino acid residues, protons, water molecules, and even single hydrogen bonds, which is discussed in detail in Section 9.6.1. The development of time-resolved step-scan FTIR spectroscopy [7] provided a means to access the kinetics of these molecular reactions and rendered bR the model system for biophysical studies of membrane proteins.

With the success of FTIR difference spectroscopy in resolving the mechanism of proton translocation by bR, the methodology was adapted for the application to other proteins. Some examples of this will be presented here, with a focus on membrane proteins. It is the relative simplicity and the economical price for instrumentation that renders FTIR spectroscopy very attractive. Here, a word of caution is required: While it may be relatively easy to record IR difference signals that can be correlated with a reaction, the molecular interpretation of difference bands in these spectra and the assignment to specific bonds or groups may be tedious and require model compound studies, studies on isotopes, mutants, quantum mechanical calculations, and more—again very similar to NMR spectroscopy.

9.2 BASICS OF IR SPECTROSCOPY

9.2.1 Vibrational Spectroscopy

Molecules possess vibrational degrees of freedom. In vibrational transitions, absorbed photons cause the bonds between atoms to lengthen and shorten and twist and turn, resulting in stretching, bending, or rocking motions. The bonds between atoms in a molecule are classically described as springs between masses. For a diatomic molecule, the frequencies of such vibrational motions are approximated by the harmonic oscillator model, that is,

$$v = \frac{1}{2\pi}\sqrt{\frac{k}{\mu}}$$

where k is the force constant representing the stiffness of the bond and μ is the reduced mass of the two atoms, A and B, given by the individual masses m_A and m_B as

$$\mu = \frac{m_A m_B}{m_A + m_B}$$

The independent, synchronous, excited vibrational motions of groups of atoms are called normal modes. The number of modes for nonlinear polyatomic molecules is $3N - 6$, where N is the number of atoms. Normal modes are classified according to their symmetries.

Because of their different selection rules, Raman and IR spectroscopy are complementary spectroscopic approaches for monitoring vibrational transitions in molecules. To be IR active, a vibration must produce a change in dipole moment of the molecule, whereas for Raman activity, a change in polarizability is required. Thus, together, they allow all vibrational modes of a molecule to be experimentally measured. In IR spectroscopy, polychromatic incident radiation in the IR range is absorbed by the sample to effect vibrational excitation. In Raman spectroscopy, intense monochromatic light, provided by a laser, of frequencies higher than those required for vibrational changes is scattered off the molecules in the sample. Some of the collisions between sample molecules and photons are inelastic, meaning there is an exchange of energy between photons and sample molecules. After such collisions, scattered photons have frequencies that are shifted from the initial laser frequency due to this exchange. The magnitudes of the shifts exactly match the gaps between certain vibrational energy levels of the molecule.

9.2.2 FTIR Spectroscopy

In FTIR spectrometers, polychromatic IR radiation is emitted by a globar. In the Michelson interferometer, the IR beam is split into two beams by a semitransparent beam splitter (see Fig. 9.1). One beam is perpendicularly reflected to the fixed mirror, and the other one is transmitted to the movable mirror, which induces a path length difference with respect to the other beam. Both beams are reflected back to the beam splitter where they interfere either constructively or destructively (depending on the optical path length difference δ). This combined IR beam passes through the sample and the transmitted light intensity is registered as a function of the movable mirror position $I(\delta)$. The relation between mirror position and frequency of the molecular vibration is provided by Fourier transformation. The latter transforms the recorded

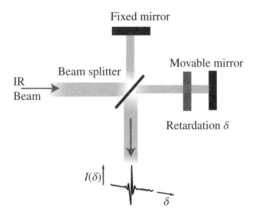

FIGURE 9.1   Schematic of the Michelson interferometer used in FTIR spectroscopy.

interferogram $I(\delta)$ into the vibrational spectrum $I(\nu)$. Unfortunately, this cannot be achieved directly and a few steps in between are required to yield the spectrum, as described in Refs. 8, 9.

9.2.3   IR Spectra of Biological Compounds

The most abundant biological macromolecules are proteins, lipids, nucleic acids, and carbohydrates. The IR spectra of these compounds exhibit characteristic vibrational bands. For proteins, the amide I (predominantly C=O stretching vibration of the peptide bond) and amide II (coupled vibrational mode of the partial double-bond C=N stretching and the N–H in-plane bending mode) are the strongest due to the fact that the peptide bond is the most frequent chemical bond in proteins. The amide I mode is used to analyze the secondary structure of proteins [10].

Lipids are comprised of the long hydrophobic hydrocarbon chain and the hydrophilic head group. The strongest dipole is the C=O in ester lipids that gives rise to a characteristic band at around $1720 \, cm^{-1}$. The weak band in the IR spectrum (see Fig. 9.2) is due to the fact that only one or two C=O bonds occur in ester lipids as compared to the more abundant C–H and the C–C bonds. The intense C–H stretching vibrational band appears at around $2900 \, cm^{-1}$. The vibrational fine structure arises from symmetric and asymmetric CH_2 stretching vibrations and from the terminal CH_3. Vinylic C=C–H, as in unsaturated fatty acids, may also contribute. The corresponding CH_2 bending vibrations absorb in the range of $1400–1500 \, cm^{-1}$. C=C double-bond vibrations of unsaturated fatty acid long chains of lipids appear around $1650 \, cm^{-1}$, the frequency of which is a very sensitive marker for *cis*- or *trans*-conformations. The vibrational frequency of the C–C stretch is in the range of $1000–1200 \, cm^{-1}$. However, as the dipole moment of the bond is very low, even when coupled with other vibrations of similar energy, a corresponding vibrational band is hardly detectable by IR spectroscopy.

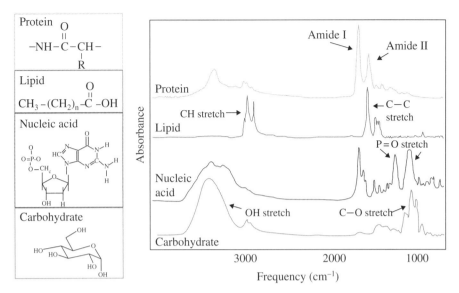

FIGURE 9.2   Mid-IR spectra of biological compounds. Arrows indicate characteristic marker bands.

In nucleic acids, strong bands appear around 1230 and 1089 cm^{-1}, which are assigned to the antisymmetric and the symmetric stretching vibrations of the phosphate groups, respectively. The frequency of the former depends on the secondary structure of the nucleic acid, whereas the latter is invariant and observed in almost all DNA or RNA spectra. Bands in the range of 1500–1700 cm^{-1} are mainly assigned to in-plane double-bond stretching vibrations of the aromatic rings of the nucleobases. The frequencies indicate different base pairing schemes in multiple-stranded DNA and RNA.

Carbohydrates (sugars) usually exhibit broad and featureless spectra. The O–H stretching vibration is of marginal diagnostic value due to the spectral overlap with the O–H stretches of water. The frequency of the C–O stretch is highly sensitive to the isomeric conformation, being at 1110 cm^{-1} when the OH ring substituent is in the *endo* position and 1064 cm^{-1} for the *exo* position. The C–O–C stretching vibration of the glycosidic linkage of oligo- and polysaccharides has been assigned to bands in the region of 1000–960 cm^{-1}.

For an extensive discussion on the vibrational bands of biological compounds, the reader is referred to Volume 5 of the *Handbook of Vibrational Spectroscopy* [11].

It is noted that the spectra displayed in Figure 9.2 are those of pure compounds. As biological material is usually in contact with water—the interaction that finally determines the structure of biological macromolecules—the spectral contribution of water has been subtracted. The bent H_2O molecule surrounded by the hydrogen-bonded network of bulk water has a large dipole moment and, thus, gives rise to strong and broad IR bands with maxima at around 3300 cm^{-1} (O–H stretching vibrations) and 1640 cm^{-1} (H–O–H bending vibration). In contrast to visible spectroscopy where water is transparent, the subtraction of solvent water in IR absorption experiments is far from trivial. For this purpose, biological material is very often dissolved in heavy water, which shifts the bands (see Section 9.2.1) down to $2500\,cm^{-1}$ (O–D stretching vibrations) and $1200\,cm^{-1}$ (D–O–D bending vibration), and thus removes the overlap with the bands of interest. Other ways to obtain a reduced water absorbance to the IR spectrum are discussed in the following.

9.2.4 Difference Spectroscopy

The manifold of IR-active vibrations of a protein leads to a broad, almost featureless absorption spectrum (see Fig. 9.2, top spectrum). Selectivity for single vibrations can be reached by applying the difference technique (for reviews, see [12, 13]). Here, the IR spectrum of the sample in one state (often the stable resting state) is subtracted from the spectrum of the protein in another state (often a metastable intermediate state). Thereby, only those vibrations that change during the action of the protein are detected, whereas all others cancel. Difference spectra with sharp bands result with difference intensities of the order of 10^{-6} absorbance units, demonstrating the superb sensitivity of the FTIR technique. Bands of single vibrations are discernible on top of the strong background absorbance of the entire protein.

9.3 REFLECTION TECHNIQUES

9.3.1 Attenuated Total Reflection

In the attenuated total reflection (ATR) technique (see Fig. 9.3), a crystal material of high refractive index is used as the internal reflection element (IRE). The probe beam passes through the IRE under an angle that exceeds the critical angle for total internal

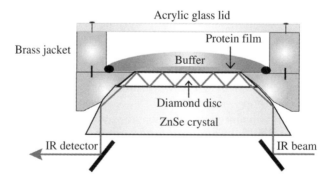

FIGURE 9.3 Schematic diagram of an ATR cell. The protein film is immersed in bulk aqueous solution. The protein sample is probed at each reflection of the IR beam by the evanescent field that decays exponentially outside the internal reflection element. Reproduced from Ref. 13 with permission from the European Society for Photobiology, the European Photochemistry Association, and The Royal Society of Chemistry.

reflection. At the interface of the highly refractive IRE and the low refractive medium, an evanescent wave evolves that penetrates into the optically rarer medium. The penetration depth is of the order of the radiation wavelength, that is, in the micrometer range when IR radiation is used [14]. If a sample is enriched at the surface of the IRE, only sample material is probed by the evanescent wave, provided the sample thickness is as large as the penetration depth of radiation. In the aqueous phase, substances such as salt, buffer, etc. can be dissolved. The strong absorbance of water and of solutes is sufficiently reduced by this approach. Moreover, the sample environment is well defined and can be conveniently modified due to the open access from above to the ATR cell [15, 16].

ZnSe and ZnS are the materials of choice for the IRE. The commonly used Ge and Si are not suitable for light-induced difference spectroscopy because both elements are indirect semiconductors that drastically change IR transmission upon absorption of visible light. The micro-ATR unit used in our work employs a diamond disk. The angle of incidence is 45°. The diameter of the active area is 4.3 mm and 5 reflections are used for probing the sample. Advantages of the micro-ATR unit are high optical throughput and low consumption of sample.

9.3.2 Surface-Enhanced IR Absorption

Surface-enhanced spectroscopy, the logical advance in ATR methodology, is used to probe monolayers of proteins. In surface-enhanced infrared absorption spectroscopy (SEIRAS), the IR-active vibrations of a sample deposited on a thin gold surface are enhanced by the strong electromagnetic field induced on the metal particles by the incident light [17]. The enhancement decays rapidly with the distance from the surface (decay length in the order of 10 nm). It is advantageous that the near-field effect of surface enhancement eliminates contributions from the bulk aqueous phase to the IR spectrum and selectively detects signals from the adsorbed monolayer of a membrane or a protein even when immersed in water (see Fig. 9.4). An important consideration for the surface-enhanced IR absorption (SEIRA) effect is the roughness of the gold layer because the enhancement factor depends on the size, shape, and density of the gold particles. Typically, the enhancement factor in SEIRAS ranges between 10 and 100, still several orders of magnitude smaller than that observed with surface-enhanced Raman scattering [18].

In order to preserve the structure and functionality of the adsorbed biomolecules, the gold film must be chemically modified [19]. Molecules that contain thiol groups (such as dithiobis-succinimidylpropionate) spontaneously form a stable self-assembled monolayer (SAM) with the bare gold surface through covalent linkage to the sulfur group [20]. The detergent-solubilized membrane protein (2–5 μM) is attached to the chemically modified metal film via an organic group connected to a His-tag. In the last step, the oriented attached protein is reconstituted in artificial lipids. Proper orientation of the protein relative to the reactant is achieved by this approach, which is crucial for the analysis of the function of the protein [21]. This is superior to the ATR technique where an arbitrary oriented

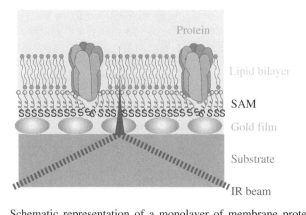

FIGURE 9.4 Schematic representation of a monolayer of membrane protein probed by surface-enhanced infrared absorption spectroscopy. The thickness of the gold layer is known to be about 50 nm in the form of a rough, nanoscaled island structure. The roughness of the gold film plays an important role for the enhancement effect. Reproduced from Ref. 13 with permission from the European Society for Photobiology, the European Photochemistry Association, and The Royal Society of Chemistry.

multilayer of proteins is formed on the surface of the IRE. All surface modification steps to tether proteins to the solid surface are monitored *in situ* by SEIRAS [20].

9.4 APPLICATION TO ELECTRON-TRANSFERRING PROTEINS

The enzymatic reactions of many biological systems, especially of membrane proteins, are driven by an electrochemical gradient across the cell membrane. Such a system can be artificially reproduced on an electrode surface to mimic the physiological properties of a biological membrane. The metal used for surface enhancement can also be employed as an electrode in such experiments.

9.4.1 Cytochrome c

Electrochemically induced oxidation and reduction processes of a monolayer of cytochrome c, a protein that mediates single-electron transfer between the integral membrane protein complexes of the respiratory chain, are a model system for electron transfer within and between proteins [22]. Electrons are directly injected to and/or retracted from cytochrome c after proper contact has been established with the electrode through a suitable surface modifier. Proteins can be adhered to such a chemically modified electrode (CME) via electrostatic attraction or covalent interaction.

As redox-active proteins are readily activated by electron injection from the electrode surface, the difference spectroscopic approach can be applied, which has been proven to be extremely useful in studying the functionality of light-driven systems (see Section 9.6). Difference spectra of cytochrome c between the fully reduced state and varying degrees of oxidation have been recorded while scanning the potential from negative to positive voltages [23]. The difference spectra represent subtle changes of the secondary structure induced by the rearrangement of the hydrogen-bonded network associated with the internal amino acid side chains surrounding the heme chromophore. The frequencies of the observed difference bands are invariant to the type of CME layer, while the relative intensities strongly depend on the CME. The former suggests that the internal conformational changes of cytochrome c are not influenced by interaction with the modifiers. The latter observation is attributed to the difference in the orientation of cytochrome c relative to the CME surface [24].

The implementation of SEIRAS is advantageous, not only with respect to monolayer sensitivity but also for providing fast electrochemical response. Pulsed electron injection from the electrode in combination with advanced time-resolved techniques enabled IR studies of monolayers of cytochrome c with microsecond time resolution [25].

9.4.2 Cytochrome c Oxidase

In mitochondria, cytochrome c carries electrons from the bc1 complex to cytochrome c oxidase (CcO). CcO represents the terminal membrane protein of the electron transfer chain where molecular oxygen is reduced to water. The binding reaction of cytochrome c to its cognate oxidase was monitored by SEIRAS.

Prior to the interaction studies, His-tagged CcO was bound to the Ni–NTA-modified gold surface (see Fig. 9.5). As the His-tag was introduced at the C-terminus

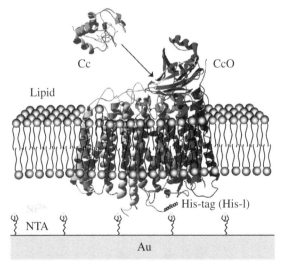

FIGURE 9.5 Solid-supported cytochrome c oxidase (CcO) is adsorbed to the Ni–NTA-terminated SAM layer via the His-tag (His-I) positioned at the C-terminus of subunit I. The binding site for the natural electron donor cytochrome c (Cc) is exposed toward the electrolyte solution. Reproduced with permission from Ref. 21. Copyright (2006) American Chemical Society.

of subunit I, CcO was oriented to expose the binding site for cytochrome c to the aqueous bulk solution [20, 21]. Binding of reduced cytochrome c led to electron injection into the CcO catalytic center where dioxygen is reduced to water. Cyclic voltammetry was used to monitor the catalytic activity of CcO. A drastic increase in the reduction current was observed at potentials lower than +260 mV (vs. NHE), which is indicative of dioxygen reduction. Solid-supported CcO did not exhibit electroactivity in the absence of cytochrome c. These results suggest that the surface-tethered CcO layer retains its catalytic properties. The preservation of protein functionality following monolayer formation is mandatory for meaningful studies and was also demonstrated for other membrane proteins [26, 27].

Electrochemically induced IR difference spectroscopy has been performed on the functional cytochrome c/CcO complex. Spectra have been recorded while the potential was repeatedly increased and decreased (cyclic voltammetry). The resulting SEIRA spectra are remarkably similar to redox-induced difference spectra of cytochrome c in solution or when adsorbed to a carboxy-terminated SAM [21]. Thus, these bands emanate from conformational changes of cytochrome c induced by electrochemical oxidation when adsorbed to the CcO surface. It is remarkable that difference spectra of cytochrome c when adsorbed to SAMs equipped with different headgroups (hydroxyl, zwitterionic, or pyridine) exhibit different relative intensities [24]. Only the carboxy-terminated surface provides similar redox-induced difference spectra of cytochrome c as if adsorbed to CcO. Thus, we concluded that the negatively charged carboxy-terminated SAM effectively mimics the (physiological) cytochrome c binding site of CcO, which is dominated by negatively charged amino acid side chains (Asp and Glu).

9.5 TIME-RESOLVED IR SPECTROSCOPY

Valuable insights into the mechanism of protein function are also obtained from techniques capable of resolving the evolution of intermediates along the reaction pathway in a non-invasive manner. For many years, time-resolved FTIR spectroscopy has proven to be superior to other IR techniques due to the multiplex advantage. However, these techniques are limited in signal-to-noise ratio by the low emissivity of the broadband light source. In recent years, tunable quantum cascade lasers (QCLs) became available whose emission power outperforms that of globars by a factor of 10,000 or more. The most relevant approaches to resolve kinetics by transient techniques are briefly introduced here.

9.5.1 The Rapid-Scan Technique

In FTIR spectroscopy, the movable mirror of the interferometer rapidly scans back and forth (rapid-scan). The time resolution depends on the maximum velocity of the movable mirror. Here, the moment of inertia of the mirror (with holders and slide) and the collinearity condition of interferometry between the movable and the fixed mirrors limit the maximum force that can be applied by a magnetic coil to accelerate the movable mirror. Thus, a typical time resolution of the order of 10 ms is achieved with commercial FTIR spectrometers at a moderate spectral resolution of $4\,cm^{-1}$. As the latter is determined by the retardation of the movable mirror, the time resolution of rapid-scan spectroscopy also depends on the spectral resolution—the higher the latter, the lower the former. Yet, the spectral range covered in a rapid-scan experiment is only limited by the optical components in the FTIR spectrometer, which represents a unique advantage of this method.

9.5.2 The Step-Scan Technique

Higher time resolution, retaining the multiplex advantage of FTIR spectroscopy, can be achieved by applying the step-scan technique (see Ref. 28 for a visual experiment). There, the movable mirror is set to a specific retardation. After triggering the kinetic process by a short perturbation such as a laser pulse, the intensity change is recorded over time by a transient recorder. Once the process has relaxed back to the initial state, the mirror is stepped to the next retardation and the reaction is initiated again (step-scan) [7, 29]. The kinetic data are collected until they cover the entire range of retardations necessary for the full interferogram. Then, the matrix of temporal changes for all retardations is rearranged to yield interferograms at specific times after triggering the reaction. Each of the serial interferograms is Fourier-transformed to provide single-channel spectra. Finally, transient absorption difference spectra are calculated by taking the ratio of single-channel spectra before and after the reaction trigger (for technical details, see Refs. 28, 30).

The advantage of the step-scan technique is that the time resolution is limited only by the detection electronics. By these means, FTIR difference spectra have been recorded with time resolutions up to a few nanoseconds [31–33]. A disadvantage of step-scan spectroscopy is the requirement of a strictly reversible process.

9.5.3 Tunable QCLs

Recently, QCLs became commercially available that exhibit high power (>200 mW on a monochromatic emission line with a FWHM of $<0.1\,cm^{-1}$) paired with broad tunability of the IR emission (up to $200\,cm^{-1}$ in continuous wave operation—maybe larger in pulsed operation) [34, 35]. In comparison, the tuning range of conventional lead salt laser diodes is smaller, and they often suffer from mode instabilities.

Flash-photolysis experiments can be performed at single frequencies with superior signal-to-noise ratio. The time resolution in these experiments is, as in the case of step-scan spectroscopy, only limited by detector electronics. An advantage over step-scan spectroscopy is that the signal-to-noise ratio is sufficient to resolve difference spectra of even small absorption changes at a time resolution of 10 ns, due to the strong emission of the QCLs. Upon tuning the emission frequency of the QCL, a segment of the spectral range is covered. A broadband spectral range can be resolved by the use of a series of QCLs.

9.6 APPLICATIONS TO RETINAL PROTEINS

This class of seven-transmembrane (TM) proteins possesses a retinal chromophore bound covalently to a lysine residue of the apoprotein. Microbial rhodopsins carry all-*trans* retinal (type I rhodopsin), whereas mammalian rhodopsins harbor 11-*cis* retinal (type II rhodopsin) in the binding pocket of the dark state of the protein. bR, the first identified microbial rhodopsin, was discovered to be a light-driven proton pump [36]. bR had a large impact on membrane structural research and bioenergetics (see Section 1.5.2) and became a model system for ion pumps. Shortly after, other rhodopsins were described as primary sensors in the phototactic responses in halobacteria [37]. More recently, the light-driven cation channel, channelrhodopsin (ChR), was discovered in *Chlamydomonas reinhardtii* [38, 39] (see Section 1.4.4). This retinal protein revolutionized neurophysiological experimentation as action potentials can be elicited by light, even in free-running animals [40]. The reader is referred to an excellent recent review on the history and the impact of these three different proteins on our current understanding of membrane proteins [41].

9.6.1 Bacteriorhodopsin

Bacteriorhodopsin (bR), the light-driven proton pump of haloarchaea, represents the best-studied membrane protein to date. It is a comparatively simple biological machinery (molecular weight 26 kDa) that converts light energy into a proton gradient across the cell membrane [42]. The difference in proton concentration is utilized to drive ATP synthesis [43]. The first three-dimensional structure of any TM protein was derived from bR by the seminal electron microscopic experiments of Richard Henderson [44]. However, it took more than 20 years to obtain an atomistic model of bR (see Fig. 9.6a). With the help of cubic lipid phase crystallization [46], the structure of bR was resolved to a resolution of 1.55Å [45], which provides a detailed picture of the structure, including the location of water molecules (see Fig. 9.6a). Crystallographic models of intermediate states of the photoreaction of bR are now also available (see Ref. 47 for a critical discussion).

A detailed picture of the sequence of proton transfer steps has emerged from time-resolved FTIR spectroscopy that links the structural data to the functional mechanism of proton translocation (see Fig. 9.6). The elucidation of the latter will now be discussed.

Absorption of light by the chromophore retinal induces a photoreaction that can be traced by molecular spectroscopy. As short laser pulses can be applied to trigger the reaction, time-resolved spectroscopy was applied very early in bR studies. The time range covers the entire chemically relevant range from femtoseconds to seconds [48].

Proton pumping and the associated conformational changes of the apoprotein occur on the microsecond to millisecond time range. Therefore, time-resolved step-scan FTIR difference spectroscopy was applied to study the kinetics of the vibrational changes after pulsed laser excitation of bR (see Fig. 9.7). Due to the broad observation range (5 μs to 100 ms), the kinetics are represented on a logarithmic

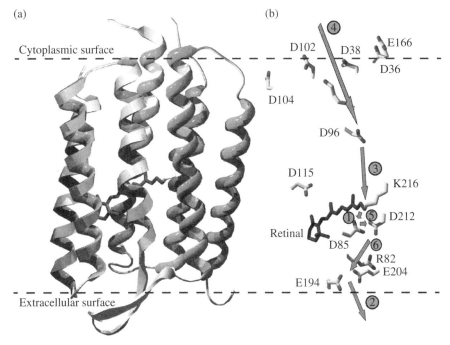

FIGURE 9.6 (a) Structure of bacteriorhodopsin (PDB entry: 1C3W [45]). (b) Proton transfer pathway across bR. The numbers 1–6 denote the temporal sequence of proton transfer steps. The dashed horizontal line was drawn to indicate the membrane surfaces on either side. Modified from Ref. 13 with permission from the European Society for Photobiology, the European Photochemistry Association, and The Royal Society of Chemistry.

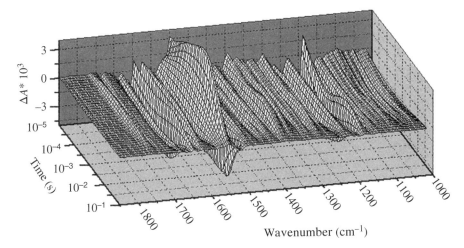

FIGURE 9.7 3D representation of the vibrational changes associated with the catalytic cycle of bacteriorhodopsin as detected by time-resolved FTIR step-scan spectroscopy in combination with the ATR technique. Negative bands are due to the (unphotolyzed) ground state of bR, and positive bands correspond to intermediate states. Data are from a multiexponential fit to the absorbance changes. The large negative difference band at 1525 cm⁻¹ is truncated to facilitate the observation of smaller band features. Experimental conditions: 1 M KCl, pH 7.4, 20°C. Reproduced from Ref. 49 with permission from Research Signpost.

timescale [50]. It is obvious that a whole variety of difference bands are observed. It is beyond the scope of this chapter to review the current status of the assignment of the numerous bands to particular molecular vibrations. For this, the reader is referred to the excellent compilation given by Maeda [51]. Nevertheless, several vibrations are discussed in the following to exemplify the power of IR spectroscopy in elucidating specific molecular steps in the function of bR.

Resolving pure intermediate state spectra is crucial to assess the sequence of functional changes in the bR photoreaction. Despite the fact that the occurrence of intermediate states is separated in time, substantial temporal overlap of rise and decay may lead to contamination in the difference spectra of intermediate states. The variation of thermodynamic parameters such as temperature and pH in the time-resolved experiments is achieved with the ATR sampling approach (see Section 9.3.1), which improved the separation of the intermediate states [50].

The large negative bands in the difference spectra arise from C=C (at $1525\,cm^{-1}$) and C–C stretching vibrations (at 1254, 1200, and $1167\,cm^{-1}$) of the conjugated double bonds of the chromophore all-*trans* retinal (see Fig. 9.8, left).

The frequency of the C–C single-bond vibrations represents a sensitive marker of the conformation of retinal. In the first observable intermediate ($10\,\mu s$ difference spectrum in Fig. 9.8, left), a positive band appears at $1189\,cm^{-1}$. This band is indicative for the all-*trans* to 13-*cis* isomerization of retinal after photon absorption. This so-called L intermediate decays with a time constant of about $100\,\mu s$ to the subsequent M state. In M, the covalent bond of retinal to the protein (a protonated Schiff base linkage between retinal and Lys-216) is deprotonated, which leads to a decrease in dipole moment. Consequently, bands of retinal lose intensity. This is obvious from the fingerprint region of the $340\,\mu s$ spectrum (see Fig. 9.8, left) where positive bands are missing. The N and O states appear after the M intermediate. During the M–N transition, the retinal Schiff base is reprotonated, that is, retinal is in a similar conformation as in the early L intermediate. This is corroborated by an almost identical band pattern in the fingerprint region. However, the positive band peaks at $1184\,cm^{-1}$ in the N intermediate, which is in contrast to $1189\,cm^{-1}$ in the L state.

The light-triggered retinal isomerization induces proton translocation across bR. The first proton transfer reaction, from the retinal Schiff base to D85, takes place with a time constant of $50\,\mu s$ (step 1 in Fig. 9.6b). This step is visualized by the appearance of the C=O stretching vibration of the carboxylic side chain of D85 (see Fig. 9.8, right). Shortly after, a proton is released to the extracellular membrane surface (step 2 in Fig. 9.6b) [52]. Because D85 remains protonated until the last stage of the reaction cycle (see Fig. 9.8, right), the released proton must originate from a different group. This proton release group (PRG) has been assigned to an excess proton distributed over a hydrogen-bonded network of water molecules and the dyad of E194 and E204 along the extracellular surface of bR [53]. Subsequent reprotonation of the retinal Schiff base is accomplished by D96 (step 3 in Fig. 9.6b, [54]). This amino acid is protonated in the ground state of bR due to the hydrophobic environment. The pK_a of D96 has been determined by pH-induced difference ATR spectroscopy to be higher than 11 [55]. We used time-resolved ATR spectroscopy to quantify the decrease in proton affinity of D96 during the M–N transition [56]. The drop in acidity from greater than 11 down to 7.1 [56] accounts for the role of D96 as the internal proton donor for the

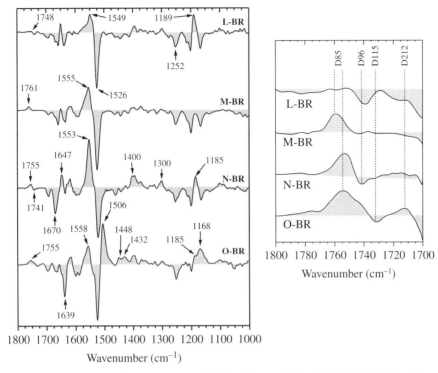

FIGURE 9.8 (Left) Difference spectra of L-BR (10 µs, pH 6.6, 20°C), M-BR (300–400 µs, pH 8.4, 20°C), N-BR (80–100 ms, pH 8.4, 20°C), and O-BR (5–10 ms, pH 4.0, 40°C). For clarity, spectra have been scaled to yield an identical difference absorbance at 1252 cm⁻¹. Frequencies of marker bands characteristic for the respective intermediate are indicated (see Ref. 30 for a discussion of the vibrational bands). (Right) Zoom-in to the difference spectra in the frequency range vibration of carboxylic acids. Vertical dashed lines indicate the C=O stretching vibration of aspartic acids that change during proton pumping activity of bR. Modified from Ref. 13 with permission from the European Society for Photobiology, the European Photochemistry Association, and The Royal Society of Chemistry.

retinal Schiff base. Evidently, time-resolved FTIR spectroscopy is the only technique able to quantify transient changes in acidity of functionally important amino acid residues over the whole time range from nanoseconds to seconds.

Reprotonation of D96 takes place from the cytoplasmic surface where protons are attracted by negatively charged residues along the surface (proton-collecting antenna) and guided to the entrance of the proton uptake pathway where D38 is located [57]. In the final stages of the photoreaction, D85 deprotonates and the released proton is transferred to the nearby residue D212 (step 5 in Fig. 9.6b) for which the C=O stretch was assigned to the positive band at 1713 cm⁻¹ (see Fig. 9.8, right). Under physiological conditions, the residence time for the proton at D212 is short and the proton is readily transferred to the initial proton release complex (step 6 in Fig. 9.6b). This reaction completes the sequence of proton transfer reactions across bR.

9.6.2   Channelrhodopsin

Microbial-type rhodopsins were identified in the eyespot of the unicellular green alga *C. reinhardtii*. The retinal binding pocket of the proteins exhibits high homology to that of bR, but the photocycle reaction exhibits distinct differences compared with other microbial rhodopsins [58].

Similar to bR, it was electron microscopy that provided the first structural evidence for the expected 7-TM arrangement of ChR2 [59]. In 2012, the atomic model of ChR was resolved [60]. Actually, a chimeric construct (C1C2) was produced by linking the last two helices (F and G) of ChR2 to the first five (A to E) of ChR1. The high-resolution structure confirmed the dimeric arrangement observed in the 2D crystals of ChR2 [59]. The C1C2 structure, depicted in Figure 9.9, is typical of microbial rhodopsins, especially for helices C–G. Yet, the most striking differences are in helices A and B, which are tilted outward by 3–4Å with respect to the bR structure. A cavity is formed by helices A, B, C, and G that connects to the extracellular medium. This half channel is rich in charged and polar residues mostly contributed by amino acids of helix B and most likely forms the open pore of the cation channel.

The major question toward resolving the functional mechanism is the link between the light-absorbing retinal chromophore and the cation channel opening. The comparison between time-resolved electrophysiological and spectroscopic data suggests that the channel is conductive during the lifetimes of the P_2^{390} and P_3^{520} states (the subscripts indicate the sequence in the linear photocycle and the superscript the wavelength of maximal absorption in the visible range of ChR2). The preceding P_1^{500} state comprises 13-*cis* retinal and structural changes in the protein backbone,

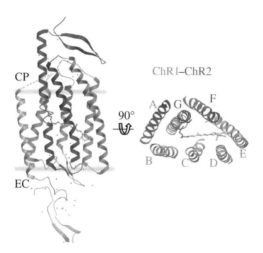

FIGURE 9.9   Structural model of the ChR1–ChR2 (C1C2) chimera of channelrhodopsin (PDB entry: 3UGA [60]). Helices A–E correspond to the amino acid sequence of ChR1, and helices F and G correspond to residues derived from ChR2. The left graph represents the in-plane view to the membrane, showing the helices and sheets including water molecules (spheres) and the retinal linked to a lysine residue in the middle of the membrane. In the right graph, the molecule is rotated by 90° for a view perpendicular to the membrane plane. Modified from Ref. 61 with permission from Elsevier.

albeit to a lesser extent than observed in the slower intermediates. The last photointermediate P_4^{480} ($\tau = 10\,\text{ms}$) is nonconductive and was associated with desensitization of the channel. The recovery of the ground state is the rate-limiting step of the photoreaction of ChR2 ($\tau = 5\,\text{s}$).

We applied static and rapid-scan FTIR spectroscopy to investigate the conformational changes of ChR2 that occur upon channel activity [62, 63]. Later, the time resolution was improved to $5\,\mu\text{s}$ by applying the step-scan technique to investigate the structural changes that occur during the opening and closing of the channel [64]. The slow turnover rate of ChR2 represented a great technical challenge and led to recording times in the order of a few weeks for a complete step-scan data set of sufficiently high signal-to-noise ratio.

The appearance of negative bands at 1240, 1200, and $1154\,\text{cm}^{-1}$ in all difference spectra (see Fig. 9.10) is in accordance with the general view that the photocycle of ChR2 originates mostly from the photoisomerization of all-*trans* retinal [58]. The negative band at $1554\,\text{cm}^{-1}$ is due to the ethylenic (C=C) vibration of ground-state retinal [65]. The frequency correlates well with the electronic absorption of retinal at 470 nm of ground-state ChR2 [66].

The time-resolved IR difference spectrum of the P_1^{500} state (top spectrum in Fig. 9.10) exhibits similar spectral changes in the amide I region (negative band at $1663\,\text{cm}^{-1}$) as the state trapped at 80 K [63, 67]. These unusually fast conformational changes of the protein backbone evolve in the femtosecond time domain [68]. The time-resolved difference spectrum of the P_1^{500} intermediate shows a positive band at $986\,\text{cm}^{-1}$ assigned to hydrogen-out-of-plane (HOOP) vibrations of the retinal, indicating a nonplanar retinal conformation in the early red-shifted intermediate, as in other microbial rhodopsins [69–71].

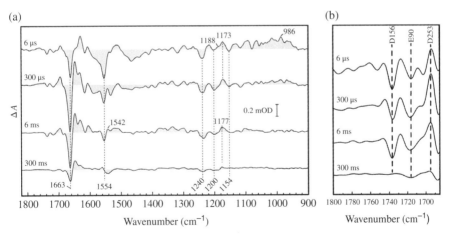

FIGURE 9.10 (a) FTIR difference spectra of ChR2 at $6\,\mu\text{s}$, $300\,\mu\text{s}$, 6 ms, and 300 ms after pulsed laser excitation. At these times, the difference spectra correspond mainly to the photocycle intermediates P_1^{500}, P_2^{390}, P_3^{520}, and P_4^{480}, respectively. (b) Zoom-in to the difference spectra in the frequency range vibration of carboxylic acids. Vertical dashed lines indicate the C=O stretching vibration of acidic amino acids that change during turnover of ChR2. Modified from Ref. 64 with permission from The National Academy of Sciences USA.

The P_2^{390} state is the predominant intermediate at 300 μs after the laser pulse (see Fig. 9.10). This intermediate is unique as it represents the only state with a deprotonated Schiff base. The acceptor of the Schiff base proton was assigned to D253 on the basis of mutants whose IR difference spectra exhibited distinct alterations of the positive band at 1695 cm^{-1} (C=O stretch of protonated D253). The rise kinetics of this band tally with the rise of the P_2^{390} state, that is, compatible with the expected kinetics of the SB proton acceptor.

At 6 ms after the laser flash, the difference spectrum of the P_3^{520} intermediate was isolated. The positive band at 1542 cm^{-1} was assigned to the ethylenic vibration of retinal as the frequency correlated well with the absorption maximum at 520 nm [66]. The negative band at 1737 cm^{-1} was assigned to the C=O stretching vibration of D156 by mutational analysis. As the kinetics of this band matched the rise and decay of the P_2^{390} state, D156 acts as the proton donor to the Schiff base. The exchange of D156 to other residues led to drastic changes in the rates of channel closure [72] rendering the reprotonation of D156 a molecular determinant for channel gating.

The difference spectrum of the P_4^{480} state was resolved at 300 ms after the laser pulse. The absorption band of the ethylenic vibration of ground-state retinal is largely canceled due to the similar frequency of this mode in the P_4^{480} state. The negative band at 1717 cm^{-1} was assigned to E90 [64, 73], that is, deprotonation occurs only during the lifetime of this intermediate that represents the desensitized (closed) state.

These FTIR experiments have been complemented by time-resolved experiments in the visible range using a pH-sensitive dye [74] and led to a comprehensive description of proton translocation in ChR2 (see Fig. 9.11). The initial proton

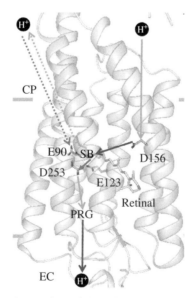

FIGURE 9.11 Proton transfer reactions of channelrhodopsin (PRG, proton release group). The proton transfer reactions are overlaid on the structural model of the C1C2 chimera (PDB 3UGA) [60]. Modified from Ref. 64 with permission from The National Academy of Sciences USA.

transfer takes place from the retinal Schiff base to D253 and proceeds with a half-life of $\tau_{1/2} = 10\,\mu s$. The Schiff base is reprotonated from D156 with $\tau_{1/2} = 2\,ms$. At the same time, a proton is released to the bulk, as deduced by transient pH changes monitored by a pH-sensitive dye [74]. The proton is released to the extracellular side, a reasonable assumption required to account for the outward proton pumping activity of ChR2 [75]. The source of the released proton is not D253, which instead stays protonated at this stage. Thus, we infer the presence of a protein release group (PRG in Fig. 9.11) that is yet to be identified [64].

D156 is reprotonated with $\tau_{1/2} = 10\,ms$ [64], concomitantly with proton uptake from the bulk [74]. Given the location of D156 in the intracellular domain of ChR2, we have proposed that D156 is reprotonated from the intracellular medium [64].

A minor fraction of ChR2 undergoes the P_3^{520} to P_4^{480} transition prior to ground-state recovery. This transition is accompanied by the deprotonation of E90 ($\tau_{1/2} = 10\,ms$), which is reprotonated upon P_4^{480} decay to the initial dark state ($\tau_{1/2} = 20\,s$). The deprotonation and reprotonation of E90 are to the aqueous bulk phase, as revealed by transient pH-change measurements of the E90A variant [64].

9.7 CONCLUSIONS

The presented results illustrate how FTIR spectroscopic techniques are able to provide a detailed understanding of the molecular dynamics of membrane proteins. The ability to vary physico-chemical parameters has been very helpful in extracting basic properties of the light-driven proton pump bR. Conformational changes of the chromophore retinal and the surrounding protein could be monitored. Probably most important is the ability of IR spectroscopy to observe proton transfer reactions of single amino acid side chains within the protein.

Based on the acquired technological knowledge on the light-driven proton pump bR, the progress in understanding of the light-gated cation channel ChR proceeded more rapidly. Yet, we are still far away from a comprehensive understanding. For instance, the functionality of an ion channel critically depends on the membrane potential. With the presented methodologies, it may be possible to address this point by combining SEIRAS with time-resolved approaches.

Another rapidly expanding field of IR spectroscopy is the analysis of very small samples by IR microspectroscopy. We succeeded in performing difference spectroscopy on microcrystals of bR [76] and halorhodopsin [77]. Even step-scan spectroscopy was applied to gauge the influence of the crystalline arrangement on the structural changes of the membrane protein [78]. For spatial resolution beyond the diffraction limit, the very exciting recent developments in scanning near-field infrared microscopy (SNIM) allow for "chemical" imaging of specimens at a spatial resolution of less than 10 nm. As an application to bio-membrane research, the native purple membrane, which is densely packed with bR, was studied by nano-FTIR spectroscopy at a lateral resolution of 30 nm [79, 80]. These studies may open an avenue for conducting single-molecule IR spectroscopy.

REFERENCES

1. F. Siebert, W. Mäntele, W. Kreutz, Flash-induced kinetic infrared spectroscopy applied to biochemical systems, *Biophys. Struct. Mech.* 6 (1980) 139–146.

2. F. Siebert, W. Mäntele, Investigations of the rhodopsin/Meta I and rhodopsin/Meta II transitions of bovine rod outer segments by means of kinetic infrared spectroscopy, *Biophys. Struct. Mech.* 6 (1980) 147–164.

3. F. Siebert, W. Mäntele, W. Kreutz, Biochemical applications of kinetic infrared spectroscopy, *Can. J. Spectrosc.* 26 (1981) 119–125.

4. F. Siebert, W. Mäntele, W. Kreutz, Evidence for the protonation of 2 internal carboxylic groups during the photocycle of bacteriorhodopsin: Investigation by kinetic infrared-spectroscopy, *FEBS Lett.* 141 (1982) 82–87.

5. K.J. Rothschild, M. Zagaeski, W.A. Cantore, Conformational changes of bacteriorhodopsin detected by Fourier-transform infrared difference spectroscopy, *Biochem. Biophys. Res. Commun.* 103 (1981) 483–489.

6. F. Siebert, W. Mäntele, Investigation of the primary photochemistry of bacteriorhodopsin by low-temperature Fourier-transform infrared spectroscopy, *Eur. J. Biochem.* 130 (1983) 565–573.

7. W. Uhmann, A. Becker, C. Taran, F. Siebert, Time-resolved FT-IR absorption spectroscopy using a step-scan interferometer, *Appl. Spectrosc.* 45 (1991) 390–397.

8. P.R. Griffiths, J.A. de Haseth, *Fourier Transform Infrared Spectroscopy*, John Wiley & Sons, Inc., New York, 1986.

9. J. Gronholz, W. Herres, *FT-IR Data Processing*, Bruker Analytische Messtechnik GmbH, Karlsruhe, 1985.

10. J.T. Pelton, L.R. McLean, Spectroscopic methods for analysis of protein secondary structure, *Anal. Biochem.* 277 (2000) 167–176.

11. J.M. Chalmers, P.R. Griffiths, *Handbook of Vibrational Spectroscopy*, John Wiley & Sons, Ltd, Chichester, 2003.

12. C. Kötting, K. Gerwert, Proteins in action monitored by time-resolved FTIR spectroscopy, *ChemPhysChem* 6 (2005) 881–888.

13. I. Radu, M. Schleeger, C. Bolwien, J. Heberle, Time-resolved methods in biophysics. 10. Time-resolved FT-IR difference spectroscopy and the application to membrane proteins, *Photochem. Photobiol. Sci.* 8 (2009) 1517–1528.

14. U.P. Fringeli, H.H. Günthard, Infrared membrane spectroscopy, in: E. Grell (Ed.), *Membrane Spectroscopy*, Springer, New York, 1981, pp. 270–332.

15. J. Heberle, C. Zscherp, ATR/FT-IR difference spectroscopy of biological matter with microsecond time resolution, *Appl. Spectrosc.* 50 (1996) 588–596.

16. R.M. Nyquist, K. Ataka, J. Heberle, The molecular mechanism of membrane proteins probed by evanescent infrared waves, *ChemBioChem* 5 (2004) 431–436.

17. M. Osawa, Surface-enhanced infrared absorption spectroscopy, in: J.M. Chalmers and P.R. Griffiths (Eds.), *Handbook of Vibrational Spectroscopy*, John Wiley & Sons, Ltd, Chichester, 2002, pp. 785–799.

18. R.F. Aroca, D.J. Ross, C. Domingo, Surface-enhanced infrared spectroscopy, *Appl. Spectrosc.* 58 (2004) 324A–338A.

19. K. Ataka, J. Heberle, Biochemical applications of surface-enhanced infrared absorption spectroscopy, *Anal. Bioanal. Chem.* 338 (2007) 47–54.

20. K. Ataka, F. Giess, W. Knoll, R. Naumann, S. Haber-Pohlmeier, B. Richter, J. Heberle, Oriented attachment and membrane reconstitution of His-tagged cytochrome c oxidase to a gold electrode: In situ monitoring of surface-enhanced infrared absorption spectroscopy, *J. Am. Chem. Soc.* 126 (2004) 16199–16206.

21. K. Ataka, B. Richter, J. Heberle, Orientational control of the physiological reaction centre of cytochrome c oxidase tethered to a gold electrode, *J. Phys. Chem. B* 110 (2006) 9339–9347.

22. M. Fedurco, Redox reactions of heme-containing metalloproteins: Dynamic effects of self-assembled monolayers on thermodynamics and kinetics of cytochrome c electron-transfer reactions, *Coord. Chem. Rev.* 209 (2000) 263–331.

23. K. Ataka, J. Heberle, Electrochemically induced surface-enhanced infrared difference absorption (SEIDA) spectroscopy of a protein monolayer, *J. Am. Chem. Soc.* 125 (2003) 4986–4987.

24. K. Ataka, J. Heberle, Functional vibrational spectroscopy of a cytochrome c monolayer: SEIDAS probes the interaction with different surface modified electrodes, *J. Am. Chem. Soc.* 126 (2004) 9445–9457.

25. N. Wisitruangsakul, I. Zebger, K.H. Ly, D.H. Murgida, S. Ekgasit, P. Hildebrandt, Redox-linked protein dynamics of cytochrome c probed by time-resolved surface enhanced infrared absorption spectroscopy, *Phys. Chem. Chem. Phys.* 10 (2008) 5276–5286.

26. A. Badura, B. Esper, K. Ataka, C. Grunwald, C. Wöll, J. Kuhlmann, J. Heberle, M. Rögner, Light-driven water splitting for (bio-)hydrogen production: Photosystem 2 as the central part of a bioelectrochemical device, *Photochem. Photobiol.* 82 (2006) 1385–1390.

27. H. Krassen, A. Schwarze, B. Friedrich, K. Ataka, O. Lenz, J. Heberle, Photosynthetic hydrogen production by a hybrid complex of photosystem I and [NiFe]-hydrogenase, *ACS Nano* 3 (2009) 4055–4061.

28. V.A. Lorenz-Fonfria, J. Heberle, Proton transfer and protein conformation dynamics in photosensitive proteins by time-resolved step-scan Fourier-transform infrared spectroscopy, *J. Vis. Exp.* (2014) e51622.

29. C.J. Manning, R.A. Palmer, J.L. Chao, Step-scan Fourier-transform infrared spectrometer, *Rev. Sci. Instrum.* 62 (1991) 1219–1229.

30. I. Radu, M. Schleeger, M. Nack, J. Heberle, Time-resolved FT-IR spectroscopy of membrane proteins, *Aust. J. Chem.* 64 (2011) 9–15.

31. O. Weidlich, F. Siebert, Time resolved step-scan FTIR investigations of the transition from KL to L in the bacteriorhodopsin photocycle: Identification of chromophore twists by assigning hydrogen-out-of-plane (HOOP) bending vibrations, *Appl. Spectrosc.* 47 (1993) 1394–1400.

32. R.A. Palmer, J.L. Chao, R.M. Dittmar, V.S. Gregoriuo, S.E. Plunkett, Investigation of time-dependent phenomena by use of step-scan FT-IR, *Appl. Spectrosc.* 47 (1993) 1297–1310.

33. R. Rammelsberg, B. Heßling, H. Chorongiewski, K. Gerwert, Molecular reaction mechanism of proteins monitored by nanosecond step-scan FT-IR difference spectroscopy, *Appl. Spectrosc.* 51 (1997) 558–562.

34. M. Brandstetter, B. Lendl, Tunable mid-infrared lasers in physical chemosensors towards the detection of physiologically relevant parameters in biofluids, *Sensors Actuators B Chem.* 170 (2012) 189–195.

35. L. Bush, B. Lendl, B. Brandstetter, Quantum cascade lasers for infrared spectroscopy: Theory, state of the art, and applications, *Spectroscopy* 28 (2013) 26–33.

36. D. Oesterhelt, W. Stoeckenius, Rhodopsin-like protein from the purple membrane of *Halobacterium halobium*, *Nat. New Biol.* 233 (1971) 149–152.

37. E. Hildebrand, N. Dencher, Two photosystems controlling behavioural responses of *Halobacterium halobium*, *Nature* 257 (1975) 46–48.

38. G. Nagel, D. Ollig, M. Fuhrmann, S. Kateriya, A.M. Musti, E. Bamberg, P. Hegemann, Channelrhodopsin-1: A light-gated proton channel in green algae, *Science* 296 (2002) 2395–2398.

39. G. Nagel, T. Szellas, W. Huhn, S. Kateriya, N. Adeishvili, P. Berthold, D. Ollig, P. Hegemann, E. Bamberg, Channelrhodopsin-2, a directly light-gated cation-selective membrane channel, *Proc. Natl. Acad. Sci. U. S. A.* 100 (2003) 13940–13945.

40. K. Deisseroth, Optogenetics, *Nat. Methods* 8 (2011) 26–29.

41. M. Grote, M. Engelhard, P. Hegemann, Of ion pumps, sensors and channels: Perspectives on microbial rhodopsins between science and history, *Biochim. Biophys. Acta-Bioenerg.* 1837 (2014) 533–545.

42. U. Haupts, J. Tittor, D. Oesterhelt, Closing in on bacteriorhodopsin: Progress in understanding the molecule, *Annu. Rev. Biophys. Biomol. Struct.* 28 (1999) 367–399.

43. E. Racker, W. Stoeckenius, Reconstitution of purple membrane vesicles catalyzing light-driven proton uptake and adenosine triphosphate formation, *J. Biol. Chem.* 249 (1974) 662–663.

44. R. Henderson, P.N. Unwin, Three-dimensional model of purple membrane obtained by electron microscopy, *Nature* 257 (1975) 28–32.

45. H. Luecke, B. Schobert, H.T. Richter, J.P. Cartailler, J.K. Lanyi, Structure of bacteriorhodopsin at 1.55 Å resolution, *J. Mol. Biol.* 291 (1999) 899–911.

46. E.M. Landau, J.P. Rosenbusch, Lipidic cubic phases: A novel concept for the crystallization of membrane proteins, *Proc. Natl. Acad. Sci. U. S. A.* 93 (1996) 14532–14535.

47. C. Wickstrand, R. Dods, A. Royant, R. Neutze, Bacteriorhodopsin: Would the real structural intermediates please stand up?, *Biochim. Biochim. Acta-Gen. Subj.* 1850 (2015), 536–553.

48. R.A. Mathies, S.W. Lin, J.B. Ames, W.T. Pollard, From femtoseconds to biology: Mechanism of bacteriorhodopsin's light-driven proton pump, *Annu. Rev. Biophys. Biophys. Chem.* 20 (1991) 491–518.

49. J. Heberle, Time-resolved ATR/FT-IR spectroscopy of membrane proteins, in: S.G. Pandalai (Ed.), *Recent Research Developments in Applied Spectroscopy*, Research Signpost, Trivandrum, 1999, pp. 147–159.

50. C. Zscherp, J. Heberle, Infrared difference spectra of the intermediates L, M, N, and O of the bacteriorhodopsin photoreaction obtained by time-resolved attenuated total reflection spectroscopy, *J. Phys. Chem. B* 101 (1997) 10542–10547.

51. A. Maeda, Application of FTIR spectroscopy to the structural study on the function of bacteriorhodopsin, *Isr. J. Chem.* 35 (1995) 387–400.

52. J. Heberle, N.A. Dencher, Surface-bound optical probes monitor proton translocation and surface potential changes during the bacteriorhodopsin photocycle, *Proc. Natl. Acad. Sci. U. S. A.* 89 (1992) 5996–6000.

53. F. Garczarek, L.S. Brown, J.K. Lanyi, K. Gerwert, Proton binding within a membrane protein by a protonated water cluster, *Proc. Natl. Acad. Sci. U. S. A.* 102 (2005) 3633–3638.

54. K. Gerwert, B. Hess, J. Soppa, D. Oesterhelt, Role of aspartate-96 in proton translocation by bacteriorhodopsin, *Proc. Natl. Acad. Sci. U. S. A.* 86 (1989) 4943–4947.

55. S. Szarez, D. Oesterhelt, P. Ormos, pH-induced structural changes in bacteriorhodopsin studied by Fourier transform infrared spectroscopy, *Biophys. J.* 67 (1994) 1706–1712.

56. C. Zscherp, R. Schlesinger, J. Tittor, D. Oesterhelt, J. Heberle, In situ determination of transient pKa changes of internal amino acids of bacteriorhodopsin by using time-resolved attenuated total reflection Fourier-transform infrared spectroscopy, *Proc. Natl. Acad. Sci. U. S. A.* 96 (1999) 5498–5503.

57. J. Riesle, D. Oesterhelt, N.A. Dencher, J. Heberle, D38 is an essential part of the proton translocation pathway in bacteriorhodopsin, *Biochemistry* 35 (1996) 6635–6643.

58. K. Stehfest, P. Hegemann, Evolution of the channelrhodopsin photocycle model, *ChemPhysChem* 11 (2010) 1120–1126.

59. M. Müller, C. Bamann, E. Bamberg, W. Kühlbrandt, Projection structure of channelrhodopsin-2 at 6 Å resolution by electron crystallography, *J. Mol. Biol.* 414 (2011) 86–95.

60. H.E. Kato, F. Zhang, O. Yizhar, C. Ramakrishnan, T. Nishizawa, K. Hirata, J. Ito, Y. Aita, T. Tsukazaki, S. Hayashi, P. Hegemann, A.D. Maturana, R. Ishitani, K. Deisseroth, O. Nureki, Crystal structure of the channelrhodopsin light-gated cation channel, *Nature* 482 (2012) 369–374.

61. V.A. Lorenz-Fonfria, J. Heberle, Channelrhodopsin unchained: Structure and mechanism of a light-gated cation channel, *Biochim. Biophys. Acta-Bioenerg.* 1837 (2014) 626–642.

62. M. Nack, I. Radu, M. Gossing, C. Bamann, E. Bamberg, G.F. von Mollard, J. Heberle, The DC gate in channelrhodopsin-2: Crucial hydrogen bonding interaction between C128 and D156, *Photochem. Photobiol. Sci.* 9 (2010) 194–198.

63. I. Radu, C. Bamann, M. Nack, G. Nagel, E. Bamberg, J. Heberle, Conformational changes of channelrhodopsin-2, *J. Am. Chem. Soc.* 131 (2009) 7313–7319.

64. V.A. Lorenz-Fonfria, T. Resler, N. Krause, M. Nack, M. Gossing, G. F. von Mollard, C. Bamann, E. Bamberg, R. Schlesinger, J. Heberle, Transient protonation changes in channelrhodopsin-2 and their relevance to channel gating, *Proc. Natl. Acad. Sci. U. S. A.* 110 (2013) E1273–E1281.

65. M. Nack, I. Radu, C. Bamann, E. Bamberg, J. Heberle, The retinal structure of channelrhodopsin-2 assessed by resonance Raman spectroscopy, *FEBS Lett.* 583 (2009) 3676–3680.

66. B. Aton, A.G. Doukas, R.H. Callendar, B. Becher, T. Ebrey, Resonance Raman studies of the purple membrane, *Biochemistry* 16 (1977) 2995–2999.

67. E. Ritter, K. Stehfest, A. Berndt, P. Hegemann, F.J. Bartl, Monitoring light-induced structural changes of Channelrhodopsin-2 by UV-visible and Fourier transform infrared spectroscopy, *J. Biol. Chem.* 283 (2008) 35033–35041.

68. M.K. Neumann-Verhoefen, K. Neumann, C. Bamann, I. Radu, J. Heberle, E. Bamberg, J. Wachtveitl, Ultrafast infrared spectroscopy on channelrhodopsin-2 reveals efficient energy transfer from the retinal chromophore to the protein, *J. Am. Chem. Soc.* 135 (2013) 6968–6976.

69. A.K. Dioumaev, M.S. Braiman, Two bathointermediates of the bacteriorhodopsin photocycle, distinguished by nanosecond time-resolved FTIR spectroscopy at room temperature, *J. Phys. Chem. B* 101 (1997) 1655–1662.

70. C. Hackmann, J. Guijarro, I. Chizhov, M. Engelhard, C. Rödig, F. Siebert, Static and time-resolved step-scan Fourier transform infrared investigations of the photoreaction of

halorhodopsin from *Natronobacterium pharaonis*: Consequences for models of the anion translocation mechanism, *Biophys. J.* 81 (2001) 394–406.

71. M. Hein, A.A. Wegener, M. Engelhard, F. Siebert, Time-resolved FTIR studies of sensory rhodopsin II (NpSRII) from *Natronobacterium pharaonis*: Implications for proton transport and receptor activation, *Biophys. J.* 84 (2003) 1208–1217.

72. C. Bamann, R. Gueta, S. Kleinlogel, G. Nagel, E. Bamberg, Structural guidance of the photocycle of channelrhodopsin-2 by an interhelical hydrogen bond, *Biochemistry* 49 (2010) 267–278.

73. K. Eisenhauer, J. Kuhne, E. Ritter, A. Berndt, S. Wolf, E. Freier, F. Bartl, P. Hegemann, K. Gerwert, In channelrhodopsin-2 Glu-90 is crucial for ion selectivity and is deprotonated during the photocycle, *J. Biol. Chem.* 287 (2012) 6904–6911.

74. M. Nack, I. Radu, B.J. Schultz, T. Resler, R. Schlesinger, A.N. Bonder, C. Del Val, S. Abbruzzetti, C. Viappiani, C. Bamann, E. Bamberg, J. Heberle, Kinetics of proton release and uptake by channelrhodopsin-2, *FEBS Lett.* 586 (2012) 1344–1348.

75. K. Feldbauer, D. Zimmermann, V. Pintschovius, J. Spitz, C. Bamann, E. Bamberg, Channelrhodopsin-2 is a leaky proton pump, *Proc. Natl. Acad. Sci. U. S. A.* 106 (2009) 12317–12322.

76. V.I. Gordeliy, J. Labahn, R. Moukhametzianov, R. Efremov, J. Granzin, R. Schlesinger, G. Büldt, T. Savopol, A.J. Scheidig, J.P. Klare, M. Engelhard, Molecular basis of transmembrane signaling by sensory rhodopsin II-transducer complex, *Nature* 419 (2002) 484–487.

77. W. Gmelin, K. Zeth, R. Efremov, J. Heberle, J. Tittor, D. Oesterhelt, The crystal structure of the L1 intermediate of halorhodopsin at 1.9 angstroms resolution, *Photochem. Photobiol.* 83 (2007) 369–377.

78. R. Efremov, V. Gordeliy, J. Heberle, G. Büldt, Time-resolved microspectroscopy on a single crystal of bacteriorhodopsin reveals lattice-induced differences in the photocycle kinetics, *Biophys. J.* 91 (2006) 1441–1451.

79. I. Amenabar, S. Poly, W. Nuansing, E.H. Hubrich, A.A. Govyadinov, F. Huth, R. Krutokhvostov, L.B. Zhang, M. Knez, J. Heberle, A.M. Bittner, R. Hillenbrand, Structural analysis and mapping of individual protein complexes by infrared nanospectroscopy, *Nat. Commun.* 4 (2013) 2890.

80. H.A. Bechtel, E.A. Muller, R.L. Olmon, M.C. Martin, M.B. Raschke, Ultrabroadband infrared nanospectroscopic imaging, *Proc. Natl. Acad. Sci. U. S. A.* 111 (2014) 7191–7196.

10

ANALYSIS OF MEMBRANE-PROTEIN COMPLEXES BY SINGLE-MOLECULE METHODS

KATIA COSENTINO[1,2], STEPHANIE BLEICKEN[1,2],
AND ANA J. GARCÍA-SÁEZ[1,2]

[1] *Max Planck Institute for Intelligent Systems, Stuttgart, Germany*
[2] *Interfaculty Institute of Biochemistry, University of Tübingen, Tübingen, Germany*

10.1 INTRODUCTION

Biomolecules and especially proteins function either as soluble or as membrane-embedded moieties, with some exceptions that are able to travel between both worlds. However, soluble proteins are far better understood in terms of their structure, function, and interaction. The reason for this imbalance is found in the better-developed methodologies for studying soluble compared to membrane-embedded proteins. Currently, relatively few methods exist to study membrane proteins in their physiological environment. Although not well understood, membrane proteins are crucial to many biological processes and changes in their structure or function are involved in many disease-related processes. Therefore, membrane proteins are very important drug targets, and understanding their function is crucial for drug development [1].

Fluorescence correlation spectroscopy (FCS) and single-particle tracking (SPT) methodologies are two powerful techniques to study the diffusion and interaction of membrane proteins in artificial or cellular membranes. The FCS method was first implemented in the 1970s by Magde, Elson, and Webb [2, 3]. Later, in the 1990s, it

Pumps, Channels, and Transporters: Methods of Functional Analysis, First Edition. Edited by
Ronald J. Clarke and Mohammed A. A. Khalid.
© 2015 John Wiley & Sons, Inc. Published 2015 by John Wiley & Sons, Inc.

was shown that FCS was able to detect single molecules, and its applications to biomolecules expanded [4, 5]. Today, the FCS technique can be considered a well-established method that has been proven capable of describing systems from simple measurements in solution [6] to soluble proteins within the cytosol or nucleus of cells [7–10], to proteins in artificial membranes [6, 11–14] or in cell membranes [15–17], and even to experiments on living zebrafish embryos [18].

Since the pioneering work of Gelles et al. [19] and Geerts et al. [20], the application of single-particle (SP) techniques to life science has experienced a rapid evolution [21–23]. The ability of these techniques to visualize individual events enlarges their applicability to all those studies where information is lost by either spatial or/ and temporal averaged measurements. This situation applies to the study of, for example, "on–off" events (e.g., the opening and closing of an ion channel); of spatial and time-resolved mechanisms, such as the stoichiometry and the assembling mechanism of proteins into oligomers; or of processes that occur only in a minority of a population, and that would be hidden if averaged with the dominant signals.

FCS and SPT are based on optical microscopy setups and on the detection of fluorophores linked to the protein of interest. The linkage between the fluorophore and the protein can be achieved genetically by introducing genes coding for fusion proteins of the protein of interest together with a fluorescent protein, like GFP (green fluorescent protein), or by covalent attachment of an organic dye or quantum dot to the protein (more information is provided in Section 10.2). In this chapter, the two methods are introduced, compared, and examples of their application to the study of membrane–protein complexes are given.

10.2 FLUOROPHORES FOR SINGLE PARTICLE LABELING

The requirement for a high signal/noise ratio in single-molecule detection depends not only on the optical setup but also on the fluorescent probe that is linked to the molecule of interest in order to allow its detection. Therefore, the choice of an appropriate probe is strategic for successful detection. Such a choice depends on several factors, such as the nature of the probe itself and its photophysical properties. The most common fluorescent probes include quantum dots (QDs), fluorescent proteins (FPs), and organic fluorophores.

QDs are crystals of semiconductor materials based on metals such as cadmium and selenium. These inorganic fluorescent probes present excellent photophysical properties, as they are extremely bright and resistant to photobleaching [24]. This means that they provide high signal/noise ratios (up to 25) and can be detected individually, even by conventional wide-field illumination. They have broad excitation spectra and a narrow emission wavelength range, which can be tuned by their size. Hence, QDs of varying sizes can be synthesized to show emission covering the entire visual spectrum. This makes QDs ideal probes for multicolor imaging. In order to be used as biological fluorescent probes, QDs must be functionalized by linkage to biomolecules, such as biotin or avidin, which specifically bind to the molecule of interest [24, 25]. They are not internalized within cells and are widely used to track proteins and lipids at the cell surface [24]. Some drawbacks in their properties are linked to their large size, ranging between 10 and 20 nm, which can affect the functionality of the molecule to which they are attached. In addition, some concerns have been raised about the toxic effect that metals, such as cadmium and selenium, from which QDs are synthesized, can produce on cells. Finally, they tend to aggregate and monovalent labeling is difficult to achieve, even though some progress has been made toward reducing multivalent conjugation [26].

If labeling specificity is a must, then FPs are the probes of choice. Absolute specificity is assured because the labeling occurs at the genetic level. FPs are also more biocompatible with many organisms and less phototoxic than other fluorescent probes. These probes have been applied successfully in many single-molecule studies [27–29]. On the other hand, they are bulky and this can affect the structure and the function of the molecule of interest. In addition, they have a low extinction coefficient, which results in very low brightness [30], and they are not as photostable as organic dyes and QDs. Moreover, FPs need to mature in order to be fluorescent, and it is very difficult to estimate the rate of conversion between fluorescent and nonfluorescent FPs in a cell, resulting in an unknown "labeling efficiency." Nonetheless, FPs can be used in combination with total internal reflection fluorescence (TIRF) microscopy, which increases the signal/noise ratio and is thus ideally suited to studies of events at the cell membrane.

Organic dye molecules are a good compromise between QDs and FPs. They offer a wider spectral range, higher stability, and quantum yield compared to FPs (but less than QDs) while offering monovalent ligation. Their most relevant advantage compared to the other fluorescent probes is their small size (<1 nm), which allows them to be linked to any kind of ligand [30]. On the other hand, like QDs, they link through chemical conjugation [30] and do not assure absolute labeling efficiency. Examples of organic dyes are rhodamines, BODIPY (a boron-containing class of fluorophores), and cyanines (the ATTO and Alexa series of dyes are derivatives of these molecules).

10.3 PRINCIPLES OF FLUORESCENCE CORRELATION SPECTROSCOPY

FCS is a powerful technique to quantitatively determine the concentration, molecular mobility (diffusion properties), and interactions of fluorescent molecules with single-molecule sensitivity. Furthermore, from FCS data, it is possible to calculate rate constants, association and dissociation constants, as well as particle sizes. The strength of FCS lies in its wide application range, as it can be used *in vitro* and *in vivo*, in solution as well as in membranes.

In most recent FCS setups the detection volume of a confocal microscope is used to detect the fluctuations of fluorescence intensity over time. Each fluorophore that enters the focal volume leads to a burst of photons recorded by the detector (see Fig. 10.1a and b). As the detection volume is tiny (~1 fl), FCS can be considered as a method with single-molecule sensitivity. The primary data of FCS are intensity traces reflecting the residence times of individual fluorescent molecules

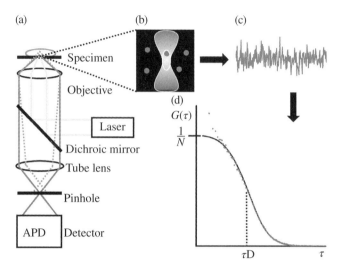

FIGURE 10.1 Schematic diagram of an experimental FCS setup and its workflow. (a) Confocal microscope with the specimen. The laser that excites the specimen is focused through an objective with high numerical aperture to a diffraction-limited spot. The excited fluorophores in the specimen emit light at a longer wavelength than the laser beam, and the emitted light can therefore be separated by a dichroic mirror and additional filters. The pinhole removes all out-of-focus photons not emitted from the focal plane in the specimen. (b) Depiction of the focal volume. The hourglass shape arises from the divergent laser beam. A confocal microscope measures a detection volume in the sub-femtoliter range, through which the molecules of interest can diffuse. (c) The diffusion of fluorescent particles is recorded as an intensity trace showing the fluorescence fluctuations in the confocal volume over time. (d) Based on the intensity traces autocorrelation curves can be calculated. Equations used are presented in Table 10.1.

in the focal volume (see Fig. 10.1c). From these traces, an autocorrelation function can be calculated, which compares the signal with itself over time using Equation 10.1 (the most important mathematical functions used for the different FCS methods are given in Table 10.1).

The autocorrelation curve (see Fig. 10.1d) allows the calculation (or comparison) of diffusion properties and concentrations of fluorescent molecules (see Figs. 10.1d and 10.2). The amplitude of the autocorrelation curve at $\tau=0$ (where τ is the lag time) is inversely proportional to the number of fluorophores in the detection volume. After calculating the size of the focal volume (V_{eff}; equation given in Table 10.1), the protein concentration can be precisely estimated. Moreover, the decay of the autocorrelation curve provides quantitative information about the average time that molecules stay in the focal volume (diffusion time, t_D). The diffusion time depends highly on the size and shape of the focal volume and on the diffusion properties and the size of the fluorescent molecule itself. By using fluorophores with well-described diffusion properties as references, the system can be calibrated, allowing the extraction of the diffusion coefficient (D) of the studied molecule (equation given in Table 10.1). Notably, the diffusion coefficient not

TABLE 10.1   Overview of the Functions used in FCS Analysis.

Equation	Title	Function
10.1	Autocorrelation function	$G(\tau) = \dfrac{\langle \delta F(t)\delta F(t+\tau)\rangle}{\langle F(t)\rangle^2}$
10.2	Cross-correlation function	$G_{RG}(\tau) = \dfrac{\langle \delta F_R(t)\,\delta F_G(t+\tau)\rangle}{\langle F_R(t)\rangle\langle F_G(t)\rangle}$
10.3	Fraction of molecules in two-color complex	$C_{RG} = \dfrac{G_{0,RG}}{G_{0,G}}$ or $C_{RG} = \dfrac{G_{0,RG}}{G_{0,R}}$
	Focal volume	$V_{eff} = \pi^{3/2}\omega_0^2 z_0$
	Structural parameter	$s = \dfrac{z_0}{\omega_0}$
	Diffusion coefficient	$D = \dfrac{\omega_0^2}{4t_D}$
	3D diffusion	$G_{3D}(\tau) = \dfrac{1}{N}\left(1+\dfrac{\tau}{\tau_D}\right)^{-1}\dfrac{1}{\sqrt{1+\dfrac{\tau}{s^2\tau_D}}}$
	2D diffusion xy (z-scan FCS)	$G_{2D}(\tau) = \dfrac{1}{N}\left(1+\dfrac{\tau}{\tau_D}\right)^{-1}$
	2D diffusion xz (sFCS)	$G(\tau) = \dfrac{1}{N}\left(1+\dfrac{\tau}{\tau_D}\right)^{-1}\left(1+\dfrac{\tau}{\tau_D s^2}\right)^{-\frac{1}{2}}$
	Dual-focus scanning FCS	$G_{12}(\tau) = \dfrac{1}{C\pi s\omega_0}\left(1+\dfrac{4D\tau}{\omega_0^2}\right)^{-\frac{1}{2}}\left(1+\dfrac{4D\tau}{\omega_0^2 s^2}\right)^{-\frac{1}{2}}$ $\exp\left(-\dfrac{d^2}{\omega_0^2+4D\tau}\right)$

Here, G is the autocorrelation and G_{RG} the cross-correlation function; τ is the lag time for correlation; $F(t)$ is the fluorescence intensity as a function of time (t); $\langle\,\rangle$ corresponds to time averaging ($\delta F(t) = F(t) - \langle F(t)\rangle$); F_R and F_G are the fluorescence intensity in the red and green channels; C_{RG} is the concentration molecules in the two-color complex; $G_{0,RG}$ is the cross-correlation amplitude; $G_{0,R}$ ($G_{0,G}$) is the amplitude of the red (or green) autocorrelation curve; V_{eff} is the focal volume; ω_0 is the radius of the focal volume in the x and y directions; z_0 is the radius of the focal volume in the z-direction; s is the structural parameter describing the shape of the focal volume; D is the diffusion coefficient; N is the average number of detected molecules; t_D is the average time the fluorophore stays in the focal volume; G_{12} is the spatial cross-correlation curve comparing the two foci; and d is the distance between the two foci.

only depends on the particle size but also on the temperature and on the viscosity of the medium, which need to be considered before experiments are performed. For detailed mathematical explanations and practical tips, see Refs. 10, 31, 32.

To obtain reliable, high-quality, quantitative FCS data, several aspects regarding the sample and the microscopy system need to be considered. (i) Photostable

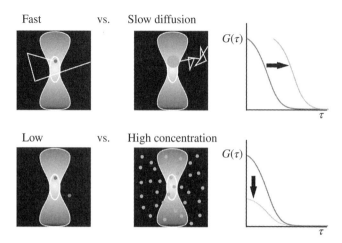

FIGURE 10.2 Effects of the diffusion rate and concentration on the autocorrelation curve. The size or, more generally, the diffusion behavior of the fluorescent molecule affects the decay of the autocorrelation curve, while the concentration of the fluorescent molecules changes the curve amplitude.

fluorophores (like organic dyes from the ATTO or Alexa series) with low bleaching or blinking (i.e., intermittent fluorescence) rates are recommended, as bleaching and blinking can both create artifacts that need to be considered in data evaluation. (ii) To reliably calculate binding constants, a high degree of labeling is needed, as the cross-correlation percentage can otherwise be too low and noisy to extract quantitative data. (iii) Quantitative FCS analysis is only suitable for a certain range of binding affinities and protein concentrations (mainly in the nanomolar range; for more information, see Refs. 10, 33). (iv) A high-quality experimental setup is necessary to have well-shaped, small detection volumes that can be adequately described by mathematical equations and that can be aligned in the two-color setups (e.g., the wavelength of the laser influences the focal volume; for two-color FCS, see Section 10.3.2). Moreover, the focal volume needs to be constant during the reference and sample measurements (e.g., the refractive index of the medium influences the shape of the focal volume and needs to be considered when reference samples are chosen). (v) Mathematical models describing the molecular motion of the studied molecules must be available in order to fit the data. Measurements in solution are mainly described by 3D diffusion models, while measurements in membranes need 2D diffusion models. Depending on the method, the membrane can cross the focal volume in the xy or xz direction, changing the shape of the detection area and the mathematical models to describe it (more information is given in later paragraphs and in Table 10.1). Moreover, for measurements in living cells, it can be quite challenging to find an adequate model to describe the molecular motion, due to the many different microcompartments that are not completely understood. (vi) The system needs to detect a reasonable and reliable "count rate." Artifacts can arise when very low fluorophore concentrations are used that are not much brighter than the detector noise. This can be counterbalanced by an increase in laser power. However, as higher laser powers increase the blinking and bleaching events, these effects need to be carefully balanced. A second error source is low particle statistics. Acquisition times at least 10^4 times longer than the diffusion time should be chosen to avoid this problem. Furthermore, photons not emitted from the fluorophore of interest but from auto-fluorescent molecules or other light sources can also create noise, which needs to be considered. More information is given in the following reviews [10, 31–33].

The following paragraphs will focus on two specialized applications of FCS: the quantitative analysis of bi-molecular complexes by two-color FCS and the study of membrane-embedded biomolecules.

10.3.1 Analysis of Molecular Complexes by Two-Color FCS

Two-color fluorescence cross-correlation spectroscopy (FCCS) is a method to quantify the interaction of two spectrally different fluorescent molecules [10, 31, 33, 34]. As shown schematically in Figure 10.3a and b, a setup with two laser lines is used to excite the two different fluorescent molecules. The emitted photons are separated by a dichroic mirror and filtered onto two photon detectors such as avalanche photodiodes (APDs), and the autocorrelation functions are calculated for both channels. Here, for simplicity, we call the channel with the shorter wavelength "green channel" and the channel with the longer wavelength "red channel." When the two fluorophores are in a complex, they will diffuse together through the focal volume (see Fig. 10.3c) and induce simultaneous fluorescence fluctuations in both channels. This is analyzed by a mathematical algorithm (see Eq. 10.2 in Table 10.1) called the cross-correlation function. By dividing the cross-correlation amplitude by one of the autocorrelation amplitudes of the two channels, the fraction of molecules in the complex can be calculated (Fig. 10.3d and Eq. 10.3).

Notably, as mentioned in Section 10.3, it is crucial to know and consider the degree of labeling of both fluorescent molecules and to align the focal volumes of both channels properly. Neglecting one or both will lead to an underestimation of the molecular interaction [10]. Moreover, setups with no or very low cross-talk between the two channels are needed to perform quantitative data analysis, as cross-talk leads to an overestimation of the molecular interaction. Under ideal conditions, the method allows the detection of the dynamic colocalization of fluorescent molecules, the following of association and dissociation processes, as well as the quantification of binding affinities (more detailed information is given in Refs. 10, 32, 33).

10.3.2 FCS Variants to Study Lipid Membranes

Generally, only very few methods allow the examination of membrane protein interactions within their physiological environment, and compared to soluble proteins, their applicability is very limited. One of the big advantages of FCS is precisely its application to study membrane proteins. However, even with FCS, these kinds of studies are challenging. Due to the higher viscosity of the membrane (compared to

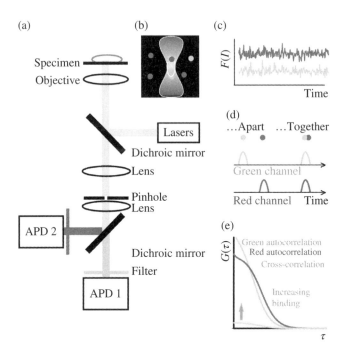

FIGURE 10.3 Schematic drawing of the two-color FCS setup and its workflow. (a) Schematic view of the confocal microscope setup for two-color FCCS. The main difference compared to the setup in Figure 10.1a is that two laser lines are used to excite the specimen; in addition, the emitted light from the two fluorophores of interest is separated by a second dichroic mirror and extra filters, and is recorded by two detectors. (b) Schematic view of the focal volume with interacting and non-interacting fluorophores of both colors. (c) Intensity traces of both channels. (d) From the intensity traces, it can be calculated whether the two fluorescent molecules interact, as shown by a cross-correlation curve with positive amplitude as plotted in (e). (e) Auto- and cross-correlation curves. The amount of complex formation is reflected by the amplitude of the cross-correlation curve.

solution), diffusion is slowed down, making longer acquisition times necessary. For statistical accuracy acquisition times of the order of minutes are necessary. Over such a long timeframe, the effects of photobleaching or phototoxicity increase, and bilayer movements in and out of the focal volume need to be considered. Moreover, reference measurements to describe the shape of the focal volume (a necessary step for data analysis) are difficult in membranes, as good reference samples are lacking. Furthermore, the correct positioning of the focus with respect to the membrane is difficult and can lead to artifacts, especially as the membrane thickness (~5 nm) is much smaller than the z-dimension of the focal volume (~1 μm). Additionally, the presence and detection of non-membrane-bound fluorescent molecules diffusing through the focal volume can create artifacts, if present in high numbers. In the following paragraphs, different FCS techniques for membranes are presented. For all of them, two-color applications exist to study molecular interactions.

The problem of accurate z-positioning of the focal volume is addressed in z-scan FCS by measuring the sample repeatedly at different z-positions (see Fig. 10.4a). Due to the divergence of the laser beam, the size of the area in which fluorophores are detected is shifted with the z-position. Thus, by moving the membrane in and out of the center of the focal volume, each measurement produces autocorrelation curves with different N and t_D values, and as the smallest detection area is found in the center of the focal plane, where N and t_D are minimized (see Fig. 10.4a). By knowing t_D and N in the different focal planes, a reference-free calculation of the particle concentration and the diffusion coefficient is possible, solving one of the major problems of FCS on membranes. However, the repeated measurements performed in z-scan FCS lengthen the total measurement time, and the method cannot correct for membrane movements during the acquisition time. Moreover, non-membrane-bound molecules concurrently diffusing through the focal volume can create artifacts that need to be considered. More information is given in Refs. 35, 36.

Total internal reflection (TIR) FCS is another method to study membrane proteins. The concept of TIRF will be described in detail in the SPT part of this chapter. Briefly, the illumination strategy creates a very small, approximately 100 nm-thick detection volume above the glass slide of the specimen. Therefore, only molecules very close to the glass are excited by the laser, making the setup very useful for studying membrane proteins in supported lipid bilayers or the plasma membrane of cells. Due to the small detection area, the influence of non-membrane-bound fluorophores is limited. However, contacts between the glass support and the membrane

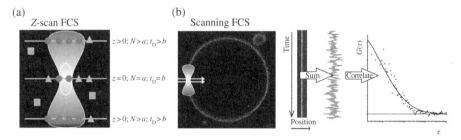

(a) Z-scan FCS   (b) Scanning FCS

$z>0; N>a; t_D>b$
$z=0; N=a; t_D=b$
$z>0; N>a; t_D>b$

Time   Sum   Correlate $G(\tau)$ Position τ

FIGURE 10.4 FCS in membranes. (a) In z-scan FCS different measurements along the z-axis are performed, moving the membrane into or out of the center of the focal volume. Due to the divergence of the laser beam, the size of the area in which fluorophores are excited is shifted with the z-position. Fluorophores within the membrane and within the laser beam are shown as circles, fluorophores in the membrane but outside the beam as triangles, and fluorophores not attached to the membrane as squares. As the smallest detection area is found in the center of the focal plane, N and t_D are minimized there. a and b are arbitrary parameters. (b) In two-focus scanning FCS, two lines are scanned perpendicular through the membrane. Notably, in contrast to z-scan FCS, the membrane passes the focal volume in the z-direction, for which reason the size/shape of the focal volume is different in both methods, and different analysis algorithms for data analysis are needed. The photon detection is presented in space and time by a kymograph, in which the membrane is visible as a bright line. From this data, the intensity traces and the autocorrelation curves are calculated.

can slow down the diffusion compared to free-standing membranes as shown in Ref. 37. More information about the method can be found in Refs. 38–40.

Scanning FCS is a method in which the focal volume is scanned repeatedly and perpendicularly through the membrane by a line scan of 32 pixels (see Fig. 10.4b). Only 2–3 of the 32 pixels excite the membrane, which limits the residence time of the laser beam on the membrane and decreases photobleaching. In a dual-focus variant of the method, two detection volumes are scanned. This allows a calibration-free calculation of the diffusion coefficient, when the distance between the two foci is known, solving the need for reference samples (see Table 10.1 and Refs. 32, 41). After detection, the photon intensities are shown in time and space as kymographs, in which the fluorescent membrane is localized by a line of high photon counts (see Fig. 10.4b). The photon intensities in the membrane are summed to an intensity trace from which the autocorrelation curve can be calculated. Thereby, the majority of the background signal from non-membrane-bound fluorophores is removed. Further advantages of this method are that corrections for membrane movements are possible and stability problems are improved. Scanning FCS can be used on freestanding membranes such as giant unilamellar vesicles (GUVs) as well as on cell membranes. More information is given in Refs. 32, 41–44.

10.3.3 FCS Applications to Membranes

In recent years the new FCS techniques have enabled several studies on membrane proteins in their physiological environments. We applied FCS to study the dynamic interplay between Bcl-2 proteins. These proteins are key players in the mitochondrial pathway of apoptosis [45, 46]. The interactions between family members in solution and in membrane are only partially understood. In two recent FCS studies, we were able to detect different modes of homo- and hetero-oligomerization in both environments (solution and membrane), showing that environmental changes can drastically affect protein interactions [6, 11]. Consistent with this idea, in another recent publication, the oligomerization of the fibroblast growth factor 2 (FGF_2) was studied in lipid environments mimicking the plasma membrane. It was shown that protein oligomerization is highly dependent on the presence of the lipid phosphatidylinositol 4,5-bisphosphate [47]. Many other FCS studies on protein oligomerization in model or plasma membranes have also been published (e.g., [12, 15, 27, 48–54]). Other FCS studies were performed on receptor–ligand or receptor–transducer interactions. One publication studied the signal transduction between the receptor–transducer pair NpSRII and HtrII of *Natronomonas pharaonis*. The interaction of both molecules had already been intensively studied in detergents, but only few methods allowed the interaction in lipid bilayers to be followed. By FCS, the diffusion coefficients of the single proteins and the complex could be determined, showing clearly their interaction in lipid bilayers [12]. Several other receptor–ligand or receptor–transducer interactions have also been studied by FCS [15, 27, 51–54]. In principle, FCS can also be used to probe the interaction of proteins, but caution is needed as technical problems, such as low labeling efficiencies, can lead to low cross-correlation amplitudes despite protein oligomerization. However, Kedrov et al. studied the oligomerization of the SecYEG complex in model membranes [49]. The SecYEG complex translocates unfolded proteins containing a pre-sequence (pre-proteins) across the inner bacterial membrane. The authors labeled the complex either with a green or a red fluorophore and reconstituted complexes of both colors into GUVs to study whether cross-correlation could be found. Neither for the SecYEG complexes alone nor in the presence of SecA or of the pre-protein could cross-correlation be detected. Thus, the authors assumed that SecYEG oligomerization is not necessary for pre-protein translocation. Another field in which FCS is used on membranes is phase separation or leaflet organization of lipid bilayers, and the diffusion and separation behavior of proteins and lipids in the different environments (e.g., [24–26, 28–30, 55, 56]). One example is the work of Chiantia et al. [26] in which FCS, atomic force microscopy, and confocal imaging were combined to study phase separation and protein/lipid diffusion in the different phases. For this purpose, supported bilayers of well-defined lipid compositions including the sphingolipid ceramide, which is proposed to modify protein interactions in membranes, were used. The authors showed that fluorescent lipid analogues and proteins normally enriched in the lipid-disordered phase were mainly excluded from ceramide-enriched regions. In contrast, proteins that were enriched in the lipid-ordered domains were enriched in the ceramide regions, implying that ceramide was either part of the lipid-ordered domains or recruited its own defined lipid domains. A very impressive example of an FCS study on receptor–ligand interaction is given by Yu et al. [18] and Ries et al. [57], who studied the interaction of FGF8 with its two receptors FGFR1 and FGFR4 in the cell membrane of living zebrafish embryos. The authors could calculate the dissociation constants between the ligand and the two receptors and demonstrate that FGF8 bound much more strongly to FGFR4 than to FGFR1.

10.4 PRINCIPLE AND ANALYSIS OF SINGLE-MOLECULE IMAGING

Among all the different approaches used in single-molecule detection and tracking (e.g., atomic force microscopy, electron microscopy, voltage-clamp measurements), optical imaging has revealed to be the technique of choice to provide combined spatially and temporally resolved information on single events [58].

In particular, TIRF microscopy is advantageous for optical imaging of membrane processes due to the capability of this technique to limit the illumination of the sample to thicknesses of only 200 nm. This considerably reduces out-of-focus fluorescence. The use of an objective with a high numerical aperture, in combination with a high-efficiency camera, allows high-resolution imaging of events where the number of photons is the limiting factor, which is often the case in single-molecule measurements. TIRF has been successfully applied to single-molecule imaging studies of membrane protein stoichiometry [59], and recent technical advances are further expanding its utility [60]. In the following sections, we will describe the principles of TIRF microscopy for single-molecule imaging and the theory behind single-particle detection and tracking, together with the methods of analysis currently used. Finally, we will discuss some key applications of single-particle imaging to the study of protein stoichiometry and kinetics in membranes.

10.4.1 TIRF Microscopy

Conventional wide-field imaging provides an axial resolution in the range of 400 nm. In the case of epifluorescence illumination, for instance, the large excitation volume in the z-direction results in the illumination of objects out of the focal plane, increasing the background and reducing the signal/noise ratio. TIRF microscopy overcomes this limitation by making use of an approach where the illumination intensity varies nonlinearly with distance from the surface. This allows selective illumination of thin areas of the sample, drastically reducing out-of-focus fluorescence, and makes TIRF an ideal technique for selective surface illumination (see Fig. 10.5), such as studies in cell membranes [61, 62].

TIRF microscopy is based on the physical principle of total internal reflection (TIR). When light travels from a medium with a high refractive index, n_1 (such as the glass of a coverslip), to the interface with a medium with a lower refractive index, n_2 (such as water solutions), two situations can occur. If the incident angle, θ (defined respect to the normal to the interface), is smaller or equal to a critical angle, θ_c, then, most of the light travels through the medium with lower refractive index, and the sample is illuminated as in conventional epifluorescence. However, if the incident angle is higher than θ_c, the light is totally (or almost) reflected (see Fig. 10.5).

The critical angle θ_c is defined by Snell's law according to the formula (Eq. 10.4):

$$\theta_c = \sin^{-1}\left(\frac{n_2}{n_1}\right) \tag{10.4}$$

In order to have TIR, the ratio n_2/n_1 must be less than 1 [63]. This effect generates a short-range electromagnetic wave at the interface, called an "evanescent wave", which propagates in the medium with lower refractive index. The intensity of this evanescent field, I_z, decreases exponentially with the distance from the coverslip. Thus, only the part of the sample that is close to the interface is efficiently illuminated. The value of the intensity at a position z is given by Equation 10.5:

$$I_z = I_0 e^{-z/d} \tag{10.5}$$

where I_0 is the intensity at position $z=0$ and d is the depth of illumination. This depth ranges between 50 and 150 nm, depending on the incident angle, and it is defined as the distance from the interface at which the intensity decays to $1/e$ of I_0 according to Equation 10.5. d can be calculated [63] from the excitation wavelength that the incident light would have in a vacuum (λ_0), the refractive index of the medium (n_2) and the incident angle of the incident light according to Equation 10.6:

$$d = \frac{\lambda_0}{4\pi}\left(n_2^2 \sin^2\theta - n_2^2\right)^{-\frac{1}{2}} \tag{10.6}$$

TIRF can be achieved by optical configurations in which light is directed to the interface either by a prism or by an objective with a high numerical aperture ($N_A > 1.4$). In the latter case the observation of emission is done by the same objective (see

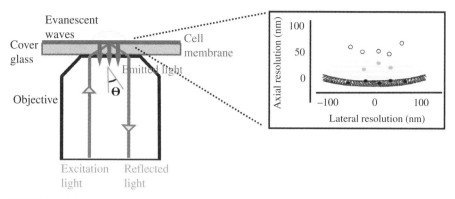

FIGURE 10.5 Principle of total internal reflection fluorescence (TIRF) microscopy. In this scheme, the excitation light travels from the cover glass (with refractive index n_1) to the interface with the water medium (and smaller refractive index n_2) of the membrane sample. TIRF occurs when the incident angle, θ (defined respect to the normal to the interface), is higher than a critical angle, θ_c. Then, the light is totally reflected and an evanescent field, which propagates in the direction of the medium with smaller refractive index, is generated at the interface. The intensity of these evanescent waves decreases with the distance from the cover glass, allowing an illumination depth of only 50–150 nm in the axial direction. This increases drastically the axial (but not the lateral) resolution. In the inset, this concept is illustrated by stripes whose thickness is reduced with distance. This allows the molecules (black circles) close to/in the membrane (membrane thickness is only 4 nm) to be completely excited, while the other particles are only weakly (gray circles) or not illuminated at all (black rings). In the set-up illustrated here, excitation, reflected, and emitted light are all collected by the same objective, but other set-ups are possible (see text).

Fig. 10.5). The use of an objective-based TIRF introduces some disadvantages, first in the illumination of the sample, because of some scattering of the excitation light within the objective, and second in the observed fluorescence, due to some luminescence arising from internal elements in the objective. On the other hand, prism-based TIRF is limited by the impossibility of using high power objectives (for a specific review on the topic, see Ref. 63).

The detectors most frequently used for recording single-molecule imaging are electron-multiplying charge-coupled devices (EM-CCDs) and silicon APDs. The latter have the advantage of a higher quantum efficiency compared to EM-CCD cameras, but they have a very limited detection area. For large samples, EM-CCD cameras are more appropriate, even though the images are limited both spatially (by the number of pixels) and temporally (by the exposure time required to collect enough photons) [64].

Summarizing, the use of TIRF microscopy not only drastically improves the sensitivity of imaging, but the selective illumination of the sample helps to reduce photodamage. This can turn out to be particularly advantageous when dealing with measurements at the cell surface. In addition, due to the use of an objective with a high numerical aperture, objective-based TIRF is particularly suitable for measurements with a low number of photons, as in the case of single-particle detection [63, 65].

Despite the obvious advantages of TIRF microscopy compared to conventional wide-field techniques, the internal reflections of polarized laser light may cause fringing that affects the quality of the image, often resulting in uneven illumination of the field of view. Consequently, if the sample is thin enough (as in the case of supported lipid bilayers), epifluorescence illumination, in combination with a high numerical aperture objective, good optics and high sensitive detectors, can be the preferred option [64].

10.4.2 Single-Molecule Detection

Standard fluorescence microscopy and TIRF are conventionally used for imaging single molecules. This allows structures to be resolved that are above the diffraction (or Abbe) limit of $r = 0.61\lambda/N_A \sim 200\,\text{nm}$, where r is the dimension of the structure of interest, λ is the wavelength of the light, and N_A is the numerical aperture of the objective.

Imaging of molecules of smaller size than the diffraction limit will result in a diffraction-limited spot. Although the size of this spot is bigger than the real object, it is possible to determine its intensity, $I(x,y)$, and its position with a positional accuracy in the order of tens of nanometers [66]. Four main algorithms are currently used for detecting and tracking single molecules: centroid, cross-correlation, sum-absolute difference and direct Gaussian fitting [67]. While the cross-correlation method seems to be the most accurate algorithm for large particles, Gaussian fitting is most appropriate for the detection of particles with sub-wavelength size and low signal/noise ratio (close to 4) [67], which is often the case in the detection of single fluorophores (see Fig. 10.6a).

In this case, the intensity profile, determined by the point spread function (PSF), is fitted to a 2D Gaussian (see Fig. 10.6a) [66, 69, 70] with a full width at half maximum (FWHM), w, according to Equation 10.7:

$$I(x,y) = N\frac{4\ln 2}{\pi w^2}\exp\left[-4\ln 2\left(\frac{(x-\mu_x)^2}{w^2} + \frac{(y-\mu_y)^2}{w^2}\right)\right] \quad (10.7)$$

where μ_x and μ_y are the position coordinates of the object. This approach requires a sufficiently high number of photons N [71], achievable by the use of fluorophores with high saturation intensity and high sensitive detectors.

Accordingly, the resolution limit for the characterization of the shape of the object is given by the precision for localizing the particle [66, 69]. The positional accuracy is affected by many factors, such as the photophysical properties of the fluorophore, the detection efficiency of the device, and the time resolution. A critical parameter is the signal-to-noise ratio [69], which can be improved by minimizing the background signal via, for example, highly sensitive CCD or EM-CCD cameras operating at low readout noise (in EM-CCD, it is reduced to sub 1 e⁻ levels), or, when possible, excitation by TIRF microscopy, which reduces out-of-focus phenomena.

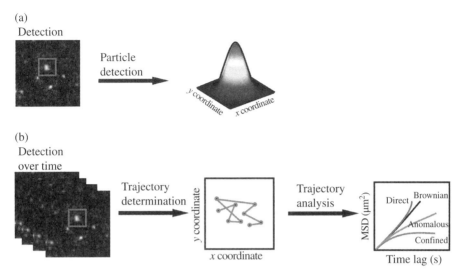

FIGURE 10.6 Single-molecule detection and tracking. (a) In order to detect a single fluorescent particle (left image, bright spot in the box), the intensity distribution of the particle is fitted with a known distribution function. This allows the determination of the center of such distribution, which corresponds to the particle's position (x and y). Here, a 2D Gaussian fit is illustrated (right), but other algorithms can be used as well (see text). (b) The detection process is repeated over time (image stack on the left) in order to determine the trajectory of the particle (central graph). The analysis of such trajectories provides important information about the motional properties of the particle. The most commonly used analysis method is the calculation of the mean square displacement (MSD) that allows the calculation of the diffusion coefficient of the particle by comparison of the MSD plot with theoretical motion models (see Table 10.2). In the right graph, different modes of motion are represented: Brownian, anomalous, direct with diffusion, and confined.

This approach presents several advantages: (i) the intensity of the spot is determined with higher accuracy than by just measuring the central pixel value and estimating the background, (ii) artifacts can be discarded based on the high width of the peak and high chi-squared values, and (iii) the spot position can be determined with an accuracy in the range of nanometers [69, 72]. Diffusional motion studies of rhodamine-labeled phospholipids in supported lipid bilayers have been carried out with a positional accuracy of 30 nm and a time resolution of 5 ms illumination [66]. Furthermore, GFP molecules have been localized with a precision of 1.5 nm but at the expense of a lower time resolution (500 ms) [73], which is not suitable for dynamic studies in living cells where the timescale of lateral diffusion is much faster (a few milliseconds). A better compromise between positional and time resolution can be achieved by using fluorophores with a high saturation intensity (e.g., Alexa647) [72], which increases the number of emitted photons N (see Eq. 10.7) when excited at high laser intensity.

Finally, in order to image single fluorophores, the density of fluorescent molecules has to be low enough (~1 μm^{-2}) to ensure a distance between two molecules higher than $3w$, or, in other words, independent intensity profiles.

10.4.3 Single Particle Tracking and Trajectory Analysis

Once the position of each particle in a given image has been properly detected, particles can be tracked over time, comparing each position in consecutive images (see Fig. 10.6b). Most of the tracking algorithms currently available are optimized for low surface density and slow diffusion [66, 74]. Although this approach allows several issues linked to high particle density (such as motion heterogeneity, particle disappearance, splitting, or merging with other particles) to be overcome, it limits the number of observed particles per experiment and does not allow the study of particle interactions. To overcome these issues, new algorithms, based on recursive detection, are used.

Schütz and coworkers have proposed a nearest-neighbor algorithm that allows single-molecule tracking up to a surface density of $1\,\mu m^{-2}$ [75]. In order to connect two particles, the molecule in image i is surrounded by a circle of radius R_{max}. If in the next image, $i+1$, a molecule is found within the circle, then it is connected to the corresponding particle in image i. Repetition of this procedure throughout the whole image sequence allows the trajectory of the particle of interest to be reconstructed [75]. When the surface density becomes too high, splitting, merging, bleaching, or interaction events make the trajectory reconstruction difficult to carry out. In order to solve this problem, global, instead of local, linking strategies can be used. An example is provided by the graph-based optimization approach proposed by Jaqaman et al. [76]. This algorithm is an implementation of the method of multiple-hypothesis tracking (MHT) [77], where the largest non-conflicting ensemble of trajectories is determined via a temporally global optimization [76].

Analysis of the trajectories can provide important quantitative biological information as, for example, the mode of motion of the object through the calculation of the mean square displacement (MSD) (see Fig. 10.6b) [68, 78, 79].

The lateral diffusion of particles in a medium with diffusion coefficient D, is described by the cumulative distribution function for the square displacement r^2 according to the Equation 10.8:

$$P\left(r^2, t_{lag}\right) = 1 - \exp\left(-\frac{r^2}{MSD\left(t_{lag}\right)}\right) \tag{10.8}$$

where $P(r^2, t_{lag})$ describes the probability that, starting at the origin at time 0, the particle will be found within a circle of radius r at time t_{lag}. For each trajectory, the two-dimensional MSD for every observation r_i and r_{i+1} separated by t_{lag} is given by Equation 10.9:

$$MSD = \frac{\sum_{i=1}^{T-t}\left(r_{i+1} - r_i\right)^2}{T - t} \tag{10.9}$$

with T being the total length of the trajectory in time. The MSD is calculated in order to reduce the noise in an experimental trajectory [37, 38]. Its value, for a given time lag, can be defined as the average over all the pairs of points in that time lag.

TABLE 10.2   Diffusion Models for Single-Particle Tracking.

Equation	Title	Function
10.10	Brownian motion	$MSD(t) = aDt$
10.11	Anomalous motion	$MSD(t) = aDt^b$
10.12	Direct motion with diffusion	$MSD(t) = aDt + (vt)^2$
10.13	Confined motion	$MSD(t) = R[1 - c_1 \exp(-c_2 \, aDt/R)]$

Here, $a = 4$ in 2D and $a = 6$ in 3D; $b < 1$; v is the speed of motion; R is the confinement size; c_1 and c_2 are positive constants related to the geometry of that region [68].

Averaging is a critical step and should not be done automatically, because it might hide a transition from diffusive to non-diffusive trajectory segments. In general, accurate MSD values (and diffusion coefficients) are obtained for short-range averaging [37]. The MSD value allows the mode of motion to be characterized and the diffusion coefficient, D, to be calculated. The different modes of motion are illustrated in Figure 10.6b, and the respective equations are reported in Table 10.2 (Eqs. 10.10–10.13).

10.5 COMPLEX DYNAMICS AND STOICHIOMETRY BY SINGLE-MOLECULE MICROSCOPY

In a single-molecule imaging experiment, fluorescence intensity is measured in order to gain information about biological processes in cells. In fact, the fluorescence intensity is related to many physical and chemical parameters of the observed molecules and their environment. Appropriate data analysis has allowed quantitative information relevant to the nature of many biological processes at the molecular level to be obtained. Here, we report some key applications where single-molecule imaging has led to a better understanding of the assembly mechanisms of proteins and the dynamics of molecules in membranes.

10.5.1 Application to Single-Molecule Stoichiometry Analysis

An important application of single-molecule imaging is the determination of the subunit stoichiometry of molecules. This approach has turned out to be particularly relevant in the study of the assembly mechanism of membrane proteins [59]. Single fluorophores can be coupled with the protein of interest and localized in the cell membranes (see Fig. 10.7a). With continuous excitation, the fluorescence signal of the observed particles decreases in discrete steps of the same amplitude. Each of these steps corresponds to the bleaching of one single fluorophore. By counting the number of discrete steps until the complete bleaching of the molecule of interest, an estimate of the stoichiometry of that molecule can be obtained (see Fig. 10.7b). Repetition of this process for a statistically significant number of molecules finally leads to Gaussian-fitted distributions of each oligomer (monomers, dimers, trimers, tetramers, etc.), from which the percentage of each species can be calculated [59].

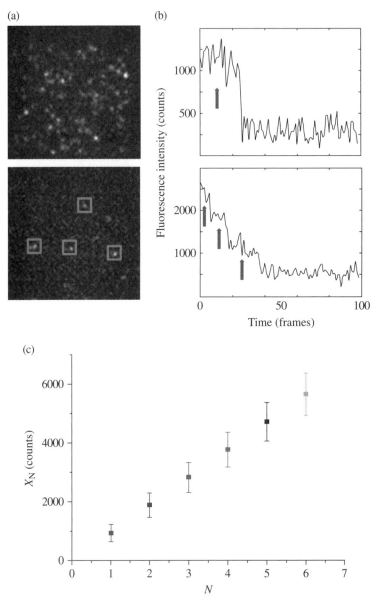

FIGURE 10.7 Determination of the subunit stoichiometry of proteins. (a) TIRF image of fluorescent particles in a model membrane before photobleaching (upper picture). After a bleaching time t, but immediately before complete bleaching, the image only includes brightness values of single fluorophores (boxes in the lower picture). (b) The fluorescent intensity of each particle is plotted over time until bleaching is complete, in order to obtain a bleaching-step graph. Each step (indicated by the arrows) corresponds to bleaching of one single fluorophore. By counting the number of discrete steps, an estimate of the stoichiometry of the molecular complex is provided. The upper graph presents a single bleaching step and corresponds to a monomer, while the lower graph, with three bleaching steps, corresponds to a trimer. (c) Graph reporting x_N and σ_N for $N=6$ species calculated according to Equation 10.16.

This method, called photobleaching-based single-molecule counting, was initially introduced by Ulbrich and Isacoff [80] to resolve the subunit composition of GFP-tagged NR1:NR3 NMDA receptor. They demonstrated a precise *in situ* determination of the stoichiometry of membrane-bound proteins, verifying the applicability of this technique to stoichiometries of up to four units [80]. Variations of this method have been successfully applied to many other systems [62, 81].

By using the same approach, Zhang et al. [62] have extended the applicability of this technique to dynamic stoichiometry of serine–threonine kinase receptors. In particular, they discovered a new model for the activation of transforming growth factor (TGF)-β type II receptor (TβRII) tagged with GFP. This membrane protein, which is present as a monomer in its resting state, was found to undergo dimerization upon stimulation with (TGF)-β1 ligand. In addition, a higher number of oligomers were observed at high expression levels of TβRII [62]. Furthermore, single-molecule imaging has also been used in combination with single-molecule force measurements to study the inhibitory effect that naringenin, a natural flavanone, exerts on the TGF-β ligand–receptor interaction [81]. Imaging analyses have proven that naringenin inhibits ligand-induced TβRII dimerization by reducing its binding affinity for TGF-β1. This result was further confirmed by force measurements [81].

In spite of the many advantages that such an approach offers, in particular for the *in situ* determination of the stoichiometries of protein complexes, its major drawback is that it is restricted to complexes composed of a relatively small number of monomers (up to 4). For higher oligomers, the detection of discrete steps becomes more difficult [80]. The same problem applies to the study of mobile particles [82]. In addition, the diffraction-limited resolution of the imaging doesn't allow a distinction between situations where molecules interact with each other or are simply in close vicinity. Dilution of the sample should reduce the likelihood of molecules being in close vicinity purely by nature of their concentration. However, in some cases, such as in *in vitro* studies of endogenous proteins in the cellular plasma membrane, where dilution is not possible and the protein surface density is too high to resolve individual clusters, alternative techniques are required.

In order to characterize the stoichiometry of mobile molecules, a different approach, called single-molecule brightness analysis, is generally used. This method, originally introduced by Schmidt et al. [83], is based on the comparison of the fluorescent intensity of a cluster of fluorophores with the brightness of a single fluorophore. The fluorescent intensity of a particle (monomer or cluster) is obtained by fitting its point spread function (PSF) to a 2D Gaussian, allowing the localization of the particle and an estimation of its brightness. In order to identify which particles are monomers, particles are photobleached and an image sequence is recorded over time. At this point, two different approaches can be used:

1. The second last image, immediately before complete bleaching, can be analyzed, as it will include only brightness values of single fluorophores (see Fig. 10.7a). In this case, the probability of observing more than one single fluorophore is negligible [84, 85].

2. For a given image sequence, the fluorescent intensity of each particle can be plotted over time until bleaching is complete in order to obtain a bleaching-step graph (see Fig. 10.7b). Only the intensity values of particles showing a single bleaching step (monomers) are then considered for further analysis.

To quantify the occurrence of each species, the theoretical intensity distribution of N colocalized independent emitters, $\rho_N(F)$, is calculated recursively as a series of convolution integrals:

$$\rho_N(F) = \int \rho_1(F') \rho_{N-1}(F - F') dF' \tag{10.14}$$

where $\rho_1(F)$ is the probability density function (pdf) characterizing the number of photons (F) detected from a single fluorophore.

Given a mixed population of monomers and higher oligomers, the overall intensity distribution results from a linear combination of the distributions according to Equation 10.15:

$$\rho(F) = \sum_{N=1}^{N_{MAX}} \beta_N \, \rho_N(F) \tag{10.15}$$

where β_N is the weight of the distribution N, corresponding to the fraction of the respective N-mer $\left(\sum_{N=1}^{N_{MAX}} \beta_N = 1\right)$ [86]. If values of $\rho(F)$ and $\rho_1(F)$ can be obtained experimentally, the β_N values can be extracted by non-linear least squares fitting [86].

An alternative approach is to approximate the intensity distribution of a high number of monomers (more than 1000) to a Gaussian, providing the intensity value, x_1, and the standard deviation, σ_1, for a single fluorophore. From the monomer intensity value, x_1, the fluorescent intensities of higher oligomers, x_N, can be calculated according to equation 10.16 [83]

$$x_N = Nx_1 \pm N^{1/2} \sigma_1 \tag{10.16}$$

where $N = 2, 3, 4$, etc., for dimers, trimers, tetramers, etc., respectively (see Fig. 10.7c). The maximum number of N-mers that can be resolved, defined as the stoichiometry resolution, is given by Equation 10.17 [83]:

$$N_{MAX} = \left(\frac{x_1}{\sigma_1}\right)^2 \tag{10.17}$$

The fluorescent intensity distribution, $\rho(F)$, of a mixed population of monomers and higher oligomers is fitted with a sum of N_{MAX} Gaussians, imposing x and σ values (from Eq. 10.16) in each fitting:

$$\rho(F) = \sum_{N=1}^{N_{MAX}} A_N \cdot \frac{1}{\sigma\sqrt{2\pi}} e^{-\frac{(i-x_N)^2}{2(\sigma_N)^2}} \tag{10.18}$$

where A_N is the area under the curve of component N. The normalized area under each fitted curve provides the percentage of occurrence for the respective species [87]. It is worth noting that the percentage of occurrence provided has to be "corrected" with respect to the labeling efficiency in order to obtain the real occurrence of the species.

Due to its applicability for the characterization of the stoichiometry of mobile particles, brightness analysis has been widely used in studies of proteins diffusing in the plasma membrane [88–90]. Kasai et al. have applied this method to study the monomer–dimer equilibrium of N-formyl peptide receptor (FPR), a chemoattractant G protein-coupled receptor (GPCR), in the plasma membrane. They revealed a fast dynamic equilibrium (of the order of ms), both in the dissociation of dimers and in their formation [89]. Teramura et al. studied the dimerization of epidermal growth factor (EGF) in the plasma membrane. The activation of EGF receptor (EGFR) requires the formation of a signaling EGFR dimer bound to a single ligand. They showed that only few EGFR molecules formed dimeric binding sites on the cell membrane, but the binding kinetics of these sites was faster than in the monomeric form [90].

In the presence of a high surface density of mobile particles, an interesting approach to carry out single molecule analysis is that of TOCCSL (Thinning Out Clusters While Conserving Stoichiometry of Labeling) [86]. In this method, the fluorophores of a small area of the sample are selectively bleached, while the molecules outside this region are unaffected. After a short recovery time, due to the diffusion of particles, the bleached area becomes populated with a few molecules/ clusters that can easily be resolved as individual diffraction-limited spots and quantified according to their stoichiometry [86]. TOCCSL has been successfully combined with single-molecule brightness analysis to study the stoichiometry of proteins in cell membranes [84–86, 88, 91, 92]. This approach has made it possible to visualize rafts in the plasma membrane [85]. By using the raft-marker glycosylphosphatidylinositol-anchored monomeric GFP in CHO cells, evidence for a cholesterol-dependent dimerization of the marker was found, as cholesterol depletion abolished this homo-association. In addition, the associated probes moved together in the plasma membrane, indicating the presence of small areas containing cholesterol [85].

Madl et al. [88] have applied brightness analysis and the TOCCSL method to study the subunit stoichiometry of the plasma membrane protein Orai1, considered responsible for the formation of store-operated calcium entry (SOCE) channels. Due to the overexpression of Orai1-mGFP, required to avoid underestimation by mixing of the labeled protein with endogenous subunits, the surface probe density was very high, preventing individual clusters from being resolved. Application of TOCCSL solved this problem and allowed the pore stoichiometry to be determined. Brightness analysis revealed a tetrameric stoichiometry for mobile Orai1 proteins in resting cells [88].

Finally, two-color TOCCSL was used for studying hetero-association of labeled cholera toxin B (CTX-B-Alexa647) with its ligand BODIPY-GM1 in model membranes. Interestingly, while free BODIPY-GM1 was present only in its monomeric form, upon addition of CTX-B-Alexa647, a distribution of colocalized monomers, dimers, and trimers was observed [84].

10.5.2 Application to Kinetics Processes in Cell Membranes

Single particle tracking (SPT) techniques have been widely used in the study of protein interactions in cell membranes. The first step in this direction was reported by Barak and Webb [93], who studied the internalization of human low-density lipoprotein (LDL) receptors in internalization-deficient human fibroblasts. The incorporation of the fluorescent probe diI (a lipophilic carbocyanine) in LDL allowed the monitoring of LDL–LDL receptor motion on living fibroblasts by standard fluorescent techniques, showing that the complex was completely immobilized on the surface [93]. Along the same lines, Smith et al. [94] studied the anomalous diffusion of major histocompatibility complex (MHC) class I molecules on the surface of HeLa cells and discussed how this mode of diffusion could affect T-cell activation [94]. More recently, SPT has been applied to the analysis of the kinetics of epidermal growth factor (EGF) binding to its receptor (EGFR, a type I receptor tyrosine kinase) on the plasma membrane of HeLa cells [90]. Rhodamine-labeled EGF, which freely diffuses in solution following Brownian motion, was found to undergo a reduction in mobility upon binding with its receptors, thus allowing the imaging of the EGF–EGFR complex. It was observed that EGF bound to preformed dimeric EGFR binding sites two orders of magnitude faster compared to its binding to monomeric receptors and the binding occurred through the formation of a kinetic intermediate, never observed before [90]. Chung et al. have delved further into the dimerization dynamics of individual EGFRs on living cells by using QD-based optical tracking of single molecules. They have found that, before ligand addition, EGFRs spontaneously formed finite-lifetime dimers and although ligand binding was not necessary for EGFR activation, it was still required for EGFR signaling [95].

Particle tracking has also been extended to the study of the mobility of biomolecules on the cell membrane, as in the case of the diffusion of dye molecules in model membranes [66, 78] and proteins in cells [96–98]. Courty et al. used this approach and provided the first live-cell characterization of the mobility of single kinesin motors [97].

An SPT study by Andrews et al. on the simultaneous high-affinity IgE receptor (FcεRI) motion and GFP-tagged actin dynamics has revealed the presence of micron-sized transient confinement zones (TCZs) in the plasma membrane. These are induced by actin filaments that confine FcεRI motion [99]. TCZs are areas of 50–700 nm diameter in which the diffusion of the observed molecules is slowed down, increasing the residence time of the particles in these domains [100]. They are thought to resemble lipid rafts and are believed to play a role in the coordination of signal transduction cascades. Lommerse et al. [98] have verified the presence of these domains in the cytoplasmic leaflet of the plasma membrane by recording the trajectories of a lipid-anchored fluorescent protein (eYFP-H-Ras). They have found that 30–40% of the molecules diffused in confined areas with a typical diameter of 200 nm. These zones, however, were not affected by the breakdown of actin, hence indicating that they can have a different chemical composition [98].

Finally, another useful method is two-color SPT, which helps significantly in the detection of differential modes of motion. Via dual-color TIRF microscopy, Pinaud et al. studied the structure and dynamics of putative raft microdomains in the membrane of HeLa cells by monitoring the correlation between the diffusion of a glycosylphosphatidylinositol-anchored avidin probe (Av-GPI) and the location of glycosphingolipid GM1-rich microdomains and caveolae. They observed a slowing down of the diffusion of Av-GPI when entering GM1-rich microdomains located in close proximity to, but distinct from, caveolae. In addition, the dynamic partitioning of Av-GPI in and out of these microdomains was cholesterol dependent, providing evidence that cholesterol/sphingolipid-rich microdomains can induce compartmentalization in the plasma membrane [96].

10.6 FCS VERSUS SPT

FCS and SPT represent two powerful techniques to gain information on the properties and motion of molecules in biological membranes. These two methods are quite different and the use of one or the other is subject to the specific experimental conditions. SPT, for example, can be used only at very low concentrations (up to 1 molecule μm^{-2}), limiting the number of observed particles for a given experiment. SPT in samples with high surface density can still be applied by using specific precautions in order to reduce the surface concentration of particles, for example, by using TOCCSL (described in Section 10.5.1). FCS allows the use of higher concentrations (between 0.1 and 100 molecules μm^{-2}). Another drawback of SPT is that it must be performed on membranes close to the glass support, while in FCS different membrane systems can be used.

FCS averages the behavior of molecules within a tiny observation volume (of the order of fL). This approach has the advantage over SPT that one can easily measure enough particles for a good statistical analysis, but this can be problematic since averaging can sometimes lead to a loss of information. This could, for example, be the case when monitoring a minority of a population or populations whose properties (such as mobility) change rapidly over space or time. SPT overcomes this issue providing crucial information on individual molecules, with spatial resolution in the range of a few nanometers and temporal resolution of milliseconds. This makes this technique especially appealing for dynamic studies of single molecules in cell membranes.

In conclusion, both FCS and SPT offer the possibility of obtaining important insights into the mechanisms occurring within cell membranes. The choice of one technique over the other is, in the end, mainly dependent on the specific biological question to be answered.

REFERENCES

1. T. Reiss, Drug discovery of the future: The implications of the human genome project, *Trends Biotechnol.* 19 (2001) 496–499.

2. D. Magde, E.L. Elson, W.W. Webb, Fluorescence correlation spectroscopy. II. An experimental realization, *Biopolymers* 13 (1974) 29–61.

3. D. Magde, E. Elson, W.W. Webb, Thermodynamic fluctuations in a reacting system— Measurement by fluorescence correlation spectroscopy, *Phys. Rev. Lett.* 29 (1972) 705–708.

4. R. Rigler, Ü. Mets, J. Widengren, P. Kask, Fluorescence correlation spectroscopy with high count rate and low background: Analysis of translational diffusion, *Eur. Biophys. J.* 22 (1993) 169–175.

5. M. Eigen, R. Rigler, Sorting single molecules: Application to diagnostics and evolutionary biotechnology, *Proc. Natl. Acad. Sci. U. S. A.* 91 (1994) 5740–5747.

6. A.J. Garcia-Saez, J. Ries, M. Orzaez, E. Perez-Paya, P. Schwille, Membrane promotes tBID interaction with BCL(XL), *Nat. Struct. Mol. Biol.* 16 (2009) 1178–1185.

7. D. Merkle, D. Zheng, T. Ohrt, K. Crell, P. Schwille, Cellular dynamics of Ku: Characterization and purification of Ku-eGFP, *ChemBioChem* 9 (2008) 1251–1259.

8. R. Broderick, S. Ramadurai, K. Toth, D.M. Togashi, A.G. Ryder, J. Langowski, H.P. Nasheuer, Cell cycle-dependent mobility of Cdc45 determined in vivo by fluorescence correlation spectroscopy, *PLoS One* 7 (2012) e35537.

9. N. Dross, C. Spriet, M. Zwerger, G. Muller, W. Waldeck, J. Langowski, Mapping eGFP oligomer mobility in living cell nuclei, *PLoS One* 4 (2009) e5041.

10. K. Bacia, P. Schwille, Practical guidelines for dual-color fluorescence cross-correlation spectroscopy, *Nat. Protoc.* 2 (2007) 2842–2856.

11. S. Bleicken, C. Wagner, A.J. García-Sáez, Mechanistic differences in the membrane activity of Bax and Bcl-xL correlate with their opposing roles in apoptosis, *Biophys. J.* 104 (2013) 421–431.

12. J. Kriegsmann, I. Gregor, I. von der Hocht, J. Klare, M. Engelhard, J. Enderlein, J. Fitter, Translational diffusion and interaction of a photoreceptor and its cognate transducer observed in giant unilamellar vesicles by using dual-focus FCS, *ChemBioChem* 10 (2009) 1823–1829.

13. F. Heinemann, S.K. Vogel, P. Schwille, Lateral membrane diffusion modulated by a minimal actin cortex, *Biophys. J.* 104 (2013) 1465–1475.

14. D. Lingwood, J. Ries, P. Schwille, K. Simons, Plasma membranes are poised for activation of raft phase coalescence at physiological temperature, *Proc. Natl. Acad. Sci. U. S. A.* 105 (2008) 10005–10010.

15. G. Vamosi, A. Bodnar, G. Vereb, A. Jenei, C.K. Goldman, J. Langowski, K. Toth, L. Matyus, J. Szollosi, T.A. Waldmann, S. Damjanovich, IL-2 and IL-15 receptor alpha-subunits are coexpressed in a supramolecular receptor cluster in lipid rafts of T cells, *Proc. Natl. Acad. Sci. U. S. A.* 101 (2004) 11082–11087.

16. P. Liu, T. Sudhaharan, R.M.L. Koh, L.C. Hwang, S. Ahmed, I.N. Maruyama, T. Wohland, Investigation of the dimerization of proteins from the epidermal growth factor receptor family by single wavelength fluorescence cross-correlation spectroscopy, *Biophys. J.* 93 (2007) 684–698.

17. D.R. Larson, J.A. Gosse, D.A. Holowka, B.A. Baird, W.W. Webb, Temporally resolved interactions between antigen-stimulated IgE receptors and Lyn kinase on living cells, *J. Cell Biol.* 171 (2005) 527–536.

18. S.R. Yu, M. Burkhardt, M. Nowak, J. Ries, Z. Petrasek, S. Scholpp, P. Schwille, M. Brand, Fgf8 morphogen gradient forms by a source-sink mechanism with freely diffusing molecules, *Nature* 461 (2009) 533–536.

19. J. Gelles, B.J. Schnapp, M.P. Sheetz, Tracking kinesin-driven movements with nanometre-scale precision, *Nature* 331 (1988) 450–453.

20. H. Geerts, M. de Brabander, R. Nuydens, Nanovid microscopy, *Nature* 351 (1991) 765–766.

21. W. Moerner, New directions in single-molecule imaging and analysis, *Proc. Natl. Acad. Sci. U. S. A.* 104 (2007) 12596–12602.

22. I. Casuso, F. Rico, S. Scheuring, High-speed atomic force microscopy: Structure and dynamics of single proteins, *Curr. Opin. Chem. Biol.* 15 (2011) 704–709.

23. A. Fürstenberg, M. Heilemann, Single-molecule localization microscopy–near-molecular spatial resolution in light microscopy with photoswitchable fluorophores, *Phys. Chem. Chem. Phys.* 15 (2013) 14919–14930.

24. S. Chiantia, N. Kahya, J. Ries, P. Schwille, Effects of ceramide on liquid-ordered domains investigated by simultaneous AFM and FCS, *Biophys. J.* 90 (2006) 4500–4508.

25. S. Chiantia, N. Kahya, P. Schwille, Raft domain reorganization driven by short- and long-chain ceramide: A combined AFM and FCS study, *Langmuir* 23 (2007) 7659–7665.

26. S. Chiantia, J. Ries, G. Chwastek, D. Carrer, Z. Li, R. Bittman, P. Schwille, Role of ceramide in membrane protein organization investigated by combined AFM and FCS, *Biochim. Biophys. Acta-Biomembr.* 1778 (2008) 1356–1364.

27. O. Meissner, H. Haberlein, Lateral mobility and specific binding to GABA(A) receptors on hippocampal neurons monitored by fluorescence correlation spectroscopy, *Biochemistry* 42 (2003) 1667–1672.

28. S. Chiantia, P. Schwille, A.S. Klymchenko, E. London, Asymmetric GUVs prepared by MbetaCD-mediated lipid exchange: An FCS study, *Biophys. J.* 100 (2011) L1–L3.

29. R. Lasserre, X.J. Guo, F. Conchonaud, Y. Hamon, O. Hawchar, A.M. Bernard, S.M. Soudja, P.F. Lenne, H. Rigneault, D. Olive, G. Bismuth, J.A. Nunes, B. Payrastre, D. Marguet, H.T. He, Raft nanodomains contribute to Akt/PKB plasma membrane recruitment and activation, *Nat. Chem. Biol.* 4 (2008) 538–547.

30. U. Golebiewska, M. Nyako, W. Woturski, I. Zaitseva, S. McLaughlin, Diffusion coefficient of fluorescent phosphatidylinositol 4,5-bisphosphate in the plasma membrane of cells, *Mol. Biol. Cell* 19 (2008) 1663–1669.

31. K. Bacia, S.A. Kim, P. Schwille, Fluorescence cross-correlation spectroscopy in living cells, *Nat. Methods* 3 (2006) 83–89.

32. J. Ries, T. Weidemann, P. Schwille, Fluorescence correlation spectroscopy, in: E.H. Egelman (Ed.), *Comprehensive Biophysics*, Elsevier, Amsterdam, 2012, pp. 210–245.

33. P. Schwille, F.J. Meyer-Almes, R. Rigler, Dual-color fluorescence cross-correlation spectroscopy for multicomponent diffusional analysis in solution, *Biophys. J.* 72 (1997) 1878–1886.

34. J. Rika, T. Binkert, Direct measurement of a distinct correlation function by fluorescence cross correlation, *Phys. Rev. A* 39 (1989) 2646–2652.

35. J. Humpolíčková, E. Gielen, A. Benda, V. Fagulova, J. Vercammen, M. Vandeven, M. Hof, M. Ameloot, Y. Engelborghs, Probing diffusion laws within cellular membranes by z-scan fluorescence correlation spectroscopy, *Biophys. J.* 91 (2006) L23–L25.

36. T. Steinberger, R. Machan, M. Hof, Z-scan fluorescence correlation spectroscopy as a tool for diffusion measurements in planar lipid membranes, *Methods Mol. Biol.* 1076 (2014) 617–634.

37. M. Przybylo, J. Sýkora, J. Humpolíčková, A. Benda, A. Zan, M. Hof, Lipid diffusion in giant unilamellar vesicles is more than 2 times faster than in supported phospholipid bilayers under identical conditions, *Langmuir* 22 (2006) 9096–9099.

38. N.L. Thompson, X. Wang, P. Navaratnarajah, Total internal reflection with fluorescence correlation spectroscopy: Applications to substrate-supported planar membranes, *J. Struct. Biol.* 168 (2009) 95–106.

39. T.E. Starr, N.L. Thompson, Total internal reflection with fluorescence correlation spectroscopy: Combined surface reaction and solution diffusion, *Biophys. J.* 80 (2001) 1575–1584.

40. J. Ries, E.P. Petrov, P. Schwille, Total internal reflection fluorescence correlation spectroscopy: Effects of lateral diffusion and surface-generated fluorescence, *Biophys. J.* 95 (2008) 390–399.

41. J. Ries, P. Schwille, New concepts for fluorescence correlation spectroscopy on membranes, *Phys. Chem. Chem. Phys.* 10 (2008) 3487–3497.

42. J. Ries, P. Schwille, Studying slow membrane dynamics with continuous wave scanning fluorescence correlation spectroscopy, *Biophys. J.* 91 (2006) 1915–1924.

43. P. Muller, P. Schwille, T. Weidemann, Scanning fluorescence correlation spectroscopy (SFCS) with a scan path perpendicular to the membrane plane, *Methods Mol. Biol.* 1076 (2014) 635–651.

44. J. Ries, S. Chiantia, P. Schwille, Accurate determination of membrane dynamics with line-scan FCS, *Biophys. J.* 96 (2009) 1999–2008.

45. A.J. Garcia-Saez, The secrets of the Bcl-2 family, *Cell Death Differ.* 19 (2012) 1733–1740.

46. R.J. Youle, A. Strasser, The BCL-2 protein family: Opposing activities that mediate cell death, *Nat. Rev. Mol. Cell Biol.* 9 (2008) 47–59.

47. J.P. Steringer, S. Bleicken, H. Andreas, S. Zacherl, M. Laussmann, K. Temmerman, F.X. Contreras, T.A.M. Bharat, J. Lechner, H.-M. Müller, J.A.G. Briggs, A.J. García-Sáez, W. Nickel, Phosphatidylinositol 4,5-bisphosphate (PI(4,5)P2)-dependent oligomerization of fibroblast growth factor 2 (FGF2) triggers the formation of a lipidic membrane pore implicated in unconventional secretion, *J. Biol. Chem.* 287 (2012) 27659–27669.

48. V. Betaneli, Eugene P. Petrov, P. Schwille, The role of lipids in VDAC oligomerization, *Biophys. J.* 102 (2012) 523–531.

49. A. Kedrov, I. Kusters, V.V. Krasnikov, A.J. Driessen, A single copy of SecYEG is sufficient for preprotein translocation, *EMBO J.* 30 (2011) 4387–4397.

50. A.J. Garcia-Saez, S.B. Buschhorn, H. Keller, G. Anderluh, K. Simons, P. Schwille, Oligomerization and pore formation by equinatoxin II inhibit endocytosis and lead to plasma membrane reorganization, *J. Biol. Chem.* 286 (2011) 37768–37777.

51. R.C. Patel, U. Kumar, D.C. Lamb, J.S. Eid, M. Rocheville, M. Grant, A. Rani, T. Hazlett, S.C. Patel, E. Gratton, Y.C. Patel, Ligand binding to somatostatin receptors induces receptor-specific oligomer formation in live cells, *Proc. Natl. Acad. Sci. U. S. A.* 99 (2002) 3294–3299.

52. T. Weidemann, R. Worch, K. Kurgonaite, M. Hintersteiner, C. Bokel, P. Schwille, Single cell analysis of ligand binding and complex formation of interleukin-4 receptor subunits, *Biophys. J.* 101 (2011) 2360–2369.

53. S.J. Briddon, J. Gandia, O.B. Amaral, S. Ferre, C. Lluis, R. Franco, S.J. Hill, F. Ciruela, Plasma membrane diffusion of G protein-coupled receptor oligomers, *Biochim. Biophys. Acta-Mol. Cell Res.* 1783 (2008) 2262–2268.

54. S.J. Briddon, R.J. Middleton, Y. Cordeaux, F.M. Flavin, J.A. Weinstein, M.W. George, B. Kellam, S.J. Hill, Quantitative analysis of the formation and diffusion of A1-adenosine receptor-antagonist complexes in single living cells, *Proc. Natl. Acad. Sci. U. S. A.* 101 (2004) 4673–4678.

55. S. Ganguly, A. Chattopadhyay, Cholesterol depletion mimics the effect of cytoskeletal destabilization on membrane dynamics of the serotonin1A receptor: A zFCS study, *Biophys. J.* 99 (2010) 1397–1407.

56. N. Kahya, D. Scherfeld, K. Bacia, B. Poolman, P. Schwille, Probing lipid mobility of raft-exhibiting model membranes by fluorescence correlation spectroscopy, *J. Biol. Chem.* 278 (2003) 28109–28115.

57. J. Ries, S.R. Yu, M. Burkhardt, M. Brand, P. Schwille, Modular scanning FCS quantifies receptor-ligand interactions in living multicellular organisms, *Nat. Methods* 6 (2009) 643–645.

58. J.K. Jaiswal, S.M. Simon, Imaging single events at the cell membrane, *Nat. Chem. Biol.* 3 (2007) 92–98.

59. R. Hallworth, M.G. Nichols, The single molecule imaging approach to membrane protein stoichiometry, *Microsc. Microanal.* 18 (2012) 771–780.

60. H. Schneckenburger, Total internal reflection fluorescence microscopy: Technical innovations and novel applications, *Curr. Opin. Biotechnol.* 16 (2005) 13–18.

61. Y. Sako, S. Minoghchi, T. Yanagida, Single-molecule imaging of EGFR signalling on the surface of living cells, *Nat. Cell Biol.* 2 (2000) 168–172.

62. W. Zhang, Y. Jiang, Q. Wang, X. Ma, Z. Xiao, W. Zuo, X. Fang, Y.-G. Chen, Single-molecule imaging reveals transforming growth factor-β-induced type II receptor dimerization, *Proc. Natl. Acad. Sci. U. S. A.* 106 (2009) 15679–15683.

63. D. Axelrod, Total internal reflection fluorescence microscopy, *Methods Cell Biol.* 89 (2008) 169–221.

64. J.R. Lakowicz, *Principles of Fluorescence Spectroscopy*, Springer, New York, 2007.

65. D. Axelrod, Selective imaging of surface fluorescence with very high aperture microscope objectives, *J. Biomed. Opt.* 6 (2001) 6–13.

66. T. Schmidt, G. Schütz, W. Baumgartner, H. Gruber, H. Schindler, Imaging of single molecule diffusion, *Proc. Natl. Acad. Sci. U. S. A.* 93 (1996) 2926–2929.

67. M.K. Cheezum, W.F. Walker, W.H. Guilford, Quantitative comparison of algorithms for tracking single fluorescent particles, *Biophys. J.* 81 (2001) 2378–2388.

68. M.J. Saxton, K. Jacobson, Single-particle tracking: Applications to membrane dynamics, *Annu. Rev. Biophys. Biomol. Struct.* 26 (1997) 373–399.

69. R.E. Thompson, D.R. Larson, W.W. Webb, Precise nanometer localization analysis for individual fluorescent probes, *Biophys. J.* 82 (2002) 2775–2783.

70. B. Zhang, J. Zerubia, J.-C. Olivo-Marin, Gaussian approximations of fluorescence microscope point-spread function models, *Appl. Optics* 46 (2007) 1819–1829.

71. N. Bobroff, Position measurement with a resolution and noise-limited instrument, *Rev. Sci. Instrum.* 57 (1986) 1152–1157.

72. S. Wieser, M. Moertelmaier, E. Fuertbauer, H. Stockinger, G.J. Schütz, (Un) confined diffusion of CD59 in the plasma membrane determined by high-resolution single molecule microscopy, *Biophys. J.* 92 (2007) 3719–3728.

73. A. Yildiz, J.N. Forkey, S.A. McKinney, T. Ha, Y.E. Goldman, P.R. Selvin, Myosin V walks hand-over-hand: Single fluorophore imaging with 1.5-nm localization, *Science* 300 (2003) 2061–2065.

74. R.N. Ghosh, W.W. Webb, Automated detection and tracking of individual and clustered cell surface low density lipoprotein receptor molecules, *Biophys. J.* 66 (1994) 1301–1318.

75. S. Wieser, G.J. Schütz, Tracking single molecules in the live cell plasma membrane— Do's and Don't's, *Methods* 46 (2008) 131–140.

76. K. Jaqaman, D. Loerke, M. Mettlen, H. Kuwata, S. Grinstein, S.L. Schmid, G. Danuser, Robust single-particle tracking in live-cell time-lapse sequences, *Nat. Methods* 5 (2008) 695–702.

77. D.B. Reid, An algorithm for tracking multiple targets, *IEEE Trans. Autom. Control* 24 (1979) 843–854.

78. G. Schütz, H. Schindler, T. Schmidt, Single-molecule microscopy on model membranes reveals anomalous diffusion, *Biophys. J.* 73 (1997) 1073–1080.

79. H. Qian, M.P. Sheetz, E.L. Elson, Single particle tracking. Analysis of diffusion and flow in two-dimensional systems, *Biophys. J.* 60 (1991) 910–921.

80. M.H. Ulbrich, E.Y. Isacoff, Subunit counting in membrane-bound proteins, *Nat. Methods* 4 (2007) 319–321.

81. Y. Yang, Y. Xu, T. Xia, F. Chen, C. Zhang, W. Liang, L. Lai, X. Fang, A single-molecule study of the inhibition effect of Naringenin on transforming growth factor-β ligand-receptor binding, *Chem. Commun.* 47 (2011) 5440–5442.

82. P.D. Simonson, H.A. DeBerg, P. Ge, J.K. Alexander, O. Jeyifous, W.N. Green, P.R. Selvin, Counting bungarotoxin binding sites of nicotinic acetylcholine receptors in mammalian cells with high signal/noise ratios, *Biophys. J.* 99 (2010) L81–L83.

83. T. Schmidt, G.J. Schütz, H.J. Gruber, H. Schindler, Local stoichiometries determined by counting individual molecules, *Anal. Chem.* 68 (1996) 4397–4401.

84. V. Ruprecht, M. Brameshuber, G.J. Schütz, Two-color single molecule tracking combined with photobleaching for the detection of rare molecular interactions in fluid biomembranes, *Soft Matter* 6 (2010) 568–581.

85. M. Brameshuber, J. Weghuber, V. Ruprecht, I. Gombos, I. Horváth, L. Vigh, P. Eckerstorfer, E. Kiss, H. Stockinger, G.J. Schütz, Imaging of mobile long-lived nanoplatforms in the live cell plasma membrane, *J. Biol. Chem.* 285 (2010) 41765–41771.

86. M. Moertelmaier, M. Brameshuber, M. Linimeier, G. Schutz, H. Stockinger, Thinning out clusters while conserving stoichiometry of labeling, *Appl. Phys. Lett.* 87 (2005) 263903.

87. D. Calebiro, F. Rieken, J. Wagner, T. Sungkaworn, U. Zabel, A. Borzi, E. Cocucci, A. Zürn, M.J. Lohse, Single-molecule analysis of fluorescently labeled G-protein–coupled receptors reveals complexes with distinct dynamics and organization, *Proc. Natl. Acad. Sci. U. S. A.* 110 (2013) 743–748.

88. J. Madl, J. Weghuber, R. Fritsch, I. Derler, M. Fahrner, I. Frischauf, B. Lackner, C. Romanin, G.J. Schütz, Resting state Orai1 diffuses as homotetramer in the plasma membrane of live mammalian cells, *J. Biol. Chem.* 285 (2010) 41135–41142.

89. R.S. Kasai, K.G. Suzuki, E.R. Prossnitz, I. Koyama-Honda, C. Nakada, T.K. Fujiwara, A. Kusumi, Full characterization of GPCR monomer–dimer dynamic equilibrium by single molecule imaging, *J. Cell Biol.* 192 (2011) 463–480.

90. Y. Teramura, J. Ichinose, H. Takagi, K. Nishida, T. Yanagida, Y. Sako, Single-molecule analysis of epidermal growth factor binding on the surface of living cells, *EMBO J.* 25 (2006) 4215–4222.

91. M. Schwarzenbacher, M. Kaltenbrunner, M. Brameshuber, C. Hesch, W. Paster, J. Weghuber, B. Heise, A. Sonnleitner, H. Stockinger, G.J. Schütz, Micropatterning for quantitative analysis of protein-protein interactions in living cells, *Nat. Methods* 5 (2008) 1053–1060.

92. A. Anderluh, E. Klotzsch, A.W.A.F. Reismann, M. Brameshuber, O. Kudlacek, A.H. Newman, H.H. Sitte, G.J. Schütz, Single molecule analysis reveals coexistence of stable serotonin transporter monomers and oligomers in the live cell plasma membrane, *J. Biol. Chem.* 289 (2014) 4387–4394.

93. L.S. Barak, W.W. Webb, Fluorescent low density lipoprotein for observation of dynamics of individual receptor complexes on cultured human fibroblasts, *J. Cell Biol.* 90 (1981) 595–604.

94. P.R. Smith, I.E. Morrison, K.M. Wilson, N. Fernandez, R.J. Cherry, Anomalous diffusion of major histocompatibility complex class I molecules on HeLa cells determined by single particle tracking, *Biophys. J.* 76 (1999) 3331–3344.

95. I. Chung, R. Akita, R. Vandlen, D. Toomre, J. Schlessinger, I. Mellman, Spatial control of EGF receptor activation by reversible dimerization on living cells, *Nature* 464 (2010) 783–787.

96. F. Pinaud, X. Michalet, G. Iyer, E. Margeat, H.-P. Moore, S. Weiss, Dynamic partitioning of a glycosyl-phosphatidylinositol-anchored protein in glycosphingolipid-rich microdomains imaged by single-quantum dot tracking, *Traffic* 10 (2009) 691–712.

97. S. Courty, C. Luccardini, Y. Bellaiche, G. Cappello, M. Dahan, Tracking individual kinesin motors in living cells using single quantum-dot imaging, *Nano Lett.* 6 (2006) 1491–1495.

98. P.H.M. Lommerse, G.A. Blab, L. Cognet, G.S. Harms, B.E. Snaar-Jagalska, H.P. Spaink, T. Schmidt, Single-molecule imaging of the H-Ras membrane-anchor reveals domains in the cytoplasmic leaflet of the cell membrane, *Biophys. J.* 86 (2004) 609–616.

99. N.L. Andrews, K.A. Lidke, J.R. Pfeiffer, A.R. Burns, B.S. Wilson, J.M. Oliver, D.S. Lidke, Actin restricts Fc[epsiv]RI diffusion and facilitates antigen-induced receptor immobilization, *Nat. Cell Biol.* 10 (2008) 955–963.

100. R. Simson, B. Yang, S.E. Moore, P. Doherty, F.S. Walsh, K.A. Jacobson, Structural mosaicism on the submicron scale in the plasma membrane, *Biophys. J.* 74 (1998) 297–308.

11

PROBING CHANNEL, PUMP, AND TRANSPORTER FUNCTION USING SINGLE-MOLECULE FLUORESCENCE

Eve E. Weatherill, John S. H. Danial, and Mark I. Wallace

Department of Chemistry, Chemical Research Laboratory, University of Oxford, Oxford, UK

11.1 INTRODUCTION

The ability of patch clamping to monitor individual ion channels revolutionized our approach to understanding the discrete functional states of these membrane proteins [1], including how ligands, point mutations, and the electrochemical potential affect channel function [2]. However, patch clamping is not without its limitations; it can only report on changes associated with distinct conducting states. Additional methods are required to explore dynamics that are related to electrically indistinguishable or inactive states. Single-molecule fluorescence (SMF) spectroscopy has emerged as a powerful biophysical tool capable of resolving the dynamics of conformational changes within individual proteins [3]. It has transformed the way we are able to understand numerous biological systems, ranging from enzyme kinetics to molecular motors [4]. SMF methods have already made a significant contribution to our understanding of ion channels, and these methods are being extended to help understand the molecular mechanisms of transporters.

Membrane proteins *can* be observed at the single-molecule level in cells. Although understanding biomolecule behavior *in situ* is obviously preferable, such experiments are not without drawbacks. In particular, inferring a molecular mechanism from single-molecule events in such a complex environment is often impossible.

Pumps, Channels, and Transporters: Methods of Functional Analysis, First Edition. Edited by Ronald J. Clarke and Mohammed A. A. Khalid.
© 2015 John Wiley & Sons, Inc. Published 2015 by John Wiley & Sons, Inc.

Therefore, *in vivo* experiments are complemented by *in vitro* measurements, where the effectors of biomolecule function can be studied in isolation.

This chapter is intended as a brief guide to introduce researchers to SMF and how these techniques have been applied to ion channels, transporters, and pumps. The main focus is on techniques that relate structure to function. Other powerful SMF techniques, including fluorescence fluctuation analysis and single-particle tracking are covered in detail in Chapter 10. By harnessing photophysical properties of fluorescent molecules and resolving transitions inaccessible by bulk measurements, we can observe the kinetics associated with structural dynamics, stoichiometry, interactions, and function over physiologically relevant timescales.

11.1.1 Basic Principles

In SMF imaging, fluorescent probes are attached to a target protein, and the spectral and positional properties corresponding to individual molecules are recorded [5]. Resolving the signal emitted by a single label is made possible by limiting either the illumination volume or the volume within which the SMF is collected that either excites or is emitted from a single molecule. While the emission of photons occurs on the nanosecond timescale, temporal resolution is effectively limited by the total rate of photon flux to around 100 µs for current fluorophores. In terms of spatial resolution, for experiments employing an imaging detector, fluorescence from a single molecule is observed as a diffraction-limited single spot of light a few hundred nanometers in diameter. As the expected distribution of fluorescence from this point object is known, this distribution can be fitted to determine the position of the center of the spot to a higher accuracy [6]. This process is limited by the number of photons detected, the background noise, and the pixel size of the charge-coupled device (CCD) used for detection [6]. For this reason, the temporal and spatial limits are interlinked, as longer acquisition times allow for the collection of more photons.

Resolving many crowded spots beyond the diffraction limit is made possible by exploiting specific photophysical properties of the fluorophores, to "turn off" the majority of the fluorophores or "turn on" only a small number of fluorophores, either at random or at specific positions [7–10]. It is only recently that such superresolution techniques are beginning to have the speed required to resolve dynamics on a timescale relevant to the majority of channels and transporters [11, 12]. Despite the current limitations, SMF has proved indispensable in deciphering the function of a wide array of important proteins with unprecedented detail.

11.2 PRACTICAL CONSIDERATIONS

In this section, we briefly outline some practical considerations for imaging channels, transporters, and pumps using SMF. The requirements for SMF imaging have been recently reviewed [5, 13]. We have attempted to provide a brief overview and would direct the reader to these reviews for more detail.

11.2.1   Observables

A number of important biological questions relating to channel structure and function can be addressed using SMF (see Fig. 11.1). Examples include (i) channel clustering: where intense spots signify the presence of localized singly labeled proteins over a dim background of isolated channels; (ii) interactions with regulatory subunits, by colocalization of differentially labeled components; (iii) stoichiometry of channel subunits and relative stoichiometry with regulatory subunits, by counting photo-bleaching steps; and (iv) structural transitions associated with gating, by single-molecule Förster resonance energy transfer (smFRET).

11.2.2   Apparatus

Two major modalities are typically used to achieve SMF detection, total internal reflection fluorescence (TIRF) microscopy and confocal fluorescence microscopy.

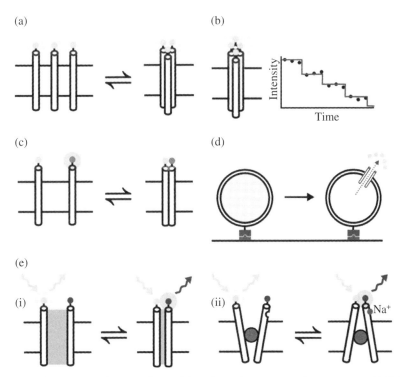

FIGURE 11.1   SMF techniques for studying channels, pumps, and transporters. (a) Subunit clustering detected by bright fluorescent spots. (b) Subunit stoichiometry by counting photo-bleaching steps. (c) Colocalization of subunits or binding partners, detected by spatially over-lapping signals in two emission channels. (d) Efflux of fluorescent substrate through a channel in a tethered liposome. (e) Single-molecule FRET for probing gating mechanisms: (i) channel gating and (ii) alternating symporter action.

TIRF microscopy relies on restricting the illumination volume by exciting only molecules that are very close to a surface using an illumination path of subwavelength depth [14]. The incident light is totally internally reflected at the glass/solution interface because the angle of incidence exceeds the critical angle governed by refractive indices either side of the interface. The intensity of the resulting evanescent field decreases exponentially from the interface with a typical decay constant of approximately 100 nm [15] (see Section 10.4.1 for further details). The setup allows for fast (ms) SMF acquisition over a wide ($\sim$10 μm $\times$ 10 μm) area. The technique is therefore well suited to imaging molecules that are confined to the plane of a bilayer and has shed much light on the dynamics of channels and transporters in both the basal plasma membranes of live cells and in artificial bilayers.

For single-molecule confocal microscopy, the incident light is focused by an objective lens onto the sample. The detection volume is restricted by placing a small pinhole aperture in a conjugate image plane in the emitted light path, thereby eliminating light from above and below the focal plane. A wider sample depth can be explored by altering the position of the objective lens permitting greater positional freedom than in a TIRF setup. Thus, objects inside cells [16] can be investigated. Single molecules can be studied by either scanning the confocal beam relative to the sample to build up an image [17] or by allowing single molecules to diffuse into a fixed beam (see Chapter 10) [18].

11.2.3 Labels

There are numerous strategies for labeling proteins for SMF imaging [19]. A persistent challenge is to detect a sufficiently strong continuous signal for a sufficiently long period of time. Few fluorophores possess the molecular brightness (the product of instrumental factors, absorption cross section, and quantum yield) required to detect a single fluorophore with a good signal-to-noise ratio [20]. At the illumination intensity required for SMF imaging, blinking due to intersystem crossing to dark triplet states and irreversible bleaching due to photochemical changes to fluorophore structure [15] are significant limitations. The size of the fluorophore and its mode of attachment to the protein of interest must also be carefully considered based on the application.

Naturally occurring fluorescent proteins, such as the *Aquorius victorius*-derived green fluorescent protein (GFP) [21] or its many derivatives, are a class of genetically encoded fluorophores widely used in SMF studies [22]. By expressing a protein as a fluorescent fusion protein, a label can be introduced without the need for an additional labeling step or purification. This is particularly useful *in vivo*, where the presence of native proteins and accessibility to the cell interior pose challenges to subsequent labeling steps. While a 100% labeling efficiency is guaranteed, there will be a proportion of fluorescent proteins that do not fold correctly and, therefore, do not fluoresce. This approach is well suited to assessing channel localization, interactions, and stoichiometry, but the large size of these fluorophores and the limited positions to which they are attached restrict their use as reporters for structural dynamics. Additionally, in comparison to other fluorophores, fluorescent proteins are relatively dim.

The small size and brightness of organic dyes afford minimal effects on the protein and allow for labeling of higher precision. Many commercially available dyes are synthesized with chemical moieties that allow for targeted labeling to an epitope tag [23, 24] or amino acid [25], such as the attachment of a maleimide-functionalized dye to the sulfhydryl group of a cysteine side chain. Fluorescent unnatural amino acids (fUAAs) provide the ultimate level of precision in targeted labeling [26] by incorporating the fluorophore at the translational level. Their use is limited by lengthy synthesis and limited commercial availability. Fluorescent ligands or substrates can provide a targeted labeling approach without the need to engineer an attachment site [27], as can antibody–fluorophore conjugates, although the antibody's size and mode of binding are not optimal [28].

Quantum dots (QDs) are large (>10 nm) semiconductor crystals with superior photophysical properties to small organic dyes. They typically afford approximately 20 min of continuous emission during SMF imaging. However, they are prone to photoblinking unless properly passivated [29, 30]. While their size limits their application as probes for changes in protein conformation, they have a dual purpose as effective labels for electron microscopy [28]. However, the possibility of toxicity of QDs in the case of measurements on live cells needs to be considered (see also Chapter 10).

11.2.4 Bilayers

Imaging in live cells is the only way to observe the activity of a protein in a truly native environment. The basal cell membrane is easily accessible by TIRF microscopy, where lateral diffusion of proteins ensures they remain within the evanescent field. As such, TIRF microscopy has been used for SMF imaging in oocytes [25, 26, 31–36], bacterial [37, 38], and mammalian cells [28, 39–43]. Oocytes are surrounded by a vitelline membrane that must be enzymatically removed in order to maneuver the plasma membrane sufficiently close to the glass surface [25].

Artificial lipid bilayers provide a more minimal environment for studying membrane proteins, affording a high degree of control over each constituent component. As such, they can be arranged to fit the needs of the technique. They can be produced using a number of different techniques, including tethered or supported lipid bilayers [44] (see Chapter 6), Montal–Mueller or "black lipid membranes" [45] (see Chapter 2), droplet interface bilayers (DIBs) [46], and giant unilamellar vesicles [47]. *In vitro* methods additionally require purification and reconstitution of membrane proteins [48, 49].

11.3 SMF IMAGING

In this section, we explore the types of experiments enabled by wide-field imaging of singly labeled proteins and the conclusions that can be drawn by observing the intensity or position of fluorescent spots over time. We present just a section of essential SMF techniques here, reserving those that exploit specific photophysical phenomena for subsequent sections.

11.3.1   Fluorescence Colocalization

One of the most basic applications of SMF is to distinguish areas where the local concentration of a protein is elevated above the background level due to the association of subunits or clustering of complexes. In a regime where individually labeled subunits are present at a sufficiently high concentration that individual spots can't be resolved, uniform fluorescence intensity denotes a homogenous distribution of monomers. A heterogeneous distribution of fluorescence, or the observation of bright spots above the background level, is indicative of multiple localized subunits or complex formation. This technique can therefore be used to monitor the formation of channels in real time [40, 50, 51]. Likewise, clustering of channels and trans- porters is exhibited by the formation of larger patches of high intensity [39, 52–54]. For example, recently, we exploited the formation of fluorescent spots to explore the assembly of the twin-arginine transporter (TAT) in live cells [50]. This complex facil- itates the transport of folded proteins across bacterial cytoplasmic and chloroplast thylakoid membranes. The transporter complex is formed on demand, to minimize ion leakage when unoccupied, in a substrate- and proton motive force (PMF)- dependent manner (see Fig. 11.2a).

For examining heterogeneous interactions, the distribution of multiple compo- nents labeled with different fluorophores can be observed simultaneously; their dual presence within a diffraction-limited spot is indicative of an interaction. Fluorescence colocalization is used to measure the stoichiometry of multiple-component complexes [31, 34, 35, 39, 43, 56] and to probe interactions with regulatory proteins [36, 53].

Ulbrich et al. recently developed a method to measure the degree of coassembly between different subunits of a tetrameric receptor in live cells [31]. The N-methyl- D-aspartate (NMDA) receptor is known to consist of an NR1 homodimer paired with either a homo- or heterodimer of NR2, NR3a, and/or NR3b [31]. It was not known whether the non-NR1 subunits coassembled in a stoichiometric fashion where each subunit is represented once or whether the assembly was random with equal chance of selecting one non-NR1 subunit as the other. C-terminal fluorescent protein tags were fused to NR2, NR3a, and NR3b, two of which were simultaneously coexpressed with unlabeled NR1 in *Xenopus* oocytes and imaged by dual-color TIRF microscopy (see Fig. 11.2b). When NR1 was coexpressed with the ttTomato-labelled NR3a and GFP labelled NR3b, significant co-localization of the two signals was observed, indi- cating a preferential rather than random coassembly of NR3a and NR3b.

The work of Waschk and coworkers provides a second example of SMF colocal- ization. They developed a method to probe the assembly of tetrameric channels in fixed cells [56]. Homotetrameric potassium hKCa3.1 channel subunits expressed in Madin-Darby canine kidney (MDCK-F) cells were singly labeled in equal proportion with red and green QDs prior to imaging. In order to achieve this labeling ratio, two constructs of the same subunit containing different epitope tags, accessible either to the extracellular or the intracellular side of the plasma membrane, were coexpressed in equal measure. The extracellular epitope tag was bound by primary and then secondary QD-conjugated antibodies prior to fixing and permeabilization of the cells for the same procedure with the intracellular epitope tag. A binomial equation was

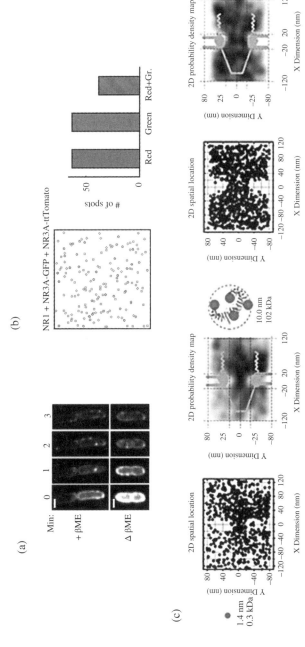

FIGURE 11.2 Single-molecule fluorescence. (a) Single-molecule fluorescence of TatA subunits shows the assembly of TatA in the presence and absence of a proton motive force, which was chemically dissipated and then restored by β-mercaptoethanol (βME). Modified from Ref. 50 with permission from the National Academy of Sciences, USA. (b) Stoichiometry of the NMDA receptor probed by single-molecule colocalization. Processed image displaying the locations of NR3A subunits labeled with either GFP (diamonds) or ttTomato (squares) *in vivo* and histogram of separate and overlapping spot populations. Data indicate that the NR3A subunit can form homodimers. Modified from Ref. 31 with permission from the National Academy of Sciences, USA. (c) Superresolution images and density maps of the interior of the nuclear pore complex demonstrating the areas occupied by fluorescently labeled cargo. The diffusion pathways of fluorescein (left) and labeled importin β1 transport receptor (right) through the complex demonstrate a distinction between passive transport through a central axial channel and facilitated transport involving interactions with Nups in the peripheral regions. Modified from Ref. 55 with permission from the National Academy of Sciences, USA.

employed to predict proportions of red, green, and yellow spots assuming subunits assembled independently and in a random manner to form a distribution of all possible homo- and heterotetramers. However, the proportions observed better matched a distribution in which tetramers containing uneven numbers of each subunit were omitted, suggesting that they are instead assembled from preexisting homodimers. While this technique is able to distinguish between stoichiometries that would have been obscured at the ensemble level, the images were obtained from fixed cells and with large labels, and so the dynamics of the observed states were inaccessible.

Yamamura and coworkers also used fluorescence colocalization to probe the stoichiometry of potassium KCa1.1 (BK) channels [39]. BK channels consist of four BKα subunits forming the pore region, and each one may be bound by an auxiliary BKβ subunit, which regulates characteristics such as Ca^{2+} sensitivity. BKβ subunits are distributed in a tissue-specific manner, and it is thought that they also facilitate trafficking of BKα to the plasma membrane. Coexpression of BKα-YFP and BKβ-CFP fusion proteins in human embryonic kidney (HEK) cells with dual-color TIRF microscopy was carried out to examine the distributions of the two subunits. While the majority of yellow spots were colocalized with blue spots, there remained a small but distinct proportion of individual spots of either color, implying that potentially a small number of BK channels are formed of uncoupled BKα subunits.

11.3.2 Conformational Changes

Site-specific attachment of fluorescent dyes to ion channels can be used to report on conformational changes by taking advantage of the sensitivity fluorophores show to their immediate environment [57]. Indeed, in recent years, environment-sensitive fluorophores (ESF) have been developed to probe such parameters as microviscosity, polarity, and hydration, which can vary dramatically within biological membranes [58]. As such, fluorescence intensity has the potential to report on conformational changes experienced by channels, pumps, and transporters, including those that arise from folding, gating, and interactions with substrates, ligands, inhibitors, or lipids. On the single-molecule level, this phenomenon has the power to resolve structural intermediates beyond those associated with distinct conductive levels. If a labeling site is selected such that a conformational change sufficiently alters its immediate environment, then the attached fluorophore will have distinct spectral properties associated with each conformation. Furthermore, a change in the location within the evanescent field and orientation of the fluorophore can further contribute to intensity changes. The magnitude and direction of the fluorescent change associated with conformational switching varies according to the attachment site.

Sonnleitner et al. first demonstrated the power of this phenomenon by observing structural dynamics of the voltage-gated potassium shaker channel *in vivo* [25]. The fluorophore was targeted to the S4 helix using highly specific maleimide chemistry by introducing a single cysteine residue into an otherwise cysteine-free channel. Cysteines on native proteins were first blocked with tetraglycine maleimide in a hyperpolarizing solution rendering the labeling site inaccessible. When washed with depolarizing solution, the free cysteine emerged from the bilayer, available for labeling. Combining TIRF microscopy with a voltage-clamp protocol, the spots corresponding to individual, fluorophore-labeled channels were identified by their voltage-responsive fluorescent fluctuations combined with eventual photobleaching in a single step. A 20-fold suppression of brightness was observed in the dim state, and the two labeling positions evaluated exhibited bright states at opposite potentials. The method revealed for the first time structural intermediates that did not directly open or close the ion channel. Conformational changes associated with potassium channel gating have also been measured using SMF in supported lipid bilayers [59].

Nguyen and coworkers used an ESF to probe the formation and insertion of staphylococcal γ-hemolysin pores into erythrocyte ghost membranes at the single-molecule level [60]. In this case, the conformational change detected was associated with insertion of the pore into the membrane. The two-component cytolysin consists of leukocidin fast fraction (LukF) and γ-hemolysin second-component (HS) monomers that assemble into ring-shaped heterooligomers of six to eight subunits each. The pre-stem of LukF, which undergoes an HS-dependent conformational change to form the transmembrane portion of the pore, was labeled with Badan, a polarity-sensitive fluorophore. An increase in fluorescence emission reflects insertion due to the increased hydrophobicity of the environment surrounding the dye. While it was known that the subunits formed heterodimers and heterotetramers prior to assembly of the ring, and that at high concentrations individual γ-hemolysin rings oligomerize, it was not clear at which stage LukF inserted into the membrane. The insertion status of intermediates was accessible thanks to a high labeling efficiency and the ability to observe individual labeled LukF monomers in ghost red blood cells by TIRF microscopy. The number of transmembrane LukF monomers in each channel was quantified by counting photobleaching steps (see Section 11.4). Only intensities corresponding to individual or multiple complete channels were observed, suggesting that LukF undergoes a conformational change when one complete ring is formed and not before.

11.3.3 Superresolution Microscopy

The organization of the nuclear pore complex (NPC) interior has been a hot topic for SMF in recent years [55, 61–63]. The approximately 125 MD channel comprising 30 different nucleoproteins (nups) regulates transport of a wide range of cargos across the double envelope surrounding the nucleus. The interior of the channel is occupied by flexible filaments of unstructured nucleoporins, which interact with cargo molecules and molecular chaperones. Ma and coworkers developed a superresolution microscopy technique to evaluate possible models for nucleocytoplasmic transport by imaging a range of fluorescent cargos traversing individual NPCs [55]. In order to limit the illumination volume, single-point edge-excitation subdiffraction (SPEED) microscopy employs an illumination beam that intercepts the focal plane at a 45° angle [61]. Thus, with illumination above and below, the focal plane is limited with this technique, rather than the detection volume. HeLa cells expressing a GFP–Nup fusion were used in order to define the edge of the pore interior. By observing the distribution of areas occupied by cargos of varying size during passive and facilitated transport, they were able to elucidate distinct but overlapping pathways throughout the channel as well as a viscous central axial pathway (see Fig. 11.2c). These findings supported an "oil-spaghetti" model to describe the NPC interior, as opposed to a sieve-like hydrogel mesh offering numerous small holes for passive diffusion.

11.4 SINGLE MOLECULE FÖRSTER RESONANCE ENERGY TRANSFER

Since the discovery of FRET over half a century ago, it has become an invaluable tool applied across a wide range of biological macromolecules [64]. FRET is the result of a nonradiative dipole–dipole interaction that occurs between a donor fluorophore (D) in the excited state and an acceptor fluorophore in the ground state (A). The donor must emit at shorter wavelengths and overlap spectrally with the absorption of the acceptor. Energy transfer occurs in a distance-dependent manner. Therefore, changes that affect D–A separation can be probed by the FRET efficiency akin to a "molecular ruler." The distance between D and A at which the energy transfer is 50% efficient is termed the Förster radius (R_o), typically in the range of 50–70Å for commercially available single-molecule fluorophores. In ensemble measurements, FRET has been used extensively to make nanoscale measurements of biological molecules, including intramolecular and intersubunit distances of proteins. The sensitivity of FRET was first extended to the single-molecule level in 1996 [65] and has since been used to observe nonequilibrium dynamics that are obscured in bulk experiments. Excellent practical guidance for smFRET can be found elsewhere [13]. In brief, the protein is labeled with donor and acceptor fluorophores at a distance reflecting the Förster radius of the pair. The sample is excited at the donor wavelength and the emitted light from donor and acceptor detected simultaneously in parallel images. The FRET efficiency of each molecule is calculated by comparing the intensity of any given spot in both channels. Cross talk is minimized by ensuring that the acceptor is not strongly excited by the laser and that the donor emission spectrum is filtered out of the acceptor channel.

The utility of smFRET is extended by using alternating laser excitation (ALEX), whereby two laser sources excite the donor and acceptor in an alternating fashion, in place of continuous donor excitation [66]. In addition to the FRET efficiency, the emission from a directly excited acceptor fluorophore is also recorded. The abundance of donors and acceptors contributing to the FRET signal can be determined, and distinct D–A ratios can be isolated. Furthermore, colocalized fluorophores with low FRET due to dye separation can be resolved from low FRET signals where the acceptor is missing or trapped in a dim state.

11.4.1 Interactions/Stoichiometry

The first smFRET study of a membrane protein was reported at the start of this century [67]. In 2003, the technique was applied to protein pores by Nguyen and coworkers [68]. They probed the mechanism of γ-hemolysin pore formation by labeling the two types of subunit it contains with donor and acceptor fluorophores.

Using FRET, they identified individual heterooligomers as spots with high efficiency (see Fig. 11.3a). The subunit composition corresponding to each FRET signal was obtained by counting the number of photobleaching steps of donor- and acceptor-only signals. The abundance of different heterooligomers was used to predict the equilibrium constants associated with each step of the assembly pathway of formation and subsequent clustering of pores. In a later study, the same group showed that two residues in the LukF subunit are essential for insertion through the membrane [71]. LukF point mutants were capable of forming intermediate heterooligomers, but the pores did not form clusters, nor did they cause K^+ efflux or lyse erythrocytes. The timing of the transmembrane step was later resolved using an ESF [60] (see Section 11.3.2). In this example and others, the smFRET signal provided a qualitative indication of subunit interaction, which was used to evaluate the stoichiometry or assembly mechanism of a channel [39, 72]. Likewise, smFRET has been used to report on channel clustering [39, 52] and interactions with other complexes [73].

More recently, Morrison and coworkers used smFRET to evaluate the topology of EmrE, the multidrug efflux transporter [74]. The antiporter is known to function as a homodimer of much debated topology. Combined with various NMR approaches, they confirmed that the dimers assume an antiparallel configuration by assessing residue accessibility to FRET labels. EmrE mutants were incorporated into liposomes such that a single cysteine became available to either the interior or exterior solution, depending on subunit orientation. The sites were labeled with donor and acceptor fluorophores, which were added either to opposite sides of the bilayer or the same side. Bicelles containing individual labeled dimers were formed and tethered to a glass surface by biotin–neutravidin linkages and imaged by TIRF microscopy. The idea is that if a FRET signal is observed as a result of the dyes accessing the same side, a parallel orientation is assumed, and if it is observed when the dyes access opposite sides of the membrane, then an antiparallel orientation is assumed. For all labeling positions examined, an antiparallel orientation was implied by the resulting smFRET efficiency. While the same results were obtained in bulk FRET measurements, confirmation that the channels were indeed dimers was only possible by counting single photobleaching steps for each fluorophore per molecule.

11.4.2 Conformational Changes

Although numerous high-resolution biophysical tools exist for obtaining detailed structural information about channels, pumps, and transporters, it is not always possible to access a full range of conformations in the protein's native environment. Using fluorescence burst imaging, in which the fluorescence intensity of molecules is obtained as they diffuse through a small confocal volume, single-molecule FRET has been used to study conformational changes of channels [75]. While this approach reflects the relative populations of molecules corresponding to distinct FRET efficiencies under particular conditions, it does not reflect kinetics associated with transitions between them. Wide-field smFRET imaging allows the observation of structural fluctuations for extended times (seconds) and has already revealed mechanistic insights into channels [76–80], pumps, and transporters [69, 70, 81, 82].

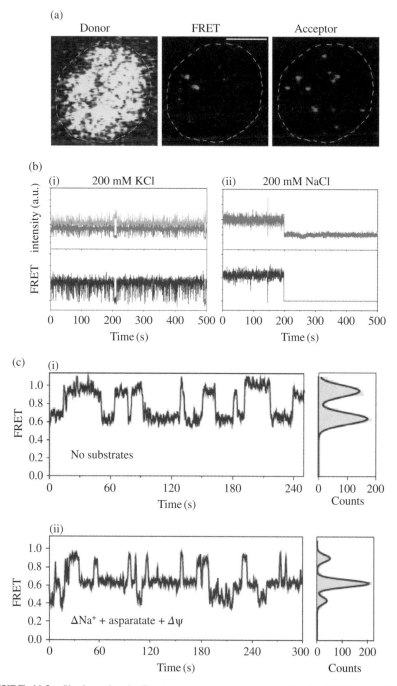

FIGURE 11.3 Single-molecule Förster resonance energy transfer (smFRET). (a) Single-molecule imaging of the assembly of gamma-hemolysin. LukF and HS labeled with TMR and IC5, respectively. Modified from Ref. 68 with permission from John Wiley & Sons.

Javitch and coworkers paved the way for applying smFRET to transporters when they published two studies involving the neurotransmitter/Na$^+$ symporter homologue, leucine transporter (LeuT) [69, 81]. Crystal structures for this transporter are misleading as they were solved in a detergent that is known to disrupt the symport mechanism. Therefore, other approaches were required to uncover precise conformational changes. Javitch and coworkers tethered LeuT to glass via biotin–streptavidin linkages in order to image a double-labeled, detergent-solubilized channel using TIRF microscopy (see Fig. 11.3b). In an initial study, they explored the FRET label positions to pinpoint which helix moves during gating [69]. They used the occupancy of distinct FRET states to explore the allosteric nature of the transporter by measuring the response to substrate and inhibitor binding at the intracellular binding site as well as point mutations to the intracellular network. In a subsequent study, they extended this to examine binding of a second substrate and inhibitor and the extracellular binding site, leading to a two-substrate-site model with a substrate stoichiometry of 2:1 at the intracellular and extracellular sites, respectively [81]. The true merit of this approach over burst measurements lies in the long recordings obtained, which revealed the relatively slow conformational switching events.

Using ALEX with smFRET, the researcher has access to the precise D–A combinations contributing to a particular FRET signal in a situation where both fluorophores are distributed randomly among the available labeling sites [70, 77]. For example, smFRET–ALEX was used quantitatively to measure conformational changes that occur in the MscL pore region during gating of the channel in tethered liposomes requiring exclusively channels labeled with a single D–A pair [77]. Each subunit of the homopentamer contained a free cysteine and was labeled with a mixture of both fluorophores, resulting in a mixture of D–A distributions among the channels. Comparing the increase in intersubunit distances on gating at two locations, their results were indicative of a helix-tilt model. While the study focused on the fully open and closed states of the channel, the technique could be used to report on structural intermediates or those corresponding to sub-conductance states, especially if combined with an electrophysiological capability.

Erkens and coworkers showed that the three subunits of a glutamate transporter homologue GltPh operate independently of one another using smFRET–ALEX [70].

FIGURE 11.3 (*Continued*) (b) Conformational changes in the transporter LeuT in the presence of different substrates. Long, single-molecule trajectories of tethered, detergent-solubilized LeuT H7C/R86C labeled with Cy3 (upper trace, top panel) and Cy5 (lower trace, top panel) and the resulting FRET efficiency (lower panel) in the presence of (i) 200 mM KCl or (ii) 200 mM NaCl. Modified from Ref. 69 with permission from the Nature Publishing Group. (c) Independent gating of subunits in excitatory amino acid transporter homologue, GltPh. Homotrimeric transporters were labeled with a mixture of donor and acceptor fluorophores, which were reconstituted into liposomes and tethered to a surface to obtain FRET efficiencies. Single-molecule FRET efficiency traces for GltPh trimers (i) in the absence of Na$^+$ and aspartate and (ii) in the presence of external Na$^+$, aspartate, and a membrane potential. In the absence of substrates, the channel alternates between two conformations in the order of seconds, whereas under conditions in which aspartate can be transported, three distinct FRET states dominate whereby every transition from low to high FRET must pass through the middle state. Modified from Ref. 70 with permission from Nature Publishing Group.

The observed FRET efficiency corresponding to triple-labeled channels (D–A ratio of 2:1 or 1:2) contained a greater number of fluctuations than channels labeled with a single D–A pair, indicating that a synchronous movement of the three subunits does not occur (see Fig. 11.3c).

While a majority of the smFRET studies have focused on structural transitions occurring in the transmembrane region of channels, pumps, and transporters, others have addressed conformational changes that occur in domains extending either side of the bilayer as a result of a binding event [78, 79] or phosphorylation [80]. Expressed in isolation, the water-soluble domains require tethering to a surface so that they remain within the focal plane for sufficiently long periods of time.

One of the limitations presented by smFRET is the range of distances accessible by this technique. To maximize the observed change in FRET efficiency, the D–A distance should fluctuate about the Förster radius. Only distances in the region of the Förster radii of available dyes can be well resolved, which are typically larger than distances of interest within channels, pumps, and transporters, such as the pore radius of a transmembrane domain. Alternative strategies such as single-molecule electron transfer [83] or a fluorophore–quencher pairing [84] could be used to probe shorter (<20Å) distances on a single-molecule level.

11.5 SINGLE-MOLECULE COUNTING BY PHOTOBLEACHING

Photobleaching is the abrupt irreversible decrease in a fluorophore's emission due to photoinduced chemical damage or covalent modification. While traditionally seen as a limitation of SMF experiments, the phenomenon can be used to evaluate the number of fluorophores, and hence subunits, in a diffraction-limited spot. The intensity profile of an illuminated fluorophore displays a single step-like decrease when photobleaching occurs. Thus, the intensity profile of a diffraction-limited spot containing multiple individually labeled subunits displays a stepwise decay whereby the number of steps corresponds to the number of subunits present. The oligomeric state of channels, transporters, and pumps in native environments is often inaccessible by high-resolution structural methods, and ensemble fluorescence measurements assume invariant stoichiometry. The detection and quantification of rare and transient stoichiometries at the single-molecule level afford the possibility of interrogating the kinetics and assembly mechanisms of channels, pumps, and transporters.

The first *in vivo* [32] and *in vitro* [85] examples of this technique emerged in 2007, in which subunits were labeled via a fluorescent fusion protein and maleimide reactive dye, respectively (see Fig. 11.4a). A step-fitting algorithm was used to determine the number of photobleaching events per spot intensity profile. The resulting distribution of photobleaching steps reflects the likelihood of detecting a particular number of fluorescent labels within the complex. The probability of detecting a fluorescent label in each subunit and ultimately the total number of subunits was obtained by fitting a binomial distribution to the data. In these initial reports, the bacterial pore leukocidin was confirmed to be an octamer containing four subunits each of LukF and LukS [85], and the NMDA receptor containing NR1 and NR3b had a tetrameric stoichiometry of 2:2 [32]. Step-fitting algorithms have since been well established

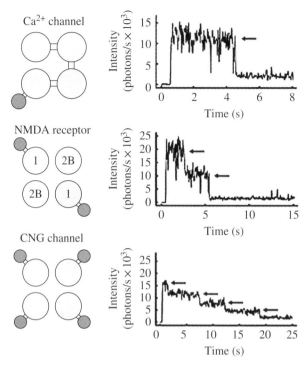

FIGURE 11.4 Subunit counting by photobleaching. Three complexes containing varying numbers of GFP-labeled subunits were imaged; the number of photobleaching steps agrees well with the number of labels per complex. Modified from Ref. 32 with permission from Nature Publishing Group.

[85] and are extensively used to probe the stoichiometry of channels [33, 35, 36, 39–41, 43] and transporters [34, 53] in live cells and in artificial bilayers [51, 59, 86, 87].

Beyond probing the oligomeric status of functional channels, pumps, and transporters, the subunit counting by photobleaching method can be used to evaluate dynamic intermediates associated with channel assembly. For γ-hemolysin, an abundance of intermediates was indicative of a modular assembly mechanism with distinct rate constants associated with each step [68], while the absence of intermediates observed for α-hemolysin suggested that the heptameric pore forms from a pool of monomers in one rapid (<5 ms) step [51].

Photobleaching step counting can be employed in conjunction with other SMF techniques, such as fluorescence colocalization [43], to facilitate a more complex level of analysis. In a smFRET experiment, it can be invaluable to know the number of donor and acceptor fluorophores contributing to a FRET signal by counting photobleaching steps in either emission channel. A more sophisticated analysis is afforded, as highlighted by the studies concerning channel assembly of a bacterial pore [68] and the gating mechanism of a mechanosensitive channel [77] (see Section 11.4).

This is by no means the only way of determining stoichiometry using SMF. As first demonstrated almost 20 years ago, the initial amplitude of the fluorescent signal divided by the signal of an individual fluorophore can be used to estimate the number of fluorophores present [88]. This approach was applied to the L-type Ca^{2+} channel [42] using YFP-labeled monomers, which were tracked in the plasma membrane to generate a time-resolved intensity profile for each diffraction-limited spot. The initial amplitude corresponding to each spot was divided by that of the individual fluorophore observed under the same conditions to obtain the stoichiometry. While a step-wise decay in the emission was reported for all samples, the stoichiometry calculation did not involve step quantification.

More recently, Leake and coworkers assigned the stoichiometry of much larger complexes using a similar approach. First, the ion flow coupler component of the bacterial flagella motor, MotB, was expressed as a GFP fusion protein, and 22 copies were observed per complex [38]. It would have been difficult to resolve this number using a step-fitting algorithm as the distributions of bleaching steps for n and $n+1$ subunits look increasingly similar with greater subunit numbers [32]. The technique was also applied to a transporter with more complex assembly dynamics [38]. The Tat complex is capable of translocating large folded proteins of variable size across the plasma membrane with high efficiency and minimal ion leakage. TatA subunits are thought to form the protein translocation element in each complex and were found to vary in abundance between 4 and 100 per complex with a median of 25. This finding suggests a highly variable structure that forms a tight channel to meet the demand of the translocated protein.

11.6   OPTICAL CHANNEL RECORDING

One limitation of patch clamping is that gating events cannot be assigned to individual channels unless a single channel is present in the bilayer at any given time. An optical approach to patch clamping, in which a fluorescent dye indicates ion flux, provides parallel readout of individually resolved channels. Typically, a fluorescent indicator is loaded into the side of the membrane devoid of that ion and a gating event, permitting ion flux, gives rise to a fluorescent signal. A steep gradient across the bilayer is required for a large and rapid concentration increase in the localized area surrounding the mouth of an open channel. The signal dissipates as ions diffuse away from the channel and by the action of chelating agents. It is very useful to know the number of channels present in an electrophysiological recording, and optical ion flux can provide this without the need for labeling steps.

Calcium indicator dyes have long been used in simultaneous fluorescence and electrophysiological recordings of calcium-signaling events in live cells [89–91]. Only in the last decade, with the ability to restrict illumination/detection volumes, have events from single channels been resolved [92, 93]. To our knowledge, the first single-channel calcium fluorescent transient (SCCaFT) recording was demonstrated by Wang et al. [94], who observed Ca^{2+} sparklets from individual L-type Ca^{2+} channels in heart cells by scanning confocal microscopy. Demuro and Parker pioneered the use of SCCaFT imaging to probe voltage-gated calcium channels and muscle nicotinic acetylcholine receptors [95–98]. They also used TIRF microscopy to interrogate inositol triphosphate (IP3) receptors on areas of the endoplasmic reticulum membrane located sufficiently close to the plasma membrane [99, 100]. Sparks from individual ryanodine receptors (RyR) have also been observed in cardiac myocytes [54].

Artificial bilayers can also be employed for SCCaFT imaging, especially those readily accessible by TIRF microscopy, and the first *in vitro* SCCaFT recording was of RyR in a vertical black lipid membrane [101]. We recently used droplet interface bilayers (DIBs) to image SCCaFTs from alamethicin pores [102] and alpha-hemolysin in the presence of a blocker [103], to identify whether an orthogonal electrical and fluorescence response could be obtained (see Fig. 11.4b). These studies confirmed that the simultaneous optical signal mirrored the electrical current recording and could therefore be directly interpreted as Ca^{2+} conductance in a purely optical experiment. Recently, arrays of DIBs capable of ensemble Ca^{2+} flux imaging in parallel have been pioneered in Oxford [104] and in the group of Takeuchi [105]. Although macroscopic ion fluxes were measured, the reports show that these devices were capable of imaging ion flux through alpha-hemolysin pores under diverse conditions.

High temporal resolution is achieved by fine-tuning the concentration and affinity of both dye and mobile calcium-buffering agents, as explored in a series of theoretical experiments by Shuai and Parker [106]; an indicator dissociation constant of $1–5\,\mu M$ showed the best compromise between signal amplitude and decay rate while calcium buffers, such as EGTA, accelerate the rise and decay of the fluorescent signal at the expense of overall amplitude. At present, the temporal resolution does not come close to that of patch clamping due to the limitations in detectors for photons as compared to ions. As with other SMF techniques, spatial resolution can surpass the diffraction limit by post-acquisition centroid fitting [100]. To our knowledge, optical patch clamping of single channels has thus far been restricted to evaluating those permeable to Ca^{2+} ions. The development and use of fluorescent indicator dyes specific to different ions could broaden the scope to channels of different ionic selectivity.

11.7   SIMULTANEOUS TECHNIQUES

So far, we have hopefully demonstrated the power of SMF techniques for probing structural dynamics of channels, pumps, and transporters. These techniques alone give only half of the picture; combining them with the single-molecule sensitivity provided by patch clamping would allow simultaneous observation of conformational and functional changes. Such an experiment, branded the "holy grail" of ion channel research, [107] would provide unprecedented insight into these biomolecules. For over two decades, the scientific community has recognized the potential of these experiments [3, 108, 109]. However, methodological challenges have proved restrictive and we wait with anticipation for these experiments to become truly tractable.

Simultaneous measurements combining ensemble fluorescence and electrical signals were achieved at least 40 years ago [110, 111]. Patch-clamp fluorometry [112], which combines epifluorescence microscopy with electrophysiology, is also largely a macroscopic technique (see Chapter 4). There are a few examples where patch clamping has been combined with single-molecule confocal microscopy [113, 114]. Lu and coworkers were able to make fluorescence lifetime measurements of FRET-labeled gramicidin and showed a degree of anticorrelation between fluorescence and electrical signals [114]. More recently, the conductance of single nAchR channels was correlated with optical signals arising from the binding of a fluorescent ligand [113]. The latter study reported for the first time a significant number of simultaneous events, where a distinct lag was reported between the electrical and optical signals.

Initial attempts at combining SMF methods with single-channel recordings in artificial bilayers were made by Ide and Yanagida, although simultaneous measurements were not initially achieved [116]. A planar bilayer was formed across an annulus and supported by a thin layer of agarose, making it accessible by TIRF microscopy. Three years later, this setup was used to make the first unambiguous simultaneous measurement of a single event, although not specifically associated with a single molecule [117]; fusion of a single vesicle containing alamethicin and fluorescent lipids with the bilayer showed that the onset of a fluorescent signal was immediately followed by electrical activity. Ten years later, the same group simultaneously recorded the conductance of alpha-hemolysin during a blocking event by DNA bound to fluorescent streptavidin [118]. Although only an isolated event is presented in the study, this is a clear demonstration of a simultaneous recording involving parallel detection of fluorescence from a single molecule and a single electrical signal. More recently, Wagner and coworkers developed a planar bilayer setup capable of both electrical measurements and confocal imaging at the single-molecule level [119]. They present an image containing hundreds of labeled PorB channels in a bilayer, individually resolved, alongside an electrical recording depicting successive insertions. While the data suggest that less than 5% of the imaged channels are active, it is impossible to assign each electrical signal with its corresponding optical counterpart. A similar setup was employed by Lu and coworkers, whereby the diffusion of colicin Ia was observed by SMF with simultaneous single-channel recording to confirm channel activity [120].

Parallel attempts were made to observe simultaneous signals from single molecules *in vivo*, initially with whole-cell currents [42]. Isacoff and coworkers provided the first truly single-molecule experiment, where SMF reported structural changes of a potassium shaker channel during a voltage-clamp protocol in oocytes [25] (see Section 11.3.2). Although the four individual channel events presented clearly show a voltage-dependent response, the lack of data in this study, and indeed almost all the studies in this section, speaks to the difficulty of these experiments.

Borisenko et al. demonstrated arguably the first true simultaneous single-molecule measurement using smFRET to report on gramicidin dimerization [115]. Two cyclic polypeptide monomers, suitably labeled for smFRET, in opposite leaflets coincide to form a bilayer-spanning channel. Hence, current and FRET signals are expected to coincide (see Fig. 11.5a). Again, the small number of events presented reflects the difficulty of such experiments, and the main challenges are discussed in a comprehensive "strategies for improvement" section.

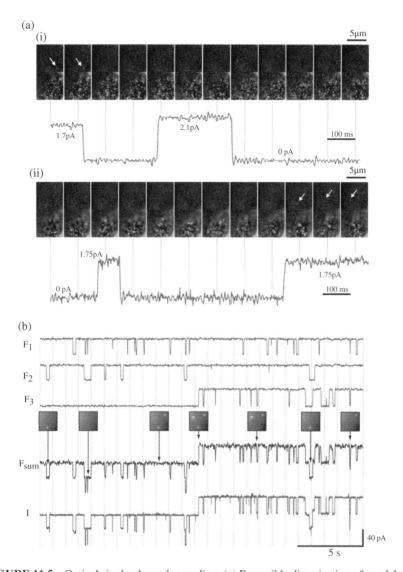

FIGURE 11.5 Optical single-channel recording. (a) Reversible dimerization of two labeled gramicidin monomers for simultaneous smFRET and single-channel recording. (i) gA-Cy3 and pF-Phe-gA-Cy5 peptides diffuse in opposite leaflets. Dimerization results in channel formation as well as bringing the fluorophores close enough for FRET to occur. (ii) Two simultaneous optical and electrical recordings of gramicidin in planar bilayers, displaying acceptor (upper panel) and donor (lower panel) emission channels, with the corresponding bilayer current below. Each panel shows one electrical event which is accompanied by a FRET signal, indicative of heterodimer formation. The two remaining electrical events that are not accompanied by a FRET signal are attributed to homodimer formation. Modified from Ref. 115 with permission from Elsevier. (b) Simultaneous fluorescence and electrical recording from alpha-hemolysin pores in droplet interface bilayers being stochastically blocked by cyclodextrin. Parallel fluorescent signals for three channels are clearly resolved, and their sum is consistent with the total current throughout the recording. Reprinted from Ref. 103 with permission from American Chemical Society.

The primary challenge presented by all of these experiments lies in the differences in requirements for both types of experiment: single-channel recordings require a single channel to be present in a bilayer maintaining a high-resistance (gigaohm) seal; SMF imaging is achieved by the photobleaching of fluorophores and so a large population is desirable for prolonged imaging. A bilayer containing a single fluorescently labeled channel would relinquish only several seconds of imaging time before photobleaching, and therefore, a large number of sequential experiments would be required in order to obtain a statistically tractable dataset. One strategy to overcome the challenge presented by the differences in requirements is to make many measurements in parallel. Bilayer arrays capable of such parallelization have been made in diverse ways, although they have thus far been limited to ensemble imaging [104, 105, 121–125]. Another strategy is to extend the lifetime of current fluorescent probes; considerable advances have recently been made to this end, including the development of new photoprotective reagents [126], coupling these reagents to fluorophores [127], the use of microsecond modulation of laser excitation [128], and the use of larger probes such as QDs [28] or nanodiamonds [129]. Alternatively, a strategy whereby ionic current and fluorescence carry orthogonal information, such as with optical channel recording (see Section 11.6), could circumvent the single-channel recording requirement of obtaining just one channel per bilayer.

11.8 SUMMARY

In this chapter, we have attempted to introduce the reader to the wide scope for SMF microscopy in understanding the role of channels, transporters, and pumps. There are clearly many impressive demonstrations, of which we have only highlighted a small fraction. However, perhaps the most striking is the future potential of these methods: although much has been promised, single-molecule methods have really just begun to provide serious mechanistic insight into the behavior of ion channels. In contrast, transporters and pumps have only very recently found the attention of scientists using these methods. There is much to be done.

ACKNOWLEDGMENTS

M.I.W is supported by the Biotechnology and Biological Sciences Research Council (BBSRC) and the European Research Council (ERC). E.E.W. is supported by the BBSRC. J.S.H.D. is supported by a Louis Dreyfus-Weidenfeld Scholarship.

REFERENCES

1. O.P. Hamill, A. Marty, E. Neher, B. Sakmann, F.J. Sigworth, Improved patch-clamp techniques for high-resolution current recording from cells and cell-free membrane patches, *Pflugers Arch.* 391 (1981), 85–100.

2. B. Hille, *Ion Channels of Excitable Membranes*, Sinauer Associates, Sunderland, MA, 2001.

3. S. Weiss, *Fluorescence spectroscopy of single biomolecules*, Science 283 (1999) 1676–1683.

4. A.E. Knight, *Single Molecule Biology*, Academic Press, San Diego, CA, 2008.

5. C. Joo, H. Balci, Y. Ishitsuka, C. Buranachai, T. Ha, Advances in single-molecule fluorescence methods for molecular biology, *Annu. Rev. Biochem.* 77 (2008) 51–76.

6. R.E. Thompson, D.R. Larson, W.W. Webb, Precise nanometer localization analysis for individual fluorescent probes, *Biophys. J.* 82 (2002) 2775–2783.

7. S.W. Hell, J. Wichmann, Breaking the diffraction resolution limit by stimulated emission: Stimulated-emission-depletion fluorescence microscopy, *Opt. Lett.* 19 (1994), 780–782.

8. S.T. Hess, T.P.K. Girirajan, M.D. Mason, Ultra-high resolution imaging by fluorescence photoactivation localization microscopy, *Biophys. J.* 91 (2006) 4258–4272.

9. M.J. Rust, M. Bates, X. Zhuang, Imaging by stochastic optical reconstruction microscopy (STORM), *Nat. Methods* 3 (2006) 793–795.

10. B. Huang, M. Bates, X. Zhuang, Super-resolution fluorescence microscopy, *Annu. Rev. Biochem.* 78 (2009) 993–1016.

11. A. Chmyrov, J. Keller, T. Grotjohann, M. Ratz, E. D'Este, S. Jakobs, C. Eggeling, S.W. Hell, Nanoscopy with more than 100,000 'doughnuts', *Nat. Methods* 10 (2013) 737–740.

12. F. Huang, T.M. Hartwich, F.E. Rivera-Molina, Y. Lin, W.C. Duim, J.J. Long, P.D. Uchil, J.R. Myers, M.A. Baird, W. Mothes, M.W. Davidson, D. Tommre, J. Bewersdorf, Video-rate nanoscopy using sCMOS camera-specific single-molecule localization algorithms, *Nat. Methods*, 10 (2013) 653–658.

13. R. Roy, S. Hohng, T. Ha, A practical guide to single-molecule FRET, *Nat. Methods* 5 (2006) 507–516.

14. M. Tokunaga, K. Kitamura, K. Saito, A.H. Iwane, T. Yanagida, Single molecule imaging of fluorophores and enzymatic reactions achieved by objective-type total internal reflection fluorescence microscopy, *Biochem. Biophys. Res. Commun.*, 235 (1997) 47–53.

15. J.R. Lakowicz, *Principles of Fluorescence Spectroscopy*, Springer, Baltimore, MD, 2010.

16. M. Minsky, Microscopy Apparatus, US Patent, vol. 3,013,467, 1961.

17. V. Vukojevic, M. Heidkamp, Y. Ming, B. Johansson, L. Terenius, R. Rigler, Quantitative single-molecule imaging by confocal laser scanning microscopy, *Proc. Natl. Acad. Sci. U. S. A.* 105 (2008) 18176–18181.

18. R. Rigler, U. Mets, J. Widengren, P. Kask, Fluorescence correlation spectroscopy with high count rate and low background: Analysis of translation diffusion, *Eur. Biophys. J.* 22 (1993) 169–175.

19. T. Ha, P. Tinnefeld, Photophysics of fluorescence probes for single molecule biophysics and super-resolution imaging, *Annu. Rev. Phys. Chem.* 63 (2013) 595–617.

20. Y. Chen, J.D. Müller, Q. Ruan, E. Gratton, Molecular brightness characterization of EGFP in vivo by fluorescence fluctuation spectroscopy, *Biophys. J.* 82 (2002) 133–144.

21. R. Y. Tsien, The green fluorescent protein, *Annu. Rev. Biochem.* 67 (1998) 509–544.

22. G.S. Harms, L. Cognet, P.H. Lommerse, G.A. Blab, T. Schmidt, Autofluorescent proteins in single-molecule research: Applications to live cell imaging microscopy, *Biophys. J.* 80 (2001) 2396–2408.

23. A. Keppler, S. Gendreizig, T. Gronemeyer, H. Pick, H. Vogel, K. Johnsson, A general method for the covalent labeling of fusion proteins with small molecules in vivo, *Nat. Biotechnol.* 21 (2003) 86–89.

24. M.W. Popp, J.M. Antos, G.M. Grotenbreg, E. Spooner, H.L. Ploegh, Sortagging: A versatile method for protein labeling, *Nat. Chem. Biol.* 3 (2007) 707–708.

25. A. Sonnleitner, L. M. Mannuzzu, S. Terakawa, E.Y. Isacoff, Structural rearrangements in single ion channels detected optically in living cells, *Proc. Natl. Acad. Sci. U. S. A.* 99 (2002) 12759–12764.

26. R. Pantoja, E.A. Rodriguez, M.I. Dibas, D.A. Dougherty, H.A. Lester, Single-molecule imaging of a fluorescent unnatural amino acid incorporated into nicotinic receptors, *Biophys. J.* 96 (2009) 226–237.

27. G.J. Schütz, V.P. Pastushenko, H.J. Gruber, H. Knaus, B. Pragl, H. Schindler, 3D imaging of individual ion channels in live cells at 40 nm resolution, *Single Mol.* 1 (2000) 25–31.

28. M. Dahan, S. Lévi, C. Luccardini, P. Rostaing, B. Riveau, A. Triller, Diffusion dynamics of glycine receptors revealed by single-quantum dot tracking, *Science* 302 (2003) 442–445.

29. I.L. Medintz, H.T. Uyeda, E.R. Goldman, H. Mattoussi, Quantum dot bioconjugates for imaging, labelling and sensing, *Nat. Mater.* 4 (2005), 435–446.

30. U. Resch-Genger, M. Grabolle, S. Cavaliere-Jaricot, R. Nitschke, T. Nann, Quantum dots versus organic dyes as fluorescent labels, *Nat. Methods* 5 (2008) 763–775.

31. M. Ulbrich, E. Isacoff, Rules of engagement for NMDA receptor subunits, *Proc. Natl. Acad. Sci. U. S. A.* 105 (2008) 14163–14168.

32. M. H. Ulbrich, E. Y. Isacoff, Subunit counting in membrane-bound proteins, *Nat. Methods* 4 (2007) 319–321.

33. J.R. Bankston, S.S. Camp, F. DiMaio, A.S. Lewis, D.M. Chetkovich, W.N. Zagotta, Structure and stoichiometry of an accessory subunit TRIP8b interaction with hyperpolarization-activated cyclic nucleotide-gated channels, *Proc. Natl. Acad. Sci. U. S. A.* 109 (2012) 7899–7904.

34. A. Reiner, R. Arant, E. Isacoff, Assembly stoichiometry of the GluK2/GluK5 kainate receptor complex, *Cell Rep.* 1 (2012) 234–240.

35. K. Nakajo, M. H. Ulbrich, Y. Kubo, E.Y. Isacoff, Stoichiometry of the KCNQ1 – KCNE1 ion channel complex, *Proc. Natl. Acad. Sci. U. S. A.* 107 (2010) 18862–18867.

36. Y. Yu, M.H. Ulbrich, M.-H. Li, Z. Buraei, X.-Z. Chen, A.C.M. Ong, L. Tong, E.Y. Isacoff, J. Yang, Structural and molecular basis of the assembly of the TRPP2/PKD1 complex, *Proc. Natl. Acad. Sci. U. S. A.* 106 (2009) 11558–11563.

37. M.C. Leake, J.H. Chandler, G.H. Wadhams, F. Bai, R.M. Berry, J.P. Armitage, Stoichiometry and turnover in single, functioning membrane protein complexes, *Nature* 443 (2006) 355–358.

38. M.C. Leake, N.P. Greene, R.M. Godun, T. Granjon, G. Buchanan, S. Chen, R.M. Berry, T. Palmer, B.C. Berks, Variable stoichiometry of the TatA component of the twin-arginine protein transport system observed by in vivo single-molecule imaging, *Proc. Natl. Acad. Sci. U. S. A.* 105 (2008) 15376–15381.

39. H. Yamamura, C. Ikeda, Y. Suzuki, S. Ohya, Y. Imaizumi, Molecular assembly and dynamics of fluorescent protein-tagged single KCa1.1 channel in expression system and vascular smooth muscle cells, *Am. J. Physiol. Cell Physiol.* 302 (2012) 1257–1268.

40. M. Tajima, J. M. Crane, S. Verkman, Aquaporin-4 (AQP4) associations and array dynamics probed by photobleaching and single-molecule analysis of green fluorescent protein-AQP4 chimeras, *J. Biol. Chem.* 285 (2010) 8163–8170.

41. P.D. Fox, R.J. Loftus, M.M. Tamkun, Regulation of Kv2.1 K^+ conductance by cell surface channel density, *J. Neurosci.* 33 (2013) 1259–1270.

42. G.S. Harms, L. Cognet, P.H. Lommerse, G.A. Blab, H. Kahr, R. Gamsjäger, H.P. Spaink, N. M. Soldatov, C. Romanin, T. Schmidt, Single-molecule imaging of L-type Ca^{2+} channels in live cells, *Biophys. J.* 81 (2001) 2639–2646.

43. W. Ji, P. Xu, Z. Li, J. Lu, L. Liu, Y. Zhan, Y. Chen, B. Hille, T. Xu, L. Chen, Functional stoichiometry of the unitary calcium-release-activated calcium channel, *Proc. Natl. Acad. Sci. U. S. A.* 105 (2008) 13668–13673.

44. E.T. Castellana, P.S. Cremer, Solid supported lipid bilayers: From biophysical studies to sensor design, *Surf. Sci. Rep.* 61 (2006) 429–444.

45. M. Montal, P. Mueller, Formation of bimolecular membranes from lipid monolayers and a study of their electrical properties, *Proc. Natl. Acad. Sci. U. S. A.* 69 (1972) 3561–3566.

46. S. Leptihn, O.K. Castell, B. Cronin, E.H. Lee, L.C.M. Gross, D.P. Marshall, J.R. Thompson, M. Holden, M.I. Wallace, Constructing droplet interface bilayers from the contact of aqueous droplets in oil, *Nat. Protoc.* 8 (2013) 1048–1057.

47. F.J. Szoka, D. Papahadjopoulos, Comparative properties and methods of preparation of lipid vesicles (liposomes), *Ann. Rev. Biophys. Bioeng.* 9 (1980) 467–508.

48. S. Demarche, K. Sugihara, T. Zambelli, L. Tiefenauer, J. Vörös, Techniques for recording reconstituted ion channels, *Analyst* 136 (2011) 1077–1089.

49. P.M. Conn, *Essential Ion Channel Methods*, Academic Press, San Diego, CA, 2010.

50. F. Alcock, M.A.B. Baker, N.P. Greene, T. Palmer, M.I. Wallace, B.C. Berks, Live cell imaging shows reversible assembly of the TatA component of the twin-arginine protein transport system, *Proc. Natl. Acad. Sci. U. S. A.* 110 (2013) 3650–3659.

51. J.R. Thompson, B. Cronin, H. Bayley, M.I. Wallace, Rapid assembly of a multimeric membrane protein pore, *Biophys. J.* 101 (2011) 2679–2683.

52. M.L. Molina, F.N. Barrera, A.M. Fernández, J.A. Poveda, M.L. Renart, J.A. Encinar, G. Riquelme, J.M. González-Ros, Clustering and coupled gating modulate the activity in KcsA, a potassium channel model, *J. Biol. Chem.* 281 (2006) 18837–18848.

53. Q. Wang, Y. Zhao, W. Luo, R. Li, Q. He, X. Fang, R.D. Michele, C. Ast, N.V. Wirén, J. Lin, Single-particle analysis reveals shutoff control of the Arabidopsis ammonium transporter AMT1;3 by clustering and internalization, *Proc. Natl. Acad. Sci. U. S. A.* 110 (2013) 13204–13209.

54. D. Baddeley, I.D. Jayasinghe, L. Lam, S. Rossberger, M.B. Cannell, C. Soeller, C. Franzini-Armstrong, Optical single-channel resolution imaging of the ryanodine receptor distribution in rat cardiac myocytes, *Proc. Natl. Acad. Sci. U. S. A.* 106 (2009) 22275–22280.

55. J. Ma, A. Goryaynov, A. Sarma, W. Yang, Self-regulated viscous channel in the nuclear pore complex, *Proc. Natl. Acad. Sci. U. S. A.* 109 (2012) 7326–7331.

56. D.E.J. Waschk, A. Fabian, T. Budde, A. Schwab, Dual-color quantum dot detection of a heterotetrameric potassium channel, *Am. J. Physiol. Cell Physiol.* 300 (2011) C843–C849.

57. L. Mannuzzu, M. Moronne, E. Isacoff, Direct physical measure of conformational rearrangement underlying potassium channel gating, *Science* 271 (1996) 213–216.

58. A.P. Demchenko, Y. Mély, G. Duportail, A.S. Klymchenko, Monitoring biophysical properties of lipid membranes by environment-sensitive fluorescent probes, *Biophys. J.* 96 (2009) 3461–3470.

59. R. Blunck, H. McGuire, H.C. Hyde, F. Bezanilla, Fluorescence detection of the movement of single KcsA subunits reveals cooperativity, *Proc. Natl. Acad. Sci. U. S. A.*, 105 (2008) 20263–20268.

60. A.H. Nguyen, V.T. Nguyen, Y. Kamio, H. Higuchi, Single-molecule visualization of environment-sensitive fluorophores inserted into cell membranes by staphylococcal gamma-hemolysin, *Biochemistry* 45 (2006) 2570–2576.

61. J. Ma, W. Yang, Three-dimensional distribution of transient interactions in the nuclear pore complex obtained from single-molecule snapshots, *Proc. Natl. Acad. Sci. U. S. A.* 107 (2010) 7305–7310.

62. M. Kahms, P. Lehrich, J. Hüve, N. Sanetra, R. Peters, Binding site distribution of nuclear transport receptors and transport complexes in single nuclear pore complexes, *Traffic* 10 (2009) 1228–1242.

63. A. Szymborska, A.D. Marco, N. Daigle, V.C. Cordes, J.A.G. Briggs, J. Ellenberg, Nuclear pore scaffold structure analyzed by super-resolution microscopy and particle averaging, *Science* 341 (2013) 655–658.

64. T. Förster, Zwischenmolekulare Energiewanderung und Fluoreszenz, *Ann. Phys.* 437 (1948) 55–75.

65. T. Ha, T. Enderle, D.F. Ogletree, D.S. Chemla, P.R. Selvin, S. Weiss, Probing the interaction between two single molecules: Fluorescence resonance energy transfer between a single donor and a single acceptor, *Proc. Natl. Acad. Sci. U. S. A.* 93 (1996) 6264–6268.

66. J. Hohlbein, T.D. Craggs, T. Cordes, Alternating-laser excitation: Single-molecule FRET and beyond, *Chem. Soc. Rev.* 43 (2014) 1156–1171.

67. Y. Sako, S. Minoghchi, T. Yanagida, Single-molecule imaging of EGFR signalling on the surface of living cells, *Nat. Cell Biol.* 2 (2000) 168–172.

68. V.T. Nguyen, Y. Kamio, H. Higuchi, Single-molecule imaging of cooperative assembly of gamma-hemolysin on erythrocyte membranes, *EMBO J.* 22 (2003) 4968–4979.

69. Y. Zhao, D. Terry, L. Shi, H. Weinstein, S.C. Blanchard, J.A. Javitch, Single-molecule dynamics of gating in a neurotransmitter transporter homologue, *Nature* 465 (2010) 188–193.

70. G.B. Erkens, I. Hänelt, J.M.H. Goudsmits, D.J. Slotboom, A.M. van Oijen, Unsynchronised subunit motion in single trimeric sodium-coupled aspartate transporters, *Nature* 502 (2013) 119–123.

71. N. Monma, V.T. Nguyen, J. Kaneko, H. Higuchi, Y. Kamio, Essential residues, W177 and R198, of LukF for phosphatidylcholine-binding and pore-formation by staphylococcal gamma-hemolysin on human erythrocyte membranes, *J. Biochem.* 136 (2004) 427–431.

72. J. Ohshiro, H. Yamamura, T. Saeki, Y. Suzuki, Y. Imaizumi, The multiple expression of Ca^{2+}-activated Cl^- channels via homo- and hetero-dimer formation of TMEM16A splicing variants in murine portal vein, *Biochem. Biophys. Res. Commun.* 443 (2014) 518–523.

73. Y. Suzuki, H. Yamamura, S. Ohya, Y. Imaizumi, Caveolin-1 facilitates the direct coupling between large conductance Ca^{2+}-activated K^+ (BKCa) and Cav1.2 Ca^{2+} channels and their clustering to regulate membrane excitability in vascular myocytes, *J. Biol. Chem.* 288 (2013) 36750–36761.

74. E.A. Morrison, G.T. DeKoster, S. Dutta, R. Vafabakhsh, M.W. Clarkson, A. Bahl, D. Kern, T. Ha, K.A. Henzler-Wildman, Antiparallel EmrE exports drugs by exchanging between asymmetric structures, *Nature* 481 (2012) 45–50.

75. G.V.D. Bogaart, I. Kusters, J. Velásquez, J.T. Mika, V. Krasinkov, A.J.M. Driessen, B. Poolman, Dual-color fluorescence-burst analysis to study pore formation and protein-protein interactions, *Methods* 46 (2008) 123–130.

76. G.S. Harms, G. Orr, M. Montal, B.D. Thrall, S.D. Colson, H.P. Lu, Probing conformational changes of gramicidin ion channels by single-molecule patch-clamp fluorescence microscopy, *Biophys. J.* 85 (2003) 1826–1838.

77. Y. Wang, Y. Liu, H.A. DeBerg, T. Nomura, M.T. Hoffman, P.R. Rohde, K. Schulten, B. Martinac, P.R. Selvin, Single molecule FRET reveals pore size and opening mechanism of a mechano-sensitive ion channel, *eLife* 3 (2014) e01834.

78. U.B. Choi, S. Xiao, L.P. Wollmuth, M.E. Bowen, Effect of Src kinase phosphorylation on disordered C-terminal domain of *N*-methyl-D-aspartic acid (NMDA) receptor subunit GluN2B protein, *J. Biol. Chem.* 286 (2011) 29904–29912.

79. U.B. Choi, R. Kazi, N. Stenzoski, L.P. Wollmuth, V.N. Uversky, M.E. Bowen, Modulating the intrinsic disorder in the cytoplasmic domain alters the biological activity of the *N*-methyl-D-aspartate-sensitive glutamate receptor, *J. Biol. Chem.* 288 (2013) 22506–22515.

80. C.F. Landes, A. Rambhadran, J.N. Taylor, F. Salatan, V. Jayaraman, Structural landscape of isolated agonist-binding domains from single AMPA receptors, *Nat. Chem. Biol.* 7 (2011) 168–173.

81. Y. Zhao, D.S. Terry, L. Shi, M. Quick, H. Weinstein, S.C. Blanchard, J.A. Javitch, Substrate-modulated gating dynamics in a Na+-coupled neurotransmitter transporter homologue, *Nature* 474 (2011) 109–113.

82. N. Akyuz, R.B. Altman, S.C. Blanchard, O. Boudker, Transport dynamics in a glutamate transporter homologue, *Nature* 502 (2013) 114–118.

83. H. Yang, G. Luo, P. Karnchanaphanurach, T.-M. Louie, I. Rech, S. Cova, L. Xun, X. S. Xie, Protein conformational dynamics probed by single-molecule electron transfer, *Science* 302 (2003) 262–266.

84. P. Zhu, J.P. Clamme A.A. Deniz, Fluorescence quenching by TEMPO: A sub-30 Å single-molecule ruler, *Biophys. J.* 89 (2005) 37–39.

85. S.K. Das, M. Darshi, S. Cheley, M.I. Wallace, H. Bayley, Membrane protein stoichiometry determined from the step-wise photobleaching of dye-labelled subunits, *Chembiochem* 8 (2007) 994–999.

86. H. McGuire, M.R.P. Aurousseau, D. Bowie, R. Blunck, Automating single subunit counting of membrane proteins in mammalian cells, *J. Biol. Chem.* 287 (2012) 35912–35921.

87. N. Groulx, H. McGuire, R. Laprade, J.L. Schwartz, R. Blunck, Single molecule fluorescence study of the *Bacillus thuringiensis* toxin Cry1Aa reveals tetramerization, *J. Biol. Chem.* 286 (2011) 42274–42282.

88. T. Schmidt, G.J. Schu, H.J. Gruber, H. Schindler, Local stoichiometries determined by counting individual molecules, *Anal. Chem.* 68 (1996) 4397–4401.

89. I. Parker, Y. Yao, Regenerative release of calcium from functionally discrete subcellular stores by inositol trisphosphate, *Proc. Biol. Sci.* 246 (1991) 269–274.

90. H. Cheng, W.J. Lederer, M.B. Cannell, Events underlying calcium sparks: Elementary events underlying excitation-contraction coupling in heart muscle, *Science* 262 (1993) 740–744.

91. S.A. Thayer, M. Sturek, R.J. Miller, Measurement of neuronal Ca^{2+} transients using simultaneous microfluorimetry and electrophysiology, *Pflugers Arch.* 412 (1988) 216–223.

92. S.-Q. Wang, C. Wei, G. Zhao, D.X.P. Brochet, J. Shen, L.S. Song, W. Wang, D. Yang, H. Cheng, Imaging microdomain Ca^{2+} in muscle cells, *Circ. Res.* 94 (2004) 1011–1022.

93. A. Demuro, I. Parker, Imaging single-channel calcium microdomains, *Cell Calcium* 40 (2006) 413–422.

94. S.Q. Wang, L.S. Song, E.G. Lakatta, H. Cheng, Ca^{2+} signalling between single L-type Ca^{2+} channels and ryanodine receptors in heart cells, *Nature* 410 (2001) 592–596.

95. A. Demuro, I. Parker, Imaging single-channel calcium microdomains by total internal reflection microscopy, *Biol. Res.* 37 (2004) 675–679.

96. A. Demuro, I. Parker, Imaging the activity and localization of single voltage-gated Ca^{2+} channels by total internal reflection fluorescence microscopy, *Biophys. J.* 86 (2004) 3250–3259.

97. A. Demuro, I. Parker, Optical single-channel recording: Imaging Ca^{2+} flux through individual N-type voltage-gated channels expressed in *Xenopus* oocytes, *Cell Calcium* 34 (2003) 499–509.

98. A. Demuro, I. Parker, "Optical patch-clamping": Single-channel recording by imaging Ca2+ flux through individual muscle acetylcholine receptor channels, *J. Gen. Physiol.* 126 (2005) 179–192.

99. I.F. Smith, I. Parker, Imaging the quantal substructure of single IP3R channel activity during Ca^{2+} puffs in intact mammalian cells, *Proc. Natl. Acad. Sci. U. S. A.* 106 (2009) 6404–6409.

100. S.M. Wiltgen, I.F. Smith, I. Parker, Superresolution localization of single functional IP_3R channels utilizing Ca^{2+} flux as a readout, *Biophys. J.* 99 (2010) 437–446.

101. S. Peng, N.G. Publicover, G.J. Kargacin, D. Duan, J.A. Airey, J.L. Sutko, Imaging single cardiac ryanodine receptor Ca^{2+} fluxes in lipid bilayers, *Biophys. J.* 86 (2004) 134–144.

102. L.M. Harriss, B. Cronin, J.R. Thompson, M.I. Wallace, Imaging multiple conductance states in an alamethicin pore, *J. Am. Chem. Soc.* 133 (2011) 14507–14509.

103. A.J. Heron, J.R. Thompson, B. Cronin, H. Bayley, M.I. Wallace, Simultaneous measurement of ionic current and fluorescence from single protein pores, *J. Am. Chem. Soc.* 131 (2009) 1652–1653.

104. O.K. Castell, J. Berridge, M.I. Wallace, Quantification of membrane protein inhibition by optical ion flux in a droplet interface bilayer array, *Angew. Chem. Int. Ed. Engl.* 51 (2012) 3134–3138.

105. T. Tonooka, K. Sato, T. Osaki, R. Kawano, S. Takeuchi, Lipid bilayers on a picoliter microdroplet array for rapid fluorescence detection of membrane transport, *Small* 10 (2014) 3275–32828.

106. J. Shuai, I. Parker, Optical single-channel recording by imaging Ca^{2+} flux through individual ion channels: Theoretical considerations and limits to resolution, *Cell Calcium* 37 (2005) 283–299.

107. P.R. Selvin, Lighting up single ion channels, *Biophys. J.* 84 (2003) 1–2.

108. T. Schmidt , P. Hinterdorfer, H. Schindler, Microscopy for recognition of individual biomolecules, *Microsc. Res. Tech.* 44 (1999) 339–346.

109. A.G. Macdonald, P.C. Wraight, Combined spectroscopic and electrical recording techniques in membrane research: Prospects for single channel studies, *Prog. Biophys. Mol. Biol.* 63 (1995) 1–29.

110. F. Conti, F. Malerba, Fluorescence signals in ANS-stained lipid bilayers under applied potentials, *Biophysik* 8 (1972) 326–332.

111. W.R. Veatch, R. Mathies, M. Eisenberg, L. Stryer, Simultaneous fluorescence and conductance studies of planar bilayer membranes containing a highly active and fluorescent analog of gramicidin A, *J. Mol. Biol.* 99 (1975) 75–92.

112. J. Kusch, G. Zifarelli, Patch-clamp fluorometry: Electrophysiology meets fluorescence, *Biophys. J.* 106 (2014) 1250–1257.

113. R. Schmauder, D. Kosanic, R. Hovius, H. Vogel, Correlated optical and electrical single-molecule measurements reveal conformational diffusion from ligand binding to channel gating in the nicotinic acetylcholine receptor, *Chembiochem* 12 (2011) 2431–2414.

114. G. Harms, G. Orr, H.P. Lu, Probing ion channel conformational dynamics using simultaneous single-molecule ultrafast spectroscopy and patch-clamp electric recording, *Appl. Phys. Lett.* 84 (2004) 1792.

115. V. Borisenko, T. Lougheed, J. Hesse, E. Füreder-Kitzmüller, N. Fertig, J.C. Behrends, G.A. Woolley, G.J. Schütz, Simultaneous optical and electrical recording of single gramicidin channels, *Biophys. J.* 84 (2003) 612–622.

116. T. Ide, T. Yanagida, An artificial lipid bilayer formed on an agarose-coated glass for simultaneous electrical and optical measurement of single ion channels, *Biochem. Biophys. Res. Commun.* 265 (1999) 595–599.

117. T. Ide, Y. Takeuchi, T. Yanagida, Development of an experimental apparatus for simultaneous observation of optical and electrical signals from single ion channels, *Single Mol.* 3 (2002) 33–42.

118. T. Ide, Simultaneous optical and electrical recording of single molecule bonding to single channel proteins, *Chemphyschem* 11 (2010) 3408–3411.

119. P. Bartsch, C. Walter, P. Selenschik, A. Honigmann, R. Wagner, Horizontal bilayer for electrical and optical recordings, *Materials* 5 (2012) 2705–2730.

120. S.P. Rajapaksha, X. Wang, H.P. Lu, Suspended lipid bilayer for optical and electrical measurements of single ion channel proteins, *Anal. Chem.* 85 (2013) 8951–8955.

121. D. Frese, S. Steltenkamp, S. Schmitz, C. Steinem, In situ generation of electrochemical gradients across pore-spanning membranes, *RSC Adv.* 3 (2013) 15752.

122. J.S. Hansen, M. Perry, J. Vogel, J.S. Groth, T. Vissing, M.S. Larsen, O. Geschke, J. Emneús, H. Bohr, C.H. Nielsen, Large scale biomimetic membrane arrays, *Anal. Bioanal. Chem.* 395 (2009) 719–727.

123. T. Osaki, K. Kamiya, R. Kawano, H. Sasaki, S. Takeuchi, Towards artificial cell array system: Encapsulation and hydration technologies integrated in liposome array, IEEE 25th Int. Conf. Micro. Electro. Mech. Syst., January 29–February 2, 2012, Paris, France, 333–336.

124. H. Bayley, B. Cronin, A. Heron, M.A. Holden, W.L. Hwang, R. Syeda, J. Thompson, M. Wallace, Droplet interface bilayers, *Mol. Biosyst.* 4 (2008) 1191–1208.

125. B.L. Jackson, J.T. Groves, Hybrid protein-lipid patterns from aluminum templates, *Langmuir* 23 (2007) 2052–2057.

126. L.A. Campos, J. Liu, X. Wang, R. Ramanathan, D.S. English, V. Muñoz, A photoprotection strategy for microsecond-resolution single-molecule fluorescence spectroscopy, *Nat. Methods* 8 (2011) 143–146.

127. R.B. Altman, D.S. Terry, Z. Zhou, Q. Zheng, P. Geggier, R.A. Kolster, Y. Zhao, J.A. Javitch, J.D. Warren, S.C. Blanchard, Cyanine fluorophore derivatives with enhanced photostability, *Nat. Methods* 9 (2012) 68–71.

128. G. Donnert, C. Eggeling, S.W. Hell, Major signal increase in fluorescence microscopy through dark-state relaxation, *Nat. Methods* 4 (2007) 81–86.

129. L.T. Hall, C.D. Hill, J.H. Cole, B. Städler, F. Caruso, P. Mulvaney, J. Wrachtrup, L.C.L. Hollenberg, Monitoring ion-channel function in real time through quantum decoherence, *Proc. Natl. Acad. Sci. U. S. A.* 107 (2010) 18777–18782.

12

ELECTRON PARAMAGNETIC RESONANCE: SITE-DIRECTED SPIN LABELING

LOUISE J. BROWN AND JOANNA E. HARE

Department of Chemistry and Biomolecular Sciences, Macquarie University, Sydney, New South Wales, Australia

12.1 INTRODUCTION

The introduction of a nitroxide spin label via cysteine substitution, as pioneered by Wayne L. Hubbell and colleagues, and advances in molecular biology in the late 1980s, established electron paramagnetic resonance (EPR) as a powerful tool for structural biology [1, 2]. Collectively defined as "site-directed spin labeling" (SDSL), today this term is best used to describe the approach of positioning spin labels at desired sites to reveal structure–function relationships by EPR. In the beginning, SDSL provided researchers with a new strategy for overcoming the limitations of relying on naturally occurring sites. Now, it has evolved into a powerful and routine method for investigating structural folds, protein dynamics, and large-scale conformational changes for biomolecules that were once considered too difficult to study. Particularly in recent years, the SDSL approaches have proved indispensable to the membrane protein structural biology community with significant contributions from SDSL studies revealing the dynamic connections between structure and mechanism for a number of pumps, channels, and transporter proteins.

In brief, SDSL typically involves individually mutating new cysteine residues at sites of interest in the protein for the subsequent covalent attachment of a

Pumps, Channels, and Transporters: Methods of Functional Analysis, First Edition. Edited by Ronald J. Clarke and Mohammed A. A. Khalid.
© 2015 John Wiley & Sons, Inc. Published 2015 by John Wiley & Sons, Inc.

sulfhydryl-reactive paramagnetic reporter group or nitroxide "spin label" (see Fig. 12.1a). Upon attachment to the engineered cysteines, the spin labels then report by means of their free unpaired electron on parameters including the mobility of the attached label, the exposure of the label to solvent, or the relative distance of the spin label from another paramagnetic center in the protein (see Fig. 12.1b). "Cysteine scanning" or "nitroxide scanning" throughout an entire region of a protein can also reveal trends in several EPR parameters as a function of the cysteine-spin-labeled position along the polypeptide chain, to provide secondary, tertiary, or quaternary structural information. With fewer restrictions placed on sample requirements compared with other structural approaches in terms of size and amounts, SDSL–EPR is

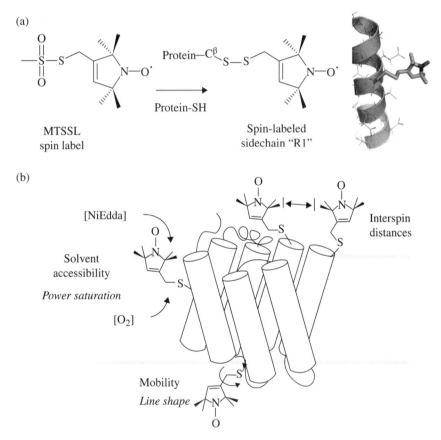

FIGURE 12.1 Site-directed spin labeling. (a) Reaction of a thiol-containing protein with a sulfhydryl-specific nitroxide reagent (MTSSL) generates a disulfide-linked nitroxide side chain "R1." The nitroxide group is tethered to the protein backbone via a flexible linker arm. (b) EPR can provide mobility, solvent accessibility, and interspin distances for spin labels introduced using site-directed mutagenesis methods. Adapted from Ref. 3 with permission from the Royal Society of Chemistry (RSC) on behalf of the Centre National de la Recherche Scientifique (CNRS).

suitable for use in a wide range of protein systems, including soluble proteins, nucleic acids, and membrane proteins. Membrane proteins can either be solubilized in detergent or reconstituted within lipid bilayers. As such, there are no real limits on the size or complexity of a system that can be examined, provided a viable background protein sequence with all reactive native cysteines replaced is achievable.

12.1.1 Development of EPR as a Tool for Structural Biology

The first EPR experiments with biological applications were performed in the 1950s and included structural studies on metalloproteins [4] and free radicals in biological tissues [5]. However, as most proteins are typically EPR silent, techniques to introduce or "label" a protein with a paramagnetic species were needed. The first use of extrinsic spin labels (nitroxide derivatives) to label a peptide was demonstrated in the 1960s by Ogawa and McConnell [6, 7]. The targeted labeling of proteins continued to develop with a number of maleimide nitroxides synthesized to specifically target cysteine residues. At the time, only native cysteines could be targeted, which was a major limitation of the technique. The game-changing moment for EPR as a structural tool occurred when molecular biology methods were first used to readily introduce new cysteine residues at desired locations [1]. In this first SDSL study [1], four labeled cysteine residues were individually introduced at desired sites into the protein bacteriorhodopsin, and comparison of EPR relaxation enhancement measurements of the mutant proteins allowed for surface-exposed and membrane-embedded protein residues to be identified. Another early pivotal study was of 30 single-spin-labeled cysteine mutants of T4 lysozyme that showed that SDSL experiments could be interpreted in terms of the protein tertiary fold, equilibrium dynamics, and time-dependent conformational changes [8]. Also included in this study was the reporting of both structural and functional perturbation effects from labeling of the 30 sites in T4 lysozyme with the nitroxide spin label. The analysis of the results from this study led to the proposal that a nitroxide spin label could be introduced onto a protein surface with little to no change in protein structure, stability, or function. This approach for engineering labeling sites at suitable and desired locations to determine local secondary and tertiary structure continued to grow in popularity during the 1990s, with several notable studies on membrane proteins appearing in the literature, including the *Streptomyces* K+ channel [9, 10], bacteriorhodopsin [2], lactose permease [11, 12], and the Mscl mechanosensitive channel [13]. The focus of these early reports was often on nitroxide-scanning techniques to reveal structural folds in the absence of high-resolution structural data. Complementary short-range interspin distance measurements were also reported which provided insights into important functional events, including gating mechanisms [10, 13].

The demonstration that SDSL could also be used as a spectroscopic ruler to measure interspin distances in proteins, initially in the range of approximately 8–20Å using continuous wave (CW) EPR, was another defining moment for SDSL as a structural tool [14, 15]. Used with sequence-specific secondary and tertiary information, these short-range interspin distances provided additional spatial restraints to better define structural folds and monitor conformational changes. There are two excellent and highly recommended reviews by Hubbell and coworkers that examine the applications of the SDSL field up to this point in time [16, 17]. Lastly, the development of pulsed EPR methods and the commercialization of pulsed EPR spectrometers at the turn of the century further revolutionized EPR as a tool for structural biology by increasing the length of the molecular ruler for measuring interspin distances up to approximately 80Å [18–20].

12.1.2   SDSL–EPR: A Complementary Approach to Determine Structure–Function Relationships

Membrane proteins play key roles in cell-to-cell communication, for moving molecules across barriers, for transferring and using energy, and for triggering the initiation of many cell signaling pathways. Many membrane protein transporters and channels are also prone to disease-causing mutations (see Chapter 1), making them highly valued drug targets. However, despite the increasing pace of protein structure determination of membrane proteins by crystallography over the past decade, a mechanistic description for how membrane proteins work is often lacking. While a static crystal structure provides a vantage point from which to begin to envision dynamics that facilitate function, a complete understanding of dynamics can only really be achieved under environmental conditions or in the "biologically relevant milieu" in which the sample exists. SDSL–EPR is well positioned as a complementary technique to other high-resolution structural techniques to achieve such mechanistic descriptions under conditions close to the physiological state of the system under investigation.

NMR spectroscopy can also provide atomic resolution structure and information on protein dynamics (see Chapter 8). However, structure determination by NMR using conventional Nuclear Overhauser Effect (NOE)-based strategies is typically only feasible for small, globular proteins smaller than 25 kDa, thus ruling out the vast majority of membrane protein systems. Even if a membrane protein of interest is of a suitable target size, detergents must often be used in place of the native lipid environment. These are often a poor representation of the external environment under which the protein operates [21]. NMR also typically requires millimolar concentrations, as opposed to EPR, which requires sample concentrations only in the nanomolar to micromolar range. This often means there are difficulties in obtaining sufficient quantities of samples for NMR analysis, even if appropriate detergent conditions can be found. Furthermore, the NMR technique is also hindered by the requirement to isotopically label the protein, a costly method in itself that often has an unfavorable impact on protein yields. When a suitable sample can be prepared, NMR can of course provide dynamic information such as discussed in Chapter 8, but in the case of proteins, this is mostly residue specific. Recent advances have led to improvements in the protein dynamic information possible by NMR, but it can still remain challenging to delineate this information, particularly when measuring dynamics on the microsecond timescale [22].

X-ray crystallography, on the other hand, cannot, by its own nature, reveal conformational flexibility required for understanding functional mechanisms. However, if two or more different crystal forms are available, such as from the crystallization of a protein with various ligands bound, then suggestions as to a mechanism of action may begin to be inferred [23, 24]. Caution, however, must be exercised when proposing an underlying mechanism for function from several crystal forms as there are many examples where crystal contacts have been shown to distort highly flexible regions or functionally critical segments. A crystal structure may also be of a minor conformer, inadvertently stabilized from the crystallization process. These issues are aside from the fact that membrane proteins per se are inherently difficult to crystallize.

SDSL–EPR is often best used as a tool to complement NMR and crystal structure information, to provide the dynamic information needed for understanding mechanisms. As a probe-based spectroscopic technique, it is not restricted by the size, complexity, or environment of the protein system. The EPR method is also highly sensitive, which can allow for a range of protein–lipid–ligand or drug ratios to be tested, enabling a more complete characterization of binding of a substrate than is practical with many other biophysical methods. More importantly, it can be used to provide dynamic information on the picosecond to nanosecond timescale and observe the time course of structural transitions on the millisecond and longer timescales [25–27]. The nature, amplitude, and timescale of conformational equilibria dictated by function or the conformational changes accompanied by functional events can, therefore, be obtained under physiological conditions such as in the required native-like environment of the lipid bilayer needed to examine pumps, transporters, and ion channels.

In the following sections, we provide an overview for performing and interpreting EPR measurements on spin-labeled proteins. We focus on SDSL methods as appropriate for monitoring dynamics and conformational changes in membrane proteins. We have aimed to do this by providing recent examples that we feel best illustrate the various SDSL–EPR techniques for revealing the connection between structure and mechanism for membrane pumps, channels, and transporters. The examples include the use of SDSL–EPR strategies for studying gating mechanisms, substrate and drug binding, and overall dynamics and structural arrangement of membrane transporters and channels accompanying their function. We also wish to direct the reader to many other recent reviews on SDSL in membrane proteins and other protein systems [16, 28–32], as well as recent reviews on technological advances in methodology, instrumentation, and data analysis [33, 34]. We conclude by discussing some current challenges of SDSL for membrane proteins and the strategies under development to address these challenges.

12.2   BASICS OF THE EPR METHOD

12.2.1   Physical Basis of the EPR Signal

The physical basis for nearly all EPR applications is the anisotropy of the signal from an unpaired electron. Thus, EPR spectroscopy is selective only for paramagnetic species, which contain the unpaired electron, and not diamagnetic species which have paired electrons. In an applied external magnetic field (B_0), the spinning electron orbiting around a nucleus of the paramagnetic species behaves as a magnetic dipole by aligning itself either parallel or antiparallel with the magnetic field (see Fig. 12.2). This spin magnetic moment is quantized, with allowed spin values (m_s) of only ±1/2. The parallel state is the low energy state, with an m_s value of −1/2, and the antiparallel state has an m_s value of +1/2. This splitting between the two energy states due to the interaction of the spinning electron with the magnetic field is called the Zeeman effect (an analogous effect on nuclear spin is discussed in Chapter 8). The difference in energy (ΔE) between these two orientations increases in proportion to the strength of the applied magnetic field.

In the simplest imaginable EPR experiment, if the application of electromagnetic (EM) radiation matches the energy gap (ΔE), then the electron is excited or "flips" from the parallel dipole orientation to an antiparallel orientation. The corresponding resonance condition for this absorption of radiation is $\Delta E = h\nu$. However, for experimental reasons, the frequency (ν) of the electromagnetic radiation is held constant in an EPR experiment, and the magnetic field is swept linearly (see Fig. 12.2c). That is, it is the magnetic field dependence of the stationary absorption of energy from the continuous EM field that is measured in a conventional CW EPR experiment. The frequency range of the exciting field for a free electron is in the microwave region of the electromagnetic spectrum, not the radiowave region as in the case of NMR (see Chapter 8). The intensity of the resulting EPR signal is, therefore, dependent on the extent of microwave absorption and is proportional to the difference in the two spin populations (see Fig. 12.2c). To complicate the situation, it is the first derivative of the CW absorption spectrum that is typically reported, and this is what is referred to as the EPR spectrum (see Fig. 12.2d). Although largely done for historical reasons to correct for linearly field-dependent baseline shifts, reporting of the first-derivative spectrum has the advantage that it enhances resolution by emphasizing rapidly changing spectral features [35].

The shape of the derivative EPR spectrum provides all the information about the magnetic properties of the sample. Although it is determined by many factors, in general, the lineshape reflects the relative orientations between the unpaired electron in the sample and the applied external magnetic field. The contributing factors to the intensity and line width of the spectral lineshape are the frequency and amplitude of any molecular motions. In other words, the EPR spectrum can be used to determine the range of orientations of the electron in the sample, in this case, the nitroxide spin label.

Further peak splitting is often seen in an EPR spectrum as a result of spin coupling of the electron spin with the nuclear spins of nearby nuclei. This coupling can either enhance or counteract the external field, depending on the orientation of the nuclear dipole (see Fig. 12.2b). The number of peaks expected in the lineshape is given by $(2I + 1)$ where I is the spin of the nucleus carrying the unpaired electron. For example, the typical EPR spectrum arising from the unpaired electron of the nitroxide radical (N–O) in a spin label shows three resonant peaks as the unpaired electron also interacts with the local magnetic field from the nearby ^{14}N nucleus ($I = 1$) (see Fig. 12.2d). This interaction is known as the hyperfine interaction. Together, it is the Zeeman effect and hyperfine interactions that define the EPR sensitivity to orientation and rotational motion. For a more comprehensive description of the basic concepts of EPR spectroscopy, the reader is referred to [35–38].

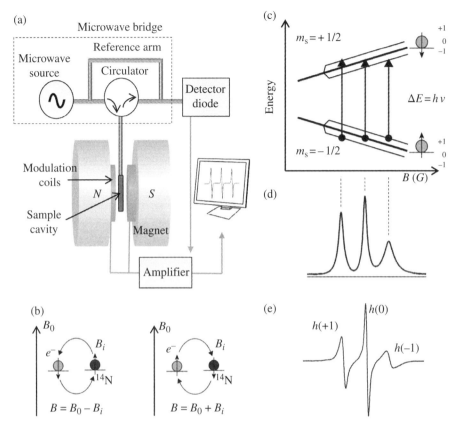

FIGURE 12.2 (a) General layout of a typical CW EPR spectrometer. (b) Hyperfine interactions between the unpaired electron (e^-) and the nitrogen (^{14}N) nuclear spin moment cause splitting of the electron spin energy levels that add or subtract from the external field (B_0). B_i is the local magnetic field at the electron due to the nearby nucleus. (c) The EPR experiment. A constant microwave radiation (ν) is applied to a sample and the magnetic field (B) is scanned. When the magnetic field for which the ΔE between the two electron states is equal to $h\nu$ is reached, there is absorption of energy by the spins, or resonance. Due to the Zeeman splitting and also the hyperfine interactions (as shown in b), there are three possible transitions indicated by the vertical arrows for a ^{14}N nitroxide spin label ($m_s = \frac{1}{2}$, $I = 1$). (d) Corresponding EPR absorption spectrum. (e) By convention, EPR spectra are presented as the first derivative of the absorbance spectrum.

12.2.2 Spin Labeling

Most proteins are diamagnetic and so do not typically contain unpaired electrons. They therefore require the introduction of a paramagnetic species, such as a spin label, to enable analysis by EPR spectroscopy. Spin labels are stable molecules that naturally contain unpaired electrons as well as a functional group for covalent attachment to the protein. It is the introduction of this spin label as an extrinsic probe into an otherwise EPR silent protein that constitutes the approach of "SDSL." The directed targeting of the protein with the spin label is typically achieved via muta-genesis, with the most commonly introduced target residue being cysteine. Prior to introducing any new cysteine residues, a functional "cys-less" construct as a background template is desired. Ideally, the target protein should not have any cys-teine residues that are essential for function and that cannot be replaced with either serine or alanine residues.

The most frequently used spin labels are nitroxide (N–O) derivatives containing a stable unpaired electron (see Fig. 12.3). The N–O group is usually enclosed in a five- or six-membered pyrrole or piperidine ring, respectively, which limits the flexibility of the N–O group. The unpaired electron of the N–O radical is localized in a highly anisotropic π bonding orbital. The radical can also be further stabilized by the presence of methyl groups on each of the neighboring carbon atoms.

In general, the nitroxide labels used for protein labeling can be grouped into two classes: thiol-reactive alkylating agents and thiol-specific labels. Both react rapidly with –SH groups, but iodoacetamide can also alkylate histidine and methionine resi-dues, depending on the pH during labeling. Likewise, maleimide labels can alkylate α-amino groups. Among the various spin labels commercially available, the most commonly used nitroxide label is 1-oxyl-2,2,5,5-tetramethylpyrroline-3-methaneth-iosulfonate (MTSSL or MTSL). This sulfhydryl spin label is covalently attached to the introduced cysteine via the formation of a disulfide bond, as shown in Figure 12.1a. The resulting spin-labeled side chain is commonly designated as "R1" and has a molecular volume similar to that of a tryptophan side chain.

Once the spin label is attached at a unique site, it then acts as a reporter molecule, with the electron sensitive to the freedom of movement of the label. However, most labels also possess an intrinsic flexibility due to rotation via several internal bonds that link the nitroxide radical group to the protein backbone [8]. For example, the "R1" MTSSL label shown in Figure 12.1 can rotate via four internal bonds. Although the high flexibility of the R1 label is sometimes viewed as an advantage as it is less likely to perturb the native protein fold, it also leads to the sampling of a large con-formational space during the time course of the experiment. This then leads to chal-lenges when interpreting structural and dynamic information from the measured EPR signals. This inherent flexibility then naturally complicates the interpretation of EPR features arising from the native side chain. Modeling methods can be used to define the conformational space sampled by the label, and these are briefly men-tioned later in Section 12.5.2. Other alternative labels are also available that have been shown to have reduced internal flexibility (see Section 12.5.1).

A further word of caution when spin labeling proteins: the disulfide bond attach-ment of the spin label to the cysteine can be readily reversed under mild reducing conditions. Therefore, exposure to reducing agents such as dithiothreitol (DTT) should be avoided at all costs. If reducing conditions are necessary for functional analysis, then the reducing agent tris(2-carboxyethyl)phosphine (TCEP) can be used as a viable alternative [43]. Additionally, spin labels that attach via a C–S bond can also be used, such as iodoacetamide (IASL) or maleimide (MSL), as shown in

Figure 12.3. However, these labels are notably less specific in their reactivity and more sterically demanding than MTSSL.

The methods described previously for the targeted labeling of cysteines are applicable to any cloned gene that can be expressed. Although cysteine residues are the most commonly targeted residue for labeling, other residues can be targeted.

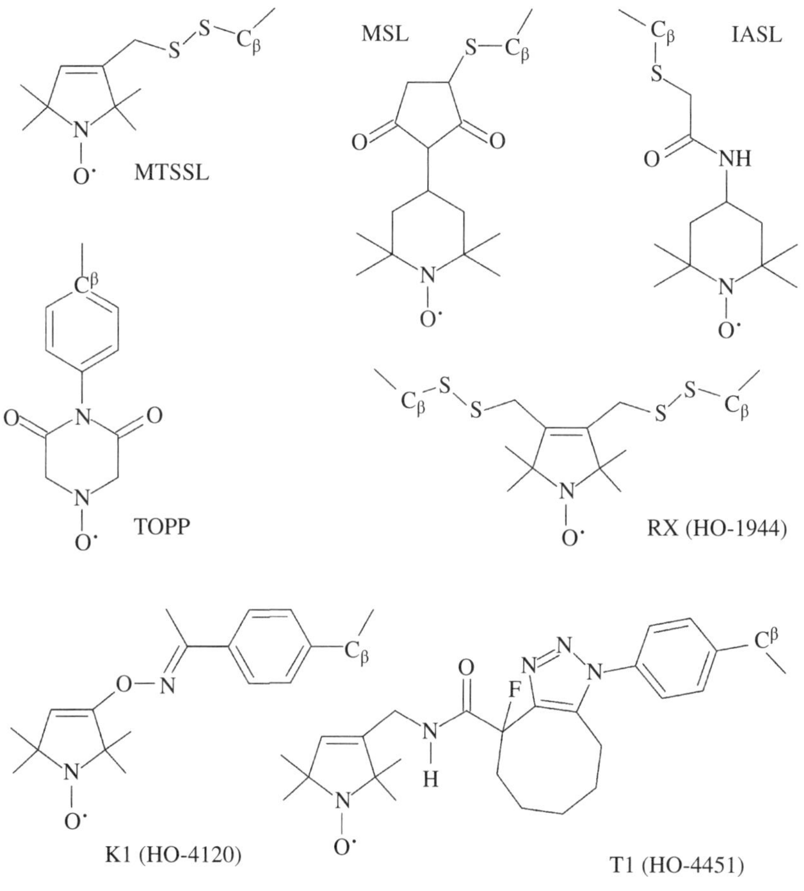

FIGURE 12.3 Structures of some common nitroxide spin labels represented as protein side chains. The stable unpaired electron is localized almost entirely on the π-orbitals of the nitroxide bond. "R1" side chains include MTSSL (1-oxyl-2,2,5,5-tetramethylpyrroline-3-methanethiosulfonate), where the nitroxide group is enclosed in a pyrrole ring that contains a double bond to limit the flexibility of the label and the adjacent methyl groups stabilize the radical through sterically hindering attack, MSL (*N*-1-oxyl-2,2,6,6-tetramethyl-4-piperidinyl-maleimide), and IASL (*N*-1-oxyl-2,2,6,6-tetramethyl-4-piperidinyl-2-iodoacetamide). TOPP is introduced by direct synthesis methods [39]. The homo-bifunctional reagent HO-1944 reacts with two cysteines to result in cross-linked side chains designated "RX" [40]. The "K1" ketoxime-linked side chain is generated by reaction of HO-4120 with the *p*-acetyl-phenylalanine unnatural amino acid [41]; and the "T4" triazole-linked side chain is generated by reaction of HO-4451 with the *p*-acetyl-phenylalanine unnatural amino acid [42].

Alternative strategies for introducing paramagnetic labels can also be explored. These alternative labeling approaches are discussed further in Section 12.5.1.

12.2.3 EPR Instrumentation

An EPR spectrometer is generally similar in design to an NMR spectrometer (see Chapter 8). The main differences are as follows: (i) the sample in an EPR spectrometer is exposed to microwave radiation and not radiowave radiation; (ii) the EPR spectrum is obtained by sweeping the magnetic field while the frequency is held constant; and (iii) it is the derivative of the EPR absorption spectrum as a function of frequency that is produced.

Although collecting and interpreting EPR spectra are still not as routine as other analytical tools, the commercial instrumentation and software development that have occurred over the past few decades have clearly aided this technique in becoming a more mainstream structural tool. Commercial EPR spectrometers are commonly available at several microwave frequencies with the majority of research groups having access to an X-band spectrometer operating at approximately 9.5 GHz. There are other commercial spectrometers available that work at lower (L-band 1.5 GHz, S-band 3.0 GHz) or higher (Q-band 36 GHz, W-band 95 GHz) frequencies.

The design of an EPR spectrometer is optimized for the measurement of the weak magnetic resonance signal. The overall layout of an EPR spectrometer has three basic components: (i) a microwave bridge and resonator, (ii) a variable field magnet, and (iii) the signal amplification circuitry (see Fig. 12.2a). The electromagnetic radiation source and the detector are located within the microwave bridge. Microwaves of desired frequency are generated from a source, either a klystron or Gunn diode. These are then directed through an attenuator to the sample, located in the microwave resonator, via a waveguide. Depending on sample availability and experimental needs, the resonator can be a standard rectangular or cylindrical cavity, split ring, or dielectric resonator. The microwave resonator enhances sensitivity by setting up a standing wave, effectively concentrating the microwave field at the sample. A circulator separates the incident and reflected microwaves, directing the incident microwaves to the resonator and the microwaves reflected back from the cavity to the detector diode. The reference arm is an additional pathway that allows for balanced detection, enhancing overall sensitivity of the system. The magnet tunes the electronic energy levels.

12.3   STRUCTURAL AND DYNAMIC INFORMATION FROM SDSL–EPR

12.3.1   Mobility Measurements

The shape of the EPR spectrum is a direct reflection of the motional rate, amplitude, anisotropy, and overall reorientational motion of the spin-labeled side chain. As such, analysis of the EPR lineshape, although complicated, provides information about various motions or steric constraints imposed on the spin label by its immediate surroundings;

including backbone fluctuations, local structure, and conformational changes. However, it must also be remembered that a spin label itself is a very flexible object, especially if the label is located on a surface-exposed region. In such a situation, it is often then the inherent spin label motion that becomes the dominant contributing feature in the EPR lineshape.

The mobility of a spin-labeled nitroxide side chain can be characterized by three correlation times: (i) the rotational correlation time due to the tumbling of the entire protein, (ii) the effective correlation time due to the rotational isomerization about the bonds linking the nitroxide to the backbone of the protein, and (iii) the effective correlation time for the segmental motion of the backbone with respect to the protein, that is, backbone fluctuation. In the case of a membrane-bound protein, the rotational correlation time due to the global tumbling of the protein can typically be ignored because the correlation time would exceed approximately 60 ns, which is beyond the sensitivity of the EPR timescale. As a comparison, for a soluble protein, such a slow correlation time would be expected for a protein with a molecular weight greater than 200 kDa, based on the Stokes–Einstein–Debye equation. The major reason for the long correlation times of membrane proteins is the high viscosity of their lipid surroundings. The effective correlation time due to rotation around the bond angles within the label is a complex function of the structure of the label, as well as the primary, secondary, tertiary, and quaternary structure of the protein. Lastly, the effective correlation time due to backbone flexibility arises from the local secondary structure of the attached site.

In general, sites labeled on loop regions or on helix surface sites result in a high degree of mobility with a small spectral line width and hyperfine splitting. Restriction on the motion of the spin label, such as from a strong interaction of the nitroxide with neighboring side chains or backbone atoms, as found for tertiary contact sites where the label makes contact with nearby secondary structures or buried sites, leads to increases in both the line width and hyperfine splitting. The motion due to the interaction of the nitroxide with neighboring atoms is, therefore, anisotropic and often shows more than one component, for example, as shown in the spectra given in Figures 12.4 and 12.5. In these cases, complicated lineshape analysis methods are needed if a description of the distribution of all motional states is desired (see Section 12.5.2).

Despite the complex challenge faced in linking lineshape back to its molecular origins, semiempirical approaches can be successfully used to identify the structure of the immediate environment of the label from its dynamic motion. These include measuring spectral parameters such as the inverse line width of the central resonance (ΔH_0^{-1}) or the inverse second moment $\langle H^2 \rangle^{-1}$. Both of these semiempirical mobility values increase in magnitude when the reorientational motion of the label also increases [8]. Originally calibrated using T4 lysozyme as a model system, the value of these two mobility parameters allows sites to be classified into three categories: exposed, buried, or tertiary contact [8]. Furthermore, plotting either mobility parameter as a function of residue number can identify secondary structure elements [9]. That is, cysteine-scanning experiments can be used to reveal a periodic variation in the mobility values if the spin label sequentially samples a surface exposed then a buried/tertiary contact site [46]. For example, if the simple visual inspection of mobility data reveals a periodicity value of 3.6 or 2.0, then the straightforward assignment of an α-helix or β-strand, respectively, can be made. If there is a lack of periodicity in the mobility parameter upon scanning, then this may suggest either the presence of random structure or a lack of an interacting tertiary contact region or buried surface. A more quantitative analysis of the magnitude and significance of such periodic behavior can also be established through frequency analysis in Fourier space [9, 47].

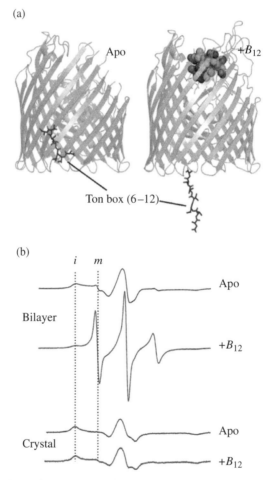

FIGURE 12.4 The conformational equilibrium in the outer membrane *E. coli* vitamin B_{12} transporter (BtuB) is modified by the crystal environment. (a) Upon binding of vitamin B_{12} (spacefill) to BtuB (+B_{12}), an unfolding event expels the N-terminal Ton box into the periplasmic space. (b) An increase in mobility for a spin label placed on the Ton box (residue 12) is observed upon binding of vitamin B_{12} but only in the context of the lipid bilayer. No change is observed in EPR mobility in the crystal form. The dashed lines indicate the positions of the EPR signal from immobilized (*i*) and mobile (*m*) components of the nitroxide side chain. Reprinted from Ref. 44 with permission from Elsevier.

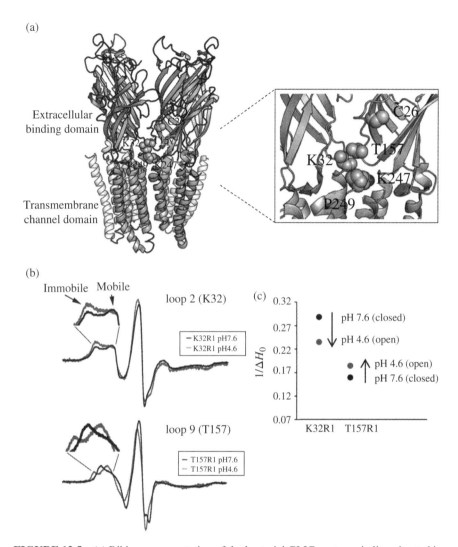

FIGURE 12.5 (a) Ribbon representation of the bacterial GLIC pentameric ligand-gated ion channel crystal structure with spin-labeled positions highlighted (zoomed view of the interface between the ECD and TMD domains). Labeled GLIC was reconstituted in liposomes and CW EPR mobility measurements performed. (b) Representative CW EPR spectra for 2 of the 5 spin-labeled mutants obtained at pH 7.6 (closed state) and pH 4.6 (desensitized open state). The immobile and mobile components of the K32R1 (loop 2) and T157R1 (loop 9) are highlighted, showing the proton-induced gating motions. (c) A plot of the inverse width of the central line (ΔH_0^{-1}) of the CW spectra shows the proton-induced decrease in mobility for loop 2 (K32R1) and increase for loop 9 (T157R1) going from a closed (pH 7.6) to open (pH 4.6) state. Adapted from Ref. 45 in accordance with the terms of Creative Commons Attribution (CC BY) license.

There are many notable examples in the literature where these simple qualitative parameters of EPR mobility have been shown to correlate with secondary structure or reveal information on local protein dynamics. One of the first membrane proteins examined by EPR mobility was rhodopsin, a G-protein-coupled receptor in vertebrate retina rod cells. Extensive cysteine scanning combined with mobility experiments was pivotal in defining the membrane boundary and tertiary contact sites [48–52], as reviewed in Ref. 17. More remarkable were the measured changes in spin label mobility upon photoactivation of rhodopsin that allowed the location, direction, and type of movements accompanying function to be described [53]. The rigid-body model describing the structural changes associated with photoactivation and G-protein binding, as proposed by EPR, was only much later confirmed by crystallography [54]. Such confirmation by a high-resolution crystal structure was a long time coming, as comparisons made with the EPR results to several earlier crystal structures obtained for both the dark- and light-activated states did not show the movements suggested by EPR [55]. This was likely a result of crystallization conditions suppressing the conformational flexibility of rhodopsin.

Another example that also provides a compelling rationale for using EPR mobility measurements to bridge the divide between static crystal structures and functional mechanisms is the study of BtuB, the outer membrane *Escherichia coli* vitamin B12 transporter (see Fig. 12.4). The structure of this transporter protein and how the substrate translocation occurs have been the subject of numerous SDSL studies over the past decade. EPR mobility spectra for a spin label positioned within the N-terminal energy-coupling motif (Ton box) of BtuB indicated that this segment of the protein undergoes a substrate-dependent conformational transition that shifts the equilibrium toward an unfolded or disordered form in lipid bilayers [56]. This substrate-dependent unfolding extends the N-terminal region of the Ton box into the periplasmic space, an observation that was also supported by EPR distance measurements [57]. However, the transition from folded to unfolded states upon substrate binding was not observed in later crystal structures. To address this controversy, researchers examined the mobility of a spin label attached to the Ton box in the presence of various crystallization solutes [44, 58]. Their study showed that the unfolding observed in the lipid environment was blocked in the crystal state and only through EPR mobility measurements could the conformational equilibria required for providing a mechanistic understanding of BtuB transport be quantified.

A more recent example of EPR mobility measurements performed on a membrane protein in its native lipid environment, which have revealed molecular mechanisms underlying channel gating, is the study of a member of the pentameric ligand-gated ion channel (pLGIC) family [45]. The pLGICs are neurotransmitter-activated receptors found in nerve and muscle cells and are involved in chemical signaling in the brain and periphery [59]. As the name implies, they consist of five homologous subunits arranged around a central ion-conducting pore structure (see Fig. 12.5a). Each subunit contains an extracellular binding domain (ECD), a β-sandwich core domain, and a transmembrane channel domain (TMD). Members of this channel family can interconvert between at least three different states: (i) the "activated" or open channel with a ligand bound, (ii) the "desensitized" or closed channel with ligand bound, and

(iii) the "resting" closed channel with ligand unbound. It is the binding of the ligand, such as a neurotransmitter, to the ECD region that causes the channel to be rapidly activated, thereby opening the ion-conducting pore. The channel then moves into the closed desensitized state following prolonged exposure to the ligand.

The current structural knowledge of the pLGIC protein family and their gating mechanism comes primarily from cryo-EM structures of liganded open and liganded closed states. Several crystal structures are also available, including the ECD domain of pLGIC and crystal structures of two whole-length bacterial homologues of pLGIC (*Erwinia chrysanthemi* ligand-gated ion channel (ELIC) and *Gloeobacter violaceus* ligand-gated ion channel (GLIC)) [60–64]. However, despite this wealth of high-resolution structural information, the mechanism by which the channel structures rearrange the equilibria between the three states is still largely unclear. Models developed from the crystal structures suggest that the opening and closing of the pore involve a twisting of the ECD and M2 α-helices lining the TMD away from the channel axis. This channel gating mechanism is suggested by the crystal structures to be coupled to ligand binding involving loops 2, 7, and 9 in the ECD and the M2–M3 loop of the TMD domain. However, it is unclear as to whether the crystal structures solved in detergent micelles accurately reflect the proposed motions of the functional channel in its native lipid environment.

EPR mobility measurements were therefore performed on the GLIC bacterial homologue as part of a larger SDSL–EPR study to determine the structural rearrangement between the resting and desensitized states of the channel [45]. Fortuitously, only a single native cysteine residue was initially required to be mutated to an alanine residue, and so individual cysteines could be readily introduced in the ECD and TMD interface for spin labeling. The changes in mobility of each labeled residue in response to proton-induced gating were measured for a fully functional reconstituted liposome system. As an example (as shown in Fig. 12.5b), the CW spectra of the spin label located in loop 2 (K32R1) showed a higher proportion of the spin labels to be mobile under closed resting state conditions (pH 7.6), whereas the spin labels located on loop 9 (T157R1) were predominantly immobile. Conversely, upon lowering the pH to 4.6 to achieve the desensitized open state of the channel, a greater proportion of the spin labels were immobile for loop 2 and mobile for loop 9. Using this mobility data and changes in interspin distances between the loops, the authors established that the surrounding loop 9 of the ECD becomes more densely packed, whereas loop 2 becomes less packed on moving from the closed resting state (pH 7.6) to the desensitized open state of the channel (pH 4.6) (see Fig. 12.5c). These results did not correlate well with observations from the crystal structures obtained in detergent micelles, demonstrating once again that EPR mobility measurements are a capable and reliable approach for testing mechanistic models of large membrane protein systems in their native-like lipid environment.

12.3.2 Solvent Accessibility

Collision of nitroxide-based spin labels with fast relaxing paramagnetic species, such as oxygen and metal ion complexes, leads to shortening of the spin lattice relaxation time (T1) of the nitroxide and broadening of the EPR signal [65].

Loosely described as "quenching" when the paramagnetic species is dissolved within the solvent, the collision frequency is directly proportional to the accessibility of the paramagnetic species to the nitroxide. Used in parallel with cysteine scanning, the sequence dependence of solvent accessibility of the attached nitroxide to paramagnetic species selectively partitioned in different environments can also, like mobility measurements, be used to define local secondary and tertiary structure and identify sites of conformational change. Revealing periodicity due to relaxation effects is best achieved if there is an unevenly solvated environment, such as a membrane surface, or if the scanned region is interacting with another polypeptide chain.

The paramagnetic species that can be used as relaxing agents include molecular oxygen, organic radicals, and metal ion complexes. In the case of membrane-embedded proteins, the concentration gradient of nonpolar oxygen increases with the distance from the surface of the membrane, peaking at the center of the bilayer. The opposite is found for polar metal ion complexes such as NiAA (nickel(II) acetylacetonate), NiEdda (nickel(II) ethylenediaminediacetate), or chromium oxalate CrOx (potassium tris(oxalatochromate)). These metal complexes all preferentially partition in the aqueous phase. The CW power saturation technique can be used to obtain the EPR accessibility parameter (Π) in the presence of the different relaxants. To achieve this, the line height of the CW EPR signal is monitored as a function of the incident microwave power at a specific relaxant concentration. The generated saturation curve provides a $P_{1/2}$ parameter that is related to the relaxation rate of the nitroxide. The difference in $P_{1/2}$ values ($\Delta P_{1/2}$) obtained in the presence and absence of the relaxing agent is, in general, proportional to the topological environment of the attached spin label. The $\Delta P_{1/2}$ for the polar agent shows minima and maxima for residues interacting with the lipid bilayer and surface-exposed residues, respectively. The nonpolar agent shows an analogous but opposite behavior. That is, residues that interact with the membrane exhibit maximum relaxation. A similar pattern is also observed for interaction of nonpolar paramagnetic agents with residues lining an aqueous channel pore.

Although a relatively straightforward EPR measurement to perform, accessibility measurements require specialized, highly gas permeable, capillary sample tubes fabricated from TPX® (polymethylpentene) [66]. A deoxygenated saturation curve is first required, and this is obtained by flowing nitrogen gas around the TPX capillary containing the sample. Air (21% oxygen) or 100% O_2 is then used to replace the nitrogen for the measurements of oxygen accessibility to the lipid phase. The neutral metal complex NiEdda at a concentration of 10 mM is usually sufficient to measure the accessibility to the water phase. Consideration must also be given when using CrOx, which is negatively charged, especially in the presence of charged membrane surfaces [67]. Used together, the ratio Φ calculated for the $\Delta P_{1/2}$ values obtained in both the polar and nonpolar relaxants can be used to measure the immersion depth of the nitroxide in the bilayer [68].

There are many examples where such accessibility measurements have been used to identify residues lining the pore of channel proteins. These include studies on annexin [69], diphtheria toxin [70], colicin [71], lactose permease [11, 12, 72, 73],

MscL [74], prokaryotic voltage-dependent K$^+$ channel [75], as well as the early impressive SDSL studies on the *Streptomyces* K$^+$ channel [10]. The "cysteine scanning" of more than 60 amino acids of transmembrane segments 1 and 2 of the KcsA K$^+$ channel from *Streptococcus lividans* set a precedent for how powerful accessibility measurements can be for providing a mechanistic understanding in the absence of complete structural data for a full-length channel in its native environment [9, 10]. The KcsA channel was also an ideal system to probe by this approach due to the lack of native cysteines, milligram expression levels in *E. coli*, a relatively straightforward reconstitution procedure to obtain functional protein, and a simple change required in pH to trigger the open channel state. The accessibility profiles, together with mobility and interspin distances, revealed the membrane topology and helical packing of the channel tetramer during the acid-induced opening of the inner channel gate [76, 77].

More recently, Perozo and coworkers used solvent accessibility measurements to investigate the structure of the voltage-sensing domain (VSD) of KvAP, a member of the voltage-gated ion channel protein family [78]. Voltage-gated ion channels are important as they are involved in a variety of biological signaling processes, including hormone secretion, osmotic balance, and electrical excit-ability. The quaternary structure of the voltage-gated ion channels is that of a functional tetramer. Within each tetramer, there are two helices (S5 and S6) that form the central pore, and another four helices (S1–S4) that form the four sur-rounding VSD. There is a highly conserved set of positively charged residues within the S4 helix that controls the gating of the pore. Opening of the pore is driven by the movement of gating charges from the resting (or "down") conforma-tion to an active (or "up") conformation state. A number of crystal structures have now been solved for several voltage-gated ion channels, with most believed to rep-resent the VSD in the "up" position [79–82]. Crystallization of the VSD in the resting "down" position has so far proved elusive due to difficulties with stabi-lizing this state for crystallography.

Previous electrophysiological experiments showed that interactions between the lipid environment and voltage-gated channels play a key role in the gating transition [83, 84]. Therefore, to gain an insight into the gating transition, Perozo and coworkers performed SDSL–EPR accessibility experiments on the reconstituted KvAP potassium channel under the influence of different lipid compositions. Spin-labeling sites were designed to distinguish between existing models for how gating is expected to occur. The gating models under investigation included the paddle model [80, 85] and the helix-screw model [86]. The authors performed the impressive task of "nitroxide scanning" throughout the entire VSD region (S1–S4) by creating 132 individual single-cysteine mutants (see Fig. 12.6a). EPR mobility (ΔH^{-1}) and acces-sibility experiments using NiEdda (ΠNi) and oxygen (ΠO_2) were performed on protein reconstituted into either nonphosphate (1,2-dioleoyl-3-trimethylammonium-propane; DOTAP) or phosphate lipids (3:1 1-palmitoyl-2-oleoyl-sn-glycero-3-phosphocholine/1-palmitoyl-2-oleoyl-sn-glycero-3-phospho-(1′-rac-glycerol); POPC/POPG), which trap the "down" and "up" states of the channel, respectively. The differences in the lipid-driven accessibility profiles for the important S4 gating

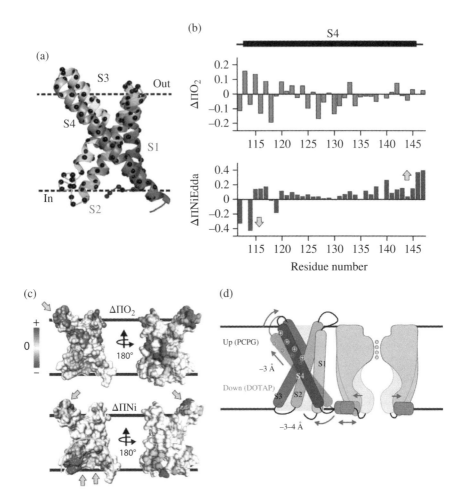

FIGURE 12.6 (a) Cartoon representation of the voltage-sensing domain (VSD) of the KvAP voltage-gated ion channel showing helices S1 to S4 and the positions of the 132 spin-labeled mutants. (b) Differences between the accessibility measurements (ΠO_2 and $\Pi NiEdda$) obtained for the S4 region of KvAP VSD reconstituted in either phosphate lipids (3:1 PC:PG or activated "up" state) or DOTAP (resting "down" state). The up and down arrows highlight the notable changes (increase and decrease, respectively) in the $\Pi NiEdda$ values for each lipid-dependent conformation (DOTAP vs. PCPG) throughout the S4 region, showing that the S4 region is important for controlling the gating reorientation in KvAP. (c) These differences are also mapped onto the crystal structure and show the opening of an intracellular water-filled crevice. (d) The "tilt–shift" gating model for KvAP where movement in the S4–S5 region, detected by accessibility measurements, is sufficient to open the pore of the channel. Reprinted from Ref. 78 with permission from Macmillan Publishers Ltd.

helix are shown in Figure 12.6b, with the differences also mapped onto the crystal structure of the KvAP VSD (Fig. 12.6c). Together with mobility and interspin distance information, the EPR accessibility data was suggestive of the S4 helix undergoing both a "tilting" and downward "shifting" motion between the different lipid environments. The authors proposed a new model called the "tilt–shift" model to account for the conformational changes observed by EPR (see Fig. 12.6d). However, it is not known if the proposed "tilt–shift" lipid-driven gating mechanism is unique for KvAP.

Each of the studies discussed previously used nitroxide scanning through extensive regions or domains to provide a description of the motion of the membrane protein in response to an environmental functional trigger. Mobility and accessibility measurements from a single labeled site are indeed valid and can often reveal the nature or magnitude of underlying structural changes connected with function. However, such interpretation of the EPR data from a single labeled site must be performed with caution, especially if a quantitative description is desired. It is likely that the protein movements that accompany the functional response of the membrane protein themselves change the local microenvironment of the targeted residue, thus altering the mobility and/or accessibility of the spin label.

12.4 DISTANCE MEASUREMENTS

12.4.1 Interspin Distance Measurements

The insertion of two cysteine residues into the protein backbone and the subsequent labeling with spin labels enables the technique of EPR to be used as a molecular spectroscopic ruler. This application was initially heralded to have the greatest potential for SDSL in structural biology as it could be used to determine structure in the absence of an NMR or X-ray crystal structure. Indeed, there are many notable examples in the early literature that show that the careful design of only a few interspin distance measurements could be used to rapidly establish structural folds in the absence of high-resolution data, such as the arrangement of α-helical or β-strand elements with respect to each other [87–89]. Other early studies also showed that interspin distances could reveal the geometry or assembly of subunits in multicomponent systems [90, 91], such as the arrangement of protein subunits in membranes [92, 93]. However, protein motions can also be directly demonstrated by changes in distances measured by EPR. In the following sections, we focus on examples where changes in interspin distances have helped develop a mechanistic understanding of channel or transport proteins in their native membrane environments.

The interspin interaction that is measured between two spin-labeled sites can be considered to comprise of (i) the exchange interaction (J), (ii) the static dipolar interaction, and (iii) the modulation of the dipolar interaction by the residual motion of the side chains of the spin label. The physical basis for the first of these interactions between coupled spins, the exchange interaction, arises from the overlap of the orbitals of unpaired electrons. In the case of the dipolar interaction, the interspin effect is a result of the magnetic moments between the two spin labels. In general, there are three EPR measurement techniques that can be used to detect these interspin interactions. The first is exchange EPR, which is used when the distance between the two labels is between 4 and 8Å and the exchange interaction is dominant [94]. The second is dipolar CW EPR, which is applicable for distances less than 25Å and typically greater than 8Å [15, 95]. Both of these EPR methods for measuring short distances are limited to strong interspin interactions and rely on detectable lineshape broadening effects in the CW EPR spectrum. The final and more advanced technique for detecting interspin interactions is the pulsed EPR method. The potential of EPR for measuring distances in labeled biomolecules significantly increased when EPR pulse protocols were developed and realized experimentally. These developments extended the interspin distance range that could be detected up to approximately 80Å. Two widely used pulsed EPR methods that each extract the weaker dipolar interactions from the spin coherence are the double electron–electron resonance (DEER) [96] and double-quantum coherence (DQC) methods [18].

Although measuring changes in interspin distances has now undoubtedly become the most popular of all the SDSL–EPR applications, the analysis of the EPR spectra to obtain the interspin distances is not always straightforward. This is because the molecular flexibility of the spin label can result in a distribution of distances. That is, the distance between spin labels is most likely not fixed at a specific value, and so it is a change in an average distance and/or the width of a distance distribution that is reported. The most desirable situation is when there is no change in the set of spin-label rotamers when comparing distances measured for two or more structural states. This is not often the case, as the protein motion under investigation may itself lead to the repacking of the label at one or both sites. Often, the best strategy to avoid this situation is to choose two label sites that are located in unconstrained positions, such as surface-exposed sites. Measurement of the mobility of the spin label at room temperature for each site can also be performed to confirm that there is no repacking of the spin label at either site due to conformational changes.

It is also worth mentioning here that the other most common analogous approach for measuring distances in biomolecules is via the measurement of fluorescence (or Förster) resonance energy transfer (FRET) efficiency. Although both are low-resolution techniques, the main advantages of using EPR over FRET are the smaller size of the spin labels relative to most fluorescent labels and the requirement for only a single type of reporter label for interspin distances, negating the need to devise complex labeling strategies. The drawback for both techniques comes when interpreting distance information in terms of molecular structure. Both the length and conformation of the spin or fluorescent label must here be taken into account during data analysis. Although the uncertainty in the precision of the distances is frequently stated to be in the order of 1–3Å, this will of course depend greatly on the method and system under investigation. The greatest uncertainty still lies in the interpretation of the distribution of distances due to label flexibility, with current efforts directed at developing better modeling tools to describe the label behavior (see Section 12.5.2).

In the next two sections, we further discuss the methodology of both CW and pulsed approaches for measuring interspin distances and give examples where each have been used to reveal structure–function relationships for a membrane protein system. Although the measurements using pulsed sequences for obtaining long interspin distances is still somewhat specialized, these methods are rapidly becoming more routine due to the commercialization of high-sensitivity pulsed

EPR spectrometers over the past decade. In contrast to pulsed EPR, the CW EPR spectrometer is still the most prevalent EPR spectrometer with simple operational requirements for the novice interested in this methodology. Even though CW methods are clearly not on par with pulsed methods for revealing interspin distances and describing distributions of distances, with careful and strategic design of EPR probe pairs, a broad range of structural questions can still be addressed using simple CW EPR methods.

12.4.2   Continuous Wave

Although there is still no theory that can fully describe the changes in the CW EPR lineshape resulting from interacting spins, the presence of line width broadening is always suggestive of two labels being within 20Å of each other. Several methods, ranging from simple semiquantitative methods to full lineshape simulation, can be used to extract the interspin distance information from such lineshape broadening, and these are described later and reviewed in more detail elsewhere [97].

Firstly, to measure the interspin dipolar coupling between two labeled sites by CW EPR from a practical viewpoint requires the preparation of a spin-labeled mutant at site i, a spin-labeled mutant at site j, and a doubly spin-labeled mutant at sites i and j. The degree of dipolar coupling is then obtained by comparing the CW EPR spectrum of the doubly labeled $(i+j)$ mutant to the sum of the spectra of the two singly labeled mutants. However, if working with homo-oligomeric protein complexes, as is often the case for membrane proteins, the dipolar coupling between the same residue on different monomers can be used when the goal is to determine the assembly of sub-units in the complex. This can be achieved by comparing the spectrum of the fully labeled protein with a mixture of labeled protein with either unlabeled or wild-type protein in excess. That is, the single-labeled reference sample is represented by a "spin-diluted" oligomer. The symmetry of the interface can then be determined from the number of interacting spins [69].

The line width broadening resulting from interacting spins is due to the large electron magnetic dipole and is a steep function of the interspin distances (r). Semiquantitative methods, therefore, often offer the best strategy to rapidly address simple structural questions. These methods include the relative intensity of the half-field transition [98, 99] and the ratio of peak heights ("d_1/d" ratio) [100, 101]. The relative intensity of the half-field transitions is best suited for very short interspin distances of around 4–12Å with a 20% error in the ratio at a length of 7–8Å, resulting in a reported error of only approximately 0.2Å. As the intensity of the half-field transition decreases proportional to r^6, distances longer than this become difficult to measure. This r^6 distance dependence also means that if a distribution of conformations is present, the average value measured will be strongly biased toward shorter values. Furthermore, this analysis requires that motion is slow relative to the anisotropy of the half-field transition and so carried out on samples in a glass phase at cryogenic temperatures.

The d_1/d peak height ratio of the center and outer lines (labeled as $h(+1)$ and $h(-1)$ in Fig. 12.2e) can be used to provide an estimate of the interspin distance if broadening of the double-labeled spectrum is evident [100, 101]. A d_1/d value of less than 0.4 is expected for single-spin-labeled proteins or when the interspin distance is greater than 20Å. The value of this ratio increases as the strength of the interspin distance also increases. For example, a value of 0.55–0.60 typically represents a distance of between approximately 7 and 8Å. As distance information is extracted directly from the d_1/d ratio, this method is particularly sensitive to interference from singly labeled proteins as the sharper lines from the noninteracting label will dominate the peak height ratios [102]. The lower limit of this approach, as well as of the half-field transition approach, is also dictated by the increasing influence of the exchange interactions. This is due to the partial overlap of the nitrogen π-orbitals of each spin label if in close proximity (<8Å). Therefore, analysis of the extent of broadening from either method is only really beneficial if one is looking for qualitative agreement for a distance obtained by another method, or when wishing to reveal secondary structural elements by observing a "pattern of proximities" for a set of double mutant pairs in nearby residues. An example of an appropriate set that can be used to distinguish local structure is $(i, i+2)$, $(i, i+3)$, and $(i, i+4)$; where i represents the first labeled site.

The Fourier deconvolution analysis method is a superior quantitative approach that can be used to estimate distances between two spin labels that adopt a static distribution of distances and orientations relative to each other [15]. Suitable for an interspin separation of between 8 and approximately 20Å, a broadening function is obtained by deconvoluting the broadened spectrum of the doubly labeled sample with the spectrum of a corresponding singly labeled or noninteracting spin sample. A Gaussian curve is then fitted to the broadening function, and that fit function is reverse Fourier transformed to provide the function from which the interspin distance is calculated [15]. An advantage of this method is that the broadening function can identify and subtract the amount of any singly labeled protein that may be contaminating the sample. Unlike half-field transition measurements, deconvolution measurements do not need to be done in the glass phase but still require conditions where motion is extensively restricted, such as by the addition of approximately 50% glycerol or 40% sucrose and cooling to −30°C. The Fourier deconvolution approach is, therefore, based on the assumption that there is a random distribution of relative orientations of the spin-label axis relative to the interspin vector. In most cases, this appears to be a sound assumption to make, as confirmed by crystal structures of several spin-labeled constructs [103].

Lastly, computer simulation of lineshape changes due to broadening effects can be performed. In principle, these simulation methods aim to define both the interspin distance and the orientation of the interspin vector relative to the magnetic axes of both spin labels, as well as separating the dipolar and exchange interactions. These methods employ automated search routines to determine best-fit parameters and have been shown to be effective for interspin distances in the range of 6–20Å [102, 104]. However, to do this is a complex process as the complete spin Hamiltonian includes an incredible number of terms, especially if expressed in its full vector form [105].

Returning to a simpler level, it is important to remember that the main observation arising from spin–spin interactions is an increase in the spectral line width. The relative magnitude of this interaction can, therefore, be quantified by empirical parameters, such as the relative intensities of spectra normalized to the same number of spins. Accordingly, Hubbell and coworkers proposed an analysis structure whereby interspin distances can be classified into three broad categories: close (<10Å), intermediate (10–15Å), or far apart (>15Å) [46]. Another alternative simplified analysis method of note is the measurement of a broadening parameter termed the "spin–spin interaction parameter" (Ω). The Ω parameter is defined as the ratio of the amplitude of the central resonance line ($h(0)$) of the normalized singly and doubly labeled spectra [9]. For distances greater than 20Å, the Ω value is approximately equal to one. For spin labels less than 13Å apart, the spectrum is significantly broadened such that the central line amplitude becomes negligible compared to the non-broadened amplitude of the singly labeled spectrum. The Ω value, although a poor estimate of distances in small proteins (<20kDa), can be used successfully to detect changes in spatial proximity accompanying protein function, such as channel gating [9, 10, 74] or, as in the example described next, ligand-binding events.

In a recent study, changes in values for the interaction parameter (Ω) were used to examine drug binding to the influenza A M2 protein channel [106]. Examining conformational changes due to drug binding in the pore region of the M2 protein is central for understanding drug disruption of the life cycle of influenza A [107]. Although the channel activity of this model ion channel can be inhibited by amino-adamantyl drugs, over 90% of circulating influenza A strains have now developed resistance [108]. This emerging resistance, as well as the ongoing threat of future pandemics, makes development of new antiviral drugs a priority.

In Figure 12.7, the CW EPR spectra are shown for several examples of M2 constructs labeled with a nitroxide spin label positioned in the transmembrane pore region [106]. In the drug-free form, the values of the interaction parameter Ω can be seen to decrease as the site of the label is positioned away from the middle of the bilayer (residue 36) toward the C-terminal end of the transmembrane domain (residue 43). This widening of the channel at the C-terminal end is also seen in several high-resolution X-ray crystal structures of the M2 channel [109]. Significant increases in the value of the interaction parameter Ω for all labeled positions were also reported upon binding of an adamantyl drug [106], indicating a narrower transmembrane region tetrameric bundle. These well-designed CW interspin distance measurements, performed in parallel with accessibility and mobility measurements, demonstrate the power of the SDSL toolbox to begin to fully and quantitatively understand the structural effects of drug binding on physiologically relevant lipid reconstituted samples.

12.4.3   Pulsed Methods: DEER

Pulsed EPR for measuring interspin distances has very much come of age over the last decade. Whereas CW methods depend upon the observation of significant broadening of the EPR lineshape to measure the dipolar interaction, pulsed techniques measure the smaller dipolar interactions that cannot be observed in the lineshape. By separating the dipolar term in the spin Hamiltonian for exclusive detection, pulsed EPR techniques have extended the measurable range of interspin

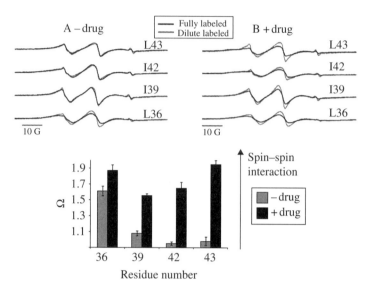

FIGURE 12.7   CW EPR spectra for four spin-labeled mutants located within the transmembrane region of the influenza A M2 protein. Spin-labeled proteins were reconstituted in liposomes, and drug-induced changes in the interaction parameter (Ω) obtained. Broadening of the CW spectrum due to interacting spins in the fully labeled sample (black line) is compared to dilute labeled sample (single spin, gray line). Upon addition of the drug, the Ω values increase, consistent with a narrower transmembrane tetrameric bundle. Reprinted from Ref. 106 with permission from Wiley.

interactions up to approximately 80Å. There are now several pulse methods available to measure interspin distances with the two most common being the DEER method [96], also known as the pulsed electron–electron double resonance (PELDOR or ELDOR) method, and the double quantum coherence (DQC) [18, 110, 111] method. Both these pulsed programs can be used not only to obtain interspin distances with great accuracy but also to provide a description of the distribution of distances. DQC will not be discussed further here, although it should be noted that it has the reported advantage over DEER of being able to measure slightly longer distances with higher sensitivity [112]. Compared to DQC, applications using DEER appear to be more prevalent in the literature, most probably because of simpler instrumentation requirements and pulse sequences. An extensive review of the DEER pulsed technique to measure interspin distances and analyze data is beyond the scope of this chapter. The interested reader is directed to excellent and comprehensive reviews of the DEER technology that further discuss its strengths and limitations as a spectroscopic ruler [20, 113, 114].

The DEER pulse method reveals distance information by producing a spin echo that is modulated by the frequency of the dipolar interaction [96, 115]. That is, the spin-echo decay of one spin label is modulated by intra- or intermolecular dipolar interactions with a second spin label or even sometimes by undesired "background" dipolar interactions between closely positioned labeled constructs. Such unwanted background contributions are often most prominent in membrane protein systems, as the two-dimensional environment of the proteoliposome can result in a high effective local concentration of the spin labels on neighboring protein molecules or subunits. These background contributions must be subtracted before the desired dipolar interactions are extracted by Fourier transformation deconvolution methods to provide distance and distance distribution information. There are many programs freely available to perform the analysis, including the DeerAnalysis program developed by Jeschke [116]. These programs provide a description of the interspin distance distribution in terms of a weighted average distance (r_{avg}) and a standard deviation (σ).

As the pulsed EPR experiments are conducted in the solid state at cryogenic temperatures (typically between 50 and 80 K), the conformational sampling of each spin label as a result of its flexibility results in static disorder. It is this static disorder that contributes directly to the width of the distance distribution (σ), as each label conformation produces a different average interspin distance. To a certain degree, it always remains somewhat undetermined as to whether the freezing process itself potentially biases certain distributions of conformational states. Therefore, the interpretation of stand-alone distances to build or support mechanistic models should be performed with caution.

Using a commercial X-band pulsed EPR spectrometer, the practical measurement range achieved by DEER is typically between approximately 18 and 60Å. An improved sensitivity in terms of smaller samples and shorter data collection times can be achieved with a Q-band spectrometer [117]. Longer distances, up to 80Å and potentially 100Å, may also be analyzed using more advanced pulsed sequences such as the five-pulse DEER program [118] or the DEER-Stitch program that combines three- and four-pulse DEER measurement [119]. Deuteration of the protein as well as the solvent can also be used to extend the range of distance measurements and improve the sensitivity [120].

In recent years, there has been a notable increase in the number of outstanding examples using DEER to measure the amplitude and direction of conformational changes in soluble proteins, as well as movements in membrane proteins accompanying function. Often performed in parallel with mobility and accessibility measurements, these include studies on the ATP-binding cassette (ABC) transporters such as the lipid flippase MsbA from *E. coli* [121, 122], the bacterial ABC maltose importer [123], the sodium-coupled amino acid transporter LeuT [124], and the outer membrane *E. coli* vitamin B$_{12}$ BtuB transporter [57], to list a few. In the case of the MsbA ABC transporter, the early spin-labeling distance studies contributed pivotal evidence that the original crystal structures were distorted [122]. These high-resolution structures were subsequently retracted and corrected [125]. Since then, further DEER and also accessibility studies have confirmed the main structural features observed in the revised crystal structures and have also provided a description for the geometric conformational transformation in MsbA that accompanies ATP hydrolysis [126, 127].

A recent outstanding example that illustrates the power of the pulsed DEER method for revealing a mechanistic description of a membrane protein is the study on the LmrP multidrug (MDR) transporter from *Lactococcus lactis* [128]. Multidrug efflux systems mediate the extrusion of toxic compounds from the cells in a coupled exchange with protons. LmrP belongs to the major facilitator superfamily (MFS) of multidrug transport proteins that are capable of transporting a range of different molecules such as anticancer drugs, sugars, amino acids, and various organic and inorganic ions from the cells [129]. Understanding the mechanism by which multidrug transporters can efficiently extrude a broad range of molecules is of vital importance to unraveling antibiotic resistance in bacterial strains.

Members of the MFS family share a conserved common architecture, consisting of two symmetry-related bundles of six, or occasionally seven, transmembrane helices. Each bundle of helices cradles the substrate at a binding site located at their interface. Over the past few years, crystal structures have captured several bacterial secondary transporters in different states along their transport cycle, providing insight into possible molecular mechanisms. This has included crystal structures of several MFS symporter proteins, such as LacY and FucP, in several conformations that are best described as an "inward-open" [130, 131], an "outward-open" [132], and an "occluded" state [101]. Together, this collection of structures led to the proposal of a rigid-body "rocker-switch" mechanistic model of symporter transport [132]. In contrast, there is only one available crystal structure of an MFS exporter, that is, EmrD in an occluded conformation [133]. The number of functional studies available to help elucidate the mechanism for how multidrug antiporters such as LmrP release substrates from the cell are also limited.

In the DEER study of Masureel et al., the authors mutated the only native cysteine in LmrP to an alanine residue (C270A) before introducing 22 pairs of spin labels in total on both the intra- (8 pairs) and extracellular sides (14 pairs) of each of the transmembrane helices (see Fig. 12.8a) [128]. The spin-labeled pairs most likely to best report on substrate-binding movement by DEER measurements were first identified by modeling the LmrP sequence onto the EmrD crystal structure. DEER spectra and distance distributions for two representative pairs from each side of the membrane are shown in Figure 12.8b. In general, distances on the extracellular side of LmrP were shown to decrease at low pH, while most distances on the intracellular side were observed to increase. The simple interpretation of the r_{avg} distance changes is that the LmrP transporter switches between an outward-open and an outward-closed conformation, depending on the protonation state of certain acidic residues, Asp68 and Glu327. These two residues were identified by mutagenesis methods. Substrate binding was also shown to initiate the transport cycle by first stabilizing the outward-open conformation (Fig. 12.8b). Another important finding from this study that could only be revealed from the distance distribution information possible from the DEER technique was that substrate transport does not involve rigid-body motions. Rather, the DEER distance distribution profiles showed that the two sides of the LmrP transporter have structural domains that can rearrange independently. The proposed LmrP transport cycle based on the EPR DEER distances is shown in Figure 12.8c. As the authors also note, it is critical to now test the proposed model in a membrane-like system.

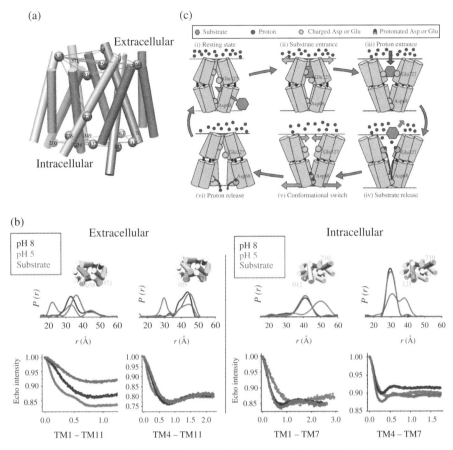

FIGURE 12.8 (a) Position of 22 interspin distance pairs for DEER analysis of the LmrP multidrug transporter. (b) DEER data analysis for two representative spin-label pairs on both the extra- and intracellular sides of LmrP. DEER distance distributions (top) obtained from time-dependent decay of DEER signals (below). DEER was measured in n-dodecyl-β-D-maltoside (DDM) micelles at pH 5, pH 8, and upon binding of the LmrP substrate (Hoechst 33342). Low pH triggers the closure (shorter distances) for interspin distances positioned on the extracellular side and opening (longer distances) for the intracellular side. Substrate binding stabilizes the open conformation on the extracellular side (similar distance profiles to pH 8). (c) Cartoon representation for the proposed LmrP transport cycle based on DEER distances. Reprinted from Ref. 128 with permission from Macmillan Publishers Ltd.

12.5   CHALLENGES

12.5.1   New Labels

The foremost limitation of SDSL–EPR is the need to modify a protein with an exogenous spin label through the introduction of new cysteines at desired locations. This assumes that a functional "cys-less" construct can first be obtained. Modification of proteins by either mutagenesis or with any extrinsic label has the potential to perturb structure and function. Described by some as the "price of peeking," such approaches continue to be routinely criticized as perturbing the "real" structure. However, what may be somewhat surprising to some is that the vast majority of cysteine-scanning and cysteine-labeling experiments described in the literature report that labeling is well tolerated by most protein systems [8, 9, 103, 128]. Even an unsuccessful modification that may perturb structure or function can be considered to be a successful experiment, helping identify key functional regions of proteins. However, while the nitroxide label attached to a cysteine is generally small, similar in magnitude to a tryptophan, a degree of caution should still be exercised as it is difficult to make generalizations about the potential perturbations arising from SDSL. That is, careful functional characterization is always necessary when the goal of a study is to correlate structural information to function.

When native cysteine residues cannot be mutated or introduced for labeling, then other strategies for introducing the unpaired electron can be explored. One developing method is the use of bioorthogonal chemical reactions, that is, reactions based on functional groups not found in biology. This includes the orthogonal spin-labeling strategy in which unnatural amino acids are used to incorporate a nitroxide group. Two recent studies report the selective modification of a genetically encoded unnatural amino acid. One of the amino acids used was p-acetyl-L-phenylalanine, which could be modified with a hydroxylamine-specific nitroxide reagent to produce a ketoxime-linked (K1) spin-labeled side chain [41]. In another study, p-azido-L-phenylalanine was modified with an azido-specific nitroxide reagent to generate a triazole-linked (T1) spin-labeled side chain [41, 42] (see Fig. 12.3). Not surprisingly, the K1 side-chain label is reported to be more flexible than the most commonly used MTSSL label, although the spatial distributions are not prohibitively broad to prevent their use for interspin distance measurements [41].

Another alternative labeling strategy that does result in a substantially more rigid spin-labeled side chain is via direct synthesis methods. Although presently of limited application for large membrane proteins, a spin-labeled side chain such as amino acid 4-(3,3,5,5-tetramethyl-2,6-dioxo-4-oxylpiperazin-1-yl)-L-phenylglycine (TOPP) or 2,2,6,6-tetramethylpiperidine-1-oxyl-4-amino-4-carboxylic acid (TOAC) can be incorporated by direct synthesis at a fixed position in peptides or small proteins [39]. The nitroxide group in these labels is fixed in the same spatial location, as all the bonds connecting the nitroxide to the protein are colinear [39]. Being rigid labels with a defined orientation of the nitroxide in space with respect to the peptide backbone, this class of labels is more effective at directly reporting on backbone or domain motion [134] and is also well suited for the accurate reporting of interspin distances [39].

Lastly, a more recent promising labeling strategy, especially if measuring very small changes of a few ångströms in interspin distances, is the development of bifunctional spin labels [40]. Bifunctional spin labels possess two functional groups for attachment to two sites on the protein. Thus, they reduce the label mobility with respect to the protein and the conformational space it can sample. For a doubly spin-labeled sample, this restricted label motion due to the cross-linked side chain gives the advantage that interspin distance distributions are significantly narrower [78].

These new labels are also a very promising prospect for measuring protein motions on the functionally important microsecond timescale [40].

12.5.2 Spin-Label Flexibility

Throughout this chapter, we have often mentioned the necessity to untangle the intrinsic contribution of the rotations of the spin label from that of protein dynamics. That is, if the accurate interpretation of EPR data is to be achieved, then the spatial distribution of the spin label needs to be described so that its contribution to mobility can be accounted for. The flexibility of the commonly used MTSSL label results in a large conformational space accessible for the nitroxide with the distance between the Cβ atom and the NO group varying between 4 and 8Å, depending on the label conformation. This flexibility of both labels in a doubly labeled sample can dramatically change the distance between the labels, which then contributes to the width of the distance distributions (σ). This drawback of using highly flexible labels can be addressed by collecting additional EPR spectra for a number of different spin-labeled sites to confirm or refine the structural data or mechanism under investigation. In another approach, direct information on preferred conformations can be obtained from diffraction of crystallized spin-labeled proteins [103]. Alternatively, if a crystal structure of the native protein is available, sophisticated methods such as atomistic molecular dynamics (MD) simulations can be performed to model the spin-label behavior when attached to a specific residue.

MD simulation methods, which sample all rotamer distributions, have been developed extensively over the past two decades (see Chapter 15) to provide details of the restriction of motion placed on the spin label due to neighboring protein atoms or the label's "microenvironment" [135–139]. This topic has also been recently reviewed in Ref. 34. Naturally, increasing the simulation time leads to better results [140], but this becomes more difficult when considering multiple label pairs. Despite several successful reports [137–139], MD simulations are still not widely or routinely used in many laboratories. A simpler "rotamer library"-based method shows potential for overcoming the need for repeating complicated and lengthy MD simulations for all label sites [141, 142]. The continued development of rotamer library-based techniques should lead to an improved agreement of the rotamer distributions with experimental data in the near future.

12.5.3 Production and Reconstitution Challenges: Nanodiscs

EPR studies of purified spin-labeled membrane proteins are traditionally performed either in detergent micelles or in proteoliposome vesicles. Each has their own weakness. Detergent micelles can be a poor membrane mimetic system, and reconstitution of proteins into liposomes can generate heterogeneity in orientation and in accessibility to ligands. Furthermore, the two-dimensional environment of the proteoliposome may also result in a high effective local concentration of the spin labels. This can accentuate background interdipolar contribution as a result of intersubunit interactions between labeled proteins being in close vicinity. As a consequence, background contributions place severe limits on EPR sensitivity, the practical distance range that can be measured, and, subsequently, the experimental output [143]. In order to minimize the background contribution arising from spins present at high concentration in the bilayer, either the lipid–protein molar ratios can be increased, or unlabeled protein can be included during the reconstitution process [144]. A more recent promising approach is to reconstitute the spin-labeled membrane proteins into "nanoscale bilayers" or nanodiscs.

Nanodiscs are lipids surrounded by a belt of amphipathic protein derived from apolipoprotein A1 [145]. Ranging in size from 9.8 to 17.0 nm in diameter, they essentially represent a bilayer system for which the desired level of protein insertion can be controlled. Not only are nanodiscs relatively easy to prepare at the concentration levels needed for EPR analysis (up to 100 µM), but by careful manipulation of the molar ratios, it is possible to achieve reconstitution of a single-spin-labeled membrane protein per bilayer disk. That is, the bilayer-mimicking environment is conserved, and the dimensionality of the background signal is also reduced. This simplifies the analysis of the DEER signal for extracting interspin distance and distribution information. A recent DEER analysis that examined the packing of transmembrane helices in the mechanosensitive channel MscS provides an interesting comparative study of the channel embedded in bicelles and nanodiscs to the detergent-solubilized crystal structure [146]. The researchers reported the extraction of highly accurate interspin distances using the lipid bilayer mimetic systems as a result of the substantial reduction in molecular crowding effects associated with the traditionally used liposome systems. Interestingly, the authors also concluded from the DEER data that the detergent-solubilized crystal structures of MscS were indeed an accurate model of the structure of the channel in the lipid bilayer.

12.6 CONCLUSIONS

The use of SDSL–EPR techniques has rapidly increased, in particular over the last decade, for examining some of the more challenging biological systems, including membrane proteins. Recent developments of SDSL techniques, including pulsed EPR methods for obtaining long-range interspin distance and distribution information, have helped spur this on, allowing structural information to be translated to a mechanism of action. Applicable to a wide range of pumps, channels, and transporter proteins, we hope that the examples we have described illustrate the strength of the SDSL approach for connecting structure with mechanism. Movements such as opening of gates, transport of substrates across bilayers, and drug binding events can all be probed by SDSL techniques. Most importantly, the movements associated with function can be probed in the native-like membrane environment of a lipid bilayer. While SDSL can be performed when there is limited structural information, it is best used as a complementary approach to crystallography, allowing static images to be brought to life. Possibly one day, the SDSL technique may also achieve its original goal of building new structures in the absence of crystal structures. This vision will of course depend on the continuing development of computational tools that can better model the dynamic behavior of the flexible spin label and thus allow more accurate distance descriptions to be obtained. Future developments to improve the sensitivity and routine nature of the tools within the SDSL toolkit should see these methods being performed in whole cells and tissues, that is, environments that are currently inaccessible to most other biophysical methods.

ACKNOWLEDGMENTS

The authors would like to thank Dr Nicole Cordina for assistance in figure preparation and critical reading of our manuscript.

REFERENCES

1. C. Altenbach, S.L. Flitsch, H.G. Khorana, W.L. Hubbell, Structural studies on transmembrane proteins. 2. Spin labeling of bacteriorhodopsin mutants at unique cysteines, *Biochemistry* 28 (1989) 7806–7812.

2. C. Altenbach, T. Marti, H.G. Khorana, W.L. Hubbell, Transmembrane protein structure: Spin labeling of bacteriorhodopsin mutants, *Science* 248 (1990) 1088–1092.

3. D.L. Farrens, What site-directed labeling studies tell us about the mechanism of rhodopsin activation and G-protein binding, *Photochem. Photobiol. Sci.* 9 (2010) 1466–1474.

4. D.J.E. Ingram, J.E. Bennett, Paramagnetic resonance in phthalocyanine, haemoglobin, and other organic derivatives, *Discuss. Faraday Soc.* 19 (1955) 127–134.

5. B. Commoner, J. Townsend, G.E. Pake, Free radicals in biological materials, *Nature* 174 (1954) 689–691.

6. T.J. Stone, T. Buckman, P.L. Nordio, H.M. McConnell, Spin-labeled biomolecules, *Proc. Natl. Acad. Sci. U. S. A.* 54 (1965) 1010–1017.

7. S. Ogawa, H.M. McConnell, Spin-label study of hemoglobin conformations in solution, *Proc. Natl. Acad. Sci. U. S. A.* 58 (1967) 19–26.

8. H.S. McHaourab, M.A. Lietzow, K. Hideg, W.L. Hubbell, Motion of spin-labeled side chains in T4 lysozyme. Correlation with protein structure and dynamics, *Biochemistry* 35 (1996) 7692–7704.

9. E. Perozo, D.M. Cortes, L.G. Cuello, Three-dimensional architecture and gating mechanism of a K^+ channel studied by EPR spectroscopy, *Nat. Struct. Biol.* 5 (1998) 459–469.

10. E. Perozo, D.M. Cortes, L.G. Cuello, Structural rearrangements underlying K^+-channel activation gating, *Science* 285 (1999) 73–78.

11. J. Voss, M.M. He, W.L. Hubbell, H.R. Kaback, Site-directed spin labeling demonstrates that transmembrane domain XII in the lactose permease of *Escherichia coli* is an α-helix, *Biochemistry* 35 (1996) 12915–12918.

12. J. Voss, W.L. Hubbell, J. Hernandez-Borrell, H.R. Kaback, Site-directed spin-labeling of transmembrane domain VII and the 4B1 antibody epitope in the lactose permease of *Escherichia coli*, *Biochemistry* 36 (1997) 15055–15061.

13. E. Perozo, A. Kloda, D.M. Cortes, B. Martinac, Site-directed spin-labeling analysis of reconstituted Mscl in the closed state, *J. Gen. Physiol.* 118 (2001) 193–206.

14. C. Altenbach, K.J. Oh, R.J. Trabanino, K. Hideg, W.L. Hubbell, Estimation of inter-residue distances in spin labeled proteins at physiological temperatures: Experimental strategies and practical limitations, *Biochemistry* 40 (2001) 15471–15482.

15. M.D. Rabenstein, Y.K. Shin, Determination of the distance between two spin labels attached to a macromolecule, *Proc. Natl. Acad. Sci. U. S. A.* 92 (1995) 8239–8243.

16. W.L. Hubbell, H.S. McHaourab, C. Altenbach, M.A. Lietzow, Watching proteins move using site-directed spin labeling, *Structure* 4 (1996) 779–783.

17. W.L. Hubbell, D.S. Cafiso, C. Altenbach, Identifying conformational changes with site-directed spin labeling, *Nat. Struct. Biol.* 7 (2000) 735–739.

18. P.P. Borbat, H.S. McHaourab, J.H. Freed, Protein structure determination using long-distance constraints from double-quantum coherence ESR: Study of T4 lysozyme, *J. Am. Chem. Soc.* 124 (2002) 5304–5314.

19. G. Jeschke, A. Bender, H. Paulsen, H. Zimmermann, A. Godt, Sensitivity enhancement in pulse EPR distance measurements, *J. Magn. Reson.* 169 (2004) 1–12.

20. G. Jeschke, Y. Polyhach, Distance measurements on spin-labelled biomacromolecules by pulsed electron paramagnetic resonance, *Phys. Chem. Chem. Phys.* 9 (2007) 1895–1910.

21. S.Y. Lee, A. Lee, J. Chen, R. MacKinnon, Structure of the KvAP voltage-dependent K^+ channel and its dependence on the lipid membrane, *Proc. Natl. Acad. Sci. U. S. A.* 102 (2005) 15441–15446.

22. A.K. Mittermaier, L.E. Kay, Observing biological dynamics at atomic resolution using NMR, *Trends Biochem. Sci.* 34 (2009) 601–611.

23. H. Krishnamurthy, C.L. Piscitelli, E. Gouaux, Unlocking the molecular secrets of sodium-coupled transporters, *Nature* 459 (2009) 347–355.

24. D.C. Rees, E. Johnson, O. Lewinson, ABC transporters: The power to change, *Nat. Rev. Mol. Cell Biol.* 10 (2009) 218–227.

25. Y.K. Shin, C. Levinthal, F. Levinthal, W.L. Hubbell, Colicin E1 binding to membranes: Time-resolved studies of spin-labeled mutants, *Science* 259 (1993) 960–963.

26. K. Qu, J.L. Vaughn, A. Sienkiewicz, C.P. Scholes, J.S. Fetrow, Kinetics and motional dynamics of spin-labeled yeast iso-1-cytochrome c: 1. Stopped-flow electron paramagnetic resonance as a probe for protein folding/unfolding of the C-terminal helix spin-labeled at cysteine 102, *Biochemistry* 36 (1997) 2884–2897.

27. A. Sienkiewicz, A.M. da Costa Ferreira, B. Danner, C.P. Scholes, Dielectric resonator-based flow and stopped-flow EPR with rapid field scanning: A methodology for increasing kinetic information, *J. Magn. Reson.* 136 (1999) 137–142.

28. G.E. Fanucci, D.S. Cafiso, Recent advances and applications of site-directed spin labeling, *Curr. Opin. Struct. Biol.* 16 (2006) 644–653.

29. J.P. Klare, H.J. Steinhoff, Spin labeling EPR, *Photosynth. Res.* 102 (2009) 377–390.

30. H.S. McHaourab, P.R. Steed, K. Kazmier, Toward the fourth dimension of membrane protein structure: Insight into dynamics from spin-labeling EPR spectroscopy, *Structure* 19 (2011) 1549–1561.

31. E. Bordignon, Site-directed spin labeling of membrane proteins, *Top. Curr. Chem.* 321 (2012) 121–157.

32. M. Drescher, EPR in protein science: Intrinsically disordered proteins, *Top. Curr. Chem.* 321 (2012) 91–119.

33. W.L. Hubbell, C.J. Lopez, C. Altenbach, Z. Yang, Technological advances in site-directed spin labeling of proteins, *Curr. Opin. Struct. Biol.* 23 (2013) 725–733.

34. G. Jeschke, Conformational dynamics and distribution of nitroxide spin labels, *Prog. Nucl. Magn. Reson. Spectrosc.* 72 (2013) 42–60.

35. G.R. Eaton, S.S. Eaton, D.P. Barr, R.T. Weber (Eds.), *Quantitative EPR*, Springer, New York, 2010.

36. J.A. Weil, J.R. Bolton (Eds.), *Electron Paramagnetic Resonance: Elementary Theory and Practical Applications*, John Wiley & Sons, Inc., Hoboken, NJ, 2007.

37. M. Brustolon, E. Giamello (Eds.), *Electron Paramagnetic Resonance: A Practitioner's Toolkit*, John Wiley & Sons, Inc., Hoboken, NJ, 2009.

38. W.R. Hagen (Eds.), *Biomolecular EPR Spectroscopy*, CRC Press, Taylor & Francis Group, Boca Raton, FL, 2009.

39. S. Stoller, G. Sicoli, T.Y. Baranova, M. Bennati, U. Diederichsen, TOPP: A novel nitroxide-labeled amino acid for EPR distance measurements, *Angew. Chem. Int. Ed. Engl.* 50 (2011) 9743–9746.

40. M.R. Fleissner, M.D. Bridges, E.K. Brooks, D. Cascio, T. Kalai, K. Hideg, W.L. Hubbell, Structure and dynamics of a conformationally constrained nitroxide side chain and applications in EPR spectroscopy, *Proc. Natl. Acad. Sci. U. S. A.* 108 (2011) 16241–16246.

41. M.R. Fleissner, E.M. Brustad, T. Kalai, C. Altenbach, D. Cascio, F.B. Peters, K. Hideg, S. Peuker, P.G. Schultz, W.L. Hubbell, Site-directed spin labeling of a genetically encoded unnatural amino acid, *Proc. Natl. Acad. Sci. U. S. A.* 106 (2009) 21637–21642.

42. T. Kálaia, M.R. Fleissner, J. Jekő, W.L. Hubbell, K. Hideg, Synthesis of new spin labels for Cu-free click conjugation, *Tetrahedron Lett.* 52 (2011) 2747–2749.

43. E.B. Getz, M. Xiao, T. Chakrabarty, R. Cooke, P.R. Selvin, A comparison between the sulfhydryl reductants tris(2-carboxyethyl)phosphine and dithiothreitol for use in protein biochemistry, *Anal. Biochem.* 273 (1999) 73–80.

44. D.M. Freed, P.S. Horanyi, M.C. Wiener, D.S. Cafiso, Conformational exchange in a membrane transport protein is altered in protein crystals, *Biophys. J.* 99 (2010) 1604–1610.

45. C.D. Dellisanti, B. Ghosh, S.M. Hanson, J.M. Raspanti, V.A. Grant, G.M. Diarra, A.M. Schuh, K. Satyshur, C.S. Klug, C. Czajkowski, Site-directed spin labeling reveals pentameric ligand-gated ion channel gating motions, *PLoS Biol.* 11 (2013) e1001714.

46. W.L. Hubbell, A. Gross, R. Langen, M.A. Lietzow, Recent advances in site-directed spin labeling of proteins, *Curr. Opin. Struct. Biol.* 8 (1998) 649–656.

47. D. Eisenberg, R.M. Weiss, T.C. Terwilliger, The hydrophobic moment detects periodicity in protein hydrophobicity, *Proc. Natl. Acad. Sci. U. S. A.* 81 (1984) 140–144.

48. R. Langen, K. Cai, C. Altenbach, H.G. Khorana, W.L. Hubbell, Structural features of the C-terminal domain of bovine rhodopsin: a site-directed spin-labeling study, *Biochemistry* 38 (1999) 7918–7924.

49. C. Altenbach, J. Klein-Seetharaman, J. Hwa, H.G. Khorana, W.L. Hubbell, Structural features and light-dependent changes in the sequence 59-75 connecting helices I and II in rhodopsin: A site-directed spin-labeling study, *Biochemistry* 38 (1999) 7945–7949.

50. C. Altenbach, K. Cai, H.G. Khorana, W.L. Hubbell, Structural features and light-dependent changes in the sequence 306-322 extending from helix VII to the palmitoylation sites in rhodopsin: A site-directed spin-labeling study, *Biochemistry* 38 (1999) 7931–7937.

51. Z.T. Farahbakhsh, K.D. Ridge, H.G. Khorana, W.L. Hubbell, Mapping light-dependent structural changes in the cytoplasmic loop connecting helices C and D in rhodopsin: A site-directed spin labeling study, *Biochemistry* 34 (1995) 8812–8819.

52. C. Altenbach, K. Yang, D.L. Farrens, Z.T. Farahbakhsh, H.G. Khorana, W.L. Hubbell, Structural features and light-dependent changes in the cytoplasmic interhelical E-F loop region of rhodopsin: A site-directed spin-labeling study, *Biochemistry* 35 (1996) 12470–12478.

53. Z.T. Farahbakhsh, K. Hideg, W.L. Hubbell, Photoactivated conformational changes in rhodopsin: A time-resolved spin label study, *Science* 262 (1993) 1416–1419.

54. H.W. Choe, J.H. Park, Y.J. Kim, O.P. Ernst, Transmembrane signaling by GPCRs: Insight from rhodopsin and opsin structures, *Neuropharmacology* 60 (2011) 52–57.

55. D. Salom, D.T. Lodowski, R.E. Stenkamp, I. Le Trong, M. Golczak, B. Jastrzebska, T. Harris, J.A. Ballesteros, K. Palczewski, Crystal structure of a photoactivated deprotonated intermediate of rhodopsin, *Proc. Natl. Acad. Sci. U. S. A.* 103 (2006) 16123–16128.

56. G.E. Fanucci, J.Y. Lee, D.S. Cafiso, Membrane mimetic environments alter the conformation of the outer membrane protein BtuB, *J. Am. Chem. Soc.* 125 (2003) 13932–13933.

57. Q. Xu, J.F. Ellena, M. Kim, D.S. Cafiso, Substrate-dependent unfolding of the energy coupling motif of a membrane transport protein determined by double electron-electron resonance, *Biochemistry* 45 (2006) 10847–10854.

58. R.H. Flores Jimenez, M.A. Do Cao, M. Kim, D.S. Cafiso, Osmolytes modulate conformational exchange in solvent-exposed regions of membrane proteins, *Protein Sci.* 19 (2010) 269–278.

59. P.S. Miller, T.G. Smart, Binding, activation and modulation of Cys-loop receptors, *Trends Pharmacol. Sci.* 31 (2010) 161–174.

60. N. Unwin, Refined structure of the nicotinic acetylcholine receptor at 4A resolution, *J. Mol. Biol.* 346 (2005) 967–989.

61. N. Unwin, Y. Fujiyoshi, Gating movement of acetylcholine receptor caught by plunge-freezing, *J. Mol. Biol.* 422 (2012) 617–634.

62. C.D. Dellisanti, Y. Yao, J.C. Stroud, Z.Z. Wang, L. Chen, Crystal structure of the extracellular domain of nAChR α1 bound to α-bungarotoxin at 1.94Å resolution, *Nat. Neurosci.* 10 (2007) 953–962.

63. R.J. Hilf, R. Dutzler, X-ray structure of a prokaryotic pentameric ligand-gated ion channel, *Nature* 452 (2008) 375–379.

64. N. Bocquet, H. Nury, M. Baaden, C. Le Poupon, J.P. Changeux, M. Delarue, P.J. Corringer, X-ray structure of a pentameric ligand-gated ion channel in an apparently open conformation, *Nature* 457 (2009) 111–114.

65. J.S. Hyde, W.K. Subczynski, Spin-label oximetry, in: L. Berliner (Eds.), *Spin Labeling Theory and Applications*, Plenum Press, New York, 1998, pp. 399–425.

66. W.K. Subczynski, C.C. Felix, C.S. Klug, J.S. Hyde, Concentration by centrifugation for gas exchange EPR oximetry measurements with loop-gap resonators, *J. Magn. Reson.* 176 (2005) 244–248.

67. R.D. Nielsen, S. Canaan, J.A. Gladden, M.H. Gelb, C. Mailer, B.H. Robinson, Comparing continuous wave progressive saturation EPR and time domain saturation recovery EPR over the entire motional range of nitroxide spin labels, *J. Magn. Reson.* 169 (2004) 129–163.

68. C. Altenbach, D.A. Greenhalgh, H.G. Khorana, W.L. Hubbell, A collision gradient method to determine the immersion depth of nitroxides in lipid bilayers: Application to spin-labeled mutants of bacteriorhodopsin, *Proc. Natl. Acad. Sci. U. S. A.* 91 (1994) 1667–1671.

69. R. Langen, J.M. Isas, H. Luecke, H.T. Haigler, W.L. Hubbell, Membrane-mediated assembly of annexins studied by site-directed spin labeling, *J. Biol. Chem.* 273 (1998) 22453–22457.

70. K.J. Oh, H. Zhan, C. Cui, K. Hideg, R.J. Collier, W.L. Hubbell, Organization of diphtheria toxin T domain in bilayers: A site-directed spin labeling study, *Science* 273 (1996) 810–812.

71. L. Salwinski, W.L. Hubbell, Structure in the channel forming domain of colicin E1 bound to membranes: The 402-424 sequence, *Protein Sci.* 8 (1999) 562–572.

72. M.L. Ujwal, H. Jung, E. Bibi, C. Manoil, C. Altenbach, W.L. Hubbell, H.R. Kaback, Membrane topology of helices VII and XI in the lactose permease of *Escherichia coli* studied by lacY-phoA fusion analysis and site-directed spectroscopy, *Biochemistry* 34 (1995) 14909–14917.

73. M. Zhao, K.C. Zen, J. Hernandez-Borrell, C. Altenbach, W.L. Hubbell, H.R. Kaback, Nitroxide scanning electron paramagnetic resonance of helices IV and V and the intervening loop in the lactose permease of *Escherichia coli*, *Biochemistry* 38 (1999) 15970–15977.

74. E. Perozo, D.M. Cortes, P. Sompornpisut, A. Kloda, B. Martinac, Open channel structure of MscL and the gating mechanism of mechanosensitive channels, *Nature* 418 (2002) 942–948.

75. L.G. Cuello, D.M. Cortes, E. Perozo, Molecular architecture of the KvAP voltage-dependent K$^+$ channel in a lipid bilayer, *Science* 306 (2004) 491–495.

76. J.F. Cordero-Morales, V. Jogini, A. Lewis, V. Vasquez, D.M. Cortes, B. Roux, E. Perozo, Molecular driving forces determining potassium channel slow inactivation, *Nat. Struct. Mol. Biol.* 14 (2007) 1062–1069.

77. L.G. Cuello, V. Jogini, D.M. Cortes, E. Perozo, Structural mechanism of C-type inactivation in K$^+$ channels, *Nature* 466 (2010) 203–208.

78. Q. Li, S. Wanderling, P. Sompornpisut, E. Perozo, Structural basis of lipid-driven conformational transitions in the KvAP voltage-sensing domain, *Nat. Struct. Mol. Biol.* 21 (2014) 160–166.

79. X. Zhang, W. Ren, P. DeCaen, C. Yan, X. Tao, L. Tang, J. Wang, K. Hasegawa, T. Kumasaka, J. He, J. Wang, D.E. Clapham, N. Yan, Crystal structure of an orthologue of the NaChBac voltage-gated sodium channel, *Nature* 486 (2012) 130–134.

80. Y. Jiang, A. Lee, J. Chen, V. Ruta, M. Cadene, B.T. Chait, R. MacKinnon, X-ray structure of a voltage-dependent K$^+$ channel, *Nature* 423 (2003) 33–41.

81. S.B. Long, X. Tao, E.B. Campbell, R. MacKinnon, Atomic structure of a voltage-dependent K$^+$ channel in a lipid membrane-like environment, *Nature* 450 (2007) 376–382.

82. J. Payandeh, T. Scheuer, N. Zheng, W.A. Catterall, The crystal structure of a voltage-gated sodium channel, *Nature* 475 (2011) 353–358.

83. H. Zheng, W. Liu, L.Y. Anderson, Q.X. Jiang, Lipid-dependent gating of a voltage-gated potassium channel, *Nat. Commun.* 2 (2011) 250.

84. Y. Xu, Y. Ramu, Z. Lu, Removal of phospho-head groups of membrane lipids immobilizes voltage sensors of K$^+$ channels, *Nature* 451 (2008) 826–829.

85. Y. Jiang, V. Ruta, J. Chen, A. Lee, R. MacKinnon, The principle of gating charge movement in a voltage-dependent K+ channel, *Nature* 423 (2003) 42–48.

86. S.R. Durell, H.R. Guy, Atomic scale structure and functional models of voltage-gated potassium channels, *Biophys. J.* 62 (1992) 238–247; discussion 247–250.

87. C.S. Klug, W. Su, J.B. Feix, Mapping of the residues involved in a proposed beta-strand located in the ferric enterobactin receptor FepA using site-directed spin-labeling, *Biochemistry* 36 (1997) 13027–13033.

88. H.A. Koteiche, H.S. McHaourab, Folding pattern of the α-crystallin domain in αA-crystallin determined by site-directed spin labeling, *J. Mol. Biol.* 294 (1999) 561–577.

89. H.S. Mchaourab, E. Perozo, Determination of protein folds and conformational dynamics using spin-labeling EPR spectroscopy, in: L.J. Berliner, S.S. Eaton and G.R. Eaton (Eds.), *Distance Measurements in Biological Systems by EPR*, Plenum Press, New York, 2000, pp. 185–247.

90. L.J. Brown, K.L. Sale, R. Hills, C. Rouviere, L. Song, X. Zhang, P.G. Fajer, Structure of the inhibitory region of troponin by site directed spin labeling electron paramagnetic resonance, *Proc. Natl. Acad. Sci. U. S. A.* 99 (2002) 12765–12770.

91. D. Hilger, H. Jung, E. Padan, C. Wegener, K.P. Vogel, H.J. Steinhoff, G. Jeschke, Assessing oligomerization of membrane proteins by four-pulse DEER: pH-dependent dimerization of NhaA Na+/H+ antiporter of *E. coli*, *Biophys. J.* 89 (2005) 1328–1338.

92. G. Jeschke, A. Bender, T. Schweikardt, G. Panek, H. Decker, H. Paulsen, Localization of the N-terminal domain in light-harvesting chlorophyll a/b protein by EPR measurements, *J. Biol. Chem.* 280 (2005) 18623–18630.

93. S. Steigmiller, M. Borsch, P. Graber, M. Huber, Distances between the b-subunits in the tether domain of F_0F_1-ATP synthase from *E. coli*, *Biochim. Biophys. Acta-Bioenerg.* 1708 (2005) 143–153.

94. S.M. Miick, G.V. Martinez, W.R. Fiori, A.P. Todd, G.L. Millhauser, Short alanine-based peptides may form 3_{10}-helices and not α-helices in aqueous solution, *Nature* 359 (1992) 653–655.

95. H.S. McHaourab, K.J. Oh, C.J. Fang, W.L. Hubbell, Conformation of T4 lysozyme in solution, Hinge-bending motion and the substrate-induced conformational transition studied by site-directed spin labeling, *Biochemistry* 36 (1997) 307–316.

96. A. Milov, K. Salikhov, M. Shirov, Application of ELDOR in electron-spin echo for paramagnetic center space distribution in solids, *Fiz. Tverd. Tela* 23 (1981) 957–982.

97. L.J. Berliner, S.S. Eaton, G.R. Eaton (Eds.), *Distance Measurements in Biological Systems by EPR, Biological Magnetic Resonance*, Vol. 19, Plenum Press, New York, 2000.

98. S.S. Eaton, K.M. More, B.M. Sawant, G.R. Eaton, Use of the EPR half-field transition to determine the interspin distance and the orientation of the interspin vector in systems with two unpaired electrons, *J. Am. Chem. Soc.* 105 (1983) 6560–6567.

99. J.C. McNulty, G.L. Millhauser, TOAC the rigid nitroxide side chain, in: L.J. Berliner, S.S. Eaton and G.R. Eaton (Eds.), *Distance Measurements in Biological Systems by EPR*, Plenum Press, New York, 2000, pp. 277–307.

100. A.I. Kokorin, K.I. Zamaraev, G.L. Grigorian, V.P. Ivanov, E.G. Rozantsev, [Measurement of the distance between paramagnetic centers in solid solutions of nitrosyl radicals, biradicals and spin-labelled proteins], *Biofizika* 17 (1972) 34–41.

101. J. Sun, J. Voss, W.L. Hubbell, H.R. Kaback, Proximity between periplasmic loops in the lactose permease of *Escherichia coli* as determined by site-directed spin labeling, *Biochemistry* 38 (1999) 3100–3105.

102. M. Persson, J.R. Harbridge, P. Hammarstrom, R. Mitri, L.G. Martensson, U. Carlsson, G.R. Eaton, S.S. Eaton, Comparison of electron paramagnetic resonance methods to determine distances between spin labels on human carbonic anhydrase II, *Biophys. J.* 80 (2001) 2886–2897.

103. R. Langen, K.J. Oh, D. Cascio, W.L. Hubbell, Crystal structures of spin labeled T4 lysozyme mutants: Implications for the interpretation of EPR spectra in terms of structure, *Biochemistry* 39 (2000) 8396–8405.

104. E.J. Hustedt, A.H. Beth, Structural information from CW-EPR spectra of dipolar coupled nitroxide spin labels, in: L.J. Berliner, S.S. Eaton and G.R. Eaton (Eds.), *Distance Measurements in Biological Systems by EPR*, Plenum Press, New York, 2000, pp. 155–184.

105. E.J. Hustedt, A.H. Beth, Nitroxide spin-spin interactions: Applications to protein structure and dynamics, *Annu. Rev. Biophys. Biomol. Struct.* 28 (1999) 129–153.

106. J.L. Thomaston, P.A. Nguyen, E.C. Brown, M.A. Upshur, J. Wang, W.F. DeGrado, K.P. Howard, Detection of drug-induced conformational change of a transmembrane protein in lipid bilayers using site-directed spin labeling, *Protein Sci.* 22 (2013) 65–73.

107. L.H. Pinto, R.A. Lamb, The M2 proton channels of influenza A and B viruses, *J. Biol. Chem.* 281 (2006) 8997–9000.

108. H. Leonov, P. Astrahan, M. Krugliak, I.T. Arkin, How do aminoadamantanes block the influenza M2 channel, and how does resistance develop?, *J. Am. Chem. Soc.* 133 (2011) 9903–9911.

109. J. Wang, J.X. Qiu, C. Soto, W.F. DeGrado, Structural and dynamic mechanisms for the function and inhibition of the M2 proton channel from influenza A virus, *Curr. Opin. Struct. Biol.* 21 (2011) 68–80.

110. S. Saxena, J.H. Freed, Double quantum two-dimensional Fourier transform electron spin resonance: Distance measurements, *Chem. Phys. Lett.* 251 (1996) 102–110.

111. P.P. Borbat, J.H. Freed, Multiple-quantum ESR and distance measurements, *Chem. Phys. Lett.* 313 (1999) 145–154.

112. Y.W. Chiang, P.P. Borbat, J.H. Freed, The determination of pair distance distributions by pulsed ESR using Tikhonov regularization, *J. Magn. Reson.* 172 (2005) 279–295.

113. G. Jeschke, DEER distance measurements on proteins, *Annu. Rev. Phys. Chem.* 63 (2012) 419–446.

114. P.G. Fajer, L.J. Brown, L. Song, Practical pulsed dipolar ESR (DEER), in: M.A. Hemminga and L.J. Berliner (Eds.), *ESR spectroscopy in membrane biophysics*, Springer, New York, 2007, pp. 95–128.

115. M. Pannier, S. Veit, A. Godt, G. Jeschke, H.W. Spiess, Dead-time free measurement of dipole-dipole interactions between electron spins, *J. Magn. Reson.* 142 (2000) 331–340.

116. G. Jeschke, V. Chechik, P. Ionita, A. Godt, H. Zimmermann, J. Banham, C.R. Timmel, D. Hilger, H. Jung, DeerAnalysis2006: A comprehensive software package for analyzing pulsed ELDOR data, *Appl. Magn. Reson.* 30 (2006) 473–498.

117. H. Ghimire, R.M. McCarrick, D.E. Budil, G.A. Lorigan, Significantly improved sensitivity of Q-band PELDOR/DEER experiments relative to X-band is observed in measuring the intercoil distance of a leucine zipper motif peptide (GCN4-LZ), *Biochemistry* 48 (2009) 5782–5784.

118. P.P. Borbat, E.R. Georgieva, J.H. Freed, Improved sensitivity for long-distance measurements in biomolecules: Five-pulse double electron-electron resonance, *J. Phys. Chem. Lett.* 4 (2013) 170–175.

119. J.E. Lovett, B.W. Lovett, J. Harmer, DEER-Stitch: Combining three- and four-pulse DEER measurements for high sensitivity, deadtime free data, *J. Magn. Reson.* 223 (2012) 98–106.

120. R. Ward, A. Bowman, E. Sozudogru, H. El-Mkami, T. Owen-Hughes, D.G. Norman, EPR distance measurements in deuterated proteins, *J. Magn. Reson.* 207 (2010) 164–167.

121. K.M. Schultz, J.A. Merten, C.S. Klug, Characterization of the E506Q and H537A dysfunctional mutants in the *E. coli* ABC transporter MsbA, *Biochemistry* 50 (2011) 3599–3608.

122. J. Dong, G. Yang, H.S. McHaourab, Structural basis of energy transduction in the transport cycle of MsbA, *Science* 308 (2005) 1023–1028.

123. C. Orelle, F.J. Alvarez, M.L. Oldham, A. Orelle, T.E. Wiley, J. Chen, A.L. Davidson, Dynamics of α-helical subdomain rotation in the intact maltose ATP-binding cassette transporter, *Proc. Natl. Acad. Sci. U. S. A.* 107 (2010) 20293–20298.

124. K. Kazmier, S. Sharma, M. Quick, S.M. Islam, B. Roux, H. Weinstein, J.A. Javitch, H.S. McHaourab, Conformational dynamics of ligand-dependent alternating access in LeuT, *Nat. Struct. Mol. Biol.* 21 (2014) 472–479.

125. A. Ward, C.L. Reyes, J. Yu, C.B. Roth, G. Chang, Flexibility in the ABC transporter MsbA: Alternating access with a twist, *Proc. Natl. Acad. Sci. U. S. A.* 104 (2007) 19005–19010.

126. P. Zou, H.S. McHaourab, Alternating access of the putative substrate-binding chamber in the ABC transporter MsbA, *J. Mol. Biol.* 393 (2009) 574–585.

127. P.P. Borbat, K. Surendhran, M. Bortolus, P. Zou, J.H. Freed, H.S. McHaourab, Conformational motion of the ABC transporter MsbA induced by ATP hydrolysis, *PLoS Biol.* 5 (2007) e271.

128. M. Masureel, C. Martens, R.A. Stein, S. Mishra, J.M. Ruysschaert, H.S. McHaourab, C. Govaerts, Protonation drives the conformational switch in the multidrug transporter LmrP, *Nat. Chem. Biol.* 10 (2014) 149–155.

129. M. Putman, H.W. van Veen, W.N. Konings, Molecular properties of bacterial multidrug transporters, *Microbiol. Mol. Biol. Rev.* 64 (2000) 672–693.

130. L. Guan, O. Mirza, G. Verner, S. Iwata, H.R. Kaback, Structural determination of wild-type lactose permease, *Proc. Natl. Acad. Sci. U. S. A.* 104 (2007) 15294–15298.

131. Y. Huang, M.J. Lemieux, J. Song, M. Auer, D.N. Wang, Structure and mechanism of the glycerol-3-phosphate transporter from *Escherichia coli*, *Science* 301 (2003) 616–620.

132. S. Dang, L. Sun, Y. Huang, F. Lu, Y. Liu, H. Gong, J. Wang, N. Yan, Structure of a fucose transporter in an outward-open conformation, *Nature* 467 (2010) 734–738.

133. Y. Yin, X. He, P. Szewczyk, T. Nguyen, G. Chang, Structure of the multidrug transporter EmrD from *Escherichia coli*, *Science* 312 (2006) 741–744.

134. C.B. Karim, T.L. Kirby, Z. Zhang, Y. Nesmelov, D.D. Thomas, Phospholamban structural dynamics in lipid bilayers probed by a spin label rigidly coupled to the peptide backbone, *Proc. Natl. Acad. Sci. U. S. A.* 101 (2004) 14437–14442.

135. I. Stoica, Solvent interactions and protein dynamics in spin-labeled T4 lysozyme, *J. Biomol. Struct. Dyn.* 21 (2004) 745–760.

136. K. Sale, L. Song, Y.S. Liu, E. Perozo, P. Fajer, Explicit treatment of spin labels in modeling of distance constraints from dipolar EPR and DEER, *J. Am. Chem. Soc.* 127 (2005) 9334–9335.

137. I.V. Borovykh, S. Ceola, P. Gajula, P. Gast, H.J. Steinhoff, M. Huber, Distance between a native cofactor and a spin label in the reaction centre of Rhodobacter sphaeroides by a two-frequency pulsed electron paramagnetic resonance method and molecular dynamics simulations, *J. Magn. Reson.* 180 (2006) 178–185.

138. M.I. Fajer, H. Li, W. Yang, P.G. Fajer, Mapping electron paramagnetic resonance spin label conformations by the simulated scaling method, *J. Am. Chem. Soc.* 129 (2007) 13840–13846.

139. P. Sompornpisut, B. Roux, E. Perozo, Structural refinement of membrane proteins by restrained molecular dynamics and solvent accessibility data, *Biophys. J.* 95 (2008) 5349–5361.

140. D. Sezer, J.H. Freed, B. Roux, Using Markov models to simulate electron spin resonance spectra from molecular dynamics trajectories, *J. Phys. Chem. B* 112 (2008) 11014–11027.

141. Y. Polyhach, E. Bordignon, G. Jeschke, Rotamer libraries of spin labelled cysteines for protein studies, *Phys. Chem. Chem. Phys.* 13 (2011) 2356–2366.

142. D. Klose, J.P. Klare, D. Grohmann, C.W. Kay, F. Werner, H.J. Steinhoff, Simulation vs. reality: A comparison of in silico distance predictions with DEER and FRET measurements, *PLoS One* 7 (2012) e39492.

143. P. Zou, H.S. McHaourab, Increased sensitivity and extended range of distance measurements in spin-labeled membrane proteins: Q-band double electron-electron resonance and nanoscale bilayers, *Biophys. J.* 98 (2010) L18–L20.

144. P. Zou, M. Bortolus, H.S. McHaourab, Conformational cycle of the ABC transporter MsbA in liposomes: Detailed analysis using double electron-electron resonance spectroscopy, *J. Mol. Biol.* 393 (2009) 586–597.

145. T.H. Bayburt, S.G. Sligar, Membrane protein assembly into Nanodiscs, *FEBS Lett.* 584 (2010) 1721–1727.

146. R. Ward, C. Pliotas, E. Branigan, C. Hacker, A. Rasmussen, G. Hagelueken, I.R. Booth, S. Miller, J. Lucocq, J.H. Naismith, O. Schiemann, Probing the structure of the mechanosensitive channel of small conductance in lipid bilayers with pulsed electron-electron double resonance, *Biophys. J.* 106 (2014) 834–842.

13

RADIOACTIVITY-BASED ANALYSIS OF ION TRANSPORT

INGOLF BERNHARDT[1] AND J. CLIVE ELLORY[2]

[1] *Laboratory of Biophysics, Saarland University, Saarbrücken, Germany*
[2] *Department of Physiology, Anatomy and Genetics, Oxford University, Oxford, UK*

13.1 INTRODUCTION

Up until around 80 years ago, it was thought that biological cell membranes (at that time still called "plasmalemma" although the terminology "membrane" was introduced by Carl Wilhelm von Nägeli in 1855 [1]) are impermeable to ions. This assumption was supported by the finding of Carl Schmidt who discovered in the 1850s that K^+ is predominantly present in the human red blood cells (RBCs), whereas Na^+ dominates in the plasma [2]. The thinking changed after two important findings. Van Slyke was able to show that RBCs have a relatively high Cl^- permeability [3]. Fenn and Cobb demonstrated that a muscle cell is permeable for Na^+ and K^+ [4]. Between 1939 and 1941, with the availability of radioisotopes for Na^+ and K^+, it was shown that RBCs are permeable for these ions [5–7]. From this time on, it became popular to investigate ion transport across biological membranes using radioisotopes. This was especially the case after Skou [8] discovered the Na^+,K^+-ATPase as the enzymatic basis of the Na^+ pump, which was hypothesized to exist in cell membranes already in the years 1941–1946 [9, 10] (see Chapter 1 for further historical detail).

Nowadays, it is well known that ion transport pathways in cell membranes can be divided into four categories: (i) active transport mediated by ATP-driven pumps, (ii) carrier-mediated transport (which can be divided in uniport and cotransport (symport

Pumps, Channels, and Transporters: Methods of Functional Analysis, First Edition. Edited by Ronald J. Clarke and Mohammed A. A. Khalid.
© 2015 John Wiley & Sons, Inc. Published 2015 by John Wiley & Sons, Inc.

or antiport) mechanisms), (iii) transport through ion channels that are more or less selective for one or more ion species, and (iv) residual or "leak" transport that remains when all specific transporters of the categories (i)–(iii) of the corresponding ion are silent or inhibited.

Radioactive isotopes played an important role in the discovery and characterization of pumps, carriers, and residual transport. Channels were identified and characterized mainly by the patch-clamp technique developed by Neher and Sakmann [11].

The application of radioisotopes for investigation of membrane transport has significantly declined over the last two decades. This is due to the fact that other techniques, especially the use of fluorescent dyes, are now available. For instance, a great variety of Ca^{2+}-sensitive fluorescent dyes, with high selectivity and sensitivity, are available, and there is no need to use the isotope $^{45}Ca^{2+}$ exclusively. However, for the investigation of Na^+ and K^+ transport across biological membranes (except through channels), the application of radioisotopes is the method of choice. This is due to the fact that the fluorescent dyes available are not sensitive enough to measure small changes in the Na^+ and K^+ concentrations inside or outside a cell. Part of this chapter focuses on techniques for ion flux measurements. Another part reviews investigations of selected ion transport pathways of RBCs. In addition, we draw attention to the fact that a combination of radioactive and fluorescent measurements might sometimes be useful for the identification of so far unknown ion transporters in membranes.

We focus on ion transport across the RBC membrane because these cells have been used to investigate cell membrane phenomena for more than 100 years. In some respect, they served as model systems for the discovery of general principles, including the physiology of membrane transport. RBCs are simple cells lacking intracellular organelles. They are easy to obtain, and there is no need to work under sterile conditions.

13.2   MEMBRANE PERMEABILITY FOR ELECTRONEUTRAL SUBSTANCES AND IONS

The flux of an ion or a substance is defined on the basis of the linear thermodynamic equation as

$$J_i = L_i X_i,$$

where J_i is the flux of the substance i, L_i is the linear thermodynamic coefficient, and X_i is the driving force. In the case of an electroneutral substance, X_i is the gradient of the chemical potential.

For an electroneutral substance, the flux in one (x) direction based on a one-dimensional concentration gradient (dc_i/dx) across the area (A) can be expressed as

$$J_i = -\frac{DAdc_i}{dx},$$

where D is the diffusion coefficient and is related to the average distance (x) over which a substance diffuses in one dimension per unit time: $x^2 = 2Dt$. To handle this equation, one assumes a constant concentration gradient within the barrier (e.g., the membrane). In addition, one takes the thickness of the barrier as constant. It follows

$$J_i = -\frac{DA\Delta c_i}{\Delta x}.$$

Often, the diffusion coefficient (D) is replaced by the permeability constant (P).
With $P_i = D_i \gamma / \Delta x$ (γ being the partition coefficient), one arrives at

$$J_i = -\frac{P_i A \Delta c_i}{\gamma}.$$

The partition coefficient (unitless parameter), which is relatively difficult to measure, is often neglected, and if this is done, the equation simplifies to

$$J_i = -P_i A \Delta c_i.$$

The unit of the flux is therefore $mol\,s^{-1}$. In biology, for example, for the description of fluxes across membranes, the flux is normalized to the area of the membrane. In other words, the flux is divided by the area and given in $mol\,cm^{-2}\,s^{-1}$.

For an ion flux, as opposed to that of a neutral molecule, X_i (see earlier text) corresponds to the electrochemical potential gradient. The corresponding solution under constant field conditions in the membrane results in the Goldman flux equation [12]

$$J_i = -\frac{P_i \alpha \left(c_i - c_0 \exp\alpha \right)}{\left(1 - \exp\alpha \right)}, \quad \text{with } \alpha = \left(\frac{z_i F}{RT} \right) \Delta\psi,$$

where c_i and c_0 are the inner and outer membrane surface concentration of ion i, $\Delta\psi$ is the transmembrane potential, and z_i, F, R, T have the usual meaning of charge valency, Faraday constant, gas constant, and absolute temperature, respectively. This equation describes an ion flux across a membrane as electrodiffusion. Both the transmembrane potential $(\Delta\psi = \psi_i - \psi_0)$ and the concentration difference at the inner and outer membrane surface $(c_i - c_0)$ are of importance for the ion flux (see Chapter 1 for the contribution of $\Delta\psi$ and the concentration difference to the energetics of transport).

It has to be taken into consideration that the flux equation for electroneutral substances as well as the equation for the diffusion of an ion can only be applied in the case of free diffusion of the substance or ion across the biological membrane. This is rarely the case. Pumps and carriers bind the substances or ions. For some carrier-mediated transport, not the substances but the substrate–carrier complex controls diffusion across the membrane. Transport through channels involves interactions of the transported ions with the channel structure. Each channel has a gating mechanism and contains a selectivity filter. Even the residual or "leak" ion flux cannot be explained by simple electrodiffusion. This would require stable (in time) water-filled pores in membranes that allow diffusion without interaction with the surrounding membrane material (for more detailed discussion of this point, see the following text).

For these reasons, the classical thermodynamic view of substance or ion fluxes across biological membranes is not very useful. However, thermodynamic flux analysis can be used to investigate the velocity of the movement of substances or ions across membranes. In addition, by modifying the surrounding conditions, one can measure the affinity of a transport protein for its substrate as well as parameters involved in the regulation of its activity.

13.3 KINETIC CONSIDERATIONS

Kinetic studies of substance or ion transport across membranes are the best way to investigate different transport processes. Fundamental reviews of this topic have been presented by Schultz [13] and Stein [14].

One of the classical (and simple) methods of flux measurements is the determination of radioactive tracer uptake, also called influx. The initial linear rate of influx declines with time until the intra- and extracellular concentrations become equal. After that time, the uptake equals the loss of the substance or ion under investigation. The equation for the uptake process can be written as

$$C_t = C_\infty \left(1 - \exp(-kt)\right),$$

where C_t is the radioactivity in the cells at time t, C_∞ is the radioactivity after equilibration, and k is the rate constant of the influx.

Alternatively, the loss of radioactivity from cells, also called efflux, can be determined by

$$C_t = C_0 \exp(-kt),$$

where C_t is again the radioactivity in the cells at time t and C_0 is the intracellular radioactivity at time zero. One can plot $\ln((C_\infty - C_t)/C_\infty)$ against t or $\ln(C_t/C_0)$ against t for influx and efflux, respectively, and obtain a straight line. The slope of the line represents the rate constant (k) of the influx or efflux.

The kinetic determination of the rate constant of influx or efflux assumes *trans*-zero conditions at the beginning of the measurements, that is, at time zero. This means that at time zero, the internal tracer concentration for influx determination and the external tracer concentration for efflux determination should be negligible compared to the tracer concentration on the other (*cis*) side. Such conditions require special cell preparations and measuring techniques, which are described in the next paragraph. The rate constant (k) has the dimensions of time^{-1}. The flux, with the units of moles per unit volume per unit time, can be obtained by multiplying the rate constant by the concentration of the transported substance or ion on the *cis* side at time zero. In addition, the apparent permeability of the membrane for the moving substance or ion can be calculated by multiplying the rate constant with the ratio of the cell volume to the cell surface area. If the cells were spherical, this would correspond to multiplying by $r/3$, where r is the radius of the cell. The apparent permeability then has the units of velocity, for example, $cm\,s^{-1}$.

To analyze carrier-mediated transport, the situation is a bit more complicated. In this case, saturation can occur. A plot of flux versus substance or ion concentration can be analyzed based on simple Michaelis–Menten kinetics or more complex kinetic models. In the first case, the flux (J) of the substance or ion is related to the apparent dissociation constant of the substance or ion (K_d) described by the following equation:

$$J = \frac{J_{max}C}{\left(C + K_d\right)},$$

where J_{max} is the maximal flux and C is the substance or ion concentration.

13.4 TECHNIQUES FOR ION FLUX MEASUREMENTS

13.4.1 Conventional Methods

As outlined previously, methods for determining influx or efflux require *trans*-zero conditions for the radioactive tracer at time zero and separation of the cells from the surrounding medium. Three possibilities have been principally described in the literature for measuring fluxes into or out of cells. Cells can be separated from the extracellular medium either by washing by centrifugation several times with a large volume of medium, or separated from the surrounding medium by centrifugation through oil or over an ion-exchange column, or by filtration. The advantages and disadvantages of the different techniques have been reviewed in detail recently [15].

To investigate fluxes that are relatively slow, the washing procedure is the method of choice. It has been applied for many cell types and has been described in several publications (see, e.g., [16]). In detail, the final cells should be washed by centrifugation into appropriate flux medium and the RBCs suspended in the flux medium at a hematocrit (i.e., the volume percentage of RBCs) of about 5% in a total volume of 1 ml contained in a 1.5 ml polypropylene microcentrifuge Eppendorf tube. To inhibit transport pathways, more or less specific inhibitors can be added. For instance, to investigate the residual K^+ influx (see following text), ouabain (0.1 mM), bumetanide (0.1 mM), and EGTA (0.1 mM) should be present in the flux medium to inhibit the K^+ transport mediated by the Na^+ pump, the $Na^+,K^+, 2Cl^-$ cotransporter (NKCC), and the Ca^{2+}-activated K^+ channel (EGTA is used to prevent a possible Ca^{2+} uptake), respectively. In addition, the composition of the flux medium can be changed. Such a strategy is of importance if no specific inhibitors are available for a certain transporter. To rule out the possibility that the K^+, Cl^- cotransporter (KCC) is involved in the measured K^+ influx, the Cl^--containing flux medium should be replaced by a medium that does not contain Cl^-. One possibility is to replace NaCl and KCl by

$NaCH_3SO_4$ and KCH_3SO_4, respectively, taking care that the tonicity of the medium remains constant (e.g., replacing 145 mM NaCl by 165 mM $NaCH_3SO_4$). If the chosen temperature for the flux measurement is different from room temperature, the tube with the cell suspension for flux measurement should be equilibrated at the flux temperature for at least 5 min. After that time, a small amount of radioactive substance can be added to start the influx measurement. For K^+ uptake, isotonic KCl or $KCH_3SO_4 + {}^{86}RbCl$ (Rb^+ as a congener for K^+) can be used at a final radioactivity of about 50 kBq/ml. ^{86}Rb is normally used as a tracer for K^+ because available K^+ isotopes, for example, ^{42}K or ^{43}K, are not very practical due to their short half-lives. The total concentration of KCl or $KCH_3SO_4 + {}^{86}RbCl$ can be chosen as required. However, for classical K^+ uptake assays, 7.5 mM has been described as appropriate (e.g., [17]). After shaking the tubes with the cell suspension and the radioisotope, the tubes are placed in a rack that is put in a water bath at the required temperature. Incubation time (i.e., the flux time) should be short enough to measure the initial rate of uptake, avoiding a significant back flux (i.e., <10% of the final equilibrium concentration). For K^+ influx at 37°C or 20°C, 30 min is adequate, whereas at lower temperature 1–2 h is required to accumulate comparable radioactivity in the cells.

The isotope uptake is stopped by quick (10 s) centrifugation at high speed (15,000 g, e.g., in an Eppendorf microcentrifuge) and the supernatant removed by aspiration. It is convenient to handle 12 Eppendorf tubes in a rack at the same time, which means centrifuging them together and removing the supernatant from one after the other, using a pump to suck the supernatant into a separate flask for waste. The cells should then be washed free of extracellular radioactivity by several successive resuspensions and centrifugations (15,000 g for 10 s). A careful vortexing of the tubes is important after each resuspension, before centrifugation. It has been established that for RBCs at 5% hematocrit (in a 1 ml cell suspension in the Eppendorf tubes) four washes, that is, five spins, are practicable. In this case, the packed cell volume in the tubes will be about 50 μl and that of the trapped extracellular medium about 10 μl. Diluted 100-fold after each wash, it will result in a dilution of 10^8 fold after 4 washes.

From the activity of the starting reaction medium and the rate of uptake of the radioactively labeled substance, one is able to calculate whether the chosen amount of washing steps will be sufficient [15]. The washing procedure can be refined by applying strategies to further limit the uptake or loss of radioactively labeled substance during the washing procedure. One possibility is to use ice-cold isotonic $MgCl_2$ solutions for washing the cells. Other possibilities are the addition of chelators, for example, EGTA for Ca^{2+}, or specific transport inhibitors.

Comparable experiments can be carried out for Na^+ influx measurements. In this case, however, more specific transport pathways involving Na^+ have to be taken into consideration (see following text). Either ^{22}Na or ^{24}Na (in NaCl) can be applied. In the case of measuring Ca^{2+} influx using ^{45}Ca, the inhibition of the Ca^{2+} pump should be considered. This can be achieved either by using a permeable Ca^{2+} chelator, an inhibitor such as orthovanadate, or by ATP depletion, as has been described by Lew [18].

In the case of relatively fast fluxes, the washing procedure to separate the cells from the surrounding medium is not applicable. In these cases, the cells have to be separated from the medium using oil or filtration. For the first method, one has to select an oil type, not toxic to the cells, which has a density greater than that of the flux solution but less than that of the cells. Diphthalate esters and silicone oils in particular have been reported as being useful. For instance, for RBCs, n-dibutyl phthalate with a density of $1.047 \, g \, ml^{-1}$ is often used (e.g., [19]). First, an oil layer is added to 1.5 ml Eppendorf tubes. An appropriate solution for stopping the flux, for example, an ice-cold solution, is placed on top of the layer. The influx measurement is achieved by adding the radioactive substance to the cell suspension prepared in a separate flux–flask at time zero. A sample of the cell suspension from the flux–flask is taken and added to the upper layer of the tubes at appropriate time intervals. The tubes are rapidly centrifuged using a microcentrifuge until the cells have moved through the oil to the bottom of the tubes. After that time, the oil and the stop solution, as well as the extracellular medium, can be removed by aspiration and the cells analyzed.

A comparable method is the filtration technique. The application of this technique for RBCs has been described in detail by Dalmark and Wieth [20]. A great variety of filter materials with different pore sizes are available; originally, a cellulose ester filter with a mean pore size of 1.2 μm from Millipore was used for RBCs.

Rapid centrifugation as well as filtration can also be applied for efflux measurements. In this case, an initial loading of the cells is required, taking up to several hours. This is a clear disadvantage of the efflux compared to influx measurements, although it must be stated that for most cell types, it is easier to obtain cell-free extracellular medium rather than the cell pellet required for influx measurements.

In all cases, the influx or efflux of a substance or ion can be expressed in units of moles per unit of volume per unit of time (see earlier text), for example, in mol per volume of cells per s. Therefore, normally one has to determine the radioactivity as well as the volume of cells in the sample under investigation. The specific activity of the radioisotope in the sample can be determined by counting a suitable sample of the radioactive stock solution. The number of cells can be measured either by a cell counter or a microscope and a counting cell chamber. For RBCs, the amount of hemoglobin in the sample can be determined with a photometer and then converted into the number of liters of cells (l_{cells}) in the sample. Therefore, the flux in this case can be obtained in the unit: $mmol/(l_{cells} \times h)$ (see, e.g., [16]).

13.4.2 Alternative Method

Semiautomatic techniques for the measurement of the rate constant of Na^+ and $K^+(Rb^+)$ effluxes from RBCs, which of course can also be used for other cell types (e.g., muscle, *Xenopus* oocytes), have been established. The general idea is to measure directly the remaining radioactivity in the cells and is described in Bernhardt et al. [21]. $^{86}RbCl$ (Rb^+ as congener for K^+) is added to an RBC suspension and incubated (hematocrit about 30%) for at least 1 h to give significant ^{86}Rb loading. The RBC suspension is then put into a temperature-controlled diffusion chamber. This chamber (0.1 ml) is closed by a membrane filter with 3 μm pores, placed in the measuring 20 ml chamber, and continuously flushed with nonradioactive solution at 5 ml min^{-1}. To prevent sedimentation, the diffusion chamber is mounted on a slowly rotating horizontal axle making 12 rotations per min. The radioactivity in the measuring chamber is measured continuously by gamma scintillation counting. Thus, the time course of radioactivity release can be followed. Because the rate constant of K$^+$(Rb$^+$) efflux from the cells is small compared with the rate constant of the flux from the extracellular solution into the flushing solution (washout), in practice only the radioactivity released from the RBCs is measured. Only at the beginning of the experiment can a fast washout effect be detected. The measuring chamber is successively flushed with two different solutions in the course of one typical experiment. Within the first 100 min, the RBC physiological suspending solution is used. This solution is then replaced by a solution of a desired composition. To check the reversibility of the change in the observed rate constant, in some experiments the initial physiological solution can be used again for the final flush. The observed rate constant of the K$^+$(Rb$^+$) efflux can be determined by graphical analysis. The results of a typical experiment are shown in Figure 13.1 [21].

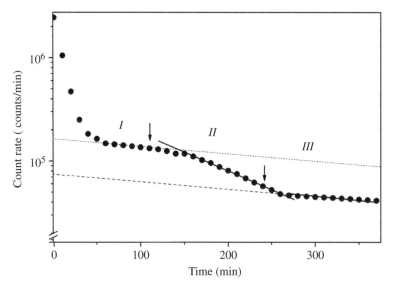

FIGURE 13.1 Radioactivity decrease of ^{86}Rb-loaded RBCs with rapid changes of the flushing solution (↓). I: Initial radioactivity decrease using the suspending solution (mM: NaCl 141.3, KCl 5.7, glucose 5.0, Na$_2$HPO$_4$/NaH$_2$PO$_4$ 5.8, pH 7.4) as flushing solution. II: Low ionic strength solution (mM: sucrose 250, KCl 5.7, glucose 5.0, Na$_2$HPO$_4$/NaH$_2$PO$_4$ 5.8, pH 7.4) as flushing solution. III: Flushing solution as in I. Modified from Ref. 21 with kind permission from Springer Science and Business Media.

13.5   KINETIC ANALYSIS OF ION TRANSPORTER PROPERTIES

Varying ion concentrations for protein-mediated transport systems provides kinetic information on the transporter properties. For example, measuring K^+ influx via the Na^+ pump as a function of external K^+ concentration gives a sigmoid curve with a Hill coefficient of about 2, which is consistent with the presence of two K^+ binding sites on the external face of the pump. A refinement of this approach can be seen in Figure 13.2 [22]. Using thallium (Tl^+) as a K^+ substitute in the external medium, Na^+ efflux through the pump (ouabain sensitive) is measured as a function of the external Tl^+ concentration. Two important observations can be made. Firstly, the sigmoid curve mentioned earlier for K^+ where Tl^+ is substituting for K^+ (see Fig. 13.2b) indicates two Tl^+ (K^+) binding sites with an affinity of 0.03 mM. Additionally, the effect of increasing external Tl^+ at a low external K^+ concentration initially activates K^+ influx by occupying the second external site and promoting K^+ influx (see Fig. 13.2a). Then, as the Tl^+ concentration is increased further, Tl^+ starts to occupy both sites, inhibiting K^+ uptake.

By contrast, fluxes via channels, which do not have binding as part of the transport process, are linear with substrate concentration up to very high substrate concentration. This difference in kinetic behavior has been taken in the past as one piece of evidence distinguishing the two classes of transporter proteins, namely, carriers and channels. Other evidence includes *trans*-stimulation (e.g., [23]) and temperature dependence, although attention must be paid to the change in pH with temperature.

To alter *internal* concentrations for kinetic studies is clearly more difficult than simple replacement of the external media. This is reflected by the power of early experiments using perfused squid giant axons [24]. Other procedures include reversible cation replacement methods, including the use of *p*-chloromercuribenzene sulfonate (PCMBS) or nystatin [22, 25, 26], or the lysis techniques for preparing resealed red cell ghosts [27].

Ion substitution experiments can identify cotransport activity. For example, NKCC function can be identified as coupled fluxes by replacing K^+ or Na^+ with an inert cation, for example, *N*-methyl D-gluconate (NMDG) or choline, and measuring the uptake of radioactive Na^+ or K^+. Similarly, Cl^- can be replaced with an inert anion, such as methyl sulfate or nitrate (see Section 13.4.1). It must be taken into consideration that altering the ionic concentrations on either side of the membrane may lead to changes in the transmembrane potential, leading to a change of the electrochemical driving force and/or the osmotic pressure and hence cell volume.

Inhibitors remain an important method for defining particular transport systems. Some examples such as the use of ouabain to inhibit the Na^+ pump are extremely specific, while furosemide, as an inhibitor of Cl^--associated transport systems (anion exchanger, band 3 (AE1), KCC, NKCC), is very promiscuous. Metabolic inhibitors such as azide, cyanide, and iodoacetic acid can indirectly reduce transport via ATP-driven systems, and nonspecific chemical modification, for example, cross-linking agents or SH group inhibitors, can reduce many transport functions.

(a)

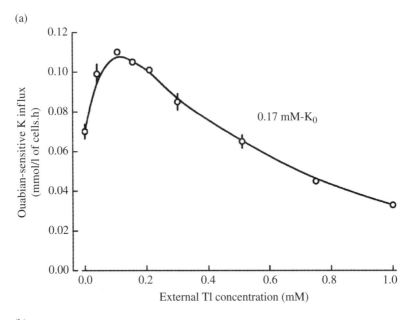

(b)

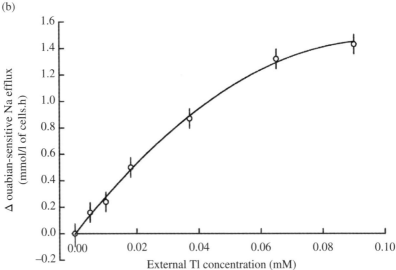

FIGURE 13.2 The effect of extracellular Tl$^+$ on Na$^+$ pump activity. For more details, see text. Reproduced from Ref. 22 with permission from Wiley.

13.6 SELECTED CATION TRANSPORTER STUDIES ON RED BLOOD CELLS

After the discovery of the Na$^+$ pump, it became possible to explain the nonsymmetrical and nonequilibrium distribution of Na$^+$ and K$^+$ between the intra- and extracellular medium. The so-called "pump-leak" concept was developed for these ions by

Tosteson and Hoffman [28]. Originally, the "leak" was thought to be simple electro-diffusion of Na^+ and K^+. Later, it became evident that the residual transport, that is, "leak," consists of a variety of carriers and channels. Such findings were made for nearly all cells. For instance, it was shown in RBCs from horse, human, and rat that the K^+ uptake consists of a linear and a saturable component [28–31]. Interestingly, a saturable component of the K^+ uptake could also be seen in the presence of ouabain [32]. Beginning with the work of Wiley and Cooper [33], who first discussed the presence of a cotransport system for Na^+ and K^+ that is sensitive to diuretic drugs (e.g., furosemide) but insensitive to ouabain, a large variety of specific passive, protein-mediated ion transport pathways have been discovered in the RBC membrane. The presently known transport pathways for monovalent cations in the human RBC membrane are presented in Figure 13.3 [34]. All of them, except the nonselective, voltage-dependent cation (NSVDC) channel, which was first demonstrated to exist in RBCs by the patch-clamp technique, have been found and investigated in more detail using radioactive isotopes.

In addition to the transport pathways for monovalent cations shown in Figure 13.3, present in the RBC membrane of various species, there is a Na^+/Ca^{2+} exchanger found in the RBC membrane of some mammalian carnivores (dog, cat, seal, bear, ferret). RBC membranes from these species do not have a Na^+ pump [35].

In parallel, a great variety of carriers and channels for other ions and specific substances such as amino acids, sugars, lipids, etc. have been described in biological cell membranes, including RBCs. We now focus on two specific examples of applications of radioactive flux measurements on RBCs, that is, the activity of the K^+,Cl^- cotransporter (KCC) and the investigation of the residual fluxes of Na^+ and K^+ after the known specific pumps, carriers, and channels for these ions have been turned off. The $K^+(Na^+)/H^+$ exchanger is discussed in detail later because its discovery in the

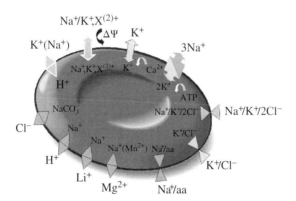

FIGURE 13.3 Summary of the principal transport pathways for Na^+ and K^+ in the human RBC membrane: Na^+ pump, $Na^+,K^+,2Cl^-$ symporter (NKCC), K^+,Cl^- symporter (KCC), Na^+-dependent amino acid (aa) transport (several discrete transporters), $Na^+(Mn^{2+})/Mg^{2+}$ antiporter, Na^+/Li^+ antiporter, Na^+/H^+ antiporter, $NaCO_3^-/Cl^-$ exchange (via band 3), $K^+(Na^+)/H^+$ antiporter, Ca^{2+}-activated K^+ channel, nonselective voltage-dependent cation (NSVDC) channel. Reproduced from Ref. 34 with kind permission from Springer Science and Business Media.

human RBC membrane is directly related to the explanation of the membrane residual or "leak" transport over the last century.

13.6.1 K⁺,Cl⁻ Cotransport (KCC)

KCC isoforms 1–4 represent K^+ and Cl^- symporters, with KCC1 being the housekeeping form ubiquitously expressed. The transporter operates in either influx or efflux mode. In addition to radioactive tracer measurements, K^+ efflux as the external appearance of K^+ or Rb^+ can be determined also by flame and atomic absorption spectroscopy (see Chapter 14 for uptake studies). In this case, the cells must be preloaded with Rb^+ by incubation in isotonic RbCl, and efflux measured into an appropriate medium containing a low K^+ concentration (0.2 mM) in a high Na^+ medium. Working at a hematocrit of 5%, K^+ or Rb^+ fluxes are typically measured over 10 min to 1 h at 37°C. Hemolysis must be kept low under the experimental conditions and hemoglobin (Hb) in the external medium measured by a spectrophotometer if any lysis is detectable by eye. Recently, the K^+ congener Tl^+, detectable via a fluorescence assay, has been proposed for measuring Tl^+ (K^+) transport (efflux) in a high-throughput system [36].

KCC activity is normally measured in the presence of ouabain (50 μM) and bumetanide (20 μM) to inhibit the Na^+ pump and NKCC transporter. The fraction of residual flux via KCC can be defined either using the inhibitor (dihydroindenyl)oxy acetic acid (DIOA; 100 μM) or Cl^- substitution. Anion replacement should be via an inert monovalent anion, for example, methyl sulfate, rather than nitrate, which might produce reactive oxygen species and interact with Hb. It is also important that cell volume is kept constant, so microhematocrit measurements should be made to check cell volume in ionic replacement media. This is particularly critical if isosmotic sucrose media are used to substitute K^+ and/or Na^+. A Good buffer ([37]; e.g., morpholinopropane sulfonic acid (MOPS)) should keep pH nearly constant with temperature changes (4, 20, or 37°C), and if cation substitution is required, N-methyl D-gluconate (NMDG) should be used (see earlier text).

KCC is normally inhibited by phosphorylation and can be stimulated with either N-ethylmaleimide (NEM) or staurosporine as kinase inhibitors. It is reduced by phosphatase inhibitors, calyculin or okadaic acid. Fluxes are also stimulated by low pH (6–6.5), cell swelling, and low Mg^{2+} concentration.

A variety of commercial antibodies are available to identify KCC isoforms via western blotting of solubilized membranes. However, in the case of RBCs, the low membrane abundance of KCC makes these experiments technically challenging.

13.6.2 Residual Transport

Originally, passive transport was envisaged as a simple electrodiffusive "leak" that could be described by the classical Goldman flux equation ([12], see also Section 13.2). However, it is necessary to consider and eliminate transport via the large variety of specific pathways in the RBC membrane given in Figure 13.3 using inhibitors and ion substitution. One should realize that in many publications calculations of the cation permeability coefficients have been made without taking into account the existence of these specific transport pathways. The problem of identifying the membrane permeability coefficients for ions is related directly to the estimation of the electrical conductance of the membrane. A detailed overview in this respect is given by Hoffman [38].

In general, the problem of how a monovalent cation passes (diffuses) through a biological membrane in the absence of a specific transport system is still unsolved. The obvious model system for ion transport studies, including studies on the electrodiffusive mechanism, appeared to be the RBC, which offered many experimental advantages for ion flux measurements. It is much easier to measure the residual K^+ transport in these cells in comparison to the residual Na^+ transport because of the greater variety of specific transport systems for Na^+ compared to K^+. However, one has to consider that even in the presence of specific inhibitors there are still two K^+ transport pathways present in the human RBC membrane, which are not affected. These are the NSVDC channel and the $K^+(Na^+)/H^+$ exchanger. The first is activated at positive transmembrane potentials only [39–42] and, therefore, should not play a substantial role under physiological conditions (i.e., negative transmembrane potentials). The contribution of the second pathway, however, is more complicated to analyze (see Section 13.7). The $K^+(Na^+)/H^+$ exchanger has been identified in the human RBC membrane by Richter et al. [43] and Kummerow et al. [44] and cloned in humans [45].

Several conditions, summarized in Ref. 34, have been described under which the assumed residual K^+ and Na^+ fluxes can be altered, not taking into consideration all existing specific transport pathways. One important parameter is the reduction of the ionic strength of the extracellular solution at constant isotonic conditions realized by the replacement of extracellular NaCl by sucrose (low ionic strength (LIS) effect), which leads to an enhancement of the (ouabain + bumetanide + EGTA)-insensitive Na^+ and K^+ influx as well as efflux [46]. It has also been demonstrated that the NSVDC channel and the KCC do not participate in this LIS effect [47].

There is an important increase in K^+ efflux of a variety of mammalian RBCs, including human RBCs, in isotonic LIS solutions (NaCl replaced by sucrose). For the history, see Bernhardt and Weiss [34]. However, the replacement of NaCl by sucrose results not only in a change of the ionic strength but also in a change of the transmembrane potential [48–50]. Therefore, attempts have been made to explain the increased K^+ efflux of human RBCs solely on the basis of electrodiffusion [49], although, because the K^+ influx is also enhanced, this invalidates this approach [34, 46]. Further evidence arises from the observation that no enhanced K^+ efflux was seen in bovine RBCs under the same experimental conditions when the transmembrane potential changes in the same way as for human RBCs [51, 52].

13.7   COMBINATION OF RADIOACTIVE ISOTOPE STUDIES WITH METHODS USING FLUORESCENT DYES

There are several possible explanations for the LIS effect (see Section 13.6.2) on residual transport that have been discussed in detail by Bernhardt and Weiss [34]. However, the most realistic was to assume an electroneutral cation carrier mechanism to explain the LIS-stimulated residual K^+ and Na^+ fluxes. The first attempt in this respect was made by Denner et al. [46]. Model calculations of possible carrier mechanisms and tracer-kinetic K^+ flux measurements at different pH's of the extracellular solutions finally lead to the idea put forward by Richter et al. [43] that a $K^+(Na^+)/H^+$ exchanger is involved in the LIS-stimulated residual K^+ and Na^+ fluxes. A fundamental idea in the work of Richter et al. [43] is the assumption that the local K^+ and Na^+ concentration near the binding site of the hypothetical carrier, that is, near the membrane surface, are of importance for the carrier-mediated ion fluxes. As described by the linearized Gouy–Chapman equation, a reduction in the ionic strength of the extracellular solution at a constant negative surface charge density leads to an enhancement of the absolute value of the negative outer membrane surface potential. This in turn results in a higher cation concentration near the surface in comparison to the free solution, as expressed by the Boltzmann equation. Combining these two equations, and taking the ion flux to be proportional to the ion concentration, as demonstrated in separate experiments, a final equation is obtained showing that the logarithm of the apparent rate constant of the ion flux is inversely proportional to the square root of the ionic strength (for more details, see Ref. 43). When the experimental results of the logarithm of the flux rate constants for the residual K^+ and Na^+ influxes or effluxes are plotted versus 1/square root of the extracellular ionic strength straight lines are in fact observed for all four fluxes, as theoretically predicted (see Fig. 13.4 [43]).

In addition to the theoretical investigations based on Na^+ and K^+ flux measurements, it was necessary to demonstrate experimentally a 1:1 relation of the residual transport, that is (ouabain + bumetanide + EGTA)-insensitive, K^+ efflux and the H^+ influx. Therefore, Kummerow et al. [44] investigated the change of intracellular pH of RBCs under different experimental conditions using the pH-sensitive fluorescent dye 2,7-biscarboxyl-5(6)-carboxyfluorescein (BCECF). The net H^+ influx was calculated from intracellular pH (pH_i) measurements taking into account the buffer capacity of hemoglobin ($10\,mmol/(mmol\ hemoglobin \times pH$ unit); [53]). To convert the fluorescence ratio of BCECF-loaded RBCs into pH_i values, a calibration was carried out by equalizing pH_i and pH_o using the K^+/H^+ ionophore nigericin. In addition to the pH_i and K^+ efflux (^{86}Rb as tracer) measurements, the Cl^- efflux (^{36}Cl as tracer) from RBCs in LIS media in the absence of extracellular Cl^- was determined.

When human RBCs were suspended in a physiological NaCl solution ($pH_o = 7.4$), the measured pH_i was 7.19 ± 0.04, which remained constant for 30 min. When RBCs were transferred into an LIS solution (NaCl replaced by sucrose), an immediate alkalinization increased the pH_i to 7.70 ± 0.15, which was followed by a slower cell acidification.

It was demonstrated that the effect of immediate alkalinization of cells in LIS media occurs via the anion transport system (band 3) [44]. The main focus in the investigations of Kummerow et al. [44], however, was the process of cell acidification after the initial pH_i increase. The movement of H^+ (or H^+ equivalents) was found to increase with decreasing ionic strength of the solution. As for K^+ and Na^+

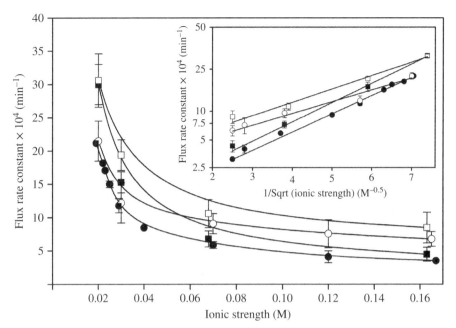

FIGURE 13.4 Effect of gradual NaCl or KCl (for Na$^+$ influx) replacement (0–145 mM) by sucrose (decreased ionic strength, constant osmolarity) on the rate constant of Na$^+$ (O, □) and K$^+$ (●, ■) efflux (□, ■) and influx (O, ●) of human RBCs. Na$^+$ and K$^+$ fluxes were measured in the presence of 0.1 mM ouabain, bumetanide, and EGTA. The insert shows the same experimental results plotted as logarithm versus reciprocal of the square root of ionic strength of the flux solutions. Reproduced from Ref. 43 with permission from Elsevier.

fluxes, the logarithm of the H$^+$ influx rate constant showed a straight line when plotted versus 1/square root of ionic strength. The comparison of the calculated H$^+$ influx with the measured unidirectional K$^+$ efflux at different extracellular ionic strengths showed a correlation with a stoichiometry nearly 1:1 (see Fig. 13.5 [44]). For bovine and porcine RBCs, in LIS media, H$^+$ influx and K$^+$ efflux were of comparable magnitude, but only about 10% of the fluxes observed in human RBCs under LIS conditions. Quinacrine, a known inhibitor of the mitochondrial K$^+$(Na$^+$)/H$^+$ exchanger [54], inhibited the K$^+$ efflux in LIS solution by about 80% and inhibited the H$^+$ influx nearly completely (demonstrated by pH measurements of cell lysates).

Based on the effect of three classical anion transport inhibitors (4,4′-diisothiocyanato stilbene-2,2′-disulfonate (DIDS), 4,4′-dinitro stilbene-2,2′-disulfonate (DNDS), and niflumic acid) on the K$^+$ efflux and H$^+$ influx in LIS solution, the involvement of the anion exchanger (band 3) in the process of acidification could be ruled out (for details, see [44]).

Alternatives to a K$^+$(Na$^+$)/H$^+$ exchanger for explaining the described findings are carefully discussed by Bernhardt and Weiss [34]. The authors were able to conclude that the K$^+$(Na$^+$)/H$^+$ exchanger is much more likely compared to the alternatives.

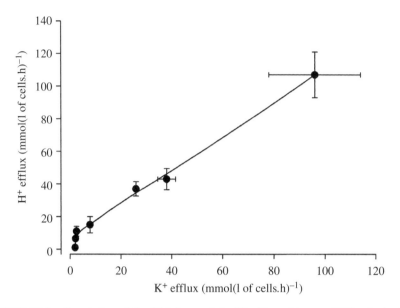

FIGURE 13.5   Correlation of the H^+ influx and the K^+ efflux of human RBCs obtained at different NaCl concentrations of the extracellular solution. NaCl concentration was reduced from 145 to 0.3 mM (sucrose replacement). Results (mean ± S.D.) are pooled data from at least four independent experiments. Where not shown, the error bar is smaller than the symbol. Reproduced from Ref. 44 with kind permission from Springer Science and Business Media.

In summary, based on combination of flux measurements using radioactive isotopes and fluorescent dyes, it was possible to describe long-existing and recently described experimental findings with a so far not-identified cation transporter in the human RBC membrane.

13.8   CONCLUSIONS

The use of radioisotopes for the characterization of ion transport pathways in biological membranes has been a useful tool for over 70 years. It is still the method of choice for the investigation of Na^+ and K^+ transport because effective fluorescent indicators to determine small changes in Na^+ and K^+ concentration are lacking. A variety of techniques to measure fluxes of Na^+ and K^+ across biological membranes have been developed, which are relatively well characterized and readily deployed if some limitations are taken into consideration. In addition, the combination of flux measurements using radioisotopes along with other techniques such as flux measurements using fluorescent dyes offers a suitable approach to detect novel ion transport pathways.

REFERENCES

1. C. Nägeli, *Pflanzenphysiologische Untersuchungen, Heft 1*, Friedrich Schulthess, Zürich, 1855.

2. S.S. Zaleski, Carl Schmidt, *Chem. Ber.* 27 (1894), 963–978.

3. D.D. Van Slyke, H. Wu, F.C. McLean, Studies of gas and electrolyte equilibria in the blood. V. Factors controlling the electrolyte and water distribution in the blood, *J. Biol. Chem.* 56 (1923) 765–849.

4. W.O. Fenn, D.M. Cobb, Electrolyte changes in muscle during activity, *Am. J. Physiol.* 115 (1936) 345–356.

5. W.E. Cohn, E.T. Cohn, Permeability of red corpuscles of the dog to sodium ion, *Prog. Soc. Exp. Biol. Med.* 41 (1939) 445–449.

6. A.J. Eisenman, L. Ott, P.K. Smith, A.W. Winkler, A study of the permeability of human erythrocytes to potassium, sodium and inorganic phosphate by the use of radioactive isotopes, *J. Biol. Chem.* 135 (1940) 165–173.

7. R.B. Dean, T.R. Noonan, L. Haege, W.O. Fenn, Permeability of erythrocytes to radioactive potassium, *J. Gen. Physiol.* 24 (1941) 353–365.

8. J.C. Skou, The influence of cell cations on adenosine triphosphatase from peripheral nerves, *Biochim. Biophys. Acta* 23 (1957) 394–401.

9. R.B. Dean, Theories of electrolyte equilibrium in muscle, *Biol. Symp.* 3 (1941) 331–348.

10. A. Krogh, The active and passive exchange of inorganic ions through the surface of living cells and through living membranes generally, *Proc. R. Soc. (Lond.) Biol. Sci.* 133 (1946) 140–200.

11. E. Neher, B. Sackmann, Single-channel currents recorded from membrane of denervated frog muscle fibres, *Nature* 260 (1976) 799–802.

12. D.E. Goldman, Potential, impedance and rectification in membranes, *J. Gen. Physiol.* 27 (1943) 37–60.

13. S.G. Schultz, *Basic Principles of Membrane Transport*, Cambridge University Press, Cambridge, 1980.

14. W.D. Stein, *Transport and Diffusion Across Cell Membranes*, Academic Press, Orlando, FL, 1986.

15. A. Hannemann, U. Cytlak, R.J. Wilkins, J.C. Ellory, D.C. Rees, J.S. Gibson, The use of radioisotopes to characterize the abnormal permeability of red blood cells from sickle cell patients, in: N. Singh (Ed.), *Radioisotopes: Applications in Bio-medical Science*, InTech, Rijeka, 2011.

16. J.D. Young, J.C. Ellory, Flux measurements, in: J.C. Ellory and J.D. Young (Eds.), *Red Cell Membranes a Methodological Approach*, Academic Press, London, 1982, pp. 119–133.

17. J.C. Ellory, P.W. Flatman, G.W. Stewart, Inhibition of human red cell sodium and potassium transport by divalent cations, *J. Physiol.* 340 (1983) 1–17.

18. V.L. Lew, On the ATP dependence of the Ca^{2+}-induced increase in K^+ permeability observed in human red cells, *Biochim. Biophys. Acta-Biomembr.* 333 (1971) 827–830.

19. K. Kirk, H.Y. Wong, B.C. Elford, C.I. Newbold, J.C. Ellory, Enhanced choline and Rb^+ transport in human erythrocytes infected with the malaria parasite *Plasmodium falciparum*, *Biochem. J.* 278 (1991) 521–525.

20. M. Dalmark, J.O. Wieth, Temperature dependence of chloride, bromide, iodide, thiocyanate and salicylate transport in human red cells, *J. Physiol.* 224 (1972) 583–610.

21. I. Bernhardt, E. Donath, R. Glaser, Influence of surface charge and transmembrane potential on rubidium-86 efflux of human red blood cells, *J. Membr. Biol.* 78 (1984) 249–255.

22. J.D. Cavieres, J.C. Ellory, Thallium and the sodium pump in human red cells, *J. Physiol.* 243 (1974) 243–266.

23. W.R. Lieb, A kinetic approach to transport studies, in: J.C. Ellory and J.D. Young (Eds.), *Red Cell Membranes a Methodological Approach*, Academic Press, London, 1982, pp. 135–164.

24. T. Begenisich, P. De Weer, Potassium flux ratio in voltage-clamped squid giant axons, *J. Gen. Physiol.* 76 (1980) 83–98.

25. P.J. Garrahan, A.F. Rega, Cation loading of red blood cells, *J. Physiol.* 193 (1967) 459–466.

26. J.D. Cavieres, Alteration of red cell Na and K using pCMBS and nystatin, in: J.C. Ellory and J.D. Young (Eds.), *Red Cell Membranes a Methodological Approach*, Academic Press, London, 1982, pp. 179–185.

27. D.E. Richards, D.A. Eisner, Preparation and use of resealed red cell ghosts, in: J.C. Ellory and J.D. Young (Eds.), *Red Cell Membranes a Methodological Approach*, Academic Press, London, 1982, pp. 165–177.

28. D.C. Tosteson, J.F. Hoffman, Regulation of cell volume by active cation transport in high and low potassium sheep red cells, *J. Gen. Physiol.* 44 (1960) 169–194.

29. T.I. Shaw, K movements in washed erythrocytes, *J. Physiol.* 129 (1955) 464–475.

30. I.M. Glynn, Sodium and potassium movement in human red cells, *J. Physiol.* 134 (1956) 278–316.

31. L.A. Beaugé, O. Ortiz, Rubidium, sodium and ouabain interactions on the influx of rubidium in red blood cells, *J. Physiol.* 210 (1970) 519–533.

32. I.M. Glynn, The action of cardiac glycoside on sodium and potassium movements in human red cells, *J. Physiol.* 136 (1957) 148–173.

33. J.S. Wiley, R.A. Cooper, Inhibition of cation cotransport by cholesterol enrichment of human red cell membranes, *Biochim. Biophys. Acta-Biomembr.* 413 (1975) 425–431.

34. I. Bernhardt, E. Weiss, Passive membrane permeability for ions and the membrane potential, in: I. Bernhardt and J.C. Ellory (Eds.), *Red Cell Membrane Transport in Health and Disease*, Springer, Berlin, 2003, pp. 83–109.

35. I. Bernhardt, A.C. Hall, J.C. Ellory, Transport pathways for monovalent cations through erythrocyte membranes, *Studia Biophys.* 126 (1988) 5–21.

36. D. Zhang, S.M. Gopalakrishnan, G. Freiberg, C.S. Surowy, A thallium transport FLIPRR-based assay for the identification of KCC2-positive modulators, *J. Biomol. Screen.* 15 (2010) 177–184.

37. N.E. Good, G.D. Winget, W. Winter, T.N. Conolly, S. Izawa, R.M.M. Singh, Hydrogen ion buffers for biological research, *Biochemistry* 5 (1966) 467–477.

38. J.F. Hoffman, Estimates of the electrical conductance of the red cell membrane, in: E. Bamberg and H. Passow (Eds.), *Progress in Cell Research*, vol. 2, Elsevier, Amsterdam, 1992, pp. 173–178.

39. P. Bennekou, P. Christophersen, A human red cell cation channel showing hysteresis like voltage activation/inactivation, *Acta Physiol. Scand.* 146 (1992) 608.

40. P. Bennekou, The voltage-gated non-selective cation channel from human red cells is sensitive to acetylcholine, *Biochim. Biophys. Acta-Biomembr.* 1147 (1993) 165–167.

41. L. Kaestner, C. Bollensdorff, I. Bernhardt, Non-selective voltage-activated cation channel in the human red blood cell membrane, *Biochim. Biophys. Acta-Biomembr.* 1417 (1999) 9–15.

42. L. Kaestner, P. Christophersen, I. Bernhardt, P. Bennekou, The non-selective voltage-activated cation channel in the human red blood cell membrane: Reconciliation between two conflicting reports and further characterisation, *Bioelectrochemistry* 52 (2000) 117–125.

43. S. Richter, J. Hamann, D. Kummerow, I. Bernhardt, The monovalent cation "leak" transport in human erythrocytes: An electroneutral exchange process, *Biophys. J.* 73 (1997) 733–745.

44. D. Kummerow, J. Hamann, J.A. Browning, R. Wilkins, J.C. Ellory, I. Bernhardt, Variations of intracellular pH in human erythrocytes via $K^+(Na^+)/H^+$ exchange under low ionic strength conditions, *J. Membr. Biol.* 176 (2000) 207–216.

45. M. Numata, J. Orlowski, Molecular cloning and characterization of a novel $(Na^+, K^+)/H^+$ exchanger localized to the trans-Golgi network, *J. Biol. Chem.* 276 (2001) 17387–17394.

46. K. Denner, R. Heinrich, I. Bernhardt, Carrier-mediated residual K^+ and Na^+ transport of human red blood cells, *J. Membr. Biol.* 132 (1993) 137–145.

47. I. Bernhardt, E. Weiss, H.C. Robinson, R. Wilkins, P. Bennekou, Differential effect of HOE642 on two separate monovalent cation transporters in the human red cell membrane, *Cell. Physiol. Biochem.* 20 (2007) 601–606.

48. W. Wilbrandt, H.J. Schatzmann, Changes in the passive cation permeability of erythrocytes in low electrolyte media, *Ciba Found. Study Group* 5 (1960) 34–52.

49. J.A. Donlon, A. Rothstein, The cation permeability of erythrocytes in low ionic strength media of various tonicities, *J. Membr. Biol.* 1 (1969) 37–52.

50. I. Bernhardt, A.C. Hall, J.C. Ellory, Effects of low ionic strength media on passive human red cell monovalent cation transport, *J. Physiol.* 434 (1991) 489–506.

51. I. Bernhardt, Untersuchungen zur Regulation des Ouabain-insensitiven Membrantransports monovalenter Kationen an Erythrozyten, D.Sc. thesis, Humboldt University, Berlin, 1986.

52. I. Bernhardt, A. Erdmann, R. Vogel, R. Glaser, Factors involved in the increase of K^+ efflux of erythrocytes in low chloride media, *Biomed. Biochim. Acta* 46 (1987) 36–40.

53. M. Dalmark, Chloride and water distribution in human red cells, *J. Physiol.* 250 (1975) 65–84.

54. K.D. Garlid, D.J. DiResta, A.D. Beavis, W.H. Martin, On the mechanism by which dicyclohexylcarbodiimide and quinine inhibit K^+ transport in rat liver mitochondria, *J. Biol. Chem.* 261 (1986) 1529–1535.

14

CATION UPTAKE STUDIES WITH ATOMIC ABSORPTION SPECTROPHOTOMETRY (AAS)

THOMAS FRIEDRICH

Institute of Chemistry, Technical University of Berlin, Berlin, Germany

14.1 INTRODUCTION

This chapter describes a method to quantify the activity of cation-transporting membrane proteins, with a particular focus on electroneutrally operating cation exchange transporters that are inaccessible to conventional voltage or patch clamping. It particularly focuses on two closely related K^+-countertransporting P-type ATPases, the net electrogenic Na^+,K^+-ATPase and the electroneutrally operating H^+,K^+-ATPase. Upon heterologous expression of these enzymes in *Xenopus laevis* oocytes, the uptake of monovalent alkali metal cations such as Li^+, Rb^+, or Cs^+ into individual cells can be assayed by atomic absorption spectrophotometry (AAS) with high accuracy, offering in addition the promising perspective of measuring ion fluxes under accurate membrane voltage control. The method is a sensitive and safe alternative to radioisotope flux experiments (see Chapter 13) and facilitates structure–function investigations as well as complex kinetic studies.

The investigation of ion transport by transmembrane proteins is still a challenging task in biophysical research. In the case of transporters with net electrogenic activity or ion channels, powerful electrophysiological techniques such as the two-electrode voltage-clamp or the patch-clamp technique (see Chapter 3) can be applied, which allows the analysis of ion transport under accurate transmembrane potential control.

Pumps, Channels, and Transporters: Methods of Functional Analysis, First Edition. Edited by Ronald J. Clarke and Mohammed A. A. Khalid.
© 2015 John Wiley & Sons, Inc. Published 2015 by John Wiley & Sons, Inc.

For this to be successful, either native cells expressing the protein of interest are used or heterologous expression in electrophysiological model cell lines like *X. laevis* oocytes or mammalian cell culture (HEK, CHO, COS, HeLa, and others) is attempted. However, problems arise if the achievable expression level in the plasma membrane is limited, or if the natural locus of expression is the membrane of some intracellular organelle such as mitochondria, lysosomes, endosomes, or the nuclear envelope, or if the host cell is phylogenetically far distant from the native organism (e.g., bacteria, cyanobacteria, algae). In such cases, protein purification and reconstitution into artificial membrane systems (planar lipid bilayers, either freely suspended as in the BLM technique (see Chapter 2) or with solid support as in the SSM configuration (see Chapter 6) or proteoliposomes can yield access to electrogenic activity. In these cases, it is often difficult to gain control over the preferential orientation of the protein in the artificial membrane system. This problem is a concern for both electrogenic and electroneutral transporters. However, particularly in the case of electroneutral transporters, where electrophysiological methods can't be applied, alternative means of investigation are required.

The situation is easily exemplified for a special pair of proteins belonging to the large superfamily of P-type ATPases, whose members are distributed in all kingdoms of life. They have in common the property of transporting their substrates across membranes at the expense of the free energy of ATP hydrolysis. P-type ATPases exhibit two major conformational states (E_1 and E_2) that differ in substrate affinities and the sidedness of accessibility of the binding sites and they reversibly form a phosphorylated intermediate (hence the classification as "P-type" as opposed to V-type or F_0F_1-type ATPases) within a conserved DKTG motif that serves in defining the whole superfamily. The transported substrates range from protons, to alkali (Na^+, K^+) and alkaline earth (Ca^{2+}, Mg^{2+}) metal cations, to transition metal cations (e.g., Cu^+), and even to phospholipids. Furthermore, the substrate specificity has still to be defined for many P-type ATPases. From the aforementioned sibling pair, that is, the Na^+,K^+-ATPase and the H^+,K^+-ATPase, the Na^+,K^+-ATPase is a net electrogenic transporter translocating $3Na^+$ ions out of and $2K^+$ ions into the cell per ATP molecule hydrolyzed. The Na^+,K^+-ATPase can well be considered as one of the most intensively studied ion transporters on earth, and it has even frequently served as a test system for pioneering new electrophysiological techniques. In contrast, the H^+,K^+-ATPase from gastric mucosa is inaccessible by voltage or patch clamping, due to its overall net electroneutral transport mode of $2H^+$ out versus $2K^+$ into the cell. The H^+,K^+-ATPase is particularly interesting from a bioenergetics perspective because it can generate the largest known concentration gradients in living systems. Because the pH in the stomach lumen can fall below 1, whereas the pH in gastric parietal cells remains around neutral, a concentration gradient of up to 10^6 is generated by the H^+,K^+-ATPase. However, this cannot be achieved by a constant $2H^+:2K^+:1ATP$ stoichiometry, because at low pH values in the stomach the free energy of ATP hydrolysis is insufficient for the pumping of $2H^+$ ions across the membrane (see Section 1.2 for a discussion of ion transport energetics). This has invoked the suggestion of a still largely hypothetical stoichiometry change to $1H^+:1K^+:1ATP$ at very acidic pH values, discussed further in Section 14.5.3. Scrutinizing this hypothesis will challenge biochemists and biophysicists.

14.2 OVERVIEW OF THE TECHNIQUE OF AAS

AAS is a spectral analytic method for the quantitative detection of elements by measuring the absorption of optical radiation by free atoms in the gas phase. Owing to the high accuracy of the method, AAS devices are widely distributed in analytical and chemical laboratories. The principal components of an atomic absorption spectrophotometer are depicted in Figure 14.1. The technique requires radiation sources that emit light of element-specific wavelengths originating from a particular electronic transition of the element of interest. In modern AAS devices (see Fig. 14.2), the radiation source is modulated in order to discriminate element-specific absorption from any constant emission originating from the flame or atomizer (see Section 14.2.1). The resulting spectral lines have bandwidths of only a few picometers (pm), which gives the technique its element selectivity. The element-specific radiation passes through the vapor of a probe containing atoms of the respective element, which absorb the element-specific light. After passing the probe, the radiation is sent through a monochromator (to separate the element-specific radiation from any other type of radiation emitted by the radiation source) and is finally measured by a detector (e.g., a photomultiplier tube). From the measured absorbance, the analyte concentration or mass can be calculated based on the law of Lambert and Beer. For analyzing a sample for its atomic constituents, the probe has to be atomized, that is, all chemical bonds need to be broken. Atomization is critical because any chemical bonds present modify the electronic energy levels of the element of interest such that the energy of its electronic transitions no longer coincides with that of the element-specific radiation source.

The temperature is the most critical parameter of the atomization process. The temperature must be high enough to atomize the constituents of the probe, but for the sake of spectral simplicity, not too high that higher energy levels can be populated.

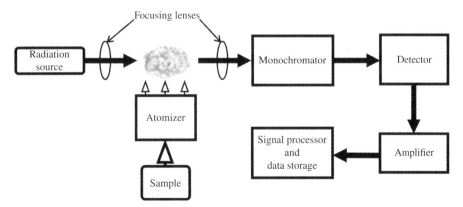

FIGURE 14.1 Principal components of an atomic absorption spectrophotometer (AAS). Picture "AASBLOCK" by K05en01, licensed under public domain via Wikimedia Commons—http://commons.wikimedia.org/wiki/File:AASBLOCK.JPG#mediaviewer/File:AASBLOCK.JPG.

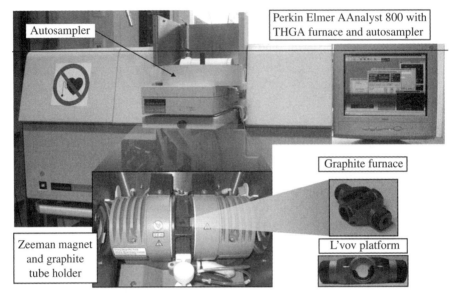

FIGURE 14.2 Overview of the AAS device used in this work. The Perkin Elmer AAnalyst 800 is equipped with a transversally heated graphite atomizer furnace, magnets for Zeeman background subtraction and autosampler.

The atomizers commonly used in modern AAS devices are flames and electro-thermal (graphite tube) atomizers. Section 14.2.1 gives a historical overview of the development of the technique and describes some of its critical components relevant for the applications described in this chapter. Most of this information is derived from the excellent textbook by Welz and Sperling [1], to which the interested reader is referred (also available in English).

14.2.1 Historical Account of AAS with Flame Atomization

The early foundations of atomic absorption spectroscopy date back to the discoveries of William Hyde Wollaston in 1802 and the later systematic studies by Josef von Fraunhofer, who described the black lines in the solar spectrum, which Sir David Brewster in 1820 proposed to result from absorption processes in the solar atmosphere. The fundamental principle was already expressed in Gustav R. Kirchhoff's radiation law: A body that, at a given temperature, exhibits a stronger absorption must also exhibit a more intensive emission, by which atomic absorption is regarded as the mirror image of atomic emission. Although Kirchhoff formulated this principle in 1859, it was not until 100 years later that the practical importance of his discovery was fully recognized for element analytics. Thus, it was in fact a rediscovery of AAS when Sir Alan Walsh in Melbourne, Australia [2], and C. Th. J. Alkemade and J. H. W. Milatz [3, 4] in the Netherlands independently described the principles of a generally applicable analytic method based on atomic absorption in 1955. Walsh, who was most influential in the field at this time, completed a patent application in 1954, but still it was not until 1963 that the first fully commercial-built AAS device (the Perkin Elmer "model 303") became available on the market.

Walsh wondered about the peculiar difference between the spectrochemical analysis of molecules and metals: Molecules were measured in absorption, whereas metals were measured by atomic emission. Because the Lambert–Beer law should apply for any absorption process, Walsh asked whether absorption of element-specific bands by the atoms of an analyte in the gas phase could provide an equally sensitive method for quantitative analysis. From his systematic studies, Walsh formulated the essentials of such a system: (i) the use of spectral line emitters as light sources, obviating the need for monochromators with picometer resolution, which are mandatory if continuum radiation sources are used; (ii) modulation of the light source that eliminates radiation from the atomizer and other nonmodulated background; and (iii) the use of transparent laminar flames ("Meker–Fisher" or "Lundegårdh" flames) that ensure comparatively low atomization temperatures of around 2000°C compared to electric arc atomization (around 5000°C). Walsh's preference for flame atomization followed the necessity of spectral simplicity, because under these circumstances only the lowest excitation state of an atom contributes to absorption. Furthermore, flame atomization avoids problems with spectral interference, which are common for electric arc atomization. Thus, Walsh set the stage for the "flame AAS" (F-AAS) format of the technique, as opposed to the "graphite furnace" ASS (GF-AAS) mode described in Section 14.2.3. The Lundegård acetylene/air flames in use today for liquid analysis employ a pneumatic nebulizer in a mixing chamber before the gas/analyte mixture exits a slit to produce a planar, laminar flame in the light path of the spectrometer.

14.2.2   Element-Specific Radiation Sources

Because the quality of the light source is critical for AAS, considerable technical efforts were undertaken to develop element-specific, line-emitting lamps with high intensity and spectral quality. Either electrical glow discharge lamps termed "hollow cathode lamps" (HCL; due to their electrode geometry) or low-pressure plasma discharge lamps known as "electrodeless discharge lamps" (EDL) are in use. HCL sources employ a hollow, mostly cylindrical cathode containing the analyte and a well-isolated anode made from tungsten or nickel within a quartz glass bulb filled with low-pressure inert gas such as Ne or Ar under 1 kPa. The cylindrical shape of the cathode ensures a "focusing" of the electrical discharge within the cylinder and, therefore, a suitable spatial intensity distribution of the emitted radiation. The EDL combine highest radiation intensity and narrow linewidth. Here, excitation power is transmitted inductively from outside the quartz bulb (which also contains inert gas at 1 kPa and micrograms of analyte) by a high-frequency electromagnetic field. Historically, the switch from gigahertz to low operation frequencies in the 100 MHz range and the use of micrograms of analyte greatly improved the long-term stability of lamp operation. Multielement lamps that emit spectral lines of several analytes are also available. Considerable effort has been expended on the use of the much cheaper continuum radiation sources such as Xe or Hg arc lamps and halogen lamps. However, as already pointed out by Walsh (reported in Ref. 2), this requires high-quality mono-chromators with 2 pm resolution such as Échelle spectrometers. Furthermore, common continuum sources have low intensity in the spectrally important region below 280 nm, and light source modulation, especially for multielement applications, imposes additional difficulties. However, continuum sources are employed for background absorption correction (see Section 14.2.4). Tunable lasers may become more important in the future, once the near UV region can also be covered by these devices at reasonable prices. Walsh's important principle of modulation is commonly achieved either by modulating the discharge current of the light source or with a rotating, sectorized aperture, with modulation frequencies in the 100 Hz range. A strong alternating magnetic field that produces a Zeeman effect can also be employed. This is an interesting option because the Zeeman effect can also be used to correct for background absorption (see Section 14.2.4).

14.2.3 Electrothermal Atomization in Heated Graphite Tubes

Besides using a flame, atomization can also be performed electrothermally, for which electrically heated graphite tubes are in use today. This so-called GF-AAS technique was introduced by the Russian chemist Boris L'vov from Leningrad State University in 1959 [5], who had access to the 1955 publication of Walsh and—as an analytical chemist—also to a graphite tube oven and HCLs. L'vov immediately rec-ognized the potential of the technique proposed by Walsh and independently worked out the principles of GF-AAS. Graphite tubes can be heated very rapidly by passing high current (resistive heating at the electrode-graphite contacts) and allow tempera-tures of 2000–2500°C to be reached under inert gas flow (Ar) to prevent burning of the tube. Modern GF-AAS devices control the inert gas flow around and within the graphite tube independently and stop the internal gas flow during the atomization phase. In modern AAS devices equipped with graphite tube atomizers, a drop of liquid sample is introduced into the graphite tube, which is then heated rapidly, thus vaporizing and atomizing the sample. The element-specific light from the radiation source passes through the vapor-filled graphite furnace to measure the atom concentration in the gas phase. The main principles of GF-AAS operation worked out by L'vov [6], which were summarized by Walter Slavin in 1981 in his "stabilized temperature platform furnace" (STPF) concept [7], are the following: (i) The atom-ization time must be short relative to the dwell time of the atom vapor in the tube, requiring very fast heating rates of $1500–2000°C\,s^{-1}$; (ii) isothermal conditions are required, meaning no significant temperature gradients between the tube walls and the site of probe atomization; and (iii) it is mandatory to measure the integrated extinction over time rather than determining only the peak amplitude, which requires a fast readout system.

To meet isothermal conditions, L'vov integrated the so-called "L'vov platform" within the graphite tubes, on which the probe is deposited for atomization. Because the platform has only a small connection with the tube wall, it is mainly heated by the surrounding tube walls, providing that it only reaches the atomization

TABLE 14.1 Temperature Profile for the Measurement of Rb in a Transversely Heated Graphite Furnace Atomic Absorption Spectrophotometer.

Step	Temperature (°C)	Ramp Time (s)	Hold Time (s)
1 (Drying)	110	1	20
2 (Drying)	130	5	30
3 (Pyrolysis)	300–600	10	20
4 (Atomization)	1700–1800	0	5
5 (Cleanout)	2400	1	2

temperature when tube and inert gas have reached local thermic equilibrium ("platform effect"). Modern graphite tubes are made from anisotropic pyrolytic graphite. In the "integrated platform" concept, the platform and the tube are sintered from a single mold, with only one suspending connection between them. The orientation of the heating direction to the optical path discriminates between longitudinally heated (LH-GFs) and transversely heated tubes (TH-GFs). As an example of a typical temperature profile of a TH-GF AAS measurement, the procedure for Rb is given in Table 14.1 [8].

The first two steps serve to evaporate water from the probe; the subsequent pyrolysis step removes any additional volatile (e.g., organic) constituents. This is followed by a short atomization step, during which the inert gas flow within the tube is stopped. During this period, the element-specific absorption is measured. The last step (cleanout) should purge the tube of any less volatile matter (compared to the analyte).

Because the ability to separate the analyte from more or less volatile probe constituents depends on the nature of the chemical bond(s) in which the analyte is engaged, much effort has been spent on the use of so-called matrix modifiers that aim to control the separation of the analyte from accompanying substances in combination with a suitable atomization temperature. For example, for the determination of copper in sea water, it is necessary to remove NaCl due to its relatively high boiling point (1413°C). If NH_4NO_3 is added, $NaNO_3$, which decomposes above 380°C, and NH_4Cl, which sublimes at 335°C, are formed, both of which are easily removed from the probe during the pyrolysis step. The literature about the best modifiers for each element is abundant. However, B. Welz showed in 1992 [9] that a mixture of palladium nitrate and magnesium nitrate works best for more than 20 elements, suggesting that the search for better modifiers might be dispensable.

F-AAS and GF-AAS commonly achieve very high and in most cases very similar sensitivities, which are expressed in terms of characteristic concentrations (F-AAS) or characteristic masses (GF-AAS). Some examples are given in Table 14.2 (data from Ref. 1).

A major difference between F-AAS and GF-AAS is the probe volume. F-AAS usually uses milliliters of solution, whereas GF-AAS employs some tens of microliters. The latter format is advantageous for the single-cell measurements described here. If single-cell equivalents were diluted, for example, to a volume of 10 ml, it would be very difficult to overcome the threshold given by the characteristic concentration (see Section 14.3).

TABLE 14.2   Characteristic Concentration and Mass Levels Detected by Different AAS Techniques for a Selection of Metals Relevant for Transport Studies.

Element	F-AAS Characteristic Concentration $(mg\,l^{-1})$	LH-GF AAS Characteristic Mass (pg)	TH-GF AAS Characteristic Mass (pg)
Rb	0.1	2.3	10
Cs	0.1	5	12
Li	0.03	1.4	5.5
Cu	0.03	8	12

14.2.4   Correction for Background Absorption

Depending on the layout of an AAS system, unspecific attenuation of radiation can occur due to the absorption of other gas molecules or scattering by other particles in the atomization zone. Methods to compensate for background absorption rely on the fact that absorption by molecules or scattering are rather broadband effects. Thus, in principle, background absorption can be measured at any wavelength different from the element-specific line. However, because spectral variations of background absorption can be large, the effect should be measured close to the element-specific wavelength. For F-AAS devices, continuum sources like deuterium lamps are used to provide an automated, quasi-simultaneous measure and compensation for background absorption. Although this procedure works well in general, background subtraction should be employed carefully, especially in the case of unknown probe composition or complex matrices such as biological extracts. In these cases, measurement with and without background subtraction is recommended.

Background absorption is particularly problematic in longitudinally heated graphite furnaces, and the use of continuum sources for background subtraction has proven unsatisfactory for these devices. For GF-AAS, background subtraction is performed on the basis of the Zeeman effect, which takes advantage of spectral splitting of degenerate emission/absorption lines in a magnetic field due to its interaction with the angular momentum of atoms. Using this principle, the absorption of a probe is first measured without external magnetic field, where total absorption is the sum of specific and background absorption, and then with magnetic field, which shifts the specific emission/absorption away from the central line so that only background absorption remains. Thus, with an alternating magnetic field, quasi-simultaneous background subtraction can be performed. At the same time, the use of an alternating magnetic field for Zeeman background subtraction allows simultaneously Walsh's modulation principle to be most straightforwardly obeyed. From all possible combinations (magnet at the light source or at the atomizer, constant or alternating magnetic field, field along or perpendicular to light path), the configuration of highest accuracy was found to be an alternating field magnet at the atomizer with magnetic force lines parallel to the light path.

14.3 THE EXPRESSION SYSTEM OF *XENOPUS LAEVIS* OOCYTES FOR CATION FLUX STUDIES: PRACTICAL CONSIDERATIONS

Xenopus laevis is a common laboratory animal because its oocytes are easy to handle as an expression system. The method described here takes advantage of large cell size (about 1.0–1.5 mm diameter; see experimental outline in the caption to Fig. 14.3) and that the protein expression level achievable exhibits low cell-to-cell variability. The high sensitivity of AAS experiments facilitates quantification of ion transport, as exemplified by a simple calculation: For the TH-GF AAS variant, the detection limit ("characteristic mass"; see Table 14.2) for Rb is 10 pg or 1.2×10^{-13} mol (Rb: $85.47 \, g \, mol^{-1}$). Expression of Na^+,K^+-ATPase in *Xenopus* oocytes results in pump currents of about 100 nA (i.e., 6.2×10^{11} elementary charges per second or $1.03 \times 10^{-12} \, mol \, s^{-1}$) yielding a charge transport of 6×10^{-6} Coulombs (C) per minute. Because the transport of one net charge per pump cycle corresponds to the uptake of two Rb^+ ions ($3Na^+:2K^+$ stoichiometry), 100 nA current for 5 min is equivalent to an uptake of 6×10^{-10} mol Rb^+. Thus, even upon a 1000-fold dilution of the sample (homogenization of a single oocyte with about 1 µl volume in 1 ml water), a typical THGA-AAS sample of 20 µl contains 1.2×10^{-11} mol Rb^+ (or 1.2 ng), which is far above the detection threshold. Therefore, even transporters with a 100-fold lower transport activity (e.g., $1 \, s^{-1}$) or with limited plasma membrane expression can reliably be assayed by appropriately adjusting the flux time of the experiment.

Because the pumping rate of an active transporter is highly temperature dependent, it is mandatory to carry out the uptake experiment under precise temperature control. Because typical activation energies for the Na^+,K^+-ATPase are in the range of $90–130 \, kJ \, mol^{-1}$, a ΔT of 1°C changes cation flux by 13% at 20°C (for $90 \, kJ \, mol^{-1}$ activation energy). Another important precaution is to select oocytes carefully for homogenous size because the membrane area scales quadratically with cell radius. This in mind, experimental errors of less than 10% with 10 cells per experimental condition can be achieved. Another advantage of the oocyte system is that ion fluxes can be performed during two-electrode voltage clamping, which assures accurate control of the membrane potential and permits the measured ion flux to be correlated with membrane potential.

14.4 EXPERIMENTAL OUTLINE OF THE AAS FLUX QUANTIFICATION TECHNIQUE

The procedure to measure cation uptake fluxes mediated by active transporters in *X. laevis* oocytes is outlined in Figure 14.3. *In vitro* cRNA transcription, cRNA injection, and protein expression in *Xenopus* oocytes proceed according to established protocols [8, 10, 11]. To discriminate the activity of the heterologously expressed Na^+,K^+-ATPase (α-subunit + β-subunit) from the endogenous Na^+,K^+-ATPase of the oocytes, the "RD" mutations (e.g., Q116R and N127D in the human α_2-subunit) need to be introduced to provide a protein with reduced ouabain sensitivity (IC_{50} in the millimolar range) [12]. Importantly, for the study of Na^+,K^+-ATPase, an additional

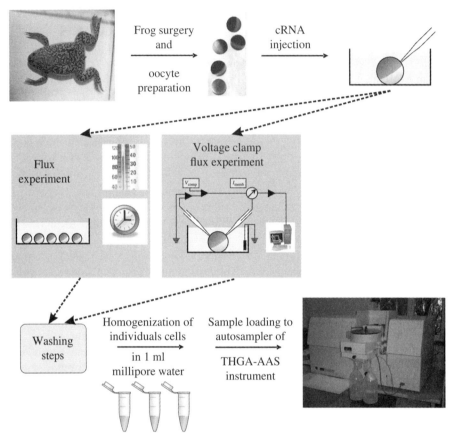

FIGURE 14.3 Experimental scheme for cation uptake experiments on *Xenopus* oocytes and subsequent analysis by atomic absorption spectrophotometry (AAS). See Sections 14.3 and 14.4 for details.

Na$^+$ loading step is required prior to activity measurements to sufficiently elevate the intracellular Na$^+$ concentration in the oocytes [13], which is a prerequisite for the enzyme to perform efficient 3Na$^+$/2K$^+$ exchange. For gastric H$^+$,K$^+$-ATPase, such a loading step is not required. Before the flux experiments, the Na$^+$,K$^+$-ATPase- or H$^+$,K$^+$-ATPase-expressing (or noninjected control) oocytes are incubated in buffer PLB (100 mM NaCl, 1 mM CaCl$_2$, 5 mM BaCl$_2$, 5 mM NiCl$_2$, and 2.5 mM MOPS, 2.5 mM TRIS, pH 7.4) supplemented with 10 μM ouabain to preblock the endogenous Na$^+$,K$^+$-ATPase. For the same reason, all solutions used for flux studies should contain 10 μM ouabain.

After the preincubation steps, groups of 5–10 oocytes are simultaneously transferred into Petri dishes filled with PLB-based flux buffers with appropriate concentrations of Rb$^+$ or other cations to be assayed. The optimal time of cation uptake should be determined for the respective transporter in preliminary experiments (e.g., 3 min for Na$^+$,K$^+$-ATPase and 15 min for H$^+$,K$^+$-ATPase). Afterward, the oocytes are simultaneously transferred into Rb$^+$-free PLB buffer, which is repeated twice, followed by one washing step in Millipore water. Subsequently, the oocytes are individually transferred into Eppendorf tubes prefilled with 1 ml Millipore water and homogenized with a narrow-gauge pipette. These probes are transferred into the sample cups of the THGA-AAS device (e.g., Perkin Elmer AAnalyst 800) and loaded by the autosampler of the instrument. Prior to analyzing the samples, a standard calibration curve is measured by analyzing a series of analyte (here: Rb$^+$) concentrations that cover the linear range of absorption for the element under study (for Rb$^+$, 10–50 µg l^{-1}). After measuring the experimental probes, the results need to be checked for samples, in which the measured amount of Rb$^+$ exceeds the maximum of the set of concentrations used for the generation of the calibration curve. In this case, the respective samples must be appropriately diluted and measured again. Probes can be stored at −20°C for several weeks for additional analyses.

Whenever cation flux determination is desired under membrane voltage control, after appropriate preincubation as described previously, the oocytes expressing the transporter of interest are transferred into the perfusion chamber of a two-electrode voltage-clamp setup. For two-electrode voltage clamping, the cells are gently impaled with sharp borosilicate capillaries (with an opening of about 20 µm, filled with, e.g., 3 M NaCl). Vibrations of the electrode holders have to be avoided to prevent damage to the cell membrane. After clamping to the desired potential, the bath solution in the perfusion chamber is rapidly exchanged for the respective flux buffer. Also in this case, precise temperature control is mandatory. For the electrogenically operating Na$^+$,K$^+$-ATPase, the pump current developing in the presence of countertransported cations is measured with appropriate electrophysiology software. After swift switching to a bath solution devoid of transported cations, the cell is removed from the recording chamber and subjected to the washing steps described earlier. The result of such a combined voltage-clamp/flux experiment is described in Section 14.5.4.

Because the plasma membrane expression of the transporter of interest is critical for the overall flux, the measured fluxes sometimes need to be cross-checked against the amount of transporter protein at the cell surface. For this purpose, a selective extraction of the plasma membrane protein fraction from the same cell preparation as used for the flux studies can be carried out [11, 14, 15]. Subsequent Western blotting using a specific antibody and densitometric analysis of protein bands helps to correlate fluxes with surface expression. The band intensities can be used to normalize the measured fluxes of various transporter constructs to the plasma membrane expression level to resolve ambiguities arising from variable numbers of transporters per cell surface area.

14.5 REPRESENTATIVE RESULTS OBTAINED WITH THE AAS FLUX QUANTIFICATION TECHNIQUE

The following section describes a series of experiments in which the AAS flux technique has been applied to the study of Na$^+$,K$^+$- and H$^+$,K$^+$-ATPases. Because the Na$^+$,K$^+$-ATPase has already been thoroughly characterized by electrophysiology, the emphasis here is on the H^+,K^+-ATPase, for which only limited data is available from intact cellular systems. Using this assay, the influence of extracellular cations, extra- and intracellular pH, and transmembrane voltage, as well as the effects of mutations on transport by the gastric H^+,K^+-ATPase, could be clarified.

14.5.1 Reaction Cycle of P-Type ATPases

To understand the considerations regarding certain reaction cycle intermediates of P-type ATPases, the cyclic reaction is briefly described (see Fig. 14.4), which is commonly termed the Post–Albers mechanism in tribute to R. L. Post and W. Albers, who first devised the scheme [16, 17], as adapted to the gastric H^+,K^+-ATPase. Upon intracellular H^+ binding to the E_1 conformation (step 1), a phosphointermediate with occluded H^+ ($E_1P(H^+)$) is formed (step 2). After a conformational change to E_2P (step 3), H^+ dissociate to the extracellular space (step 4). Subsequently, K^+ ions bind from the extracellular side (step 5) and become occluded. This stimulates dephosphorylation (step 6), and after a conformational change from E_2 to E_1 (step 7), K^+ ions are released intracellularly (step 8). The gray box indicates the reaction sequence that can be studied by voltage pulses, with the general assumption that all steps (intra- as well as extracellular H^+ release/reverse binding and the $E_1P\leftrightarrow E_2P$ conformational change) might be dependent on membrane potential.

14.5.2 Rb$^+$ Uptake Kinetics: Inhibitor Sensitivity

Figure 14.5a–c shows results from a Rb$^+$ uptake experiment with oocytes expressing the human Na$^+,$K$^+$-ATPase composed of the α_2-and the β_1-subunit, either in the presence of 100 mM extracellular Na$^+$ (see Fig. 14.5a) or upon replacement of Na$^+$ for NMDG$^+$ (see Fig. 14.5b; extracellular Na$^+$-free conditions). Rb$^+$ uptake increases

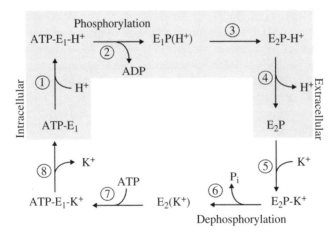

FIGURE 14.4 Reaction mechanism of gastric H^+,K^+-ATPase adapted from the Post–Albers scheme, as originally postulated for the Na$^+,$K$^+$-ATPase [16, 17]. Reproduced from Ref. 18 with permission from the Public Library of Science (PLoS One).

in a dose-dependent fashion. The Rb⁺ uptake tested on noninjected oocytes from the same preparation (in the presence of 10 mM Rb⁺ in the flux buffer) indicates that the level of unspecific Rb⁺ uptake is small. In addition, the effect of 10 mM ouabain on the Rb⁺ uptake in the presence of 1 mM Rb⁺ (also at 0.1 mM Rb⁺ in B) shows the specificity of the measured fluxes for the overexpressed enzyme. Analysis of the flux data with a Michaelis–Menten-type fit function yields the apparent K_M values for Rb⁺ transport. Figure 14.5d shows the results of similar experiments on gastric H⁺,K⁺-ATPase (α-subunit mutant S806C, used also for voltage-clamp fluorimetry experiments; see Chapter 4) at an extracellular pH of 5.5 [8]. The data on both enzymes show that the concentration for half-maximal activation of Rb⁺ uptake is substantially larger in the presence of extracellular Na⁺, which is due to the competition of Na⁺ with Rb⁺ ions for the extracellular-facing ion binding sites in the E_2P conformation of the pump molecules.

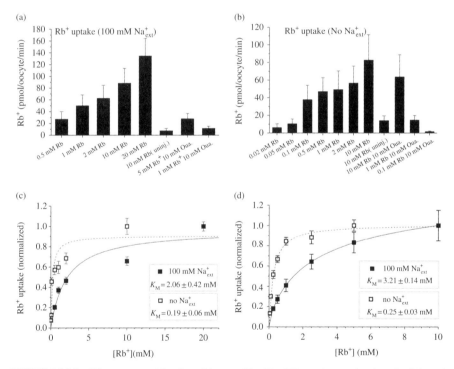

FIGURE 14.5   Rb⁺ transport kinetics of human Na⁺,K⁺-ATPase (a–c, subunits α2 + β1) and rat gastric H⁺,K⁺-ATPase (d) and dependence of Rb⁺ uptake on extracellular Na⁺ at an extracellular pH of 5.5 (c) shows fits of a Michaelis–Menten-type function to the data from (a and b), respectively. Data in (a and b) were obtained from 14 to 34 oocytes from three batches. (d) Rb⁺ uptake by rat gastric H⁺,K⁺-ATPase (α-subunit mutant S806C, which has also been used for voltage-clamp fluorimetry experiments, see Chapter 4) at extracellular pH 5.5 with 8–12 oocytes out of three batches. Maximal uptake rates at 10 mM Rb⁺ used for normalization were (in pmol/oocyte/min): 29.3/27.4/30.8 in the absence of Na⁺ and 21.1/23.4 in the presence of Na⁺. Data in panels (a–c): J. Drucker and T. Friedrich (unpublished observations). Data in (d) were reproduced from Ref. 8 with permission from the Journal of Visualized Experiments.

14.5.3 Dependence of Rb+ Transport of Gastric H+,K+-ATPase on Extra- and Intracellular pH

Acidification of the stomach lumen is performed by H+,K+-ATPase residing apically in parietal cells of the gastric mucosa. Because the gastric juice can reach pH values below 1, whereas the intracellular pH in parietal cells remains neutral, the gastric H+,K+-ATPase generates the largest ion concentration gradient (more than 10^6-fold) known in nature. To achieve this, thermodynamic considerations suggest that the free energy of hydrolysis of one ATP molecule is insufficient for extrusion of two H+ at such a steep gradient, which invokes a stoichiometry change to 1H+:1K+:1ATP at very acidic pH values. From a bioenergetics viewpoint, extracellular acidification should impose a large energetic barrier for H+ extrusion and, thus, H+/K+ exchange. Thus, the data shown in Figure 14.6a [18] were rather paradoxical because a decrease of the extracellular pH from 7.4 to 5.5 led to a more that twofold increase of Rb+ uptake by gastric H+,K+-ATPase. To resolve this enigma, again the combination of voltage-clamp fluorometry (see Chapter 4), Rb+ uptake determination by AAS, and mutagenesis was helpful. In voltage-clamp fluorometry experiments, the voltage dependence of fluorescence amplitudes also exhibited the perplexing feature that the enzyme's E_1P/E_2P distribution was shifted toward E_2P upon extracellular acidification (see Fig. 14.6b [18]). This prompted the hypothesis that extracellular acidification imposes a strong effect on the intracellularly oriented ion binding sites, which overcompensates the effect at the outward-facing binding sites. Protons are exceptional among cations because they can permeate more easily through biological membranes. If, for example, bicarbonate is present in the buffer solution, the neutral carbonic acid can be formed [19], which then passes the membrane and subsequently dissociates intracellularly to acidify the cytoplasm.

To test, whether an extracellular pH change from 7.4 to 5.5 entails intracellular acidification, the so-called "acid-bath" method was employed to achieve a fairly well-defined acidification of the cytoplasm. For this purpose, extracellular pH 7.4 buffers were used that contained 40 mM Na-butyrate. Butyric acid is a weak organic acid ($pK_a = 4.82$) that can—like the carbonic acid/bicarbonate couple—permeate through membranes in the neutral acid form and dissociate intracellularly, so that an intracellular acidification by about 0.5 units (from pH ~7.1 to ~6.6) is achieved [18, 20], whereas the extracellular solution was set to pH 7.4. Remarkably, the Rb+ flux measured in extracellular pH 7.4 buffer with 40 mM butyrate was very similar to the uptake determined with an extracellular pH 5.5 solution, and also the voltage-dependent distributions between E_1P and E_2P states from voltage-clamp fluorometry nearly superimposed for these two conditions [18]. The flux studies showed that an extracellular acidification by about two pH units essentially does not affect H+/K+ turnover of the enzyme, but rather that a slight intracellular acidification by approximately 0.5 pH units increases turnover more than twofold (compare pH 7.4 fluxes with and without butyrate in Fig. 14.6a). The shift of the E_1P/E_2P equilibrium with intracellular acidification toward E_2P was corroborated by the fact that at extracellular pH 7.4 with butyrate, the $E_2(P)$-specific inhibitor SCH28080 was more effective than at pH 7.4 without butyrate [18]. The more profound inhibition in an extracellular pH 5.5 solution is due to the additional effect that SCH28080 itself underlies a

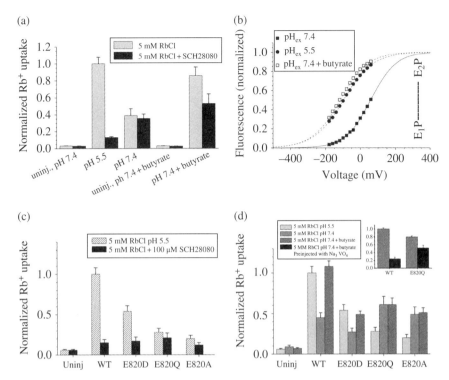

FIGURE 14.6   Intra- and extracellular pH dependence of Rb⁺ transport by gastric H⁺,K⁺-ATPase. (a) Rb⁺ uptake by H⁺,K⁺-ATPase measured in the absence (gray) or presence (black) of 10 μM SCH28080 at different extracellular pH and ionic conditions. Results from noninjected and H⁺,K⁺-ATPase-expressing oocytes (α-subunit mutant S806C + WT β-subunit) at extracellular pH 5.5, pH 7.4, and pH 7.4 + 40 mM butyrate are shown. Data are from three experiments on different cell batches with 15–20 oocytes per condition and normalized to the maximal Rb⁺ uptake at pH 5.5 (uptake rates were 15.4, 23.1, and 39.0 pmol/oocyte/min). (b) Voltage dependence of fluorescence amplitudes from voltage-clamp fluorometry at extracellular pH 7.4 (■), pH 7.4 + 40 mM butyrate (□), and pH 5.5 (•). Fits of a Boltzmann-type function are superimposed to each data set, and the fluorescence amplitudes were normalized to saturation values from the fits. Note that the normalized fluorescence amplitude indicates the voltage-dependent distribution of the enzyme between E_1P and E_2P states. (c) Rb⁺ uptake of H⁺,K⁺-ATPase (WT and α-subunit mutants E820D, -Q, and -A) at 5 mM RbCl in the absence (hatched bars) or presence of 100 μM SCH28080 (black bars). (d) H⁺,K⁺-ATPase-mediated Rb⁺ uptake at 5 mM RbCl and extracellular pH 7.4 (light gray bars), pH 5.5 (dark gray bars), or pH 7.4 + 40 mM butyrate, which causes an intracellular acidification by about 0.5 pH units (hatched dark gray bars). Data are from 15 to 20 oocytes in four individual experiments, normalized to Rb⁺ uptake of the reference construct (H⁺,K⁺-ATPase α-subunit mutant S806C + WT β-subunit; uptake rates in the experiments were 23.9, 22.5, 21.2, and 24.9 pmol/oocyte/min). The inset in (d) shows inhibition of Rb⁺ uptake upon injection of sodium orthovanadate into oocytes (~5 mM intracellular concentration) expressing either the WT or the α-subunit mutant E820Q at 5 mM RbCl and extracellular pH 7.4 + 40 mM butyrate (black bars). Panels (a) and (b) were reproduced from Ref. 18 with permission from the Public Library of Science (PLoS One). Panels (c) and (d) were reproduced from Ref. 23 with permission from the American Society for Biochemistry and Molecular Biology.

protonation equilibrium (with a pK_a around 5.5, [21, 22]) so that the pharmacologically active, protonated species is more abundant at pH 5.5.

The previous studies suggested that an intracellular H^+-dependent step in the H^+-translocating branch of the catalytic cycle is rate limiting for the turnover of gastric H^+,K^+-ATPase. This raised the question of which site in the enzyme is responsible for the intracellular interaction with protons and whether residues involved in cation coordination during transport could be involved. One important structural feature that distinguishes gastric H^+,K^+-ATPase from the highly homologous Na^+,K^+-ATPase is the presence of a lysine residue, instead of serine in Na^+,K^+-ATPase, in the otherwise highly conserved (K/S)NIPEIT motif on TM5, which is the only positively charged amino acid in the whole transmembrane domain of gastric H^+,K^+-ATPase. According to structural homology modeling and biochemical data, this lysine (K791) was suggested to form an interhelical salt bridge to a glutamate in TM6 (E820) [28], which is part of the cation binding pocket. This salt bridge could be responsible for the inherent E_2 preference of gastric H^+,K^+-ATPase [23, 24] that efficiently prevents reverse operation of the pump, in contrast to the Na^+,K^+-ATPase and Ca^{2+}-ATPase, which can be forced to undergo backward cycling under nonphysiological conditions [25, 26]. Furthermore, K791 was suggested to be responsible for the electroneutral operation in nongastric H^+,K^+-ATPases (also termed "X^+,K^+-ATPase" to reflect the fact that the nongastric pumps can also perform electroneutral Na^+/K^+ exchange). Upon neutralizing or charge-inverting mutations of K791, Na^+/K^+ transport by nongastric H^+,K^+-ATPase becomes electrogenic with positive membrane currents, and *vice versa*, mutation of the serine in TM5 of Na^+,K^+-ATPase to arginine abolished electrogenicity and pump current [27].

In contrast to the aforementioned findings for the nongastric enzyme, mutations of K791 in gastric H^+,K^+-ATPase did not give rise to electrogenic activity. Instead, Rb^+ uptake activity was largely reduced. However, destabilization of E_2- toward E_1-like states was inferred from voltage-clamp fluorometry and from the reduction of sensitivity toward the imidazo pyridine inhibitor SCH28080 (2-methyl-8-[phenylmethoxy] imidazo-(1,2a) pyridine 3-acetonitrile), which binds preferentially to the E_2P conformation of the proton pump [23]. Similarly, the effects of mutations in the salt bridge partner residue E820 were analyzed. As shown in Figure 14.6c, charge-neutralizing mutations of E820 also profoundly reduced Rb^+ uptake, and the conservative mutation E820D already led to a reduction by 50%. All mutants displayed an increased SCH28080 sensitivity, again suggesting an E_2-destabilizing effect. Notably, as shown in Figure 14.6d [23], the charge-neutralizing mutations made the enzyme essentially insensitive to intracellular acidification, because Rb^+ uptake of mutants E820Q and E820A at extracellular pH 7.4 was not dependent on the presence of butyrate, whereas the conservative E820D mutation, albeit operating with about 50% reduced activity, conserved the WT behavior. Furthermore, mutants E820Q and E820A showed a reduction of Rb^+ uptake at extracellular pH 5.5 compared to pH 7.4. As a control for the E_2-destabilizing effect, the sensitivity of the E820Q mutant against vanadate (an E_2-specific inhibitor acting from the intracellular side, which was applied by oocyte injection) was tested by Rb^+ flux studies. The inset in Figure 14.6d shows that mutant E820Q is less sensitive to vanadate than the WT enzyme, supporting the notion of an E_2-destabilizing effect of the mutation [23].

The implications suggested from these observations are twofold. Firstly, E820 might be responsible for determining K$^+$ over H$^+$ selectivity of the extracellular-facing cation binding pocket because the charge-neutralizing mutations profoundly increase the sensitivity to extracellular acidification. Secondly, E820 is apparently crucial for intracellular H$^+$ sensitivity as well, because charge-neutralizing mutations interfere with an H$^+$-dependent step in the $E_1(P) \rightarrow E_2(P)$ branch of the cycle, thus preventing the strong E_2 preference of the enzyme. These conclusions nicely conform with biochemical studies [28], which showed that mutation E820D had decreased apparent ATP affinity in ATPase assays as a consequence of a decreased H$^+$ affinity at the intracellular-facing binding site(s). The 820 position could even be the crucial determinant for the discrimination between H$^+$ and Na$^+$ because Na$^+$,K$^+$-ATPase employs an aspartate (D804 in sheep Na$^+$,K$^+$-ATPase α1-subunit) instead of a glutamate in the homologous position [23].

14.5.4 Determination of Na$^+$,K$^+$-ATPase Transport Stoichiometry and Voltage Dependence of H$^+$,K$^+$-ATPase Rb$^+$ Transport

For the electrogenic Na$^+$/K$^+$-ATPase, Rb$^+$ fluxes can be determined in conjunction with two-electrode voltage-clamp experiments in order to draw correlations between net charge transport (time integral of pump current) and Rb$^+$ transport. Figure 14.7a shows the recording of Na$^+$/K$^+$ pump current (about 40 nA stationary current amplitude) induced by perfusion of a Na$^+$,K$^+$-ATPase-expressing oocyte with a solution containing 1 mM Rb$^+$. Integration of the current signal (gray shaded area under the current trace) yielded 7.8 µC of total transported charge (equivalent to about 0.0822 nmol of charge) [8]. When the same oocyte was subsequently subjected to the AAS determination method, a total amount of 0.176 nmol of Rb$^+$ was found. The ratio between total charge and Rb$^+$ flux was 0.47, which is close to 0.5, the value expected for the 3:2 Na$^+$/K$^+$(or Rb$^+$) transport stoichiometry because net outward transport of one elementary charge by Na$^+$,K$^+$-ATPase corresponds to the uptake of two K$^+$ (or Rb$^+$ ions). Repetition of such a single-cell experiment (see Fig. 14.7b) yields a linear correlation between total charge and Rb$^+$ transport, with a slope of 0.49 determined from linear least squares fitting. Thus, reliable results are obtained even with a limited set of single cells [8].

Rb$^+$ uptake experiments combined with two-electrode voltage clamping was applied to determine the voltage dependence of H$^+$/K$^+$ exchange transport of gastric H$^+$,K$^+$-ATPase. In these experiments, Rb$^+$ flux was induced by timely, well-defined solution exchanges during clamping H$^+$,K$^+$-ATPase-expressing oocytes to a certain membrane potential. In Figure 14.7c, the Rb$^+$ uptake activity at 5 mM Rb$^+$ concentration of oocytes clamped to 100 mV was compared to that of non-voltage-clamped oocytes, for which the membrane potential was determined to be between -10 and -20 mV (in zero current clamp mode) [18]. Only a slight decrease of Rb$^+$ transport was observed at -100 mV compared to nonclamped oocytes. Importantly, the roughly twofold increase of Rb$^+$ transport at pH 5.5 compared to pH 7.4, as shown in Figure 14.6a, was independent of the membrane potential, suggesting that an intracellular pH-sensitive and only weakly voltage-dependent event does not only determine the rate of the $E_1P \rightarrow E_2P$ transition (see Section 14.5.3) but also the overall turnover rate of the pump.

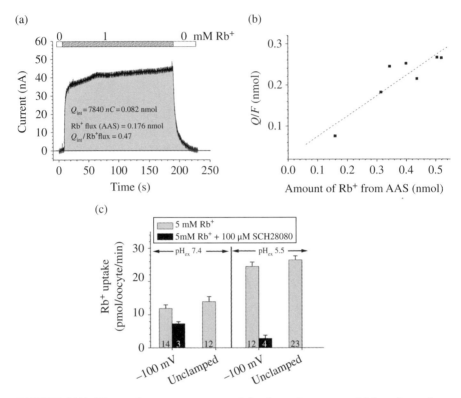

FIGURE 14.7 Rb$^+$ uptake measurements at defined membrane potential in voltage-clamp experiments. (a) Recording of Na$^+$,K$^+$-ATPase (human $\alpha 2 + \beta 1$ subunits) pump current due to a change in the perfusion solution from zero to 1 mM Rb$^+$ and back. The integral of the pump current (gray) yields the amount of transported charge. The same oocyte was subjected to Rb$^+$ quantification by AAS. (b) Correlation between transported charge from voltage-clamp experiments as in (a) and Rb$^+$ uptake determined by AAS on the same cell for determination of transport stoichiometry. (c) Rb$^+$ uptake measurements on oocytes expressing rat gastric H$^+$,K$^+$-ATPase (α-subunit mutant S806C + WT β-subunit) carried out under membrane potential control in voltage-clamp experiments to determine the voltage dependence of Rb$^+$ transport. Values represent Rb$^+$ uptake (in pmol/oocyte/min) at 5 mM Rb$^+$ and extracellular pH 7.4 or 5.5, which either were held at -100 mV or subjected to Rb$^+$ uptake without voltage control (the "resting" membrane potential was determined to be between $+10$ and -20 mV before switching on the voltage clamp). Data are from 12 to 23 oocytes of one cell batch. Panels (a) and (b) were reproduced from Ref. 8 with permission from the Journal of Visualized Experiments. Panel (c) was reproduced from Ref. 18 with permission from the Public Library of Science (PLoS One).

14.5.5 Effects of C-Terminal Deletions of the H$^+$,K$^+$-ATPase α-Subunit

One of the most outstanding features revealed by the first crystal structures of the Na$^+$,K$^+$-ATPase [29] was the intimate embedding of the α-subunit's most C-terminal segment within the transmembrane domain of the protein. Although the sequence

PGGWVEKETYY is highly conserved in Na^+,K^+-ATPase isoforms of many species and has no correlate on the sarcoplasmic reticulum Ca^{2+}-ATPase, the high structural and functional importance of this segment had not been discovered. Whereas the first part of this extension is accommodated between the transmembrane segments 7 and 10 of the α-subunit and βTM, the two C-terminal tyrosine residues are recognized by a binding pocket between TM7, TM8, and TM5 of the α-subunit. The most C-terminal Y1016 projects toward K766 of αTM5 as well as R933 in the loop connecting αTM8 and αTM9 in close vicinity to the cation binding pocket. It was shown by several laboratories that deletions of the most C-terminal residues dramatically affects the apparent Na^+ affinity and voltage dependence [30–34], suggesting that the C-terminus contributes to the formation or stabilization of the third Na^+ binding site, which is unique for the Na^+,K^+-ATPase compared to the Ca^{2+}- or H^+,K^+-ATPases. Because the C-terminal sequence of the gastric H^+,K^+-ATPase α-subunit (PGSWWDQ**EL**YY) is similar, including the terminal tyrosines, C-terminally truncated H^+,K^+-ATPases were also functionally analyzed. Deletion of one, two, or five of the last amino acids led to successively decreasing Rb^+ transport activity and, in turn, almost complete loss of sensitivity toward 10 μM SCH28080, as shown in Figure 14.8a [8]. Such dramatic losses in cation transport activity were not observed for similar truncations in the case of the Na^+,K^+-ATPase. However, in this experiment, it was important to analyze the plasma membrane expression of the truncated variants (see Fig. 14.8b).

Western blot analysis on plasma membrane preparations from oocytes expressing the truncated constructs showed that the reduction in the observed Rb^+ uptake activity is entirely due to the graded reduction of surface expression of the truncated enzymes [8]. Thus, in the case of H^+,K^+-ATPase, the primary, again presumably structural role of the α-subunit's C-terminus is already exerted on the level of plasma membrane targeting, which might be due to a structure-stabilizing effect that allows newly synthesized protein to escape posttranslational quality control in the cell. It is suggested as a general precaution to always cross-check the observed transport activities of a membrane protein with some means of plasma membrane expression quantification to avoid conceptual ambiguities.

14.5.6 Li^+ and Cs^+ Uptake Studies

Besides Rb^+, the Na^+,K^+-ATPase also accepts Cs^+ and Li^+ ions as surrogates for K^+ to stimulate electrogenic Na^+/X^+ exchange transport. It has also been shown that intracellular Li^+ stimulates electrogenic Li^+ outward transport [35]. For both cation species, HCL for AAS are available, and the detection limits for the THGA-AAS configuration are very low (see Table 14.1). Exemplary results for uptake studies involving these cations by the Na^+,K^+-ATPase are shown in Figure 14.9. Notably, Li^+ uptake measurements were more problematic due to substantial nonspecific Li^+ uptake (i.e., nonrelated to Na^+,K^+-ATPase activity) into the oocytes, which could not be blocked by ouabain or the addition of Ni^{2+} or other divalent heavy metal cations to inhibit a potential endogenous Na^+/Ca^{2+} exchanger of the oocytes. Because the nonspecific Li^+ uptake was nonlinearly dependent on the extracellular Li^+ concentration (data not shown), this cation requires that the nonspecific Li^+ uptake must be measured at every concentration used for the flux study. The Michaelis–Menten evaluation in Figure 14.9b used a linear interpolation for the subtraction of nonspecific uptake measured at 10 mM LiCl on noninjected oocytes from the same batch. In contrast, Cs+ uptake measurements were not biased by excessive nonspecific uptake (see Fig. 14.9c), so that this cation can be assayed with the same accuracy as Rb+.

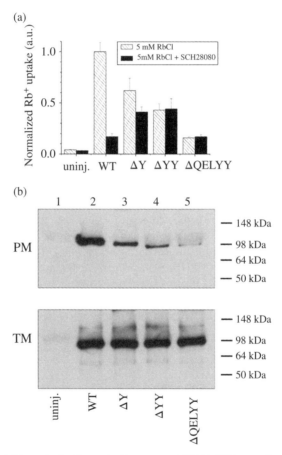

FIGURE 14.8 Rb+ uptake by C-terminally truncated H+,K+-ATPase variants. (a) Rb+ uptake mediated by H+,K+-ATPase at 5 mM RbCl in the absence (hatched bars) or presence (black bars) of 10 μM SCH28080. Results from noninjected oocytes, oocytes injected with cRNA of the HKβ-subunit, and either HKα-S806C or HKα-constructs with the indicated C-terminal truncations are shown. Data are means ± S.E.M. from 3 individual experiments with 15–20 oocytes, normalized to Rb+ uptake of the HKα-S806C/HKβ (corresponding to 20.4, 23.7, and 29.5 pmol/oocyte/min, respectively). (b) Western blot analysis of plasma membrane (PM, upper panel) and total membrane fractions (TM, lower panel) isolated from H+,K+-ATPase-expressing oocytes. Detection used anti-HKα-subunit antibody HK12.1823. One representative Western blot out of at least three from different oocyte batches is shown. Reproduced from Ref. 8 with permission from the Journal of Visualized Experiments.

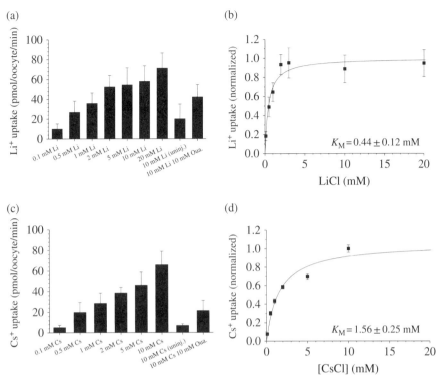

FIGURE 14.9 (a and b) Li^+ and (c and d) Cs^+ uptake kinetics of human Na^+,K^+-ATPase (subunits $\alpha2+\beta1$) in extracellular Na^+-free buffers. (b) and (d) show fits of a Michaelis–Menten-type function to the data from (a) and (c), respectively, with concentrations of half-maximal activation (K_M) as indicated. Data in (a and b) were obtained from 14 to 36 oocytes from three batches; those in (c and d) from 16 to 50 oocytes from four batches (J. Drucker and T. Friedrich, unpublished observations).

14.6 CONCLUDING REMARKS

The described AAS technique has contributed valuable data for the characterization of gastric H^+,K^+-ATPase from the viewpoint of correlations between structure and function. Its value for studying other types of transporters has still not been fully exploited, although essentially any metal ion transporter that can be expressed in the plasma membrane of *Xenopus* oocytes could be analyzed. It should be noted that cation flux measurements by AAS are not limited to alkali metal ions, but can easily be extended to other trace elements of biological relevance. Investigation of passive uptake mechanisms is possible as well. For example, the AAS technique could be used to measure the uptake of the anticancer drug cisplatin (*cis*-diamminedichloridoplatinum(II)) into cells, which has been suggested to inhibit the Na^+,K^+-ATPase and to explain the frequent nephrotoxic side effects occurring during therapeutical administration of the compound. It is even conceivable to conduct studies on endocytotic uptake mechanisms using a suitable element as a fluid phase marker.

Another interesting application would be pharmacological screening, and again, the gastric H^+,K^+-ATPase could be an ideal candidate. Gastric H^+/K^+-ATPase is an important therapeutic target for drugs known as acid suppressants or proton pump inhibitors, representing a billion dollar market [36]. However, only very few assays are available for this enzyme that meet the strict legal approval regulations of the international drug administration authorities. There is still a high demand for therapeutic proton pump agents with better bioavailability, pharmacokinetics, or reduced side effects. Although the THGA-AAS method with *Xenopus* oocytes does not support high-throughput screening, its high information content makes it an excellent technique for profiling hundreds, if not thousands of promising lead structures within a short time because robotic devices for the handling of *Xenopus* oocytes (e.g., from Multi Channel Systems, Reutlingen, Germany) are readily available.

ACKNOWLEDGMENTS

The author would like to acknowledge generous support by Prof. Ernst Bamberg (Max-Planck-Institute of Biophysics, Frankfurt, Germany), Dr. Neslihan N. Tavraz, and Dr. Katharina L. Dürr for their efforts in establishing the technique and, together with Dr. Susan Spiller (neé Meier) and Janine Drucker, accumulated data within uncounted hours of concentrated pipetting and THGA mechanics adjustment. Prof. Kazuhiro Abe (Nagoya University, Japan) encouraged us with many fruitful discussions about H,K-ATPase structure and mechanism, and Dr. Michael Kohl (Analytik-Service, Woltersdorf, Germany) provided profound technical support and advice. The work was funded by the German Research Foundation DFG (Cluster of Excellence "Unifying Concepts in Catalysis").

REFERENCES

1. B. Welz, M. Sperling, *Atomabsorptionsspektrometrie*, fourth ed., Wiley-VCH, Weinheim, 1999.
2. A. Walsh, The application of atomic absorption spectra to chemical analysis, *Spectrochim. Acta* 7 (1955) 108–117.
3. C.T.J. Alkemade, J.M.W. Milatz, Double-beam method of spectral selection with flames, *J. Opt. Soc. Am.* 45 (1955) 583–584.
4. C.T.J. Alkemade, J.M.W. Milatz, A double-beam method of spectral selection with flames, *Appl. Sci. Res. Sect. B* 4 (1955) 289–299.
5. B.V. L'vov, Russian, translated: Investigation of atomic absorption spectra by complete vaporization of a probe in a graphite cuvette, *Inzhener.-Fiz. Zhur. Akad. Nauk Beloruss. SSR (J. Eng. Phys. [USSR])* 2 (1959) 44–52.
6. B.V. L'vov, Electrothermal atomization: The way toward absolute methods of atomic absorption analysis, *Spectrochim. Acta Part B* 33 (1978) 153–193.
7. W. Slavin, D.C. Manning, G.R. Carnrick, The stabilized temperature platform furnace, *Atom. Spectrosc.* 2 (1981) 137–145.

8. K.L. Dürr, N.N. Tavraz, S. Spiller, T. Friedrich, Measuring cation transport by Na,K- and H,K-ATPase in *Xenopus* oocytes by atomic absorption spectrophotometry: An alternative to radioisotope assays, *J. Vis. Exp.* (2013) e50201.

9. B. Welz, G. Schlemmer, J.R. Mudakavi, Palladium nitrate–magnesium nitrate modifier for electrothermal atomic absorption spectrometry. Part 5. Performance for the determination of 21 elements, *J. Anal. At. Spectrom.* 7 (1992) 1257–1271.

10. R. Richards, R.E. Dempski, Examining the conformational dynamics of membrane proteins in situ with site-directed fluorescence labeling, *J. Vis. Exp.* (2011) e2627.

11. K.L. Dürr, N.N. Tavraz, D. Zimmermann, E. Bamberg, T. Friedrich, Characterization of Na,K-ATPase and H,K-ATPase enzymes with glycosylation-deficient beta-subunit variants by voltage-clamp fluorometry in *Xenopus* oocytes, *Biochemistry* 47 (2008) 4288–4297.

12. E.M. Price, J.B. Lingrel, Structure-function relationships in the Na,K-ATPase alpha subunit: Site-directed mutagenesis of glutamine-111 to arginine and asparagine-122 to aspartic acid generates a ouabain-resistant enzyme, *Biochemistry* 27 (1988) 8400–8408.

13. R.F. Rakowski, L.A. Vasilets, J. LaTona, W. Schwarz, A negative slope in the current-voltage relationship of the Na^+/K^+ pump in *Xenopus* oocytes produced by reduction of external [K^+], *J. Membr. Biol.* 121 (1991) 177–187.

14. N.N. Tavraz, T. Friedrich, K.L. Dürr, J.B. Koenderink, E. Bamberg, T. Freilinger, M. Dichgans, Diverse functional consequences of mutations in the Na^+/K^+-ATPase alpha2-subunit causing familial hemiplegic migraine type 2, *J. Biol. Chem.* 283 (2008) 31097–31106.

15. N.N. Tavraz, K.L. Dürr, J.B. Koenderink, T. Freilinger, E. Bamberg, M. Dichgans, T. Friedrich, Impaired plasma membrane targeting or protein stability by certain ATP1A2 mutations identified in sporadic or familial hemiplegic migraine, *Channels* 3 (2009) 82–87.

16. R.W. Albers, Biochemical aspects of active transport, *Ann. Rev. Biochem.* 36 (1967) 727–756.

17. R.L. Post, C. Hegyvary, S. Kume, Activation by adenosine triphosphate in the phosphorylation kinetics of sodium and potassium transporting adenosine triphosphatase, *J. Biol. Chem.* 247 (1972) 6530–6540.

18. K.L. Dürr, N.N. Tavraz, T. Friedrich, Control of gastric H,K-ATPase activity by cations, voltage and intracellular pH analyzed by voltage clamp fluorometry in *Xenopus* oocytes, *PLoS One* 7 (2012) e33645.

19. F.A. Norris, G.L. Powell, The apparent permeability coefficient for proton flux through phosphatidylcholine vesicles is dependent on the direction of flux, *Biochim. Biophys. Acta-Biomembr.* 1030 (1990) 165–171.

20. G. Nagel, D. Ollig, M. Fuhrmann, S. Kateriya, A.M. Musti, E. Bamberg, P. Hegemann, Channelrhodopsin-1: A light-gated proton channel in green algae, *Science* 296 (2002) 2395–2398.

21. B. Wallmark, C. Briving, J. Fryklund, K. Munson, R. Jackson, J. Mendlein, E. Rabon, G. Sachs, Inhibition of gastric H^+,K^+-ATPase and acid secretion by SCH 28080, a substituted pyridyl(1,2a)imidazole, *J. Biol. Chem.* 262 (1987) 2077–2084.

22. C. Briving, B.M. Andersson, P. Nordberg, B. Wallmark, Inhibition of gastric H^+/K^+-ATPase by substituted imidazo[1,2-a]pyridines, *Biochim. Biophys. Acta-Biomembr.* 946 (1988) 185–192.

23. K.L. Dürr, I. Seuffert, T. Friedrich, Deceleration of the E_1P-E_2P transition and ion transport by mutation of potentially salt bridge-forming residues Lys-791 and Glu-820 in gastric H$^+$/K$^+$-ATPase, *J. Biol. Chem.* 285 (2010) 39366–39379.

24. K.L. Dürr, I. Seuffert, T. Friedrich, Deceleration of the E_1P-E_2P transition and ion transport by mutation of potentially salt bridge-forming residues Lys-791 and Glu-820 in gastric H$^+$/K$^+$-ATPase, *J. Biol. Chem.* 285 (2012) 39366–39379.

25. M.L. Helmich-de Jong, S.E. van Emst-de Vries, J.J.H.H.M. De Pont, F.M. Schuurmans Stekhoven, S.L. Bonting, Direct evidence for an ADP-sensitive phosphointermediate of (K$^+$ + H$^+$)-ATPase, *Biochim. Biophys. Acta-Biomembr.* 821 (1985) 377–383.

26. K. Abe, K. Tani, T. Nishizawa, Y. Fujiyoshi, Inter-subunit interaction of gastric H$^+$,K$^+$-ATPase prevents reverse reaction of the transport cycle, *EMBO J.* 28 (2009) 1637–1643.

27. M. Burnay, G. Crambert, S. Kharoubi-Hess, K. Geering, J.D. Horisberger, Electrogenicity of Na,K- and H,K-ATPase activity and presence of a positively charged amino acid in the fifth transmembrane segment, *J. Biol. Chem.* 278 (2003) 19237–19244.

28. H.P.H. Hermsen, H.G.P. Swarts, J.B. Koenderink, J.J.H.H.M. De Pont, Mutagenesis of glutamate 820 of the gastric H$^+$,K$^+$-ATPase alpha-subunit to aspartate decreases the apparent ATP affinity, *Biochim. Biophys. Acta-Biomembr.* 1416 (1999) 251–257.

29. J.P. Morth, B.P. Pedersen, M.S. Toustrup-Jensen, T.L. Sørensen, J. Petersen, J.P. Andersen, B. Vilsen, P. Nissen, Crystal structure of the sodium-potassium pump, *Nature* 450 (2007) 1043–1049.

30. S. Yaragatupalli, J.F. Olivera, C. Gatto, P. Artigas, Altered Na$^+$ transport after an intracellular alpha-subunit deletion reveals strict external sequential release of Na$^+$ from the Na/K pump, *Proc. Natl. Acad. Sci. U. S. A.* 106 (2009) 15507–15512.

31. S. Meier, N.N. Tavraz, K.L. Dürr, T. Friedrich, Hyperpolarization-activated inward leakage currents caused by deletion or mutation of carboxy-terminal tyrosines of the Na$^+$/K$^+$-ATPase a-subunit, *J. Gen. Physiol.* 135 (2010) 115–134.

32. N. Vedovato, D.C. Gadsby, The two C-terminal tyrosines stabilize occluded Na/K pump conformations containing Na or K ions, *J. Gen. Physiol.* 136 (2010) 63–82.

33. M.S. Toustrup-Jensen, R. Holm, A.P. Einholm, V.R. Schack, J.P. Morth, P. Nissen, J.P. Andersen, B. Vilsen, The C terminus of Na$^+$,K$^+$-ATPase controls Na$^+$ affinity on both sides of the membrane through Arg935, *J. Biol. Chem.* 284 (2009) 18715–18725.

34. H. Poulsen, H. Khandelia, J.P. Morth, M. Bublitz, O.G. Mouritsen, J. Egebjerg, P. Nissen, Neurological disease mutations compromise a C-terminal ion pathway in the Na$^+$/K$^+$-ATPase, *Nature* 467 (2010) 99–102.

35. A.N. Hermans, H.G. Glitsch, F. Verdonck, Activation of the Na$^+$/K$^+$ pump current by intra- and extracellular Li ions in single guinea-pig cardiac cells, *Biochim. Biophys. Acta-Biomembr.* 1330 (1997) 83–93.

36. D.A. Peura, R.R. Berardi, J. Gonzalez, L. Brunetti, The value of branded proton pump inhibitors: Formulary considerations, *Pharm. Ther.* 36 (2011) 434–445.

15

LONG TIMESCALE MOLECULAR SIMULATIONS FOR UNDERSTANDING ION CHANNEL FUNCTION

BEN CORRY

Research School of Biology, Australian National University, Canberra, Australian Capital Territory, Australia

15.1 INTRODUCTION

Ion channels and transporters are large dynamic protein molecules undergoing a range of motions that are essential to their function. These movements range from relatively small vibrations and rotations of amino acid side chains essential for dictating the binding preference of the protein to large conformational changes between specific functional states. Gaining a complete understanding of how these proteins work requires an understanding of the dynamics of the protein as well as the structure at both high spatial and temporal resolutions.

In order to know how an ion is transported through a biological channel, for example, we need to know how the structure of the protein and surrounding water changes as the ion moves and how this influences the detailed interactions of the ion with its environment. Similarly, in order to know how a specific mutation influences the gating behavior of a channel, we need to know how it changes the local structure of the protein, how it alters the relative free energies of the different functional states, and how it changes the dynamic process of moving between them. All of these subjects involve aspects taking place on a range of lengths and timescales, and for this reason, they can be extremely difficult to study. Indeed, fundamental questions concerning the basic functioning of a channel or transporter require both structural and dynamic information over a range of

Pumps, Channels, and Transporters: Methods of Functional Analysis, First Edition. Edited by
Ronald J. Clarke and Mohammed A. A. Khalid.
© 2015 John Wiley & Sons, Inc. Published 2015 by John Wiley & Sons, Inc.

length scales if they are to be answered. Common questions include: How does a transporter gain substrate selectivity? Can specificity be modified by specific mutations? Can drug candidates reach potential target sites in the protein, and how well do they bind and alter the protein function?

Molecular dynamics (MD) simulations provide one way to address the challenge of gaining fundamental structural and dynamic information about protein function as they can combine both atomic spatial resolution with femtosecond time resolution. This approach was first developed as a tool for theoretical physics, where it was used to study the motion of idealized particles [1] or liquids [2], but quickly caught on as a useful tool for studying the behavior of more complex molecules and materials. It was first used to study a protein by McCammon et al. [3] and has since become an important technique for studying biological systems. A high level of resolution, however, also necessitates large amounts of computational power to simulate large molecular systems or long timescales, creating a limit on the questions that can be addressed. In addition, this simulation methodology invariably makes use of simplified descriptions of the physical interactions between particles to ease the computational burden. As a consequence, careful parameterization of the method, validation of the results, and consideration of the domain of applicability are required.

Over the last four decades, however, improvements to computer hardware, algorithms, and software have enabled many of the limitations of MD simulations to be overcome, and simulations lasting more than a millisecond have now been conducted. This has enabled physiological events taking place over long timescales to be studied as well as for the statistical reliability of shorter timescale properties to be improved.

This chapter starts by introducing the basic methodology of MD simulations, before discussing why it requires so much computational effort to conduct them and some of the algorithmic improvements utilized to ease the computational burden. Biologically relevant timescales and system sizes are introduced prior to describing the historical development of the method in order to highlight the kinds of physical questions that can be addressed and the constant methodological advances that are being made. Some of the limitations of the method are described, after which examples are given of how the method has been used to answer questions that are important to understanding the function of membrane proteins. While significant insight into the function of ion channels, transporters, and a vast range of other biological molecules has been achieved by this approach, the focus here is on results relating to ion channels.

15.2 FUNDAMENTALS OF MD SIMULATION

15.2.1 The Main Idea

The basic idea of MD simulation is to make an atomic-level representation of a set of molecules and to follow their motion as they interact over time. The first step in this process involves generating starting coordinates of all the atoms in the system. For protein simulations, the initial coordinates are assigned using a predetermined structure of the protein, such as that found by X-ray crystallography or homology modeling. To save on computational effort, very early simulations tended to simulate proteins in vacuum, but reproducing the correct behavior of the target molecule requires the inclusion of a realistic environment. For ion channels and transporters, this means placing them in a lipid bilayer solvated by water and ions. Initial velocities must also be assigned to each atom, which is usually done by assigning each atom a separate velocity randomly selected from a Maxwell–Boltzmann distribution, thus yielding the desired simulation temperature when averaged over all the atoms.

Once the initial coordinates and velocities of every atom are assigned, the forces, F, acting on each atom must be calculated. This requires knowledge of the interatomic interactions arising between each pair of atoms. Exactly how this is done is described in the next section but typically involves using a simplified empirical expression known as a force field. The acceleration of each atom, a, can then be calculated using Newton's equation of motion:

$$F_i(r_i) = m_i a_i$$

in which m, r, and F are the mass, position, and force acting on each atom, i. Even though we know the starting position and velocity and how to calculate the acceleration of each atom, analytically determining the motion of all the atoms over time is intractable. As a consequence, the problem is solved in an iterative numerical

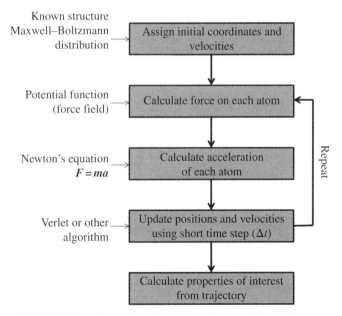

FIGURE 15.1   Basic algorithm for conducting MD simulations.

fashion, by predicting where each atom will be a short time later, moving the atoms to this new position, and then repeating the whole procedure many times (see Fig. 15.1). The numerical integration can be done in many ways, but most commonly, this is done using variants of the Verlet algorithm [4] in which the two previous positions of each atom (r_{i-1} and r_i) are used to determine the next position (r_{i+1}) after a small time step of Δt:

$$r_{i+1} = \left(2r_i - r_{i-1}\right) + a_i \left(\Delta t\right)^2$$

This expression is only theoretically accurate to third order, meaning that each iteration can introduce errors proportional to $(\Delta t)^4$. As a consequence, the accuracy of the trajectory will improve as the time step gets smaller. But, smaller time steps also mean that more steps are required to propagate the system for the same total time, which in turn means more computational effort. Factors influencing the choice of time step are discussed in Section 15.2.4.

15.2.2 Force Fields

The most time-consuming part of the MD algorithm is the calculation of the interatomic forces. Because of this, these forces are not generally calculated using high-level quantum mechanical theory (although this can be done as in "ab initio MD"), but rather one makes use of an approximate empirically parameterized expression known as a force field. The force acting on each atom is obtained from the derivative of the potential function, U, with respect to position. In the force field approach, molecules are described by the "ball and spring" model in which atoms (balls) interact with each other via covalent bonds (springs) and via nonbonded interactions:

$$F = -\nabla U$$

$$U = U_{bonded} + U_{nonbonded}$$

The bonded interactions include terms representing bond stretching, bending, torsional rotations, and out-of-plane distortions (improper dihedral angles):

$$U_{bonded} = U_{stretch} + U_{bend} + U_{torsion} + U_{improper}$$

The nonbonded interactions, on the other hand, include atom–atom repulsion and dispersion (van der Waals interactions) as well as electrostatic interactions:

$$U_{nonbonded} = U_{vdW} + U_{electrostatic}$$

A number of different force fields exist. One of the ways in which they differ is in how each of the interactions is calculated. The most common types of force fields use an expression similar to this:

$$U = \sum_{\text{bonds}} K_b \left(b - b_0 \right)^2$$

$$+ \sum_{\text{angles}} K_\theta \left(\theta - \theta_0 \right)^2$$

$$+ \sum_{\text{dihedrals}} K_\chi \left(1 + \cos \left(n\chi - \delta \right) \right) + \sum_{\text{impropers}} K_\varphi \left(\varphi - \varphi_0 \right)^2$$

$$+ \sum_i \sum_{j>i} \varepsilon_{ij} \left(\left(\frac{R_{ij}^{\text{min}}}{r_{ij}} \right)^{12} - 2 \left(\frac{R_{ij}^{\text{min}}}{r_{ij}} \right)^6 \right) + \frac{q_i q_j}{4\pi \varepsilon_0 r_{ij}}$$

Here, b is the bond length, θ is the bond angle, χ is the dihedral angle, φ is the improper angle, and r_{ij} is the distance between atoms i and j. The constants in the previous expression are known as parameters and have to be defined for every bond and atom type. These parameters include the equilibrium bond length, b_0, and force constant for bond stretching, K_b; the equilibrium bond angle, θ_0, and force constant for bending, K_θ; the torsional force constant, K_χ, multiplicity, n, and phase angle, δ; the improper force constant, K_φ, and equilibrium improper dihedral angle, φ_0; the charges on each atom, q_i; and the Lennard-Jones parameters describing the van der Waals interaction that yield the well depth, ε_{ij}, and equilibrium atomic separation, R_{ij}^{min}. While this yields an enormous number of parameters, the number can be reduced by recognizing that molecules are formed from units that are structurally similar. For example, the equilibrium bond lengths and vibrations of all C–H bonds are very similar [5]. This means that chemically similar atoms and bonds can be given the same parameter values. Having more chemically distinct units in the force field (e.g., differentiating the bonds from H atoms to singly and doubly bonded C) yields more accurate forces but requires a greater number of parameters to be determined.

As can be expected, determining the parameter values is not a straightforward task. It involves an iterative process in which the values are optimized to yield agreement with experimentally measured or quantum mechanically calculated properties. We refer the reader to an excellent review for a discussion of the different approaches that have been used for parameterization [6].

A great many force fields exist, differing in the parameter values or in the form of the potential function. The most popular force fields for biomolecular simulation are the CHARMM [7, 8], GROMOS [9, 10], and AMBER [11] families that have been optimized for use with proteins and nucleic acids and have undergone many revisions over the last two decades.

All of the force fields described so far assign fixed charges to each atom. One of the greatest deficiencies of these models is that they do not account for the fact that the charge on an atom can change in response to the environment. If an atom is in an electric field—for example, if it is near a charged molecule—a change in the electronic structure of the atom can be induced, altering its effective charge and changing the electrostatic interaction it has with other atoms in the system [12, 13].

In response to this, a number of "polarizable" force fields are being developed [14, 15], but all of these come at greater computational cost and have not been as extensively tested in biomolecular simulations as have the nonpolarizable versions.

15.2.3 Other Simulation Considerations

There are a number of additional factors that have to be considered when performing MD simulations of biological systems. Firstly, it is desirable to simulate the molecules of interest in a natural environment so that they behave in as realistic a way as possible. For biological simulations, this means solvating the biomolecules in water, while for ion channels and transporters, they should also be placed in a lipid bilayer. This increases the system size and extends the edges of the simulation system further from the region of most interest. But, we still have a problem at the edge of the system, as some of the lipid and water are unrealistically surrounded by vacuum. Although it is possible that these edge effects may not influence the behavior of the center of the simulations if the system is big enough, a more popular way to remove edge effects is to use a periodic boundary. In this approach, the simulation system is surrounded by copies of itself, so that when atoms leave one edge of the system, they appear at the opposite edge. By tiling images of the central simulation system in this way, there is effectively no edge and thus no edge effects. However, the simulation system still needs to be large enough so that molecules don't interact strongly with their own images. In addition, this means that there is an effectively infinite system and an infinite number of interatomic interactions that need to be calculated. Ways for dealing with this are described in the following section.

Unmodified simulations proceed in a microcanonical ensemble in which the number of atoms, N, volume, V, and energy of the system, E, are constant, but the temperature and pressure are not controlled. Usually, we want to compare our simulations to experiments performed at a constant temperature and pressure (NPT ensemble). Therefore, the dynamics of the system have to be modified with algorithms that act like a thermostat or barostat. There are a number of different approaches for doing this, and it has been shown that different algorithms can yield different dynamics [16–19]. We do not wish to give a detailed review of these methods here, but note that one of these algorithms will normally be applied and that some consideration should be given as to the most appropriate choice. For example, the Langevin thermostat is very stable and reproduces temperature well over long timescales and is well suited to simulations aimed at assessing protein structures. But this thermostat can influence the rates of some dynamic processes. The Berendsen thermostat, on the other hand, has less influence on the dynamics of atoms, but does not always reproduce the correct kinetic energy distribution.

15.2.4 Why Do MD Simulations Take So Much Computational Power?

MD simulations are a computationally intensive technique. The two main reasons for this are described in the following sections.

15.2.4.1 ***Force Calculations***   Each step in an MD simulation involves a large number of calculations. Most of these are related to determining the forces on each atom. The number of bonds in a molecular system scales roughly linearly with the number of atoms in the system, N. As a consequence, the number of terms that must be calculated to determine the bonded forces also scales linearly with N. The number of nonbonded interactions that must be determined is usually much larger, as these must be calculated for every pair of atoms, yielding N^2 interactions in total. Thus, as the system size (i.e., number of atoms) is increased, the number of calculations required to determine the total nonbonded force increases rapidly, scaling with N^2, and these calculations usually account for the majority of the computational cost. The situation is even worse when using periodic boundary conditions, as the periodic copies of the system create more pairs of atoms (in fact an infinite set of pairs) whose interactions have to be evaluated. Clearly, simulations of proteins surrounded by lipid molecules, water molecules, and ions contain 10s to 100s of thousands of atoms and thus represent a significant computational challenge.

A number of methods have been developed to ease the computational burden of calculating the nonbonded interactions, and these provide examples of the algorithmic improvements that have allowed for long timescale simulations to be conducted. The simplest is to use a cutoff distance, such that only interactions between atoms within this separation are determined. This approach is commonly applied to the vdW interactions as their strength drops off rapidly as $1/r^6$. Cutoff distances in the range of 10–14Å are commonly applied, as interactions over distances beyond this are small. As the force is calculated from the derivative of the potential energy with respect to distance, the interaction is adjusted to smoothly decay to zero at the cutoff distance rather than sharply removing interactions beyond this point. The strength of the electrostatic interactions decreases more slowly over space, dropping off as $1/r^2$. As a consequence, cutoffs are less commonly applied to these interactions as even those taking place over tens of Ångström units can have a significant influence on the simulation trajectories [5].

A more common approach to reducing the computational cost of the electrostatic interactions is to make use of the periodic boundary conditions described previously to calculate the interactions in Fourier space rather than doing a direct calculation in real space. The approach known as Ewald summation is now usually used for this purpose [20]. In this, the interactions are broken into a short-range component (usually interactions between atoms in the central simulation box), summed in real space, and a long-range component, summed in Fourier space. As each of these components converges quickly in the respective space, they can be truncated with little loss of accuracy (provided the periodic simulation cell is charge neutral). In most simulations, the Fourier transforms are made using a discrete grid, yielding the particle mesh Ewald summation scheme [21]. Appropriate use of cutoff and summation schemes can reduce the scaling of computational burden with system size to approximately $N\ln N$.

15.2.4.2 ***Time Step***   The other major factor influencing how much computational power is required to run simulations is related to the size of the time step, Δt. If one wants to simulate a 100 ns trajectory, for example, the effort required increases as the size of the time step decreases, simply due to the fact that more calculation steps are required. The time step should, therefore, be chosen to be as long as possible to reduce the computational demands without compromising the accuracy of the result.

The length of the time step needs to be significantly shorter than the fastest atomic motions arising in the system; otherwise, the atoms have the chance to move to unrealistic positions during each iterative step. These fast motions are typically bond stretching vibrations, with the fastest being those involving light atoms, which have frequencies in the range of 10^{11}–10^{14} s^{-1} [5]. This necessitates time steps in the range of 1 fs (10^{-15} s) or less. A simulation lasting 1 ns therefore requires one million steps, with all the interatomic interactions having to be recalculated for each step!

Methods have been developed to reduce the number of time steps required in a simulation. The simplest is to remove fast motions from the simulation, allowing for a longer time step. Because the fastest motions typically involve vibrations of bonds to hydrogen, and these vibrate so quickly over only a fraction of an Ångström, this motion does not have a large effect on the structure of proteins. One common approach, therefore, is to freeze these bonds using the SHAKE [22, 23], RATTLE [24], or LINCS [25] algorithms, allowing a time step of 1.5–2 fs to be used. While this may seem like a minimal gain, the result is that half the time is needed to complete the simulation. It is important that the motions that are removed do not influence the properties of interest, which is why it is rare to freeze any other kinds of motion that are likely to have a greater effect on molecular structures and energies.

Another common approach to improve efficiency is to use multitime step approaches in which some interactions are reevaluated every step, but others that change more slowly are recalculated less frequently. One common way of doing this is to recalculate the bonded interactions every step, the vdW and short-range electrostatic interactions every 1–2 steps, and the long-range electrostatic interactions (those calculated between atoms in different periodic cells in Fourier space) every 2–4 steps. This approach can double the speed of the simulation without compromising accuracy [26].

15.3   SIMULATION DURATION AND SIMULATION SIZE

Because the amount of effort required to conduct a simulation is related to the number of atoms in the system and the simulation duration, there is a trade-off between the size and length of the simulations that are practically feasible. Computational effort increases directly with simulation duration and with the number of atoms as roughly $N\ln N$. Thus, the timescale of important biological events and the size of relevant biological systems dictate which questions can be addressed with MD simulations.

As shown in Figure 15.2, biologically relevant phenomena take place over a very wide range of timescales [27]. The fastest motions are the vibration of covalent bonds

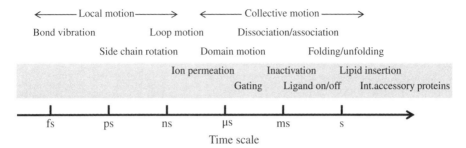

FIGURE 15.2 Timescales of important biological processes. Motions occurring in a range of proteins are shown at the top, while dynamic phenomena specific to ion channels are shown in the gray region.

taking place in the fs to ps regime, the rotations of side chains in the ps–ns regime, and the motion of flexible protein loops in the 1–100 ns regime. While all of these involve local fluctuations and do not directly dictate the overall conformation of the protein structure, they can still be critical to protein function. Such fluctuations can dictate the magnitude of local interactions and have been shown to be important, for example, in influencing the balance of forces responsible for creating selective ion binding [28]. Local flexibility is also important for allowing larger-scale con-formational changes to take place. Because these local vibrations result in a large number of possible atomic configurations, they are also critical to describing the entropy of the system [27], and this fact can make local motions one of the primary factors dictating the overall protein structure as well as the probability of transitions to nearby conformational states.

Longer timescale motions are dictated by the fluctuation of molecules between kinetically distinct states, separated by energy barriers of several kT, and taking place on timescales of 1 μs or longer. Typically, these involve the collective motion of groups of atoms between a small number of states, meaning that they play a smaller role in determining the entropy of the system. Many of the most interesting biological processes take place in this regime, including enzyme catalysis, movement between functional states of proteins, the working of molecular machines, protein folding, and protein–protein interactions.

Ion channels display dynamics over a range of these length scales (gray region in Fig. 15.2). Channels typically carry currents of 1–100 pA, meaning that ions cross individual channels at a rate of 1 per 10 ns to 1 μs. The opening and closing of channels in response to stimuli are slower, as these require the collective motion of protein domains. Such motion occurs on a millisecond timescale. Some channels are also known to inactivate, meaning that the pore becomes non-conductive to ions even though the stimulus prompting them to open remains present. Channel inactivation is about an order of magnitude slower than gating, allowing the channels to conduct for a short period before inactivating. Rates for ligands or drugs to associate/dissociate with the channels are generally slower again, while the trafficking and insertion of the proteins into the membrane can be expected to occur at even slower rates. Finally, many channels rely on interactions with other proteins to function, and this can occur over a very large range of timescales from milliseconds to seconds. The feasible duration of ion channel simulations is discussed in the next section.

Proteins and other biological molecules come in a range of sizes, and so the number of atoms needed in an MD simulation also varies significantly depending on the application. Early MD simulations focused on small proteins to reduce the computational load. For example, simulations of small, fast-folding proteins are common and can involve on the order of 10,000 atoms when including surrounding water molecules. Simulations of channels and transporters typically require larger system sizes as these are larger proteins, and placing them in a realistic environment means including lipids, water, and ions. Figure 15.3 shows an example of a small MD simulation system of a voltage-gated ion channel. In this example, parts of the protein responsible for detecting changes in membrane potential (the voltage sensors) have been removed to reduce the simulation size, as only transport through a single state of the protein was being investigated. Even so, the simulation includes about 7000 of the protein's atoms. The channel is surrounded by a small patch of lipid bilayer $(70 \times 70\text{Å})$, accounting for another 14,000 atoms. Above and below this are aqueous regions representing the extracellular space and cytosol, requiring an additional 20,000 atoms belonging to water molecules, and within this are placed a small number of ions at physiologically relevant concentrations (70 atoms). Thus, even this minimal system requires the inclusion of more than

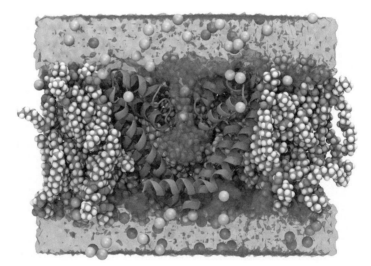

FIGURE 15.3   An example system used to conduct MD simulations on a voltage-gated ion channel. Two of the four protein subunits are shown as ribbons; atoms of the lipid molecules on each side of the protein are represented as balls. Water molecules are not shown explicitly, but the area containing them is indicated by the surrounding surface. Na^+ and Cl^- are shown as isolated balls in the water-filled regions. The front halves of the atoms are removed to allow the protein to be seen.

40,000 atoms. As every additional atom increases the computational cost of running the simulation, increasing the system size to include the entirety of the protein and the larger environment (>150,000 atoms) results in a much slower simulation.

15.4   HISTORICAL DEVELOPMENT OF LONG MD SIMULATIONS

MD simulations have progressed significantly since the first simulation of a protein by McCammon, Gelin, and Karplus in 1977 [3]. This simulation studied the dynamics of the bovine pancreatic trypsin inhibitor, chosen because of its small size and accurate X-ray structure. Although the simulation system contained only about 1000 protein atoms in a vacuum, did not include hydrogen atoms, and lasted only 8.8 ps, local fluctuations in the average structure and small concerted motions could be observed. Since then, advances in the simulation algorithms (such as the development of improved methods for propagating the system, calculating electrostatic interactions, and multi-time stepping described previously) and in computer hardware have allowed for an enormous increase in the size and duration of simulations that can be routinely conducted.

To get an idea of how the field has developed, Figure 15.4 shows the progression of long and large simulations over time. Here, the positions of a number of groundbreaking MD studies in terms of their duration and system size have been plotted on a log scale. The gray diagonal lines roughly indicate simulations requiring the same computational effort.

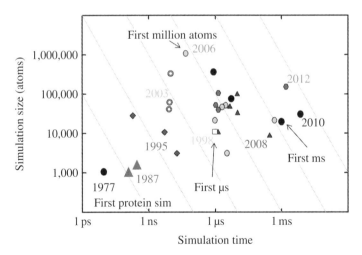

FIGURE 15.4   The historical progress of MD simulations of biological molecules. Examples of cutting edge simulations taken from the literature are plotted in terms of the simulation duration and the number of atoms in the simulation system. The year in which each simulation was conducted is indicated by the symbols (1977 closed circles, 1987 large triangles, 1995 diamonds, 1998 square, 2003 open circles, 2006 small closed circles, 2008 small triangles, 2010 closed circles, 2012 hexagons). The diagonal lines represent simulations requiring equal amounts of computational effort. The majority of the simulations described in the text are placed on this graph.

The group of Karplus pioneered the early development of MD simulations of biomolecules, and within 10 years of their initial report, they were commonly conducting simulations of proteins lasting hundreds of picoseconds—up to two orders of magnitude longer than their first published simulation [29]. For example, the work of Elber and Karplus examined dynamics in myoglobin (1400 atoms) over 300 ps to show the large number of thermally accessible minima in the neighborhood of the crystallized structure [30].

Within another 10 years, simulations had progressed by another order of magnitude, with simulations running for multiple nanoseconds [31, 32] or containing tens of thousands of atoms [33], enabling fully solvated proteins and membranes to be examined. One of the most impressive feats of brute computing power was the simulation of Duan and Kollman published in 1998 [34]. By making use of 4 months of dedicated time, improving the computational algorithms to make better use of parallel architecture, and making judicious choices in interaction cutoffs (8Å), they were able to simulate a solvated 36 amino acid peptide (a total of almost 10,000 atoms) for more than 1 μs. This was two orders of magnitude longer than any previous protein simulation and allowed them to directly observe the folding of a denatured protein.

While this record took a number of years to beat, the inevitable improvement in computing power saw simulations of 100,000 or more atoms [35, 36] (even up to a million atoms [37]) or lasting multiple microseconds [38] being achieved by the early to mid-2000s, enabling membrane proteins to be studied in a natural lipid and water environment. While multi-microsecond simulations of smaller systems also became popular, a more common approach has been to run multiple shorter simulations in order to gain a better appreciation of the reproducibility of the results [38]. This approach was exemplified by a study from the Folding@Home project [39], which distributed calculations across huge numbers of personal computers in order to gain a large number of independent trajectories (with a total simulation time of 475 μs) whose results could be collectively analyzed.

The last decade has seen a continued advance in the size and duration of MD simulations. Part of this derives from improved algorithms to make better use of the new generation of massively parallel hardware. By the late 2000s, there were numerous reports of simulations containing tens of thousands of atoms lasting for tens of microseconds [40–43]. Alternatively, others simulated hundreds of thousands of atoms for nearly a microsecond [44] or conducted hundreds of independent microsecond simulations [39]. A big step forward was made after the research group of D.E. Shaw developed custom hardware, with the individual processor architecture optimized to perform MD simulations [45]. These specialized machines have allowed for simulations to be run at the incredible rate of more than 15 μs per day. This has seen the first millisecond simulations to be run for small soluble proteins [45–47]; and more recently, even membrane-bound proteins could be simulated in a realistic environment for multiple milliseconds [48]. Although such simulations remain the domain of those with access to specialized facilities, the forefront of long timescale MD simulations continues to advance as fast as ever.

15.5 LIMITATIONS AND CHALLENGES FACING MD SIMULATIONS

Although the ever-increasing simulation durations are impressive, it is important to keep in mind some of the limitations of this method as longer simulations need not necessarily yield more accurate results.

15.5.1 Force Field and Algorithm Accuracy

A well-known limitation of MD simulations is the accuracy of the force fields. As described previously, simplified expressions for the interatomic interactions are used to reduce computational effort. This means that there are some aspects of physical reality that they cannot capture. Classical nonpolarizable force fields can never provide a description of electronic polarization and phenomena associated with this. For example, it can be difficult to study the permeation of divalent ions through ion channels or accurately model the binding of highly charged drugs to proteins [49], although refined parameterization of these ions is improving simulation results. While polarized force fields exist, these still make use of simplified expressions rather than the highest-level quantum theory. They also include more parameters and require more validation, which together with the greater computational burden has limited their use for protein simulations thus far. As the validation of these polarizable models and computer power increases, their use can be expected to become more common.

Although the classical force fields have been well parameterized and tested over the last few decades, they are only guaranteed to reproduce the properties they were parameterized against in the conditions in which they were tested. As soon as one moves away from these situations, one needs to ensure that one tests and verifies results against experimental data. Most force fields are validated at room or body temperature, for example, and one must be careful if simulating at temperatures far from this [50]. Another example of moving beyond the realm of parameterization can be seen in the application of parameters used for studying nucleic acids. These have been slowly developed over many years, with continual improvements as longer simulations showed that distortions could arise from the canonical structures [51]. The force fields now perform well for reproducing the structures of double-stranded DNA for which they are parameterized, but deviations from experimental structures can arise when simulating four-stranded DNA quadruplexes that contain a number of competing hydrogen bonds [52]. Reproducing these structures requires a careful balance of all the bonded and nonbonded force field terms, and as a consequence, quadruplex DNA is now being used to further refine the force field parameters [52]. Limitations in force fields are also apparent as it is common to find that different force fields can produce different results [52, 53].

Other algorithms used within MD can also influence the simulation accuracy. The integration algorithm introduces small errors, as noted previously, which can increase in size with the length of the time step and total simulation duration. Truncation of nonbonded interactions, multiple time stepping, thermostats, and barostats are all elements that could alter the dynamics of the system.

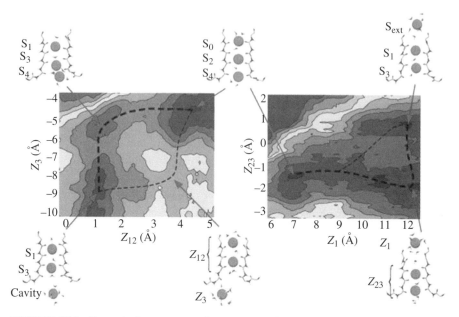

FIGURE 15.5 Example free energy surface constructed to understand ion permeation in the KcsA potassium channel. Each level of shading corresponds to an energy of $1\,\mathrm{kcal\,mol^{-1}}$. The axis coordinates represent the position of the center of mass of either a single ion (one subscript) or a pair of ions (two subscripts). The most likely conduction pathways are shown by dashed lines, and representative snapshots of the ions in the selectivity filter of the channel corresponding to each minimum are shown. Reproduced from Ref. 83 by permission from Macmillan Publishers Ltd., Copyright (2001).

15.5.2 Sampling Problems

Effectively sampling all the complex motions of proteins using MD simulations is a difficult task due to the limitations on the timescales over which simulations can be run [54]. As noted previously, channels and transporters undergo a range of dynamics across both short and long timescales. If one is trying to exhaustively study a particular motion, then one needs to not only simulate long enough to capture the event but also to sample the full range of short timescale motions occurring during the event. Furthermore, for statistical reliability, it is advisable to see events occurring multiple times before drawing conclusions from them. Even if one is studying a relatively simple and fast event, such as the energetics of ion binding to a protein, one needs to sample all the possible protein and environmental states consistent with ion binding in order to get an accurate estimation of the binding process. Not only is it computationally demanding to sample all the conformations of the protein and environments, it is often hard to assess how good one's sampling is. Thus, the so-called sampling problem remains one of the greatest limitations of MD simulations.

A number of simulation methodologies have been developed to improve conformational sampling and to speed up slow events. These can be broken into two groups. The first group works by modifying the dynamics of the system, such as high-temperature simulations [55], replica-exchange simulations [56, 57], locally enhanced sampling [58], repeated annealing [59], milestoning [60], the weighted ensemble method [61], and self-guiding MD [62]. The second group acts by modifying the potential energy landscape of the system such as in accelerated MD [63], metadynamics [64, 65], umbrella sampling [66], conformational flooding [67], potential smoothing [68], and the local elevation method [69]. Examples of the use of some of these advanced sampling methods are given in the following section, and good reviews can be found in the work of Adcock [50] or Elber [70]. Care must be taken in analyzing advanced sampling simulations as the ratio of rare to frequent events is altered, and thus, additional methods must be used to reconstruct physical quantities from the biased trajectories.

15.6 EXAMPLE SIMULATIONS OF ION CHANNELS

To gain an appreciation of how MD simulations have been applied to understanding the molecular basis of channel function, the following subsections describe a number of studies that have used simulation to address an important question relating to channel function. As described in Section 15.3, the biologically relevant dynamics of ion channels take place on a range of timescales. The examples discussed here are based around three themes, representing at least two different timescales. The first relates to understanding the steps involved in an ion passing through a channel, while the second discusses the related issue of how channels can distinguish between different ion types. Both of these aspects of channel function take place on relatively fast timescales but are still challenging to address directly with simulations. The last set of examples relates to understanding the conformational changes involved in the opening and closing of ion channels and how stimuli are translated into channel gating, that is, processes occurring on a much longer timescale.

15.6.1 Simulations of Ion Permeation

As noted previously, ion channel currents are in the order of pA, meaning that each conduction event can be expected to take approximately 1 μs. While the time per event can be less if studying a large conductance channel and applying large driving potentials, direct simulation of conduction events with MD has only become possible in the last decade. Prior to this, an early study had used a 100 ns simulation of a simplified model channel under a 1.1 V driving force to directly determine the channel conductance. But it was unclear how the results could describe conduction in a more realistic pore under physiological conditions [71]. However, water conduction through aquaporin channels occurs more rapidly. Thus, simulations of membrane-embedded channels lasting tens of nanoseconds were sufficient to determine the steps involved in water translocation [35, 36].

A number of alternatives to direct simulation of ion permeation have been followed over the years. One approach is to use simplified models, such as Brownian dynamics, in which not all atoms have to be represented explicitly.

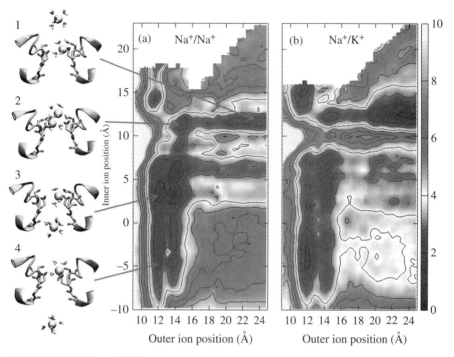

FIGURE 15.6 Example free energy surface used to understand ion permeation and selectivity in a voltage-gated sodium channel. The potential of mean force (PMF) is plotted as a function of the axial position of each of two ions assuming (a) both ions are Na$^+$ and (b) the inner ion is K$^+$ and the outer ion is Na$^+$. Energy values are shown in the scale bar in units of kcal mol^{-1} and the contour interval is 1 kcal mol^{-1}. The steps involved in conduction and their positions on the two ion PMF are shown on the left. Reprinted with permission from Ref. 84. Copyright (2012) American Chemical Society.

By doing this, much longer simulation times can be reached, and channel currents have been determined for a range of channels in many conditions [72, 73]. More recent versions of this approach have used prior MD simulations to determine the average forces on an ion at each position that are utilized in the Brownian dynamics simulation [74, 75]. Alternatively, diffusion theory can be used to approximately link the forces on ions found from detailed simulations to channel currents using an additional layer of approximation [74, 76–78].

Alternatively, a number of shorter MD simulations can be used to determine the energy landscape seen by ions in the channel, from which the steps involved in conduction and the rate at which this occurs can be deduced [79]. The free energy landscape for an ion could, in principle, be deduced from the probability of finding the ions at each position from a very long equilibrium MD simulation. But, as it can take a very long time for the ion to move out of free energy minima (one form of the sampling problem), a more common approach is to use the method of umbrella sampling [80], although other approaches such as metadynamics [65] and steered

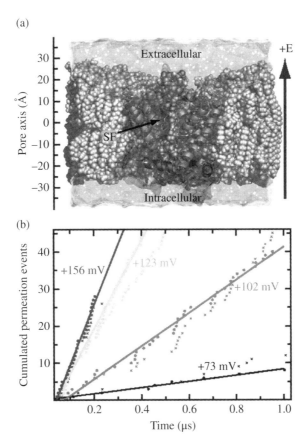

FIGURE 15.7 Results of a study directly simulating ion permeation. (a) The simulation system is shown highlighting the structure of the pore (SF, selectivity filter) and the direction of the applied electric field used to drive ion conduction. (b) Ion (•) and water (×) permeation events versus time at selected voltages; linear fits to the ion permeation events are shown. Reprinted from Ref. 86 with permission from the National Academy of Sciences USA.

MD [81] have been applied for this purpose. In umbrella sampling, the ion is held by a restraint (usually a harmonic potential) around a target location. The ion then samples positions around this target position during a simulation. If many simulations are run with different target positions (distributed, e.g., along the length of the channel), the ion effectively samples all the desired positions in the pore. The effect of the restraints can be removed to gain the unbiased probability distribution, from which the free energy landscape is derived. This last step is achieved via the weighted histogram analysis method, an iterative approach that combines the data from all the simulations to find the most likely unbiased distribution [82].

An excellent example of the energy landscape approach for understanding ion permeation was published by Berneche and Roux in 2001 [83], who determined the free energy map for multiple K+ ions passing through the KcsA potassium channel. Figure 15.5 shows the free energy map they found as a function of the position of three ions in the pore. This shows the most likely configurations of the ions as the free energy minima, as well as the most likely pathways between them. Conduction was seen to proceed in a "knock-on" manner, alternating between states with two and three ions in the selectivity filter. Notably, the barriers along this pathway are small (2–3 kcal mol^{-1}), meaning the current will be large— in the order of magnitude expected for this channel. More recently, a similar approach has been used to elucidate ion permeation in a sodium-selective channel, as seen in Figure 15.6 [84]. Conductance is also seen to involve knock-on conduction. In this case, however, it was suggested that it can arise with two or three ions in the channel. The wider pore also allows for greater independence in the motion of each ion and the potential for ions to pass each other, yielding the so-called "loosely coupled knock-on" mechanism.

The direct simulation of ion permeation in a potassium channel using MD was first demonstrated by Khalili-Araghi and Schulten in 2006 [85]. While they only witnessed three permeation events in 50 ns with a large driving force (1 V), it proved that such direct simulations were possible. More recently, Jensen et al. conducted much longer simulations enabling them to witness hundreds of conduction events and to determine current–voltage (I–V) curves under physiological conditions (see Fig. 15.7) [86]. This last point is important: Being able to predict measurable properties means that a direct comparison with experiment can be made. This study showed a knock-on conduction mechanism similar to that predicted previously and was able to highlight the rate-limiting steps as well as structural changes of the protein to transient nonconducting conformations.

The comparison with experiment has recently been taken up in more detail by the same authors [87]. Making use of extensive computational resources and more than 1 ms of simulation time, the ionic current passing through a voltage-gated potassium channel and the simpler gramicidin A channel was calculated over a range of voltages, allowing for a direct comparison with one of the most fundamental experimentally measureable properties. The simulated currents were about 40 times less and 300 times less than equivalent (highly accurate) experimental measurements on the potassium channel and gramicidin A, respectively, which was slightly disappointing from the point of view of being able to accurately reproduce experimental measurements. The authors suggest that the most likely culprit for this is the accuracy of the nonpolarizable force fields, something that has been in simulators' minds since the work of McCammon and Karplus 33 years ago. Although this has demonstrated some of the limitations in trying to directly simulate ion currents, such long MD simulations provide invaluable information on the microscopic behavior of ions in these proteins as well as help to understand how simulation force fields can be improved.

15.6.2 Simulations of Ion Selectivity

Although it may seem that using simulations to understand the selectivity of ion channels is more difficult than simply simulating ion permeation, there are many more studies addressing selectivity than permeation. The reason for this is that a large contribution to the selectivity of a channel can come from the thermodynamics of ion binding to the protein, and this can be studied by examining the relative free energies of ions at specific positions in the channel. This means that long simulations capturing ion permeation events may not be necessary and that useful information could be gained from shorter simulations examining the differences in binding of different ions. Given the large amount of literature on this topic, we do not aim to summarize all the findings for different channel types here. Instead, we focus on a few example studies to highlight the range of MD simulation methods that have been applied to understand how the discrimination between ions arises. Although selection between many different ion types has been investigated, here, we only discuss the discrimination between Na^+ and K^+ as it provides an excellent example of selection that is essential to the proper functioning of cells.

A simple way to gain a first understanding of how an ion channel discriminates between ions is to simply run two short simulations, each containing one of the ion types. By examining where the ion binds to the protein and the specific interactions involved, it is possible to ascertain which factors might influence the thermodynamics of binding. For example, studies appearing soon after the first potassium channel structure was published were able to highlight likely differences in ion hydration for Na^+ and K^+ in the pore as an origin of ion selectivity [88, 89]. The success of this method requires that either the ion-binding sites already are known or simulations are run long enough for ions to find their most likely binding locations. In addition, it is necessary to be able to sample all the protein conformations consistent with this ion-binding position in order to work out realistic structures and interaction energies. But being able to ascertain the origins of selectivity requires one to be able to predict selectivity ratios from the simulations, and so, in general, more complex approaches are employed.

The most common approaches to examine selective binding of ions in a channel (or any protein) are the methods of alchemical free energy perturbation (FEP) [90] and thermodynamic integration [91]. In these, one group of atoms slowly disappears during the simulation, while another group slowly appears. In the context of selectivity, this simply means that Na^+ is slowly replaced with K^+ within a binding site in the channel during the simulation or vice versa. The free energy change during this simulation is then determined. From a comparison of this to the free energy change associated with swapping the ions in bulk water, the total free energy representing the selectivity of a specific binding site can be estimated [92]. Again, simulations have to be run long enough for the environment of the ions (protein/water/lipid) to sample all available conformations to ensure that the final free energy value converges to a global minimum. Not only is this difficult to achieve, it is also difficult to demonstrate that adequate sampling has been obtained, which invariably leaves some room for doubt in the results. Sometimes, the protein can undergo a slow conformational change as the ions are swapped, and it can be difficult to run the simulation for long enough to capture the probability of this occurring. Further complicating the issue is the fact that many channels hold multiple ions, and the presence of additional ions can alter the selectivity of any given site. Thus, a simulation must capture all realistic ion/protein configurations during the FEP calculation to be useful.

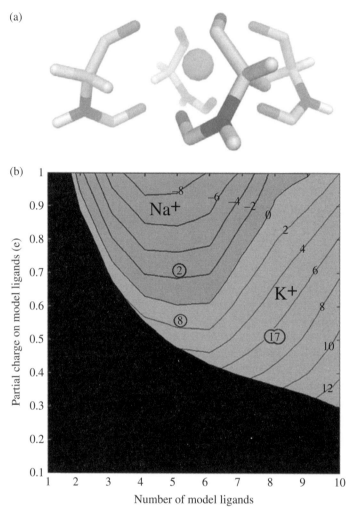

FIGURE 15.8 Examples of simulations using toy models to explain the origins of ion selectivity. (a) The toy model used by Noskov and Roux [96] to mimic a binding site in the KcsA potassium channel. The ion (sphere) is shown and sections from the four surrounding protein chains included in the model. Reproduced from Ref. 96 by permission from Macmillan Publishers Ltd., Copyright (2004). (b) A free energy "map" indicating how the selectivity of a binding site for Na^+ or K^+ is influenced by the number of coordinating ligands in the binding site and the charge on these ligands. Regions on the plot are selective for Na^+ and those selective for K^+ are indicated by two different shades. The degree of selectivity is quantified by the difference in free energy to place each ion in the site from bulk water as indicated by the contour lines. The numbers in circles represent the location of a number of selective ion binding sites from biological molecules (1 K^+ ion channel; 2 Na^+ ion channel; 7 DNA quadruplex; 8 aminoimidazole riboside kinase). Reproduced from Ref. 98 with permission from Elsevier.

Early applications of the FEP approach to understanding ion channel selectivity were conducted by using less than 10 ns to perform the ion replacement [79, 93, 94]. Although this may not be sufficient to capture any larger structural changes occurring in the protein, it was enough to indicate which positions and ion configurations in the pore would be selective for K^+ over Na^+. Once locations that are selective for the ions are identified, then the interactions at that site can be analyzed to determine what might cause selectivity. For example, it was shown that the greater desolvation penalty for Na^+ and better coordination of K^+ in the pore could generate K^+ selectivity [94].

It has become apparent that very long simulations are often required for free energy values in MD simulations to accurately converge. In a recent study examining the insertion of a peptide into a lipid bilayer, it was suggested that simulations lasting more than 4 μs may be needed at every target position due to the need to sample slow motions and conformational changes in the system [95]. While the time required differs for each system, it does mean that the quantitative validity of some of these early channel simulations in which individual simulations lasted much less than 1 ns can be questioned.

One way to allow for better sampling and to investigate the principles that can lead to selective binding of ions is to simplify the system, an approach often termed as the use of "toy models." By including fewer atoms and removing slow motions, accurate convergence can be ensured, and the ability to gain many results with different parameters enables a range of situations to be surveyed to see which yield selectivity and which do not. One of the first examples of this method was able to highlight the critical importance of the interactions between the ion and backbone carbonyl atoms in potassium channels in generating selectivity [96]. To do this, a simple model of one of the ion-binding sites in KcsA was created, as indicated in Figure 15.8a. Using this, it was shown that even in a flexible pore, the specific magnitude of the ion–carbonyl electrostatic interaction can create a preference for Na^+, as the balance of attractive ion–carbonyl and repulsive carbonyl–carbonyl interactions differs as the ion size changes. The toy model concept was extended by many groups, which has enabled the selectivity to be calculated as a function of a range of parameters such as the magnitude of charge on the coordinating protein atoms, the number of coordinating atoms in the site, and the presence or absence of water molecules as shown in Figure 15.8b [97, 98]. The toy model approach has been a great success in mapping out the factors that can create selective binding and for generating ways to rapidly predict the selectivity of a given binding site. However, the removal of the more distant parts of the protein means that this focuses on local interactions and it is difficult to take into account multi-ion effects.

One way to take account of the entirety of the protein, different binding positions, and multi-ion effects is to calculate the free energy landscape for different ions in the pore in a similar way to that described for elucidating ion permeation. By recreating maps such as shown in Figure 15.5 for multiple ion types, the location and size of energy minima and barriers for each can be ascertained. Although this can be done for single ions in the pore (as in the early examples of Allen et al. [93]), it is most informative in multi-ion channels when multiple ions are included. For example, Egwolf and Roux constructed multi-ion free energy landscapes with either K^+ in a potassium channel pore or a mixture of K^+ and Na^+ [99]. This was a computationally intensive task at the time, as the different coordinates of the three ions in the pore meant that almost 4000 unique configurations of ion positions had to be simulated, each for 0.2 ns, to generate each free energy surface. The outcome, however, shows the location of each free energy minimum, the selectivity of each, and the height of the barriers between. In this case, it was found that one ion-binding site is more selective than the others and that the barriers between the minima are greater in the presence of Na^+. A combined approach of FEP and free energy landscapes has also been recently employed to highlight the importance of kinetic barriers between multi-ion states in generating ion selection in potassium channels [100].

More recently, free energy landscapes have been used to highlight the location at which ion selectivity arises in a voltage-gated sodium channel [84], which is known to display about a 20-fold preference for Na^+ over K^+. This study, similar to that of Egwolf and Roux, created free energy landscapes with either two Na^+ in the pore or a mixture of Na^+ and K^+. As shown in Figure 15.6, with pure Na^+, sodium ions can move between the energy minima with only small barriers of height approximately $2 \, kcal \, mol^{-1}$ (roughly the accuracy of the simulations themselves). However, when K^+ is present, the height of the largest barrier to conduction increases dramatically, indicating that discrimination between the ions is likely to arise at the location of this energy barrier. Further work then focused on this location to show how differing hydration geometries created by the narrow charged pore could generate the differing energy barrier for Na^+ and K^+ [84].

15.6.3 Simulations of Channel Gating

The opening and closing of ion channels take place on the millisecond timescale, making it difficult to simulate explicitly. Until recently, MD simulations addressing gating had to either extrapolate from local fluctuations in structure or add additional forces to bias the channel toward a different functional state.

There have been a number of simulations in which direct forces were added to witness the opening of mechanosensitive channels. These are an ideal target for this kind of structural manipulation, because their physiological role is to respond to mechanical forces. The best studied of these channels is the family of mechanosensitive channels of large conductance from bacteria, MscL, which open in response to membrane tension to rescue cells from osmotic shock [101]. After the first MscL structure was published in 1998, a number of simulations were conducted in which either radial forces or membrane pressure was applied to the protein to mimic the influence of membrane tension [102–105]. However, all these simulations were of relatively short duration ($\leq$10 ns). Therefore, to observe conformational changes within a feasible time frame, the magnitude of the added forces was significantly greater than that found under physiological conditions. An alternative approach has been to bias the structure using experimentally determined restraints to force a conformational change [106, 107]. While this can provide information about alternative conformations, it cannot describe the pathway by which the protein moves from one conformation to another.

Another approach to examine the likely conformational changes involved in channel gating has been to construct free energy surfaces, similar to those described previously, as a function of a structural parameter of the protein. This requires a large number of simulations restraining the protein around different values of the structural parameter and is therefore not computationally trivial. Furthermore, it is difficult to select the most relevant structural parameter to restrain without prior knowledge. But the upside is that the pathway between conformational states can, in principle, be determined as well as the energetics of the process. While there are not as yet examples in which the energy landscape for the full structural change involved in gating has been elucidated, a recent study by Fowler et al. took a step in this direction by finding the energy landscape for opening the activation gate of a potassium channel [108]. The free energy as a function of the position of the pore-lining S6 helices indicates that in the absence of the surrounding portion of the protein responsible for sensing membrane potential, the channel prefers to reside in an open conformation. Presumably, work must be done on the pore-lining helices by the voltage sensors to close the channel, otherwise the channel will spring open. This study had a system containing 72,000 atoms simulated for $0.7\,\mu s$, but the full conformational change in the presence of the voltage sensors was neither mapped out, nor was the result assessed in the presence or absence of an activating membrane potential, as this would require significantly more atoms and more simulation time.

Due to the impressive computational advances made by the D.E. Shaw Research Institute, very long simulations now have been conducted that capture the transition of a channel between open and closed states [48]. To do this, the crystallographically determined open structure of the Kv1.2/Kv2.1 potassium channel chimera was subjected to hyperpolarizing voltages that can be expected to induce the channel to close. As can be seen in Figure 15.9, the initial structure contained a continuous pore through which water molecules could pass and which remained open under a depolarizing membrane potential. But, within $20\,\mu s$ of applying the hyperpolarizing potential, the size of the pore rapidly shrunk and water molecules became excluded from the narrow gate, indicative of the channel progressing to a closed state. Upon reversing the membrane potential, the channel was seen to open again. Analysis of the voltage sensors showed that the charged residues moved in response to an electric field during simulations lasting $260\,\mu s$ and the magnitude of this movement was in line with experimentally measured "gating charges" (see Fig. 15.9). Despite the magnitude of the membrane potentials being very large, the uncertainty about the applicability of the current force fields for such long simulations [87, 109–111], and the potential for errors in the algorithms to propagate over time, these direct simulations of channel gating allow for the possible molecular motion of the protein during gating to be studied in a way not feasible by any other approach.

15.7 CONCLUSIONS

In this chapter, we have introduced the concept of MD simulations, shown how they have progressed over time, and described a number of examples in which MD has been used to understand properties of ion channels. Notably, the scope of MD simulations has expanded greatly over the years as the algorithms and computer hardware have improved, and the amount of insight that can be gained from these simulations has increased accordingly. While it was not computationally feasible to

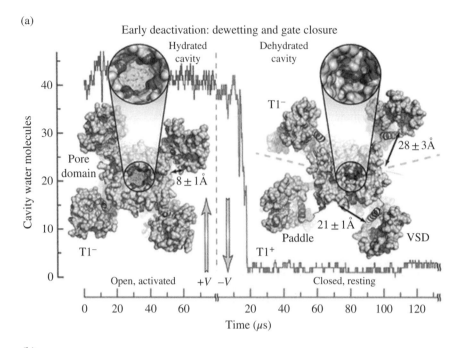

(a)

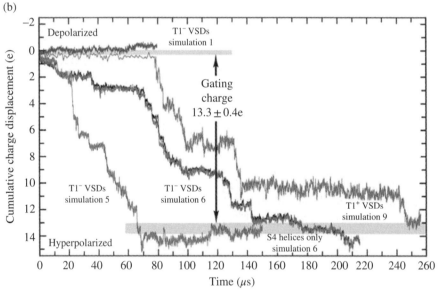

(b)

FIGURE 15.9 Examples of simulations used to directly monitor channel gating. (a) The number of water molecules occupying the narrow portion of a Kv1.2/Kv2.1 potassium channel chimera is plotted verses time. Initially, a depolarizing voltage is applied to the system (left side of graph "$+V$") that is switched to a hyperpolarizing voltage (right side of graph "$-V$"). Rapid dewetting of the channel occurs within 20 μs of switching the voltage, indicative of pore closure. Intracellular views of activated ($+V$, conducting) and resting ($-V$, nonconducting) states are also shown. (b) Displacement of charges in the voltage-sensing region as a function of time after switching to a hyperpolarizing potential. The total gating charge, $13.3 \pm 0.4e$, was estimated as the difference between the final displacements at hyperpolarizing and depolarizing potentials. Reprinted from Ref. 48 with permission from the American Association for the Advancement of Science.

directly simulate the opening of an ion channel a few years ago, simulation timescales have now reached the point where this can be done in some situations. But, while the very long simulations that have recently been published have led to insight into some important biological processes, they have also highlighted some shortcomings that continue to dog the simulation method. The inability to precisely reproduce measured currents, for example, serves as a warning that inaccuracies remain. Any bias in the simulation force fields can be exacerbated over long timescales. For example, a propensity of a force field to favor the formation of an α-helix over a β-sheet will not show up in a short simulation starting from a specified structure, but could compromise the result of a longer simulation in which there is time for the protein structure to change [110].

A common question faced by those conducting MD simulations is whether it is better to run one long simulation, many short simulations, or to conduct simulations that are biased by additional forces to explore specific questions of interest. Unfortunately, the answer to this question is not straightforward and will depend upon the system being studied and the problem being tackled. Results from individual long simulations can suffer from a lack of reproducibility—just because a certain event happened once does not mean it will always happen this way. But it may not be practical to run multiple simulations for long enough to witness slow events. For this reason, biasing methods can be applied to speed up the event, but this often requires some prior knowledge of the process being investigated. Ultimately, the design of the simulation experiment must be carefully considered to give the best chance of a useful outcome.

This chapter has touched only briefly on the use of biased simulations for speeding up events, improving conformational sampling, or quantifying free energies. Such approaches are becoming more common and probably account for the majority of simulations currently being conducted. Calculating free energies from simulations often allows for a more direct comparison with experimentally measured properties, while improved conformational sampling can allow for alternative conformational states of a protein to be found without having to conduct individual long simulations. A number of recent reviews on these topics are available for interested readers [50, 112–114].

MD simulations provide a high level of temporal and spatial resolution that is not available with experimental techniques. Because of this, they can provide detailed insight into the molecular basis of a number of important biological phenomena. While the technique has limitations, improvements are continually being made, and with the availability of free easy-to-use software, MD has become an essential technique for studying the function of a range of biological molecules, including ion channels, pumps, and transporters.

ACKNOWLEDGMENTS

The author would like to acknowledge the financial support from the Australian Research Council (Future Fellowship FT-130100781).

REFERENCES

1. B.J. Alder, T.E. Wainwright, Studies in molecular dynamics. 1. General method, *J. Chem. Phys.* 31 (1959) 459–466.
2. A. Rahman, Correlations in motion of atoms in liquid argon, *Phys. Rev. A* 136 (1964) A405–A411.
3. J.A. McCammon, B.R. Gelin, M. Karplus, Dynamics of folded proteins, *Nature* 267 (1977) 585–590.
4. L. Verlet, Computer experiments on classical fluids. I. Thermodynamical properties of Lennard-Jones molecules, *Phys. Rev.* 159 (1967) 98–103.
5. F. Jensen, *Introduction to Computational Chemistry*, second ed., John Wiley & Sons, Ltd, Chichester, 2007.
6. J.W. Ponder, D.A. Case, Force fields for protein simulations, *Adv. Protein Chem.* 66 (2003) 27–85.
7. A.D. MacKerell, D. Bashford, M. Bellott, R.L. Dunbrack, J.D. Evanseck, M.J. Field, S. Fischer, J. Gao, H. Guo, S. Ha, D. Joseph-McCarthy, L. Kuchnir, K. Kuczera, F.T.K. Lau, C. Mattos, S. Michnick, T. Ngo, D.T. Nguyen, B. Prodhom, W.E. Reiher, B. Roux, M. Schlenkrich, J.C. Smith, R. Stote, J. Straub, M. Watanabe, J. Wiorkiewicz-Kuczera, D. Yin, M. Karplus, All-atom empirical potential for molecular modeling and dynamics studies of proteins, *J. Phys. Chem. B* 102 (1998) 3586–3616.
8. A.D. Mackerell, M. Feig, C.L. Brooks, Extending the treatment of backbone energetics in protein force fields: Limitations of gas-phase quantum mechanics in reproducing protein conformational distributions in molecular dynamics simulations, *J. Comput. Chem.* 25 (2004) 1400–1415.
9. C. Oostenbrink, A. Villa, A.E. Mark, W.F. Van Gunsteren, A biomolecular force field based on the free enthalpy of hydration and solvation: The GROMOS force-field parameter sets 53A5 and 53A6, *J. Comput. Chem.* 25 (2004) 1656–1676.
10. D.Q. Wang, F. Freitag, Z. Gattin, H. Haberkern, B. Jaun, M. Siwko, R. Vyas, W.F. van Gunsteren, J. Dolenc, Validation of the GROMOS 54A7 force field regarding mixed alpha/beta-peptide molecules, *Helv. Chim. Acta* 95 (2012) 2562–2577.
11. W.D. Cornell, P. Cieplak, C.I. Bayly, I.R. Gould, K.M. Merz, D.M. Ferguson, D.C. Spellmeyer, T. Fox, J.W. Caldwell, P.A. Kollman, A 2nd generation force-field for the simulation of proteins, nucleic-acids, and organic-molecules, *J. Am. Chem. Soc.* 117 (1995) 5179–5197.
12. A. Warshel, M. Kato, A.V. Pisliakov, Polarizable force fields: History, test cases, and prospects, *J. Chem. Theory. Comput.* 3 (2007) 2034–2045.
13. P. Cieplak, F.Y. Dupradeau, Y. Duan, J.M. Wang, Polarization effects in molecular mechanical force fields, *J. Phys. Condens. Matter* 21 (2009) 333102.
14. P.E.M. Lopes, J. Huang, J. Shim, Y. Luo, H. Li, B. Roux, A.D. MacKerell, Polarizable force field for peptides and proteins based on the classical Drude oscillator, *J. Chem. Theory. Comput.* 9 (2013) 5430–5449.
15. Y. Shi, Z. Xia, J.J. Zhang, R. Best, C.J. Wu, J.W. Ponder, P.Y. Ren, Polarizable atomic multipole-based AMOEBA force field for proteins, *J. Chem. Theory. Comput.* 9 (2013) 4046–4063.
16. P. Hunenberger, Thermostat algorithms for molecular dynamics simulations, *Adv. Polym. Sci.* 173 (2005) 105–147.

17. J.E. Basconi, M.R. Shirts, Effects of temperature control algorithms on transport properties and kinetics in molecular dynamics simulations, *J. Chem. Theory. Comput.* 9 (2013) 2887–2899.

18. E. Rosta, N.V. Buchete, G. Hummer, Thermostat artifacts in replica exchange molecular dynamics simulations, *J. Chem. Theory. Comput.* 5 (2009) 1393–1399.

19. M. Thomas, B. Corry, Thermostat choice significantly influences water flow rates in molecular dynamics studies of carbon nanotubes, *Microfluid. Nanofluid.* 18 (2015) 41–47.

20. P. Ewald, Die Berechnung optischer und elektrostatischer Gitterpotentiale, *Ann. Phys.* 369 (1921) 253–287.

21. T. Darden, L. Perera, L.P. Li, L. Pedersen, New tricks for modelers from the crystallography toolkit: The particle mesh Ewald algorithm and its use in nucleic acid simulations, *Structure* 7 (1999) R55–R60.

22. J.P. Ryckaert, G. Ciccotti, H.J.C. Berendsen, Numerical-integration of Cartesian equations of motion of a system with constraints: Molecular-dynamics of N-alkanes, *J. Comput. Phys.* 23 (1977) 327–341.

23. D.J. Tobias, C.L. Brooks, Molecular-dynamics with internal coordinate constraints, *J. Chem. Phys.* 89 (1988) 5115–5127.

24. H.C. Andersen, Rattle: A velocity version of the shake algorithm for molecular-dynamics calculations, *J. Comput. Phys.* 52 (1983) 24–34.

25. B. Hess, H. Bekker, H.J.C. Berendsen, J.G.E.M. Fraaije, LINCS: A linear constraint solver for molecular simulations, *J. Comput. Chem.* 18 (1997) 1463–1472.

26. J.C. Phillips, R. Braun, W. Wang, J. Gumbart, E. Tajkhorshid, E. Villa, C. Chipot, R.D. Skeel, L. Kale, K. Schulten, Scalable molecular dynamics with NAMD, *J. Comput. Chem.* 26 (2005) 1781–1802.

27. K. Henzler-Wildman, D. Kern, Dynamic personalities of proteins, *Nature* 450 (2007) 964–972.

28. M. Thomas, D. Jayatilaka, B. Corry, An entropic mechanism of generating selective ion binding in macromolecules, *PLoS Comput. Biol.* 9 (2013) e1002914.

29. M. Karplus, Molecular-dynamics simulations of proteins, *Phys. Today* 40 (1987) 68–72.

30. R. Elber, M. Karplus, Multiple conformational states of proteins: A molecular-dynamics analysis of myoglobin, *Science* 235 (1987) 318–321.

31. A.J. Li, V. Daggett, Investigation of the solution structure of chymotrypsin inhibitor 2 using molecular dynamics: Comparison to X-ray crystallographic and NMR data, *Protein Eng.* 8 (1995) 1117–1128.

32. E. Demchuk, D. Bashford, D.A. Case, Dynamics of a type VI reverse turn in a linear peptide in aqueous solution, *Fold. Des.* 2 (1997) 35–46.

33. H. Heller, M. Schaefer, K. Schulten, Molecular-dynamics simulation of a bilayer of 200 lipids in the gel and in the liquid-crystal phases, *J. Phys. Chem.* 97 (1993) 8343–8360.

34. Y. Duan, P.A. Kollman, Pathways to a protein folding intermediate observed in a 1-microsecond simulation in aqueous solution, *Science* 282 (1998) 740–744.

35. B.L. de Groot, H. Grubmuller, Water permeation across biological membranes: Mechanism and dynamics of aquaporin-1 and GlpF, *Science* 294 (2001) 2353–2357.

36. E. Tajkhorshid, P. Nollert, M.O. Jensen, L.J.W. Miercke, J. O'Connell, R.M. Stroud, K. Schulten, Control of the selectivity of the aquaporin water channel family by global orientational tuning, *Science* 296 (2002) 525–530.

37. P.L. Freddolino, A.S. Arkhipov, S.B. Larson, A. McPherson, K. Schulten, Molecular dynamics simulations of the complete satellite tobacco mosaic virus, *Structure* 14 (2006) 437–449.

38. M.M. Seibert, A. Patriksson, B. Hess, D. van der Spoel, Reproducible polypeptide folding and structure prediction using molecular dynamics simulations, *J. Mol. Biol.* 354 (2005) 173–183.

39. D.L. Ensign, P.M. Kasson, V.S. Pande, Heterogeneity even at the speed limit of folding: Large-scale molecular dynamics study of a fast-folding variant of the villin headpiece, *J. Mol. Biol.* 374 (2007) 806–816.

40. P.L. Freddolino, F. Liu, M. Gruebele, K. Schulten, Ten-microsecond molecular dynamics simulation of a fast-folding WW domain, *Biophys. J.* 94 (2008) L75–L77.

41. P. Maragakis, K. Lindorff-Larsen, M.P. Eastwood, R.O. Dror, J.L. Klepeis, I.T. Arkin, M.O. Jensen, H.F. Xu, N. Trbovic, R.A. Friesner, A.G. Palmer, D.E. Shaw, Microsecond molecular dynamics simulation shows effect of slow loop dynamics on backbone amide order parameters of proteins, *J. Phys. Chem. B* 112 (2008) 6155–6158.

42. A. Perez, F.J. Luque, M. Orozco, Dynamics of B-DNA on the microsecond time scale, *J. Am. Chem. Soc.* 129 (2007) 14739–14745.

43. A. Grossfield, M.C. Pitman, S.E. Feller, O. Soubias, K. Gawrisch, Internal hydration increases during activation of the G-protein-coupled receptor rhodopsin, *J. Mol. Biol.* 381 (2008) 478–486.

44. F. Khalili-Araghi, V. Jogini, V. Yarov-Yarovoy, E. Tajkhorshid, B. Roux, K. Schulten, Calculation of the gating charge for the Kv1.2 voltage-activated potassium channel, *Biophys. J.* 98 (2010) 2189–2198.

45. D.E. Shaw, R.O. Dror, J.K. Salmon, J.P. Grossman, K.M. Mackenzie, J.A. Bank, C. Young, M.M. Deneroff, B. Batson, K.J. Bowers, E. Chow, M.P. Eastwood, D.J. Ierardi, J.L. Klepeis, J.S. Kuskin, R.H. Larson, K. Lindorff-Larsen, P. Maragakis, M.A. Moraes, S. Piana, Y.B. Shan, B. Towles, Millisecond-scale molecular dynamics simulations on Anton. Proceedings of the Conference on High Performance Computing Networking, Storage and Analysis, November 14–20, 2009, Oregon Convention Center, Portland, OR, p. 39.

46. D.E. Shaw, P. Maragakis, K. Lindorff-Larsen, S. Piana, R.O. Dror, M.P. Eastwood, J.A. Bank, J.M. Jumper, J.K. Salmon, Y.B. Shan, W. Wriggers, Atomic-level characterization of the structural dynamics of proteins, *Science* 330 (2010) 341–346.

47. K. Lindorff-Larsen, S. Piana, R.O. Dror, D.E. Shaw, How fast-folding proteins fold, *Science* 334 (2011) 517–520.

48. M.O. Jensen, V. Jogini, D.W. Borhani, A.E. Leffler, R.O. Dror, D.E. Shaw, Mechanism of voltage gating in potassium channels, *Science* 336 (2012) 229–233.

49. S. Mamatkulov, M. Fyta, R.R. Netz, Force fields for divalent cations based on single-ion and ion-pair properties, *J. Chem. Phys.* 138 (2013) 024505.

50. S.A. Adcock, J.A. McCammon, Molecular dynamics: Survey of methods for simulating the activity of proteins, *Chem. Rev.* 106 (2006) 1589–1615.

51. A. Perez, I. Marchan, D. Svozil, J. Sponer, T.E. Cheatham, III, C.A. Laughton, M. Orozco, Refinement of the AMBER force field for nucleic acids: Improving the description of alpha/gamma conformers, *Biophys. J.* 92 (2007) 3817–3829.

52. E. Fadrna, N.A. Spackova, J. Sarzynska, J. Koca, M. Orozco, T.E. Cheatham, III, T. Kulinski, J. Sponer, Single stranded loops of quadruplex DNA as key benchmark for testing nucleic acids force fields, *J. Chem. Theory Comput.* 5 (2009) 2514–2530.

53. S. Piana, K. Lindorff-Larsen, D.E. Shaw, How robust are protein folding simulations with respect to force field parameterization? *Biophys. J.* 100 (2011) L47–L49.

54. J.B. Clarage, T. Romo, B.K. Andrews, B.M. Pettitt, G.N. Phillips, A sampling problem in molecular dynamics simulations of macromolecules, *Proc. Natl. Acad. Sci. U. S. A.* 92 (1995) 3288–3292.

55. R.E. Bruccoleri, M. Karplus, Conformational sampling using high-temperature molecular dynamics, *Biopolymers* 29 (1990) 1847–1862.

56. A. Mitsutake, Y. Sugita, Y. Okamoto, Generalized-ensemble algorithms for molecular simulations of biopolymers, *Biopolymers* 60 (2001) 96–123.

57. K.Y. Sanbonmatsu, A.E. Garcia, Structure of Met-enkephalin in explicit aqueous solution using replica exchange molecular dynamics, *Proteins* 46 (2002) 225–234.

58. A. Roitberg, R. Elber, Modeling side chains in peptides and proteins with the locally enhanced sampling/simulated annealing method, in: K.M. Merz and S.M. Le Grand (Eds.), *The Protein Folding Problem and Tertiary Structure Prediction*, Birkhäuser, Boston, MA, 1994, pp. 1–41.

59. N. Kamiya, J. Higo, Repeated-annealing sampling combined with multicanonical algorithm for conformational sampling of bio-molecules, *J. Comput. Chem.* 22 (2001) 1098–1106.

60. A.K. Faradjian, R. Elber, Computing time scales from reaction coordinates by milestoning, *J. Chem. Phys.* 120 (2004) 10880–10889.

61. G.A. Huber, S. Kim, Weighted-ensemble Brownian dynamics simulations for protein association reactions, *Biophys. J.* 70 (1996) 97–110.

62. X.W. Wu, S.M. Wang, Self-guided molecular dynamics simulation for efficient conformational search, *J. Phys. Chem. B* 102 (1998) 7238–7250.

63. D. Hamelberg, J. Mongan, J.A. McCammon, Accelerated molecular dynamics: A promising and efficient simulation method for biomolecules, *J. Chem. Phys.* 120 (2004) 11919–11929.

64. A. Laio, F.L. Gervasio, Metadynamics: A method to simulate rare events and reconstruct the free energy in biophysics, chemistry and material science, *Rep. Prog. Phys.* 71 (2008) 126601.

65. A. Laio, M. Parrinello, Escaping free-energy minima, *Proc. Natl. Acad. Sci. U. S. A.* 99 (2002) 12562–12566.

66. G.M. Torrie, J.P. Valleau, Non-physical sampling distributions in Monte-Carlo free energy estimation: Umbrella sampling, *J. Comput. Phys.* 23 (1977) 187–199.

67. H. Grubmuller, Predicting slow structural transitions in macromolecular systems: Conformational flooding, *Phys. Rev. E.* 52 (1995) 2893–2906.

68. R.K. Hart, R.V. Pappu, J.W. Ponder, Exploring the similarities between potential smoothing and simulated annealing, *J. Comput. Chem.* 21 (2000) 531–552.

69. T. Huber, A.E. Torda, W.F. van Gunsteren, Local elevation: A method for improving the searching properties of molecular dynamics simulation, *J. Comput. Aided Mol. Des.* 8 (1994) 695–708.

70. R. Elber, Long-timescale simulation methods, *Curr. Opin. Struct. Biol.* 15 (2005) 151–156.

71. P.S. Crozier, R.L. Rowley, N.B. Holladay, D. Henderson, D.D. Busath, Molecular dynamics simulation of continuous current flow through a model biological membrane channel, *Phys. Rev. Lett.* 86 (2001) 2467–2470.

72. S.H. Chung, T.W. Allen, M. Hoyles, S. Kuyucak, Permeation of ions across the potassium channel: Brownian dynamics studies, *Biophys. J.* 77 (1999) 2517–2533.

73. B. Corry, T.W. Allen, S. Kuyucak, S.H. Chung, Mechanisms of permeation and selectivity in calcium channels, *Biophys. J.* 80 (2001) 195–214.

74. S. Berneche, B. Roux, A microscopic view of ion conduction through the K^+ channel, *Proc. Natl. Acad. Sci. U. S. A.* 100 (2003) 8644–8648.

75. D. Gordon, V. Krishnamurthy, S.H. Chung, Generalized Langevin models of molecular dynamics simulations with applications to ion channels, *J. Chem. Phys.* 131 (2009) 134102.

76. T.W. Allen, O.S. Andersen, B. Roux, Molecular dynamics: Potential of mean force calculations as a tool for understanding ion permeation and selectivity in narrow channels, *Biophys. Chem.* 124 (2006) 251–267.

77. P.W. Fowler, E. Abad, O. Beckstein, M.S. Sansom, Energetics of multi-ion conduction pathways in potassium ion channels, *J. Chem. Theory Comput.* 9 (2013) 5176–5189.

78. C. Song, B. Corry, Testing the applicability of Nernst-Planck theory in ion channels: Comparisons with Brownian dynamics simulations, *PLoS One* 6 (2011) e21204.

79. J. Aqvist, V. Luzhkov, Ion permeation mechanism of the potassium channel, *Nature* 404 (2000) 881–884.

80. G.M. Torrie, J.P. Valleau, Monte-Carlo free-energy estimates using non-Boltzmann sampling: Application to subcritical Lennard-Jones fluid, *Chem. Phys. Lett.* 28 (1974) 578–581.

81. S. Park, F. Khalili-Araghi, E. Tajkhorshid, K. Schulten, Free energy calculation from steered molecular dynamics simulations using Jarzynski's equality, *J. Chem. Phys.* 119 (2003) 3559–3566.

82. S. Kumar, D. Bouzida, R.H. Swendsen, P.A. Kollman, J.M. Rosenberg, The weighted histogram analysis method for free-energy calculations on biomolecules. 1. The method, *J. Comput. Chem.* 13 (1992) 1011–1021.

83. S. Berneche, B. Roux, Energetics of ion conduction through the K^+ channel, *Nature* 414 (2001) 73–77.

84. B. Corry, M. Thomas, Mechanism of ion permeation and selectivity in a voltage gated sodium channel, *J. Am. Chem. Soc.* 134 (2012) 1840–1846.

85. F. Khalili-Araghi, E. Tajkhorshid, K. Schulten, Dynamics of K^+ ion conduction through Kv1.2, *Biophys. J.* 91 (2006) L2–L4.

86. M.O. Jensen, D.W. Borhani, K. Lindorff-Larsen, P. Maragakis, V. Jogini, M.P. Eastwood, R.O. Dror, D.E. Shaw, Principles of conduction and hydrophobic gating in K^+ channels, *Proc. Natl. Acad. Sci. U. S. A.* 107 (2010) 5833–5838.

87. M.O. Jensen, V. Jogini, M.P. Eastwood, D.E. Shaw, Atomic-level simulation of current-voltage relationships in single-file ion channels, *J. Gen. Physiol.* 141 (2013) 619–632.

88. L. Guidoni, V. Torre, P. Carloni, Potassium and sodium binding to the outer mouth of the K^+ channel, *Biochemistry* 38 (1999) 8599–8604.

89. I.H. Shrivastava, D.P. Tieleman, P.C. Biggin, M.S.P. Sansom, K^+ versus Na^+ ions in a K channel selectivity filter: A simulation study, *Biophys. J.* 83 (2002) 633–645.

90. R.W. Zwanzig, High-temperature equation of state by a perturbation method. 1. Nonpolar gases, *J. Chem. Phys.* 22 (1954) 1420–1426.

91. T.P. Straatsma, J.A. Mccammon, Multiconfiguration thermodynamic integration, *J. Chem. Phys.* 95 (1991) 1175–1188.

92. Y.Q. Deng, B. Roux, Computation of binding free energy with molecular dynamics and grand canonical Monte Carlo simulations, *J. Chem. Phys.* 128 (2008) 115103.

93. T.W. Allen, A. Bliznyuk, A.P. Rendell, S. Kuyucak, S.H. Chung, The potassium channel: Structure, selectivity and diffusion, *J. Chem. Phys.* 112 (2000) 8191–8204.

94. V.B. Luzhkov, J. Aqvist, K^+/Na^+ selectivity of the KcsA potassium channel from microscopic free energy perturbation calculations, *Biochim. Biophys. Acta-Protein Struct. Molec. Enzym.* 1548 (2001) 194–202.

95. C. Neale, J.C.Y. Hsu, C.M. Yip, R. Pomès, Indolicidin binding induces thinning of a lipid bilayer, *Biophys. J.* 106 (2014) L29–L31.

96. S.Y. Noskov, S. Berneche, B. Roux, Control of ion selectivity in potassium channels by electrostatic and dynamic properties of carbonyl ligands, *Nature* 431 (2004) 830–834.

97. M. Thomas, D. Jayatilaka, B. Corry, The predominant role of coordination number in potassium channel selectivity, *Biophys. J.* 93 (2007) 2635–2643.

98. M. Thomas, D. Jayatilaka, B. Corry, Mapping the importance of four factors in creating monovalent ion selectivity in biological molecules, *Biophys. J.* 100 (2011) 60–69.

99. B. Egwolf, B. Roux, Ion selectivity of the KcsA channel: A perspective from multi-ion free energy landscapes, *J. Mol. Biol.* 401 (2010) 831–842.

100. A.N. Thompson, I. Kim, T.D. Panosian, T.M. Iverson, T.W. Allen, C.M. Nimigean, Mechanism of potassium-channel selectivity revealed by Na^+ and Li^+ binding sites within the KcsA pore. *Nat. Struct. Mol. Biol.* 16 (2009) 1317–1324.

101. B. Corry, B. Martinac, Bacterial mechanosensitive channels: Experiment and theory, *Biochim. Biophys. Acta-Biomembr.* 1778 (2008) 1859–1870.

102. J. Gullingsrud, D. Kosztin, K. Schulten, Structural determinants of MscL gating studied by molecular dynamics simulations, *Biophys. J.* 80 (2001) 2074–2081.

103. J. Gullingsrud, K. Schulten, Lipid bilayer pressure profiles and mechanosensitive channel gating, *Biophys. J.* 86 (2004) 3496–3509.

104. G. Colombo, S.J. Marrink, A.E. Mark, Simulation of MscL gating in a bilayer under stress, *Biophys. J.* 84 (2003) 2331–2337.

105. L.E. Bilston, K. Mylvaganam, Molecular simulations of the large conductance mechano-sensitive (MscL) channel under mechanical loading, *FEBS Lett.* 512 (2002) 185–190.

106. B. Corry, A.C. Hurst, P. Pal, T. Nomura, P. Rigby, B. Martinac, An improved open-channel structure of MscL determined from FRET confocal microscopy and simulation, *J. Gen. Physiol.* 136 (2010) 483–494.

107. E. Deplazes, M. Louhivuori, D. Jayatilaka, S.J. Marrink, B. Corry, Structural investigation of MscL gating using experimental data and coarse grained MD simulations, *PLoS Comput. Biol.* 8 (2012) e1002683.

108. P.W. Fowler, M.S.P. Sansom, The pore of voltage-gated potassium ion channels is strained when closed, *Nat. Commun.* 4 (2013) 1872.

109. A. Raval, S. Piana, M.P. Eastwood, R.O. Dror, D.E. Shaw, Refinement of protein structure homology models via long, all-atom molecular dynamics simulations, *Proteins* 80 (2012) 2071–2079.

110. K. Lindorff-Larsen, P. Maragakis, S. Piana, M.P. Eastwood, R.O. Dror, D.E. Shaw, Systematic validation of protein force fields against experimental data, *PLoS One* 7 (2012) e32131.

111. S. Piana, J.L. Klepeis, D.E. Shaw, Assessing the accuracy of physical models used in protein-folding simulations: Quantitative evidence from long molecular dynamics simulations, *Curr. Opin. Struct. Biol.* 24 (2014) 98–105.

112. C. Abrams, G. Bussi, Enhanced sampling in molecular dynamics using metadynamics, replica-exchange, and temperature-acceleration, *Entropy* 16 (2014) 163–199.

113. C. Chipot, Frontiers in free-energy calculations of biological systems, *Comput. Mol. Sci.* 4 (2014) 71–89.

114. C.D. Christ, A.E. Mark, W.F. van Gunsteren, Basic ingredients of free energy calculations: A review, *J. Comput. Chem.* 31 (2010) 1569–1582.

INDEX

Note: Italic page numbers indicate reference to a figure.

Action potential(s), 4, 9, 14, 15, 52
  elicited by light, 256
Active transport
  definition, 7
  history, 3
  primary, definition, 7, *8,* 51
  secondary, definition, 7, 51, 136
Alanine serine cysteine (ASC)
      transporter(s), 122, 137
  electrogenicity, 137–139
  mechanism, 138
  patch-clamp analysis using caged alanine,
    137–139
  patch-clamp analysis via voltage-jump,
    138–139
Amino acid exchanger, *see* Alanine serine
    cysteine transporter
Aquaporin(s), 1
  MD simulations, 425
  NMR measurements, 230–233
Aspartate transporter,
  crystal structure, 136

Atomic absorption spectrophotometry (AAS)
  atomization, 389–394
    flame(s), 390–391
    graphite furnace, 390–394
      L'vov platform, *390,* 392
      temperature profile, 392–393
  Beer-Lambert law, *see* Atomic absorption
      spectrophotometry (AAS),
      Lambert-Beer law
  calibration curve, 397
  characteristic concentration(s), 393–394
  correction for background absorption, 394
  element selectivity, 389
  history of development, 390–391
  instrument components, *389, 390*
  Lambert-Beer law, 389, 391
  light source modulation, 391, 392
  matrix modifiers, 393
  radiation source(s), 391
    electrodeless discharge lamp(s)
      (EDL), 391
    hollow cathode lamp(s) (HCL), 391, 405

Pumps, Channels, and Transporters: Methods of Functional Analysis, First Edition. Edited by
Ronald J. Clarke and Mohammed A. A. Khalid.
© 2015 John Wiley & Sons, Inc. Published 2015 by John Wiley & Sons, Inc.

CHEMICAL ANALYSIS

A SERIES OF MONOGRAPHS ON ANALYTICAL CHEMISTRY
AND ITS APPLICATIONS

Editor
MARK F. VITHA

Editorial Board
STEPHEN C. JACOBSON
STEPHEN G. WEBER

Pumps, Channels, and Transporters: Methods of Functional Analysis, First Edition. Edited by Ronald J. Clarke and Mohammed A. A. Khalid.
© 2015 John Wiley & Sons, Inc. Published 2015 by John Wiley & Sons, Inc.